MW01600384

Semiconductor and Metal Nanocrystals

OPTICAL ENGINEERING

Founding Editor

Brian J. Thompson

University of Rochester
Rochester, New York

Additional Volumes in Preparation

Semiconductor and Metal Nanocrystals

Synthesis and Electronic and Optical Properties

edited by

Victor I. Klimov

Los Alamos National Laboratory
Los Alamos, New Mexico, U.S.A.

MARCEL DEKKER, INC.　　　　　　　　　　NEW YORK · BASEL

To my loving wife, Tatiana, and my sons, Peter and Paul

Cover artist: Howard Coe.

Library of Congress Cataloging-in-Publication Data
A catalog record for this book is available from the Library of Congress.

ISBN: 0-8247-4716-X
This book is printed on acid-free paper.

Headquarters
Marcel Dekker, Inc., 270 Madison Avenue, New York, NY 10016, U.S.A.
tel: 212-696-9000; fax: 212-685-4540

Distribution and Customer Service
Marcel Dekker, Inc., Cimarron Road, Monticello, New York 12701, U.S.A.
tel: 800-228-1160; fax: 845-796-1772

World Wide Web
http://www.dekker.com

The publisher offers discounts on this book when ordered in bulk quantities. For more information, write to Special Sales/Professional Marketing at the headquarters address above.

Current printing (last digit):
10 9 8 7 6 5 4 3 2 1

PRINTED IN THE UNITED STATES OF AMERICA

Preface

This book consists of a collection of review chapters that summarize the recent progress in the areas of metal and semiconductor nanosized crystals (nanocrystals). The interest in the optical properties of nanoparticles dates back to Faraday's experiments on nanoscale gold. In these experiments, Faraday noticed the remarkable dependence of the color of gold particles on their size. The size dependence of the optical spectra of semiconductor nanocrystals was first discovered much later (in the 1980s) by Ekimov and coworkers in experiments on semiconductor-doped glasses. Nanoscale particles (islands) of semiconductors and metals can be fabricated by a variety of means, including epitaxial techniques, sputtering, ion implantation, precipitation in molten glasses, and chemical synthesis. This book concentrates on nanocrystals fabricated via chemical methods. Using colloidal chemical syntheses, nanocrystals can be prepared with nearly atomic precision, having sizes from tens to hundreds of angstroms and size dispersions as narrow as 5%. The level of chemical manipulation of colloidal nanocrystals is approaching that for standard molecules. Using suitable surface derivatization, colloidal nanoparticles can be coupled to each other or can be incorporated into different types of inorganic or organic matrices. They can also be assembled into close-packed ordered and disordered arrays that mimic naturally occurring solids. Because of their small dimensions, size-controlled electronic properties, and chemical flexibility, nanocrystals can be viewed as tunable "artificial" atoms with properties that can be engineered to suit either a particular technological application or the needs of a particular experiment designed to address a specific research problem. The large technological potential of these materials, as well as new appealing physics, have led to an explosion in nanocrystal research over the past several years.

This book covers several topics of recent, intense interest in the area of nanocrystals: synthesis and assembly, theory, spectroscopy of interband and intraband optical transitions, single-nanocrystal optical and tunneling spectroscopy, transport properties, and nanocrystal applications. It is written by experts who have contributed pioneering research in the nanocrystal field and whose work has led to numerous, impressive advances in this area over the past several years.

This book is organized into two parts: Semiconductor Nanocrystals (Nanocrystal Quantum Dots) and Metal Nanocrystals. The first part begins with a review of progress in the synthesis and manipulation of colloidal semiconductor nanoparticles. The topics covered in this first chapter by Hollingsworth and Klimov include size and shape control, surface modification, doping, phase control, and assembly of nanocrystals of such compositions as CdSe, CdS, PbSe, HgTe, and so forth. The second chapter, by Norris, overviews results of spectroscopic studies of the interband (valence-to-conduction band) transitions in semiconductor nanoparticles with a focus on CdSe nanocrystals. Because of a highly developed fabrication technology, these nanocrystals have long been model systems for studies on the effects of three-dimensional quantum confinement in semiconductors. As described in this chapter, the analysis of absorption and emission spectra of CdSe nanocrystals led to the discovery of a "dark" exciton, a fine structure of band-edge optical transitions, and the size-dependent mixing of valence-band states. This topic of electronic structures and optical transitions in CdSe nanocrystals is continued in Chapter 3 by Efros. This chapter focuses on the theoretical description of electronic states in CdSe nanoparticles using the effective mass approach. Specifically, it reviews the "dark/bright" exciton model and its application for explaining the fine structure of resonantly excited photoluminescence, polarization properties of spherical and ellipsoidal nanocrystals, polarization memory effects, and magneto-optical properties of nanocrystals. Chapter 4, by Guyot-Sionnest, Shim, and Wang, reviews studies of intraband optical transitions in nanocrystals performed using methods of infrared spectroscopy. It describes the size-dependent structure and dynamics of these transitions as well as the control of intraband absorption using charge carrier injection. In Chapter 5, Klimov concentrates on the underlying physics of optical amplification and lasing in semiconductor nanocrystals. The chapter provides a description of the concept of optical amplification in "ultrasmall," sub-10-nm particles, discusses the difficulties associated with achieving the optical gain regime, and gives several examples of recently demonstrated lasing devices based on CdSe nanocrystals. Chapter 6, by Shimizu and Bawendi, overviews the results of single-nanocrystal (single-dot) emission studies with focus on CdSe nanoparticles. It discusses such phenomena as spectral diffusion and fluorescence intermittency ("blinking"). The studies of these effects provide important insights into the dynamics of charge carriers in a single nanoparticle and interactions between the nanocrystal internal and interface states. The focus in Chapter 7, written by Ginger and Greenham, switches from spectroscopic to electrical and transport properties of semiconductor nanocrystals. This chapter surveys studies of carrier injection into nanocrystals and carrier transport in nanocrystal assemblies and between nanocrystals and organic molecules. It also describes the potential applications of these phenomena in electronic and optoelectronic devices. In Chapter 8, Banin and Millo review the work on tunneling and optical spectroscopy of colloidal InAs nanocrystals. Single-electron tunneling experiments discussed in this chapter provide unique information on electronic states and the spatial distribution of electronic wave functions in a single nanoparticle. These data are further compared with results of more traditional optical spectroscopic studies. Nozik and Mićić provide a comprehensive overview of the synthesis, structural, and optical properties of semiconductor nanocrystals of III–V compounds (InP, GaP, $GaInP_2$, GaAs, and GaN) in Chapter 9. This chapter discusses such unique properties of nanocrystals and nanocrystal assemblies as efficient anti-Stokes photoluminescence, photoluminescence intermittency, anomalies between the absorption and the photoluminescence excitation spectra, and long-range energy transfer. Furthermore, it reviews photogenerated carrier dynamics in nanocrystals, including the issues and controversies related to the cooling of hot carriers in "ultrasmall" nanoparticles. Finally, it discusses the potential applications of nanocrystals in novel photon conversion devices, such as quantum-dot solar cells and photoelectrochemical systems for fuel production and photocatalysis.

The next three chapters, which comprise Part II of this book, examine topics dealing with the chemistry and physics of metal nanoparticles. In Chapter 10, Doty et al. describe methods for fabricating metal nanocrystals and manipulating them into extended arrays (superlattices). They also discuss microstructural characterization and some physical properties of these metal nanoassemblies, such as electron transport. Chapter 11, by Link and El-Sayed, reviews the size/shape-dependent optical properties of gold nanoparticles with a focus on the physics of the surface plasmons that leads to these interesting properties. In this chapter, the issues of plasmon relaxation and nanoparticle shape transformation indued by intense laser illumination are also discussed. A review of some recent studies on the ultrafast spectroscopy of mono-component and bicomponent metal nanocrystals is presented in Chapter 12 by Hartland. These studies provide important information on timescales and mechanisms for electron–phonon coupling in nanoscale metal particles.

Of course, the collection of chapters that comprises this book cannot encompass all areas in the rapidly evolving science of nanocrystals. As a result, some exciting topics were not covered here, including silicon-based nanostructures, magnetic nanocrystals, and nanocrystals in biology. Canham's discovery of efficient light emission from porous silicon in 1990 has generated a widespread research effort on silicon nanostructures (including that on silicon nanocrystals). This effort represents a very large field that could not be comprehensively reviewed within the scope of this book. The same reasoning applies to magnetic nanostructures and, specifically, to magnetic nanocrystals. This area has been strongly stimulated by the needs of the magnetic storage industry. It has grown tremendously over the past several years and probably warrants a separate book project. The connection of nanocrystals to biology is relatively new, however, it already shows great promise. Semiconductor and metal nanoparticles have been successfully applied to tagging biomolecules. On the other hand, biotemplates have been used for the assembly of nanoparticles into complex, multiscale structures. Along these lines, a very interesting topic is bio-inspired assemblies of nanoparticles that efficiently mimic various biofunctions (e.g., light harvesting and photosynthesis). "Nanocrystals in Biology" may represent a fascinating topic for some future review by a group of experts in biology, chemistry, and physics.

I would like to thank all contributors to this book for finding time in their busy schedules to put together their review chapters. I gratefully acknowledge M. A. Petruska and Jennifer A. Hollingsworth for help in editing this book. I would like to thank my wife, Tatiana, for her patience, tireless support, and encouragement during my research career and specifically during the work on this book.

Victor I. Klimov

Contents

Contributors

Uri Banin   The Hebrew University, Jerusalem, Israel

M. G. Bawendi   Massachusetts Institute of Technology, Cambridge, Massachusetts, U.S.A.

R. Christopher Doty   The University of Texas, Austin, Texas, U.S.A.

Al. L. Efros   Naval Research Laboratory, Washington, D.C., U.S.A.

Mostafa A. El-Sayed   Georgia Institute of Technology, Atlanta, Georgia, U.S.A.

David S. Ginger   Cavendish Laboratory, Cambridge, England

Neil C. Greenham   Cavendish Laboratory, Cambridge, England

Phillipe Guyot-Sionnest   James Franck Institute, Chicago, Illinois, U.S.A.

Gregory V. Hartland   University of Notre Dame, Notre Dame, Indiana, U.S.A.

Jennifer A. Hollingsworth   Los Alamos National Laboratory, Los Alamos, New Mexico, U.S.A.

Victor I. Klimov   Los Alamos National Laboratory, Los Alamos, New Mexico, U.S.A.

Brian A. Korgel   The University of Texas, Austin, Texas, U.S.A.

Stephan Link   Georgia Institute of Technology, Atlanta, Georgia, U.S.A.

Olga I. Mičić   The National Renewable Energy Laboratory, Golden, Colorado, U.S.A.

Oded Millo   The Hebrew University, Jerusalem, Israel

David J. Norris   University of Minnesota, Minneapolis, Minnesota, U.S.A.

Arthur J. Nozik   The National Renewable Energy Laboratory, Golden, Colorado, and University of Colorado, Boulder, Colorado, U.S.A.

Aaron E. Saunders   The University of Texas, Austin, Texas, U.S.A.

Parag S. Shah   The University of Texas, Austin, Texas, U.S.A.

Moonsub Shim   James Franck Institute, Chicago, Illinois, U.S.A.

K. T. Shimizu   Massachusetts Institute of Technology, Cambridge, Massachusetts, U.S.A.

Michael B. Sigman, Jr.   The University of Texas, Austin, Texas, U.S.A.

Cynthia A. Stowell   The University of Texas, Austin, Texas, U.S.A.

Congjun Wang   James Franck Institute, Chicago, Illinois, U.S.A.

1

"Soft" Chemical Synthesis and Manipulation of Semiconductor Nanocrystals

Jennifer A. Hollingsworth and Victor I. Klimov
Los Alamos National Laboratory, Los Alamos, New Mexico, U.S.A.

I. INTRODUCTION

An important parameter of a semiconductor material is the width of the energy gap that separates the conduction from the valence energy bands (Fig. 1a, left). In semiconductors of macroscopic sizes, the width of this gap is a fixed parameter, which is determined by the material's identity. However, the situation changes in the case of nanoscale semiconductor particles with sizes smaller than ~10 nm (Fig. 1a, right). This size range corresponds to the regime of quantum confinement for which electronic excitations "feel" the presence of the particle boundaries and respond to changes in the particle size by adjusting their energy spectra. This phenomenon is known as the *quantum size effect*, whereas nanoscale particles that exhibit it are often referred to as quantum dots (QDs).

As the QD size decreases, the energy gap increases, leading, in particular, to a blue shift of the emission wavelength. In the first approximation, this effect can be described using a simple "quantum box" model. For a spherical QD with radius R, this model predicts that the size-dependent contribution to the energy gap is simply proportional to $1/R^2$ (Fig. 1b). In addition to increasing energy gap, quantum confinement leads to a collapse of the continuous energy bands of the bulk material into discrete, "atomic" energy levels. These well-separated QD states can be labeled using atomic-like notations (1S, 1P, 1D, etc.), as illustrated in Fig. 1a. The discrete

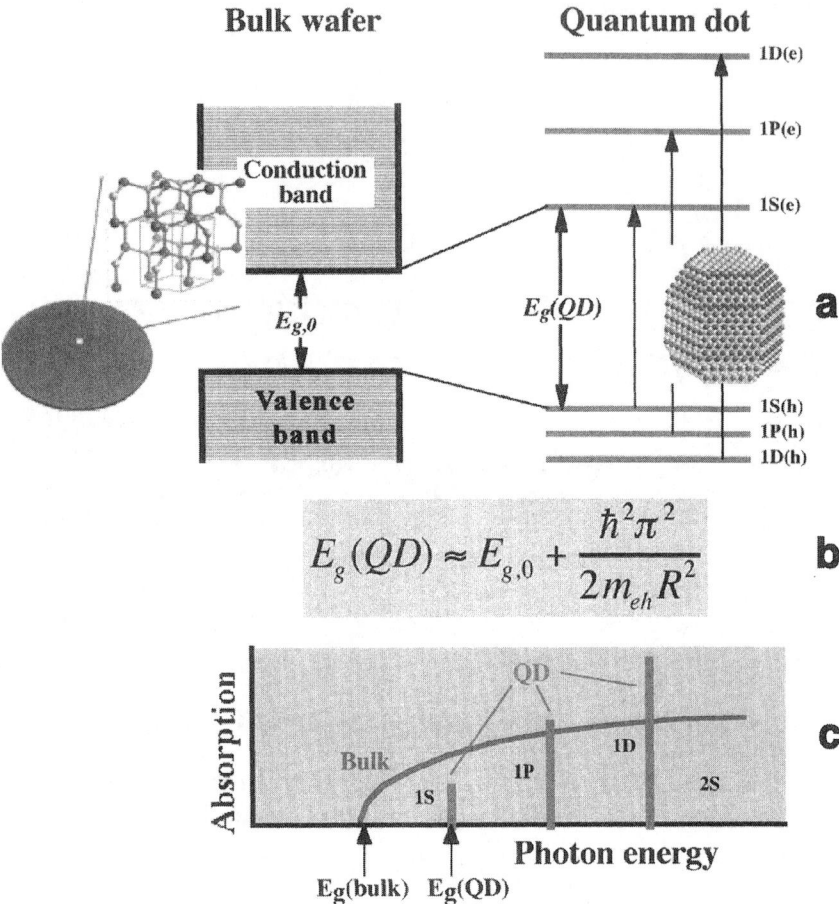

$$E_g(QD) \approx E_{g,0} + \frac{\hbar^2\pi^2}{2m_{eh}R^2}$$

Figure 1 (a) A bulk semiconductor has continuous conduction and valence energy bands separated by a "fixed" energy gap, E_{g0} (left), whereas a quantum dot (QD) is characterized by discrete atomiclike states with energies that are determined by the QD radius R (right). (b) The expression for the size-dependent separation between the lowest electron [1S(e)] and hole [1S(h)] QD states (QD energy gap) obtained using the "quantum box" model [$m_{eh} = m_e m_h/(m_e + m_h)$, where m_e and m_h are effective masses of electrons and holes, respectively]. (c) A schematic representation of the continuous absorption spectrum of a bulk semiconductor (curved line), compared to the discrete absorption spectrum of a QD (vertical bars).

structure of energy states leads to the discrete absorption spectrum of QDs (schematically shown by vertical bars in Fig. 1c), which is in contrast to the continuous absorption spectrum of a bulk semiconductor (Fig. 1c).

Semiconductor QDs bridge the gap between cluster molecules and bulk materials. The boundaries among molecular, QD, and bulk regimes are not well defined and are strongly material dependent. However, a range from ~100 to ~10,000 atoms per particle can been considered as a crude estimate of sizes for which the nanocrystal regime occurs. The lower limit of this range is determined by the stability of the bulk crystalline structure with respect to isomerization into molecular structures. The upper limit corresponds to sizes for which the energy level spacing is approaching the thermal energy kT, meaning that carriers become mobile inside the QD.

Semiconductor QDs have been prepared by a variety of "physical" and "chemical" methods. Some examples of physical processes, characterized by high-energy input, include molecular-beam-epitaxy (MBE) and metalorganic-chemical-vapor-deposition (MOCVD) approaches to quantum dots [1–3] and vapor–liquid solid (VLS) approaches to quantum wires [4,5]. High-temperature methods have also been applied to chemical routes, including particle growth in glasses [6,7]. Here, however, we emphasize "soft" (low-energy input) colloidal chemical synthesis of crystalline semiconductor nanoparticles that we will refer to as nanocrystal quantum dots (NQDs). NQDs comprise an inorganic core overcoated with a layer of organic ligand molecules. The organic capping provides electronic and chemical passivation of surface dangling bonds, prevents uncontrolled growth and agglomeration of the nanoparticles, and allows NQDs to be chemically manipulated like large molecules with solubility and reactivity determined by the identity of the surface ligand. In contrast to substrate-bound epitaxial QDs, NQDs are "freestanding." In this discussion, we concentrate on the most successful synthesis methods, where success is determined by high crystallinity, adequate surface passivation, solubility in nonpolar or polar solvents, and good size monodispersity. Size monodispersity permits the study and, ultimately, the use of materials size effects to define novel materials properties. Monodispersity in terms of colloidal nanoparticles (1–15-nm size range) requires a sample standard deviation of $\sigma \le 5\%$, which corresponds to $\pm$ one lattice constant [8]. Because colloidal monodispersity in this strict sense remains relatively uncommon, preparations are included in this chapter that achieve approximately $\sigma \le 20\%$, in particular where other attributes, such as novel compositions or shape control, are relevant. In addition, we discuss "soft" approaches to NQD chemical and structural modification, as well as to NQD assembly into artificial solids or artificial molecules.

II. COLLOIDAL NANOSYNTHESIS

The most successful NQD preparations in terms of quality and monodispersity entail pyrolysis of metal–organic precursors in hot coordinating solvents (120–360°C). Understood in terms of La Mer and Dinegar's studies of colloidal particle nucleation and growth [8,9], these preparative routes involve a temporally discrete nucleation event followed by relatively rapid growth from solution-phase monomers and finally slower growth by Ostwald ripening (referred to as recrystalization or aging) (Fig. 2). Nucleation is achieved by quick injection of a precursor into the hot coordinating solvents, resulting in thermal decomposition of the precursor reagents and supersaturation of the formed "monomers" that is partially relieved by particle generation. Growth then proceeds by the addition of monomer from solution to the NQD nuclei. Monomer concentrations are below the critical concentration for nucleation; thus, these species only add to existing particles, rather than form new nuclei [10]. Once monomer concentrations are sufficiently depleted,

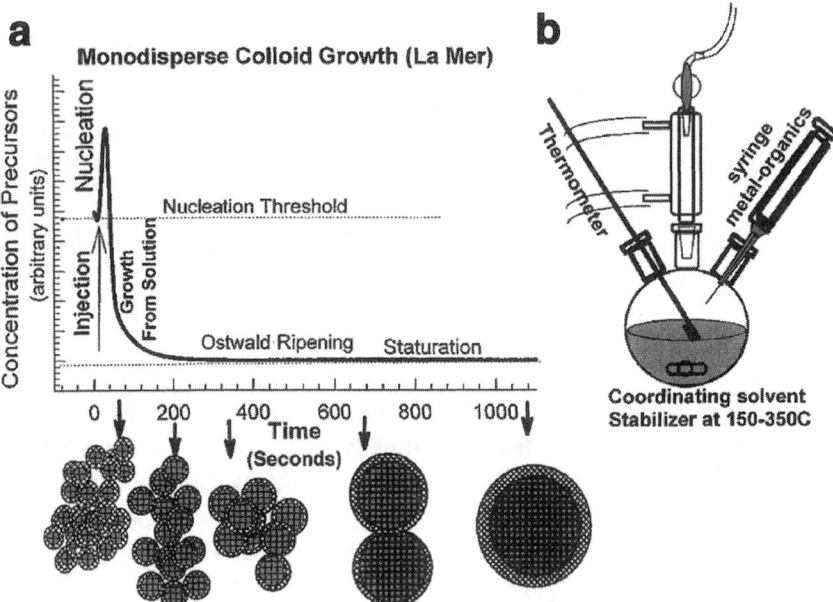

Figure 2 (a) Schematic illustrating La Mer's model for the stages of nucleation and growth for monodisperse colloidal particles. (b) Representation of the synthetic apparatus employed in the preparation of monodisperse NQDs. (From Ref. 8, reprinted with permission.)

growth can proceed by Ostwald ripening. Here, sacrificial dissolution of smaller (higher-surface-energy) particles results in growth of larger particles and, thereby, fewer particles in the system [8].

Alternatively, supersaturation and nucleation can be triggered by a *slow* ramping of the reaction temperature. Precursors are mixed at low temperature and slowly brought to the temperature at which precursor reaction and decomposition occur sufficiently quickly to result in supersaturation [11]. Supersaturation is again relieved by a "nucleation burst," after which temperature is controlled to avoid additional nucleation events, allowing monomer addition to existing nuclei to occur more rapidly than new nuclei formation. Thus, nucleation does not need to be instantaneous, but it must be a single, temporally discreet event to provide for the desired nucleation-controlled narrow size dispersions [10].

Size and size dispersion can be controlled during the reaction, as well as postpreparatively. In general, time is a key variable; longer reaction times yield a larger average particle size. Nucleation and growth temperatures play contrasting roles. *Lower* nucleation temperatures support lower monomer concentrations and can yield larger-size nuclei, whereas *higher* growth temperatures can generate larger particles as the rate of monomer addition to existing particles is enhanced. Also, Ostwald ripening occurs more readily at higher temperatures. Precursor concentration can influence both the nucleation and the growth processes, and its effect is dependent on the surfactant/precursor-concentration ratio and the identity of the surfactants (i.e., the strength of interaction between the surfactant and the NQD or between the surfactant and the monomer species). All else being equal, higher precursor concentrations promote the formation of fewer, larger nuclei and, thus, larger NQD particle size. Similarly, low stabilizer/precursor ratios yield larger particles. Also, weak stabilizer–NQD binding supports growth of large particles and, if too weakly coordinating, agglomeration of particles into insoluble aggregates [10]. Stabilizer–monomer interactions may influence growth processes as well. Ligands that bind strongly to monomer species may permit unusually high monomer concentrations that are required for very fast growth (see Sect. III) [12], or they may promote reductive elimination of the metal species (see below) [13].

The steric bulk of the coordinating ligands can impact the rate of growth subsequent to nucleation. Coordinating solvents typically comprise alkylphosphines, alkylphosphine oxides, alkylamines, alkylphosphates, alkylphosphites, alkylphosphonic acids, alkylphosphoramide, alkylthiols, fatty acids, and so forth of various alkyl chain lengths and degrees of branching. The polar head group coordinates to the surface of the NQD, and the hydrophobic tail is exposed to the external solvent/matrix. This interaction permits solubility in common nonpolar solvents and hinders aggregation of individual nanocrystals by shielding the van der Waals attractive forces between NQD cores that would otherwise lead to aggregation and flocculation. The NQD–surfactant connection is dynamic, and monomers can add or subtract relatively unhindered to the crystallite surface. The ability of component atoms to reversibly come on and off of the NQD surface provides a necessary condition for high crystallinity; particles can anneal while particle aggregation is avoided. Relative growth rates can be influenced by the steric bulk of the coordinating ligand. For example, during growth, bulky surfactants can impose a comparatively high steric hindrance to approaching monomers, effectively reducing growth rates by decreasing diffusion rates to the particle surface [10].

The two stages of growth (the relatively rapid first stage and Ostwald ripening) differ in their impact on size dispersity. During the first stage of growth, size distributions remain relatively narrow (dependent on the nucleation event) or can become more focused, whereas during Ostwald ripening, size tends to defocus as smaller particles begin to shrink and eventually dissolve in favor of growth of larger particles [14]. The benchmark preparation for CdS, CdSe, and CdTe NQDs [15], which dramatically improved the total quality of the nanoparticles prepared until that point, relied on Ostwald ripening to generate size series of II–VI NQDs. For example, CdSe NQDs from 1.2 to 11.5 nm in diameter were prepared [15]. Size dispersions of 10–15% were achieved for the larger-size particles and had to be subsequently narrowed by size-selective precipitation. The size-selective process simply involves first titrating the NQDs with a polar "nonsolvent," typically methanol, to the first sign of precipitation plus a small excess, resulting in precipitation of a small fraction of the NQDs. Such controlled precipitation preferentially removes the largest dots from the starting solution, as these become unstable to solvation before the smaller particles do. The precipitate is then collected by centrifugation, separated from the liquids, redissolved, and precipitated again. This iterative process separates larger from smaller NQDs and can generate the desired size dispersion of $\leq 5\%$.

More recently, preparations for II–VI semiconductors have been developed that specifically avoid the Ostwald-ripening growth regime. One such method maintains the regime of relatively fast growth (the "size-focusing" regime) by adding additional precursor monomer to the reaction solution after nucleation and before Ostwald growth begins. The additional monomer is not sufficient to nucleate more particles (i.e., not sufficient to again surpass the nucleation threshold). Instead, monomers add to existing particles and promote relatively rapid particle growth. Sizes focus as monomer preferentially adds to smaller particles rather than to larger ones [14]. The high monodispersity is evident in transmission electron micrograph (TEM) images (Fig. 3). Alternatively, growth can be stopped during the fast-growth stage

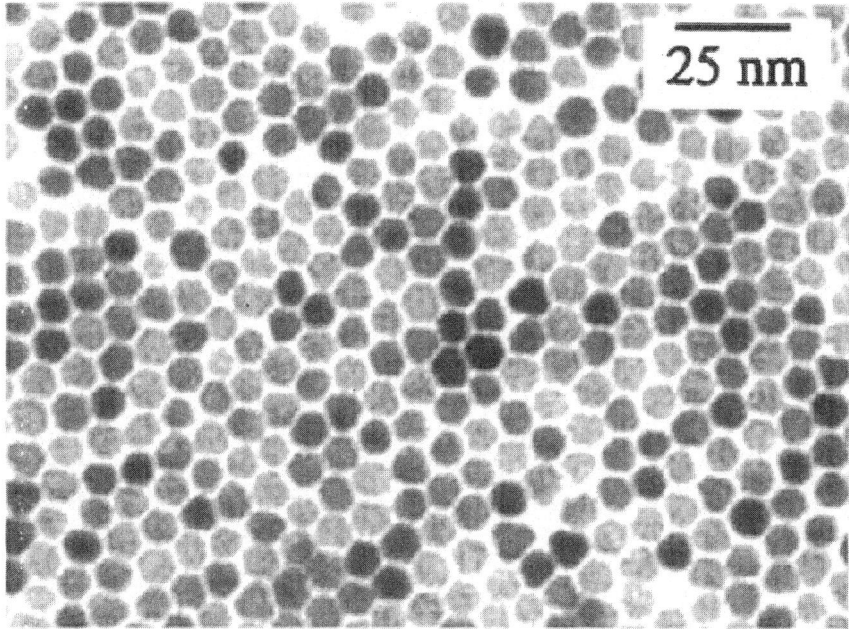

Figure 3 Transmission electron micrograph of 8.5 nm diameter CdSe nanocrystals demonstrating the high degree of size monodispersity achieved by the "size-focusing" synthesis method. (From Ref. 14, reprinted with permission.)

(by removing the heat source), and sizes are limited to those relatively close to the initial nucleation size. Because nucleation size can be manipulated by changing precursor concentration or reaction injection temperature, narrow size dispersions of controlled average particle size can be obtained by this method of simply stopping the reaction shortly following nucleation, during the rapid-growth stage.

Because of the ease with which high-quality samples can be prepared, the II–VI compound CdSe has compromised the "model" NQD system and been the subject of much basic research into the electronic and optical properties of NQDs. CdSe NQDs can be reliably prepared from pyrolysis of a variety of cadmium precursors, including alkyl cadmium compounds (e.g., dimethylcadmium [15]) and various cadmium salts (cadmium oxide, cadmium acetate, and cadmium carbonate [16]), combined with a selenium precursor prepared simply from Se powder dissolved in trioctylphosphine (TOP) or tributylphosphine (TBP). Initially, the surfactant–solvent combination, technical-grade trioctylphosphine oxide (tech-TOPO) and TOP, was used, where tech-TOPO performance was batch-specific due to the relatively random presence of adventitious impurities [15]. More recently, tech-TOPO has been replaced with "pure" TOPO to which phosphonic acids have been added to controllably mimic the presence of the tech-grade impurities [17]. In addition, TOPO has been replaced with various fatty acids, such as stearic and lauric acid, where shorter alkyl chain lengths yield relatively faster particle growth. The fatty acid systems are compatible with the full range of cadmium precursors, but they are most suited for growth of larger NQDs (>6 nm in diameter), compared to the TOPO–TOP system, as growth proceeds relatively more quickly [16]. The cadmium precursor is typically dissolved in the fatty acid at moderate temperatures, converting the Cd compound to cadmium stearate, for example. Alkyl amines were also successfully employed as CdSe growth media [16]. Incompatible systems are those that contain the anion of a strong acid (present as the surfactant ligand or as the cadmium precursor) and thiol-based systems [16]. Perhaps the most successful system, in terms of producing high quantum yields (QYs) in emission and monodisperse samples, uses a complex mixture of surfactants: stearic acid, TOPO, hexadecylamine, TBP, and dioctylamine [18].

High QYs are indicative of a well-passivated surface. NQD emission can suffer from the presence of unsaturated, "dangling" bonds at the particle surface which act as surface traps for charge carriers. Recombination of trapped carriers leads to a characteristic emission band ("deep-trap" emission) on the low-energy side of the "band-edge" photoluminescence (PL) band. Band-edge emission is associated with recombination of carriers in NQD "interior" quantized states. Coordinating ligands help to passivate surface trap sites, enhancing the relative intensity of band-edge emission compared to the deep-trap emission. The complex mixed-solvent system, described earlier, has been used to generate NQDs having QYs as high as 70–80%. These remarkably high PL efficiencies are comparable to the best achieved by inorganic epitaxial-shell surface-passivation techniques (see Sect. III). They are attributed to the presence of a primary amine ligand, as well as to the use of excess selenium in the precursor mixture (ratio Cd : Se of 1 : 10). The former alone (i.e., coupled with a "traditional" Cd:Se ratio of 2 : 1 or 1 : 1) yields PL QYs that are higher than those typically achieved by organic passivation (40–50% compared to 5–15%). The significance of the latter likely results from the unequal reactivities of the cadmium and selenium precursors. Accounting for the relative precursor reactivities by using such concentration-biased mixed precursors may permit improved crystalline growth and, hence, improved PL QYs [18]. Further, in order to achieve the very high QYs, reactions must be conducted over limited time spans from 5 to 30 mins. PL efficiencies reach a maximum in the first half of the reaction and decline thereafter (Fig. 4). Optimized preparations yield rather large NQDs,

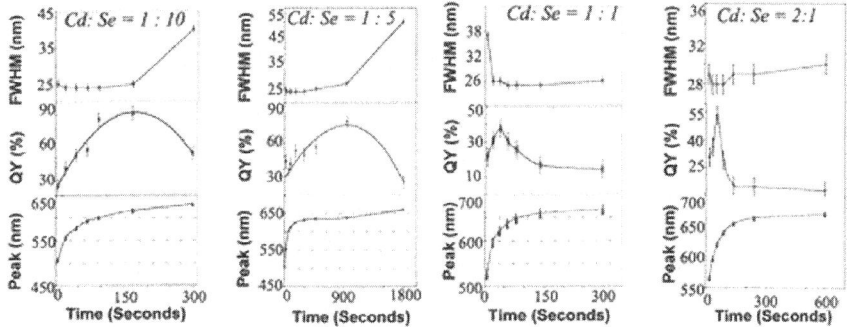

Figure 4 Temporal evolution of the full width at half-maximum (FWHM) of the PL band, the PL QY, and the PL peak position for identical reactions differing only with respect to their initial Cd : Se precursor ratios (see top graphs). (From Ref. 18, reprinted with permission.)

emitting in the orange-red. However, high-QY NQDs representing a variety of particle sizes are possible. By controlling precursor identity, total precursor concentrations, the identity of the solvent system, the nucleation and growth temperatures, and the growth time, NQDs emitting with >30% efficiency from ~510 to 650 nm can be prepared [18]. Finally, the important influence of the primary amine ligands may result from their ability to pack more efficiently on the NQD surfaces. Compared to TOPO and TOP, primary amines are less sterically hindered and may simply allow for a higher capping density [19]. However, the amine–CdSe NQD linkage is not as stable as for other more strongly bound CdSe ligands [20]. Thus, growth solutions prepared from this procedure are highly luminescent, but washing or processing into a new liquid or solid matrix can dramatically impact the QY. Multidentate amines may provide both the desired high PL efficiencies and the necessary chemical stabilities [18].

High-quality NQDs are no longer limited to cadmium-based II–VI compounds. Preparations for III–V semiconductor NQDs are well developed and are discussed in Chapter 9. Exclusively band-edge ultraviolet (UV) to blue-emitting ZnSe NQDs ($\sigma = 10\%$) exhibiting QYs from 20% to 50% have been prepared by pyrolysis of diethylzinc and TOPSe at high temperatures (nucleation: 310°C; growth: 270°C). Successful reactions employed hexadecylamine (HDA)–TOP as the solvent system (elemental analysis indicating that bound surface ligands comprised two-thirds HDA and one-third TOP), whereas the TOPO–TOP combination did not work for this material. Indeed, the nature of the reaction product was very sensitive to the TOPO/TOP ratio.

Too much TOPO, which binds strongly to Zn, generated particles so small that they could not be precipitated from solution by the addition of a non-solvent. Too much TOP, which binds very weakly to Zn, yielded particles that formed insoluble aggregates. As somewhat weaker bases compared to phosphine oxides, primary amines constitute as ligands of intermediate strength and in addition may provide enhanced capping density (as discussed earlier) [19]. HDA, in contrast with shorter-chain primary amines (octylamine and dodecylamine), provided good solubility properties and permitted sufficiently high growth temperatures for reasonably rapid growth of highly crystalline ZnSe NQDs [19].

High-quality NQDs absorbing and emitting in the infrared have also been prepared by way of a surfactant-stabilized pyrolysis reaction. PbSe colloidal quantum dots can be synthesized from the precursors: lead oleate- {prepared in situ from lead(II) acetate trihydrate and oleic acid [21]} and TOPSe [10,21]. TOP and oleic acid are present as the coordinating solvents, whereas phenyl ether, a noncoordinating solvent, provides the balance of the reaction solution. Injection and growth temperatures were varied (injection: 180–210°C; growth: 110–130°C) to control particle size from ~3.5 to ~9 nm in diameter [21]. The particles respond to "traditional" size-selection precipitation methods, allowing the narrow as-prepared size dispersions ($\sigma \leq 10\%$) to be further refined ($\sigma = 5\%$) (Fig. 5) [10]. Oleic acid provides excellent capping properties as PL quantum efficiencies, relative to infrared (IR) dye

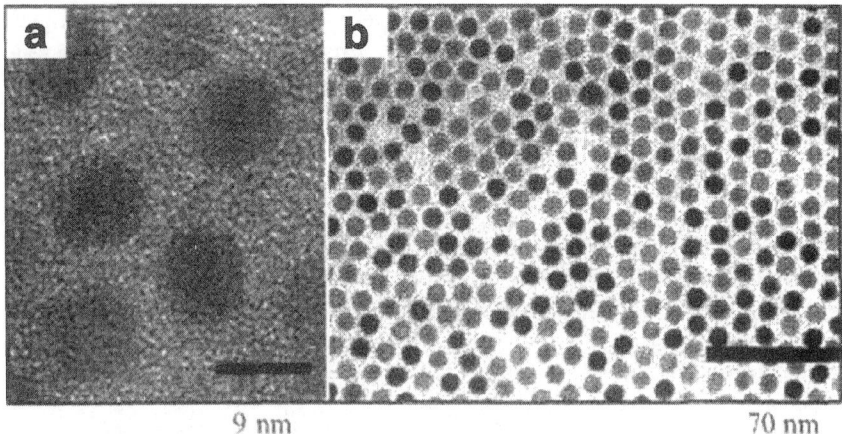

9 nm   70 nm

Figure 5   (a) High-resolution TEM of PbSe NQDs, where the internal crystal lattice is evident for several of the particles. (b) Lower-magnification imaging reveals the nearly uniform size and shape of the PbSe NQDs. (From Ref. 10, reprinted with permission.)

No. 26, are ~85% (Fig. 6) [21]. Although absolute PL QY measurements are required to confirm this remarkable figure, PbSe NQDs are, in the very least, substantially more efficient IR emitters than their organic dye counterparts. Further, the inorganic semiconductor offers enhanced stability compared to existing IR fluorophores.

In addition to the moderate (~150°C) and high-temperature (>200°C) preparations discussed earlier, many room-temperature reactions have been developed. The two most prevalent schemes entail thiol-stabilized aqueous-phase growth and inverse-micelle methods. We will briefly consider these approaches here, and the former is discussed in some detail in Section III as it pertains to core–shell nanoparticle growth, whereas the latter is revisited in Section VI with respect to its application to NQD doping. In general, the low-temperature methods suffer from relatively poor size dispersions (σ > 20%) and often exhibit significant, if not exclusively, trap-state PL. The latter is inherently weak and broad compared to band-edge PL, and it is less sensitive to quantum-size effects and particle-size control. Further, low-temperature aqueous preparations are limited in their applicability to relatively ionic materials. Higher temperatures are required to prepare *crystalline* covalent compounds [barring reaction conditions that may reduce the energetic barriers to crystalline growth (e.g., catalysts and templating structures)].

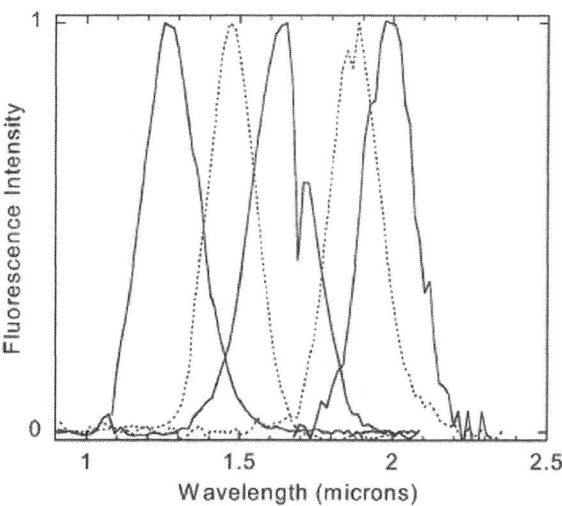

Figure 6 PbSe NQD size-dependent room-temperature fluorescence (excitation source: 1.064 μm laser pulse). Sharp features at ~1.7 and 1.85 μm correspond to solvent (chloroform) absorption. (From Ref. 21, reprinted with permission.)

Thus, II–VI compounds, which are more ionic compared to III–V compounds, have been successfully prepared at low temperatures (room temperature or less), whereas III–V compound semiconductors have not [22].

The processes of nucleation and growth in aqueous systems are conceptually similar to those observed in their higher-temperature counterparts. Typically, the metal perchlorate salt is dissolved in water, and the thiol stabilizer is added (commonly, 1-thioglycerol). After the pH is adjusted to > 11 (or from 5 to 6 if the ligand is a mercaptoamine [23]) and the solution is deaerated, the chalcogenide is added as the hydrogen chalcogenide gas [23–25]. The addition of the chalcogenide induces particle nucleation. The nucleation process appears not to be an ideal, temporally discrete event, as the initial particle-size dispersion is large, as evidenced, for example in broad PL spectra. Growth, or "ripening," is allowed to proceed over several days, after which a red shift in the PL spectrum is observed, and the spectrum is still broad [22]. For example, fractional precipitation of an aged CdTe growth solution yields a size series exhibiting emission spectra centered from 540 to 695 nm, where the full width at half-maximum (FWHM) of the size-selected samples are at best 50 nm [22], compared to ~25 nm for the best high-temperature reactions. In Cd-based systems, the ripening process can be accelerated by warming the solution; however, in Hg-based systems, heating the solution results in particle instability and degradation [22]. The initial particle size can be roughly "tuned" by changing the identity of the thiol ligand. The thiol binds to metal ions in solution prior to particle nucleation, and extended x-ray absorption fine structure (EXAFS) studies have demonstrated that the thiol stabilizer binds exclusively to metal surface sites in the formed particles [26]. By changing the strength of this metal–thiol interaction, larger or smaller particle sizes can be obtained. For example, decreasing the bond strength by introducing an electron-withdrawing group adjacent to the sulfur atom leads to larger particles [23,26].

The advantage of room-temperature, aqueous-based reactions lies in their ability to produce nanocrystal compositions that are, as yet, unattainable by higher-temperature pyrolysis methods. Of the II–VI compounds, Hg-based materials are thus far restricted to the temperature–ligand combination afforded by the aqueous thiol-stabilized preparations. The nucleation and growth of mercury chalcogenides have proven difficult to control in higher-temperature, nonaqueous reactions. Relatively weak ligands, fatty acids and amines (stability constant $K < 10^{17}$), yield fast growth and precipitation of the mercury chalcogenide, whereas stronger ligands, polyamines, phosphines, phosphine oxides, and thiols (stability constant $K > 10^{17}$), promote reductive elimination of metallic mercury at elevated temperatures [13]. Very high PL efficiencies (up to 50%) are reported for HgTe NQDs prepared in water [25]. However, the as-prepared samples yield approximately featureless absorption spectra and broad PL spectra. Further, the PL QYs for NQDs that emit at >1 μm have been determined in comparison with Rhodamine 6G, which has a PL maximum at ~550 nm. Typically, spectral overlap between the NQD emission signal and the reference organic dye is required to ensure reasonable QY values by taking into account the spectral response of the detector.

An alternative low-temperature approach that has been applied to a variety of systems, including mercury chalcogenides, is the inverse-micelle method. In general, the reverse-micelle approach entails preparation of a surfactant–polar solvent–nonpolar solvent microemulsion, where the content of the spontaneously generated spherical micelles is the polar-solvent fraction and that of the external matrix is the nonpolar solvent. The surfactant is commonly dioctyl sulfosuccinate, sodium salt (AOT). Precursor cations and anions are added and enter the polar phase. Precipitation follows and particle size is controlled by the size of the inverse-micelle "nanoreactors," as determined by the water content, W, where $W = [H_2O]/[AOT]$. For example, in an early preparation, AOT was mixed with water and heptane, forming the microemulsion. Cd^{2+}, as $Cd(ClO_4)_2 \cdot 6H_2O$, was stirred into the microemulsion, allowing it to become incorporated into the interior of the reverse micelles. The selenium precursor was subsequently added and, upon mixing with cadmium, nucleated colloidal CdSe. Untreated solutions were observed to flocculate within hours, yielding insoluble aggregated nanoparticles. The addition of excess water quickened this process. However, promptly evaporating the solutions to dryness, removing micellar water, yielded surfactant-encased colloids that could be redissolved in hydrocarbon solvents. Alternatively, surface passivation could be provided by first growing a cadmium shell via further addition of the Cd^{2+} precursor to the microemulsion, followed by addition of phenyl(trimethylsilyl)selenium (PhSeTMS). PhSe-surface passivation prompted precipitation of the colloids from the microemulsion. The colloids could then be collected by centrifugation or filtering and redissolved in pyridine [27].

More recently, the inverse-micelle technique has been applied to mercury chalcogenides as a means to control the fast growth rates characteristic of this system (discussed earlier) [13]. The process employed is similar to traditional micelle approaches; however, the metal and chalcogenide precursors are phase segregated. The mercury precursor [e.g., mercury(II) acetate] is transferred to the aqueous phase, whereas the sulfur precursor [bis (trimethylsilyl) sulfide, $(TMS)_2S$] is introduced to the nonpolar phase. Additional control over growth rates is provided by the strong mercury ligand, thioglycerol, similar to thiol-stabilized aqueous-based preparations. Growth is arrested by replacing the sulfur solution with aqueous or organometallic cadmium or zinc solutions. The Cd or Zn add to the surface of the growing particles and sufficiently alter surface reactivity to effectively halt growth.

Interestingly, the addition of the *organometallic* metal sources results in a significant increase in PL QY to 5–6%, whereas no observable increase accompanies passivation with the aqueous sources. Wide size dispersions are reported ($\sigma = 20$–30%). Nevertheless, absorption spectra are sufficiently well developed to clearly demonstrate that associated PL spectra, red-shifted with respect to the absorption band edge, derive from band-edge luminescence and not deep-trap-state emission. Finally, ligand exchange with thiophenol permits isolation as aprotic polar-soluble NQDs, whereas exchange with long-chain thiols or amines permits isolation as nonpolar-soluble NQDs [13].

The inverse-micelle approach may also offer a generalized scheme for the preparation of monodisperse metal-oxide nanoparticles [28]. The reported materials are ferroelectric oxides and, thus, stray from our emphasis on optically active semiconductor NQDs. Nevertheless, the method demonstrates an intriguing and useful approach: the combination of sol-gel techniques with inverse-micelle nanoparticle synthesis (with *moderate*-temperature nucleation and growth). Monodisperse barium titanate ($BaTiO_3$) nanocrystals, with diameters controlled in the range from 6 to 12 nm, were prepared. In addition, proof-of-principle preparations were successfully conducted for TiO_2 and $PbTiO_3$. *Single-source* alkoxide precursors are used to ensure proper stoichiometry in the preparation of complex oxides (e.g., bimetallic oxides) and are commercially available for a variety of systems. The precursor is injected into a stabilizer-containing solvent (oleic acid in diphenyl ether; "moderate" injection temperature: 140°C). The hydrolysis-sensitive precursor is, up to this point, protected from water. The solution temperature is then reduced to 100°C (growth temperature), and 30 wt% hydrogen peroxide solution (H_2O/H_2O_2) is added. The addition of the H_2O/H_2O_2 solution generates the microemulsion state and prompts a vigorous exothermic reaction. Control over particle size is exercised either by changing the precursor/stabilizer ratio or the amount of H_2O/H_2O_2 solution that is added. Increasing either results in an increased particle size while decreasing the precursor/stabilizer ratio leads to a decrease in particle size. Following growth over 48 h, the particles are extracted into nonpolar solvents such as hexane. By controlled evaporation from hexane, the $BaTiO_3$ nanocrystals can be self-assembled into ordered superlattices exhibiting periodicity over several microns, confirming the high monodispersity of the sample (see Sect. VII) [28].

III. INORGANIC SURFACE MODIFICATION

Surfaces play an increasing role in determining nanocrystal structural and optical properties as particle size is reduced. For example, due to an increasing surface-to-volume ratio with diminishing particle size, surface trap states exert an enhanced influence over photoluminescence properties, including emission efficiency, and spectral shape, position, and dynamics. Further, it is often through their surfaces that semiconductor nanocrystals interact with "their world," as soluble species in an organic solution, reactants in common organic reactions, polymerization centers, biological tags, electron-hole donors/acceptors, and so forth. Controlling inorganic and organic surface chemistry is key to controlling the physical and chemical properties that make NQDs unique compared to their epitaxial QD counterparts. In Section II, we discussed the impact of organic ligands on particle growth and particle properties. In this section, we review surface modification techniques that utilize *inorganic* surface treatments.

Overcoating highly monodisperse CdSe with epitaxial layers of either ZnS [29,30] or CdS (Fig. 7) [20] has become routine and typically provides almost an order-of-magnitude enhancement in PL efficiency compared to the exclusively organic capped starting nanocrystals [5–10% efficiencies can yield 30–70% efficiencies (Fig 8)]. The enhanced quantum efficiencies result from enhanced coordination of surface unsaturated, or dangling, bonds, as well as from improved confinement of electrons and holes to the particle core. The latter effect occurs when the bandgap of the shell material is larger than that of the core material, as is the case for (CdSe)ZnS and (CdSe)CdS (core) shell particles. Successful overcoating of III–V semiconductors has also been reported [31–33].

The various preparations share several synthetic features. First, the best results are achieved if initial particle size distributions are narrow, as some size-distribution broadening occurs during the shell-growth process. Because absorption spectra are relatively unchanged by surface properties, they can be used to monitor the stability of the nanocrystal core during and following growth of the inorganic shell. Further, if the conduction band offset between the core and the shell materials is sufficiently large (i.e., large compared to the electron confinement energy), then significant red-shifting of the absorption band edge should not occur, as the electron wave function remains confined to the core (Fig. 9). A large red shift in (core)shell systems, having sufficiently large offsets (determined by the identity of the core/shell materials and the electron and hole effective masses), indicates growth of the core particles during shell preparation. A small broadening of absorption features is common and results from some broadening of the particle size dispersion (Fig. 9). Alloying, or mixing of the shell components into the interior of the core, would also be evident in absorption spectra if it were to occur. The band edge would shift to some intermediate energy between the band energies of the respective materials comprising the alloyed nanoparticle.

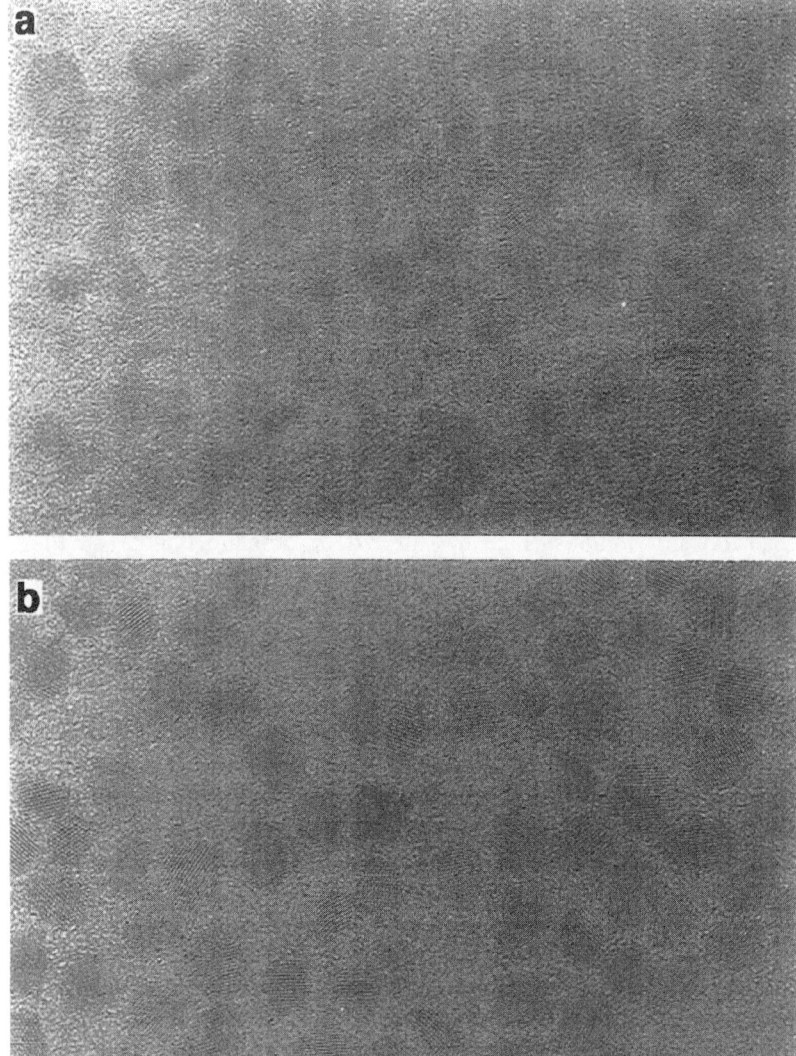

100 Å

Figure 7 Wide-field high-resolution TEMs of (a) 3.4-nm-diameter CdSe core particles and (b) (CdSe)CdS (core)shell particles prepared from the core NQDs in (a) by overcoating with a 0.9-nm-thick CdS shell. Where lattice fringes are evident, they span the entire nanocrystal, indicating epitaxial (core) shell growth. (From Ref. 20, reprinted with permission.)

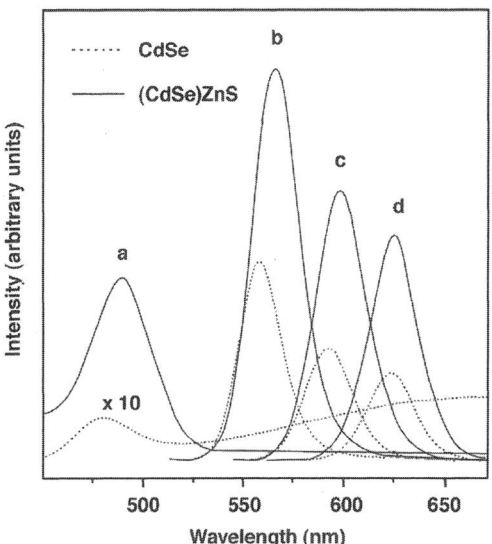

Figure 8 Photoluminescence spectra for CdSe NQDs and (CdSe)ZnS (core)shell NQDs. Core diameters are (a) 2.3, (b) 4.2, (c) 4.8, and (d) 5.5 nm. (Core)shell PL QYs are (a) 40%, (b) 50%, (c) 35%, and (d) 30%. Trap-state emission is evident in the (a) core–particle PL spectrum as a broad band to the red of the band-edge emission and absent in the respective (core)shell spectrum. (From Ref. 30, reprinted with permission.)

   Photoluminescence spectra can be used to indicate whether effective passivation of surface traps has been achieved. In poorly passivated nanocrystals, deep-trap emission is evident as a broad tail or hump to the red of the sharper band-edge emission spectral signal. The broad trap signal will disappear and the sharp band-edge luminescence will increase following successful shell growth (Fig. 8a). Note: The trap-state emission signal contribution is typically larger in smaller (higher relative surface area) nanocrystals than in larger nanoparticles (Fig. 8a).

   Homogeneous nucleation and growth of shell material as discrete nanoparticles may compete with *heterogeneous* nucleation and growth at core–particle surfaces. Typically, a combination of relatively low precursor concentrations and reaction temperatures is used to avoid particle formation. Low precursor concentrations support undersaturated-solution conditions and, thereby, shell growth by *heterogeneous* nucleation. The precursors, diethylzinc and bis(trimethylsilyl) sulfide in the case of ZnS shell growth, for example, are added dropwise at relatively low temperatures to prevent the buildup and supersaturation of unreacted precursor monomers in the

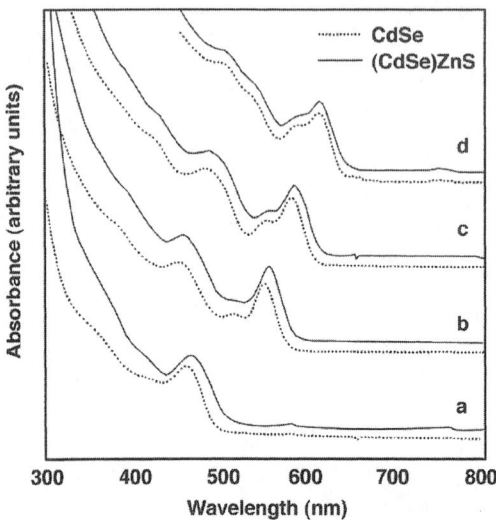

Figure 9 Absorption spectra for bare (dashed lines) and one- to two-monolayer ZnS-overcoated (solid lines) CdSe NQDs. (Core)shell spectra are broader and slightly red-shifted compared to the core counterparts. Core diameters are (a) 2.3, (b) 4.2, (c) 4.8, and (d) 5.5 nm. (From Ref. 30, reprinted with permission.)

growth solution. Further, employing relatively low reaction temperatures avoids growth of the starting core particles [20,30]. ZnS, for example, can nucleate and grow as a crystalline shell at temperatures as low as 140°C [30], and CdS shells have been successfully prepared from dimethylcadmium and bis(trimethylsilyl) sulfide at 100°C [20], thereby avoiding complications due to homogeneous nucleation and core-particle growth. Additional strategies for preventing particle growth of the shell material include using organic capping ligands that have a particularly high affinity for the shell metal. The presence of a strong binding agent seems to lead to more controlled shell growth; for example, TOPO is replaced with TOP in CdSe shell growth on InAs cores, where TOP (softer Lewis base) coordinates more tightly than TOPO (harder Lewis base) with cadmium (softer Lewis acid) [33]. Finally, the ratio of the cationic to anionic precursors can be used to prevent shell-material homogeneous nucleation. For example, increasing the concentration of the chalcogenide in a cadmium–sulfur precursor mixture hinders formation of unwanted CdS particles [20].

Successful overcoating is possible for systems where relatively large lattice mismatches between core and shell crystal structures exist. The most commonly studied (core)shell system, (CdSe)ZnS, is successful despite a 12%

lattice mismatch. Such a large lattice mismatch could not be tolerated in flat heterostructures, where strain-induced defects would dominate the interface. It is likely that the highly curved surface and reduced facet lengths of nanocrystals relax the structural requirements for epitaxy. Indeed, two types of epitaxial growth are evident in the (CdSe)ZnS system: coherent (with large distortion or strain) and incoherent (with dislocations), the difference arising for thin (approximately one to two monolayers, where a monolayer is defined as 3.1 Å) versus thick (>two monolayers) shells, respectively [30]. High-resolution (HR) TEM images of thin-shell ZnS-overcoated CdSe QDs reveal lattice fringes that are continuous across the entire particle, with only a small "bending" of the lattice fringes in some particles indicating strain. TEM imaging has also revealed that thicker shells (>two monolayers) lead to the formation of deformed particles, resulting from uneven growth across the particle surface. Here, too, however, the shell appeared epitaxial, oriented with the lattice of the core (Fig. 10). Nevertheless, wide-angle x-ray scattering (WAXS) data showed reflections for both CdSe and ZnS, indicating that each was exhibiting its own lattice parameter in the thicker-shell systems. This type of structural relationship between the core and the shell was described as *incoherent* epitaxy. It was speculated that at low coverage, the epitaxy is coherent (strain is tolerated), but that at higher coverages, the high lattice mismatch can no longer be sustained without the formation of dislocations and low-angle grain boundaries. Such defects in the core–shell

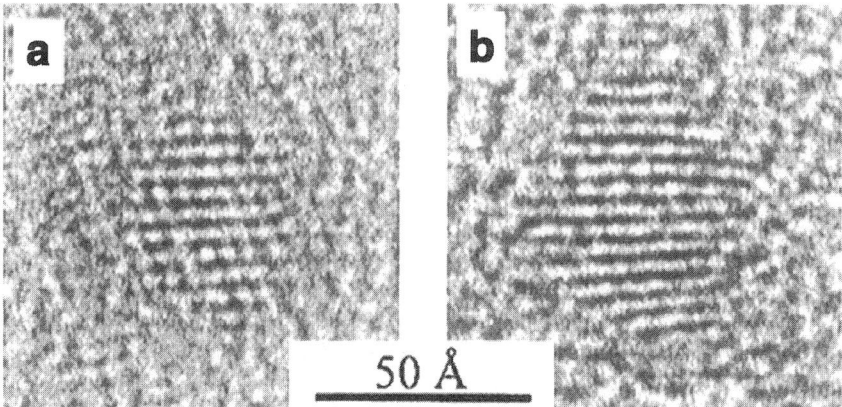

Figure 10 High-resolution TEMs of (a) CdSe core particle and (b) a (CdSe)ZnS (core)shell particle (2.6-monolayer ZnS shell). Lattice fringes in (b) are continuous throughout the particle, suggesting epitaxial (core)shell growth. (From Ref. 30, reprinted with permission.)

boundary provide nonradiative recombination sites and lead to diminished PL efficiency compared to coherently epitaxial thinner shells. Further, in all cases studied where more than a single monolayer of ZnS was deposited, the shell appeared to be continuous. X-ray photoelectron spectroscopy (XPS) was used to detect the formation of SeO_2 following exposure to air. The SeO_2 peak was observed only in bare TOPO–TOP-capped dots and dots having less than one monolayer of ZnS overcoating. Together, the HR TEM images and XPS data suggest complete epitaxial shell formation in the highly lattice-mismatched system of (CdSe)ZnS.

The effect of lattice mismatch has also been studied in III–V semiconductor core systems. Specifically, InAs has been successfully overcoated with InP, CdSe, ZnS, and ZnSe [33]. The degree of lattice mismatch between InAs and the various shell materials differed considerably, as did the PL efficiencies achieved for these systems. However, no direct correlation between lattice mismatch and QY in PL was observed. For example, (InAs)InP produced quenched luminescence, whereas (InAs)ZnSe provided up to 20% PL QYs, where the respective lattice mismatches are 3.13% and 6.44%. CdSe shells, providing a lattice *match* for the InAs cores, also produced up to 20% PL QYs. In all cases, shell growth beyond two monolayers (where a monolayer equals the d_{111} lattice spacing of the shell material) caused a decrease in PL efficiencies, likely due to the formation of defects that could provide trap sites for charge carriers {as observed in (CdSe)ZnS [30] and (CdSe)CdS [20] systems}. The perfectly lattice-matched CdSe shell material should provide the means for avoiding defect formation; however, the stable crystal structures for CdSe and InAs are different under the growth conditions employed. CdSe prefers the wurtzite structure, whereas InAs prefers cubic. For this reason, this "matched" system may succumb to interfacial defect formation with thick shell growth [33].

The larger contributor to PL efficiency in the (InAs)shell systems was found to be the size of the energy offset between the respective conduction and valence bands of the core and shell materials. Larger offsets provide larger potential energy barriers for the electron and hole wave functions at the (core)shell interface. For InP and CdSe, the conduction band offset with respect to InAs is small. This allows the electron wave function to experience the surface of the nanoparticle. In the case of CdSe, fairly high PL efficiencies can still be achieved because native trap sites are less prevalent than they are on InP surfaces. Both ZnS and ZnSe provide large energy offsets. The fact that the electron wave function remains confined to the core of the (core)shell particle is evident in the absorption and PL spectra. In these confined cases, no red-shifting was observed in the optical spectra following shell growth [33]. The observation that PL enhancement to only 8% quantum yield was possible using ZnS as the shell material may have been due to the large lattice mismatch between InAs and ZnS of ~11%. Otherwise, ZnS and ZnSe should behave similarly as shells for InAs cores.

Shell chemistry can be precisely controlled to achieve unstrained (core) shell epitaxy. For example, the zinc–cadmium alloy, $ZnCdSe_2$ was used for the preparation of (InP)$ZnCdSe_2$ nanoparticles having essentially zero lattice mismatch between the core and the shell [31]. High-resolution TEM images demonstrated the epitaxial relationship between the layers, and very thick epilayer shells were grown—up to 10 monolayers, where a monolayer was defined as 5 Å. The shell layer successfully protected the InP surface from oxidation, a degradation process to which InP is particularly susceptible (see Chapter 9).

Optoelectronic devices comprising two-dimensional quantum-well structures are generally limited to material pairs that are well lattice matched due to the limited strain tolerance of such planar systems; otherwise, very thin well layers are required. In order to access additional quantum-well-type structures, more strain-tolerant systems must be employed. As already alluded to, the highly curved quantum-dot nanostructure is ideal for lattice-mismatched systems. Several quantum-dot/quantum-well (QD/QW) struc-tures have been successfully synthesized, ranging from the well lattice-matched CdS(HgS)CdS [34–36] (quantum dot, quantum well, cladding) to the more highly strained ZnS(CdS)ZnS [37]. The former provides emission color tunability in the infrared spectral region, whereas the latter yields access to the blue-green spectral region. In contrast to the very successful (core)shell preparations discussed earlier in this section, the QD/QW structures have been prepared using ion-displacement reactions, rather than heterogeneous nucleation on the core surface (Fig. 11). These preparations have been either aqueous or polar-solvent based and conducted at low temperatures (room temperature to $-77°C$). They entail a series of steps that first involves the preparation of the nanocrystal cores (CdS and ZnS, respectively). Core preparation is followed by ion-exchange reactions in which a salt precursor of the "well" metal ion is added to the solution of "core" particles. The solubility product constant (K_{sp}) of the metal sulfide corresponding to the added metal species is such that it is significantly less than that of the metal sulfide of the core metal species. This solubility relationship leads to precipitation of the added metal ions and dissolution of the surface layer of core metal ions via ion exchange. Analysis of absorption spectra during addition of "well" ions to the nanoparticle solution revealed an apparent concentration threshold, after which the addition of the "well" ions produced no more change in the optical spectra. Specifically, in the case of the CdS (HgS)CdS system, ion exchange of Hg^{2+} for Cd^{2+} produced a red shift in absorption until a certain amount of "well" ions had been added. According to inductively coupled plasma–mass spectrometry (ICP-MS), which was used

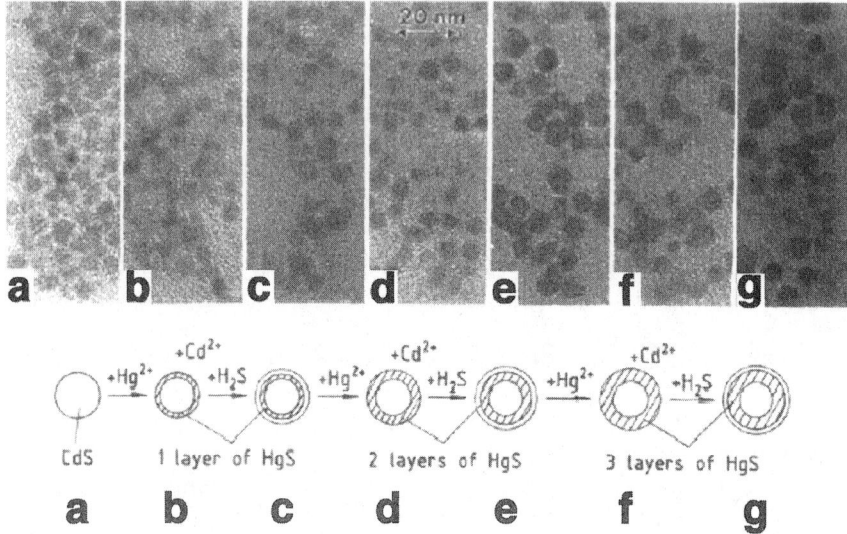

Figure 11 Transmission electron micrographs of CdS(HgS)CdS at various stages of the ion-displacement process, where the latter is schematically represented in the figure. (From Ref. 35, reprinted with permission.)

to measure the concentration of free ions in solution for both species, up until this threshold concentration was reached, the concentration of free Hg^{2+} ion was essentially zero, whereas the Cd^{2+} concentration increased linearly. After the threshold concentration was reached, the Hg^{2+} concentration increased linearly (with each externally provided addition to the system), whereas the Cd^{2+} concentration remained approximately steady. These results agree well with the ion-exchange reaction scenario, and, perhaps more importantly, suggest a certain natural limit to the exchange process. It was determined that in the example of 5.3-nm CdS starting core nanoparticles, approximately 40% of the Cd^{2+} was replaced with Hg^{2+}. This value agrees well with the conclusion that one complete monolayer has been replaced, because the surface-to-volume ratio in such nanoparticles is 0.42. Further dissolution of Cd^{2+} core ions is prevented by formation of the complete monolayer-thick shell, which also precludes the possibility of island-type shell growth [35].

Subsequent addition of H_2S or Na_2S causes the precipitation of the off-cast core ions back onto the particles. The ion-replacement process, requiring the sacrifice of the newly redeposited core metal ions, can then be repeated in order to increase the thickness of the "well" layer. This process has been successfully repeated for up to three layers of well material. The "well" is then capped with a redeposited layer of core metal ions to generate the full QD/QW structure. The thickness of the cladding layer could be increased by addition in several steps (up to five) of the metal and sulfur precursors [35].

The nature of the QD/QW structure and its crystalline quality have been analyzed by HR TEM. In the CdS(HgS)CdS system, evidence has been presented for both approximately spherical particles, as well as faceted particle shapes such as tetrahedrons and twinned tetrahedrons. In all cases, well and cladding growth is epitaxial, as evidenced by the absence of amorphous regions in the nanocrystals and in the smooth continuation of lattice fringes across particles. Analysis of HR TEM micrographs also reveals that

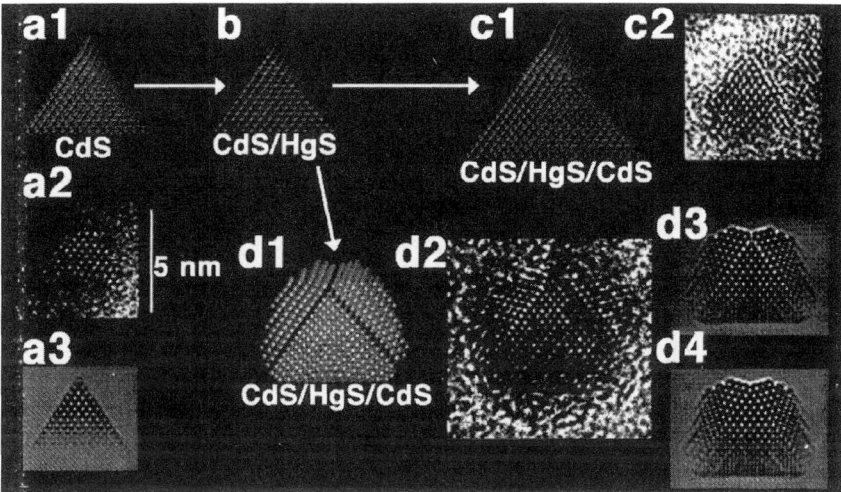

Figure 12 High-resolution TEM study of the structural evolution of a CdS core particle to a (CdS)(core)shell particle to the final CdS(HgS)CdS nanostructure. (a1) Molecular model showing that all surfaces are cadmium terminated (111); (a2) TEM of a CdS core that exhibits tetrahedral morphology; (a3) TEM simulation agreeing with (a2) micrograph. (b) Model of the CdS particle after surface modification with Hg. (c1) Model of a tetrahedral CdS(HgS)CdS nanocrystal; (c2) a typical TEM of a tetrahedral CdS(HgS)CdS nanocrystal. (d1) Model of a CdS(HgS)CdS nanocrystal after twinned epitaxial growth, where the arrow indicates the interfacial layer exhibiting increased contrast due to the presence of HgS; (d2) TEM of a CdS (HgS)CdS nanocrystal after twinned epitaxial growth; (d3) simulation agreeing with model (d1) and TEM (d2) showing increased contrast due to presence of HgS; (d4) simulation assuming all Hg is replaced by Cd—no contrast is evident. (From Ref. 36, reprinted with permission.)

the tetrahedral shapes are terminated by (111) surfaces that can be *either* cadmium *or* sulfur faces [36]. The choice of stabilizing agent—an anionic polyphosphate ligand—favors cadmium faces and likely supports the faceted tetrahedral structure that exposes exclusively cadmium-dominated surfaces (Fig. 12). In addition, both the spherical particles and the twinned tetrahedral particles provide evidence for an embedded HgS layer in the presumed QD/QW structure. Due to differences in their relative abilities to interact with electrons (HgS more strongly than CdS), contrast differences are evident in HR TEM images as bands of HgS surrounded by layers of CdS (Fig. 12).

Size dispersions in these low-temperature, ionic-ligand-stabilized reactions are reasonably good (~20%), as indicated by absorption spectra, but poor compared to those achieved using higher-temperature pyrolysis and amphiphilic coordinating ligands (4–7%). Nevertheless, the polar-solvent-based reactions give us access to colloidal materials, such as mercury chalcogenides, thus far difficult to prepare using pyrolysis-driven reactions (Sect. II). Further, the ion-exchange method provides the ability to grow well and shell structures that appear to be precisely one, two, or three monolayers deep. Heterogeneous nucleation provides less control over shell thicknesses, resulting in incomplete and variable multilayers (e.g., 1.3 or 2.7 monolayers on average, etc.). The stability of core–shell materials against solid-state alloying is an issue, at least for the CdS(HgS)CdS system. Specifically, cadmium in a CdS–HgS structure will, within minutes, diffuse to the surface of the nanoparticle, where it is subsequently replaced by a Hg^{2+} solvated ion [35]. This process is likely supported by the substantially greater aqueous solubility of Cd^{2+} compared to Hg^{2+}, as well as the structural compatibility between the two lattice-matched CdS and HgS crystal structures.

IV. SHAPE CONTROL

The nanoparticle growth process described in Section II, where fast nucleation is followed by slower growth, leads to the formation of spherical or approximately spherical particles. Such essentially isotropic particles represent the thermodynamic, lowest-energy shape for materials having relatively isotropic underlying crystal structures. For example, under this growth regime, the wurtzite crystal structure of CdSe, having a *c/a* ratio of ~1.6, fosters the growth of slightly prolate particles, typically exhibiting aspect ratios of ~1.2. Furthermore, even for materials whose underlying crystal structure is more highly anisotropic, nearly spherical nanoparticles typically result due to the strong influence of the surface in the nano-size regime. Surface energy is minimized in spherical particles, compared to more anisotropic morphologies.

Under a different growth regime, one that promotes fast, kinetic growth, more highly anisotropic shapes, such as rods and wires, can be obtained. In semiconductor nanoparticle synthesis, such growth conditions have been achieved using high precursor, or monomer, concentrations in the growth solution. As discussed previously (Sect. II), particle-size distributions can be "focused" by maintaining relatively high monomer concentrations that prevent the transition from the fast-growth to the slow-growth (Ostwald ripening) regime [14]. Even higher monomer concentrations can be used to effect a transition from thermodynamic to kinetic growth. Access to the regime of very fast kinetic growth allows control over particle shape. The system is essentially put into "kinetic overdrive," where dissolution of particles is minimized as the monomer concentration is maintained at levels higher than the solubility of all of the particles in solution. Growth of all particles is thereby promoted [14]. Further, in this regime, the rate of particle growth is not limited by diffusion of the monomer to the growing crystal surface, but, rather, by how fast atoms can add to that surface. In this way, the relative growth rates of different crystal faces have a strong influence over the final particle shape [38]. Specifically, in systems where the underlying crystal lattice structure is anisotropic (e.g., the wurtzite structure of CdSe), simply the presence of high monomer concentrations (kinetic growth regime) at and immediately following nucleation can accentuate the differences in relative growth rates between the unique c axis and the remaining lattice directions, promoting rod growth. The monomer-concentration-dependent transition from slower-growth to fast-growth regimes coincides with a transition from diffusion-controlled to reaction-rate-controlled growth and from dot to rod growth. In general, longer rods are achieved with higher initial monomer concentrations, and rod growth is sustained over time by maintaining high monomer concentrations using multiple-injection techniques. Finally, the relative rates of different crystallographic faces can be further controlled by the judicious choice of organic ligands [12,17].

In order to more precisely tune the growth rates controlling CdSe rod formation, high monomer concentrations are used in conjunction with appropriate organic ligand mixtures. In this way, a wide range of rod aspect ratios has been produced (Fig. 13) [12,17,38]. Specifically, the "traditional" TOPO ligand is supplemented with alkyl phosphonic acids. The phosphonic acids are strong metal (Cd) binders and may influence rod growth by changing the relative growth rates of the different crystal faces [38]. CdSe rods form by enhanced growth along the crystallographically unique c axis (taking advantage of the anisotropic wurtzite crystal structure). Interestingly, the fast growth has been shown to be unidirectional—exclusively on the $(00\bar{1})$ face [38]. The $(00\bar{1})$ facets comprise alternating Se and Cd layers, where the Cd atoms are relatively unsaturated (three dangling bonds per atom). In contrast,

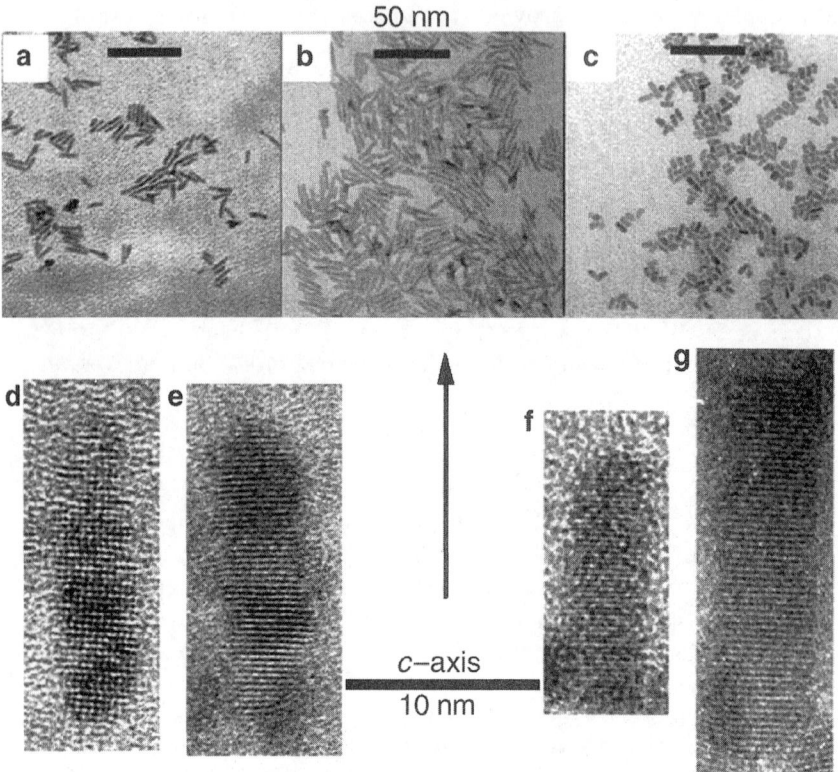

Figure 13 (a–c) Transmission electron micrographs of CdSe quantum rods demonstrating a variety of sizes and aspect ratios. (d–g) HR TEMs of CdSe quantum rods revealing lattice fringes and rod growth direction with respect to the crystallographic c axis. (From Ref. 17, reprinted with permission.)

the related (001) facet exposes relatively saturated Cd faces having one dangling bond per atom (Fig. 14). Thus, relative to (00$\bar{1}$), the (001) face (opposite c-axis growth) and {110} faces (*ab* growth), for example, are slow growing, and *unidirectional* rod growth is promoted. The exact mechanism by which the phosphonic acids alter the relative growth rates is not certain. Their influence may be in inhibiting the growth of (001) and {110} faces or it may be in directly promoting growth of the (00$\bar{1}$) face by way of interactions with surface metal sites [38]. Alternatively, it has been proposed that a more important contribution to the formation of rod-shaped particles by the strong metal ligands is their influence on "monomer" concentrations, where mono-

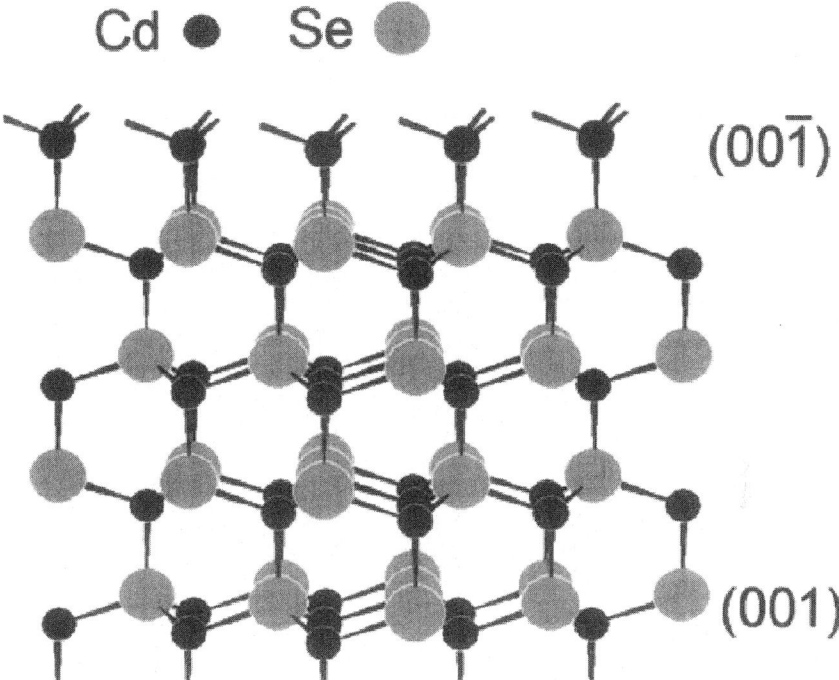

Figure 14 Atomic model of the CdSe wurtzite crystal structure. The (00$\bar{1}$) and the (001) crystal faces are emphasized to highlight the different number of dangling bonds associated with each Cd atom (three and one, respectively). (From Ref. 38, reprinted with permission.)

mer again refers to various molecular precursor species. Specifically, the phosphonic acids may simply *permit* the high monomer concentrations that are required for kinetic, anisotropic growth. As strong metal binders, they may coordinate Cd monomers, stabilizing them against decomposition to metallic Cd [12].

More complex shapes, such as "arrows," "pine trees," and "teardrops," have also been prepared in the CdSe system, and the methods used are an extension of those applied to the preparation of CdSe rods. Once again, CdSe appears to be the "proving ground" for semiconductor nanoparticle synthesis. Several factors influencing growth of complex shapes have been investigated, including the time evolution of shape and the ratio of TOPO to phosphonic acid ligands [38], as well as the steric bulk of the phosphonic acid [12]. Predictably, reaction temperature also influences the character of the growth regime [12,38]. In the regime of rod growth (i.e., fast kinetic growth), complex shapes can evolve over time. Rods and "pencils" transform into "arrows" and "pine-tree-shaped" particles (Fig. 15). The sides of the arrow or tree points comprise wurtzite (101) faces. As predicted by traditional crystal growth theory, these slower growing faces have replaced the faster growing $(00\bar{1})$ face, permitting the evolution to more complex structures [38]. Shape and shape evolution dynamics were also observed to be highly dependent on phosphonic acid concentrations. Low concentrations (<10 mol%)of hexyl-phosphonic acid (HPA), for example, relative to TOPO produced approximately spherical particles, whereas moderate amounts (20 mol%) yielded rods, and high concentrations (60 mol%) resulted in arrow-shaped particles. As discussed previously, HPA appears to enhance the growth of the $(00\bar{1})$ face *relative* to other crystallographic faces, and higher concentrations simply permit even higher relative growth rates. Therefore, shape evolution to the arrow and tree morphologies proceeds more quickly in the presence of high HPA concentrations. The growth of single-headed arrows, as opposed to double-headed ones, results from the characteristic unidirectional growth [i.e., growth from the $(00\bar{1})$ face only, not from the (001) face].

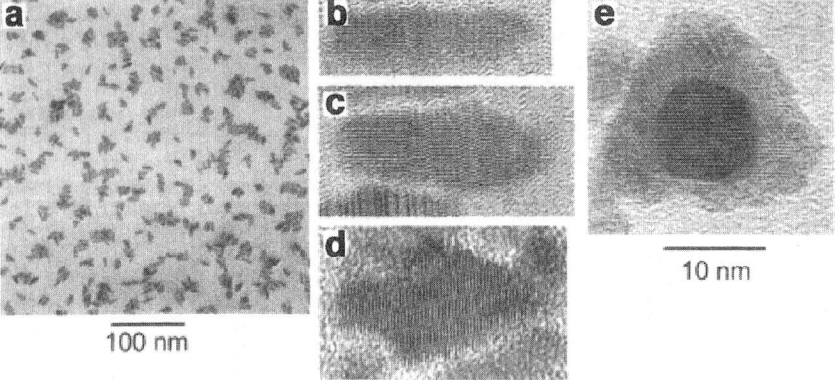

Figure 15 (a) Transmission electron micrographs of a CdSe NQD sample dominated by arrow-shaped particles (60% hexylphosphonic acid (HPA) reaction]. (b–d) HR TEMs demonstrating the shape evolution from (b) pencil-shaped to (c) arrow-shaped to (d) pine-tree-shaped CdSe NQDs. (e) Pine-tree-shaped particle looking down the [001] direction (i.e., the long axis). Analysis of lattice spacings obtained by HR TEM imaging revealed that wurtzite is the dominate phase for each shape and that the angled facets of the arrows comprise the (101) faces. (From Ref. 38, reprinted with permission.)

"Teardrop-shaped" particles also arise from the tendency toward unidirectional growth. In this case, rod-shaped crystals are exposed to growth conditions favoring spherical particle shapes (i.e., equilibrium slow growth and low monomer concentrations), causing the rods to become rounded. Monomer concentration is then quickly increased to force elongation of the "droplet" from one end into particles resembling tadpoles [38]. The growth regime governing the evolution of rods to spherical particles has been termed "1D to 2D intraparticle ripening" [12]. Nanoparticle volumes and total numbers remain approximately constant (as do monomer concentrations), whereas nanoparticle shape changes dramatically. Intraparticle diffusion of c-axis atoms to other crystal faces may explain this transformation. The process is distinguished from "interparticle ripening," or Ostwald ripening, which is observed at even lower monomer concentrations. Intraparticle ripening is thought to occur when a "diffusion equilibrium" exists between the nanoparticles and the monomers in the bulk solution [12]. Alternatively, it has also been shown that nanodots can be used to "seed" the growth of nanorods. Here, the spherical particles are exposed to high monomer concentrations that promote one-dimensional (1D) growth from the template particles. Improved short-axis and aspect-ratio distributions have been reported for these rods (Fig. 16) [12].

Rod-growth dynamics also depend on the identity of the phosphonic acid. The effectiveness of the phosphonic acid in promoting rod growth depends critically on its steric bulk, or the length of its alkyl chain. Shorter-chain phosphonic acids, such as HPA, more effectively promote rod growth compared to longer-chain phosphonic acids, such as tetradecylphosphonic acid (TDPA). Combinations of longer- and shorter-chain phosphonic acids can be used to readily tune rod aspect ratios [12] and control shape evolution dynamics.

The above morphologies reflect the underlying wurtzite crystal structure of CdSe. Occasionally, however, CdSe nucleates in the zinc-blende phase. When this occurs, a different type of morphology, the tetrapod, is observed. Here, the zinc-blende nuclei expose four equivalent (111) faces that comprise the crystallographic equivalent of the wurtzite (001) faces (alternating planes of Cd or Se). From these (111) surfaces, four wurtzite "arms" grow unidirectionally. Further addition of monomer either lengthens the wurtzite arms, in the case of "purely" wurtzite arms, or generates dendriticlike wurtzite branches, when zinc-blende stacking faults are present in the arm ends (Fig. 17) [38].

CdSe rod QYs in PL are typically relatively low, ~1–4%. Like their spherical counterparts, however, rods can be overcoated with a higher bandgap inorganic semiconductor, increasing QYs to 14–20% [39,40]. Lattice mismatch requirements for rods are somewhat more severe than for spherical

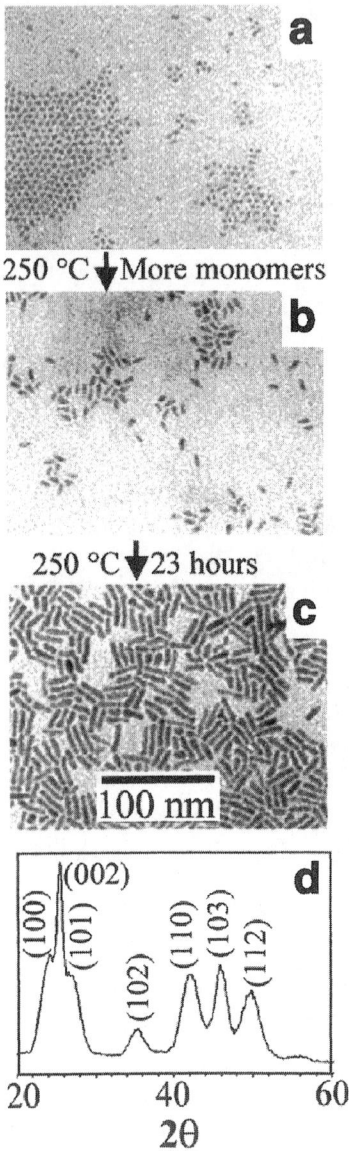

Figure 16 "Seeding" CdSe NQD rod growth for improved size monodispersity. (a) TEM of CdSe dots prepared in 13% tetradecylphosphonic acid (TDPA) using an injection temperature of 360°C and a growth temperature of 250°C; (b) TEM of NQD rods grown from the dot seeds following injection of additional monomer; (c) TEM of NQD rods after 23 h of growth; (d) x-ray diffraction (XRD) pattern for CdSe rods from (c). (From Ref. 12, reprinted with permission.)

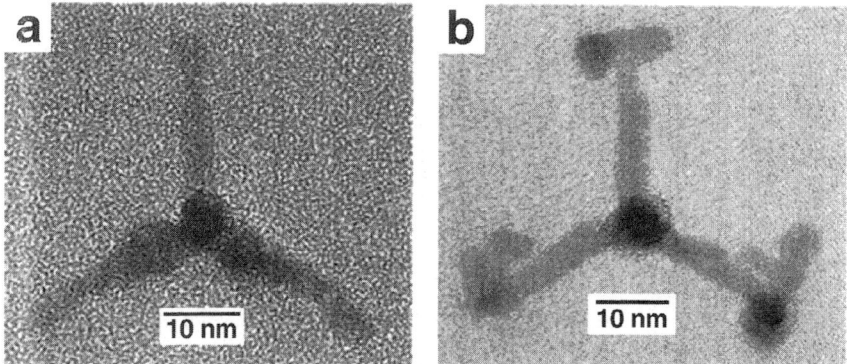

Figure 17 High-resolution TEMs of CdSe tetrapods. (a) Image down the [001] direction of one of the four arms. All arms are wurtzite phase, as confirmed by analysis of lattice spacings. (b) "Dendritic" tetrapod, where branches have grown from each arm. Some stacking faults are present in the branches, and zinc-blende layers are present at the ends of the original four arms. (From Ref. 38, reprinted with permission.)

particles, and synthetic steps unique to rod overcoating have been employed in the most successful preparations [40]. As discussed previously (Sect. III), spherical nanoparticle systems benefit from having highly curved surfaces, compared to less strain-tolerant planar systems. Nanorods provide an intermediate case. The average curvature of rods lies between that of dots and films, and, due to their larger size/surface area compared to dots, more interfacial strain can accumulate, leading to the formation of dislocations. The 12% lattice mismatch between ZnS and CdSe is, therefore, less well tolerated in rods. CdS can be used as a lattice-mismatch "buffer layer" between CdSe and ZnS (only ~4% lattice mismatch with CdSe and ~8% with ZnS). The addition of a small amount of Cd precursor to the shell precursor solution (Cd : Zn of 0.12 : 1.0) appears to lead to the preferential formation of CdS at the surface of the rods. ZnS growth then proceeds on the CdS. High-resolution TEM images demonstrate uniform and epitaxial growth. Interestingly, QYs remain low, and the benefits of inorganic over-coating in the graded epitaxial approach are only fully realized following photochemical annealing (via laser irradiation) of the rod particles (Fig. 18) [40].

Solution-phase preparations of unusually shaped and highly aniso-tropic particles that are soluble, relatively monodisperse, and sufficiently small to exhibit quantum-confinement effects are thus far limited, but not exclusive to CdSe. CdS and CdTe rods can be prepared using phosphonic-

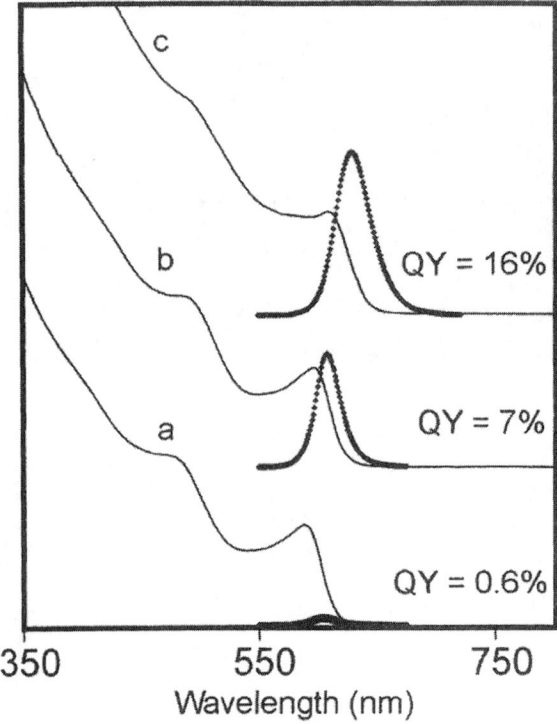

Figure 18 Absorption (solid line) and PL (dashed line) spectra for medium-length (3.3 × 21 nm) CdSe nanorods; (a) core nanorods without ZnS shell; (b) (core)shell nanorods with thin CdS–ZnS shells (~2 monolayers of shell material, where the CdS "buffer" shell comprises ~35% of the total shell); (c) (core)shell nanorods with medium CdS–ZnS shells (~4.5 monolayers of shell material, where the CdS "buffer" shell comprises ~22% of the total shell). PL spectra were recorded following photoannealing of the samples. (From Ref. 40 reprinted with permission.)

acid-controlled reactions. In addition, CdS rods and multipods can be prepared in a monosurfactant system in which hexadecylamine (HDA) serves both as the stabilizing ligand and as the shape-determining ligand [41]. Here, rod and multipod formation is temperature dependent. Rods form at high temperatures (~300°C), whereas bipods, tripods, and tetrapods dominate at lower temperature (120–180°C). The dependence of shape on temperature likely results from preferential formation of wurtzite CdS nuclei (thermodynamic phase) at high temperatures and zinc-blende nuclei (kinetic phase) at lower temperatures. As in the CdSe system, the zinc-blende {111} faces can support fast growth of (00$\bar{1}$) wurtzite "arms." Significantly, this method allows isolation of tetrapods in ~82% yield (compared to 15–40% by the HPA method) at 120°C. Size control is less well developed, and particles are relatively large compared to their HPA-derived counterparts. Nevertheless, it is the first significant report of solution-based growth of bipod and tripod morphologies in the II–VI system and provides a more predictable method of producing tetrapods [41].

The same monosurfactant system can be applied to shape-controlled preparation of the magnetic semiconductor MnS [42]. At low solution-growth temperatures (120–200°C), MnS prepared from the single-source precursor $Mn(S_2CNEt_2)_2$ can nucleate in either the zinc-blende or the wurtzite phase, whereas at high temperatures (>200°C), MnS nucleates in the rock-salt phase. Low-temperature growth yields a variety of morphologies: highly anisotropic nanowires, bipods, tripods, and tetrapods (120°C), nanorods (150°C), and spherical particles (180°C). The "single pods" comprise wurtzite cores with wurtzite-phase arms. In contrast, the multipods comprise zinc-blende cores with wurtzite arms, where the arms grow in the [001] direction from the zinc-blende {111} faces, as discussed previously with respect to the Cd–chalcogenide system. Dominance of the isotropic spherical particle shape in reactions conducted at moderate temperatures (180°C) implies a shift from predominantly kinetic control to predominantly thermodynamic control over the temperature range from 120°C to 180°C [42]. Formation of 1D particles at low temperatures results from kinetic control of relative growth rates. At higher temperatures, differences in activation barriers to growth of different crystal faces are more easily surmounted, equalizing relative growth rates. Finally, high-temperature growth supports only the thermodynamic rock-salt structure—large cubic crystals. Also, by combining increased growth times with low growth temperatures, shape evolution to "higher-temperature" shapes is achieved [42].

Extension of the ligand-controlled shape methodology to highly symmetric cubic crystalline systems is also possible. Specifically, PbS, having the rock-salt structure, can been prepared as rods, tadpole-shaped monopods, multipods (bipods, tripods, tetrapods, and pentapods), stars, truncated octahedra, and cubes [43]. The rod-based particles, including the monopods and multipods, retain short-axis dimensions that are less than the PbS Bohr exciton radius (16 nm) and, thus, can potentially exhibit quantum-size effects. These highly anisotropic particle shapes represent truly metastable morphologies for the inherently isotropic PbS system. The underlying PbS crystal lattice is the symmetric rock-salt structure, the thermodynamically stable manifestations of which are the truncated octahedra and the cubic nanocrystals. The PbS particles are prepared by pyrolysis of a single-source precursor, $Pb(S_2CNEt_2)_2$, in hot phenyl ether in the presence of a large excess of either a long-chain alkyl thiol or amine. The identity of the coordinating

ligand and the solvent temperature determine the initial particle shape
following injection (Fig. 19). Given adequate time, particle shapes evolve
from the metastable rods to the stable truncated octahedron and cubes, with
star-shaped particles comprising energetically intermediate shapes [43].

As in the CdSe system, the particle shape in the cubic PbS system
depends intimately on the ligand concentration and its identity. The highest

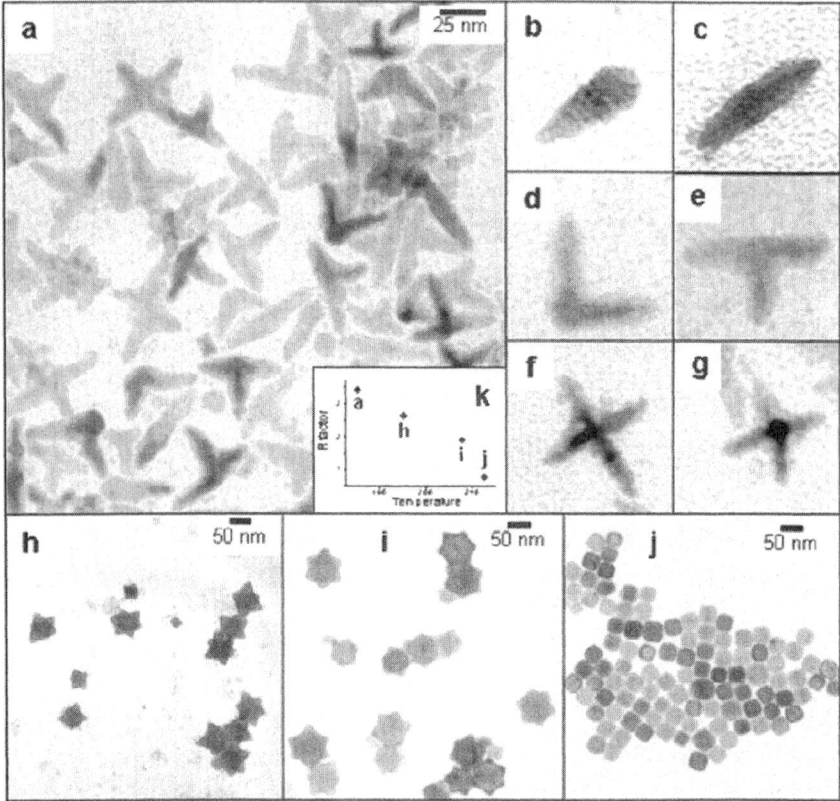

Figure 19 Transmission electron micrographs showing the variety of shapes
obtained from the PbS system grown from the single-source precursor $Pb(S_2CNEt_2)_2$
at several temperatures: (a) Multipods prepared at $140°C$; (b) tadpole-shaped
monopod ($140°C$); (c) I-shaped bipod ($140°C$); (d) L-shaped bipod ($140°C$); (e) T-
shaped tripod ($140°C$); (f) cross-shaped tetrapod ($140°C$); (g) pentapod ($140°C$); (h)
star-shaped nanocrystals prepared at $180°C$; (i) Rounded star-shaped nanocrystals
prepared at $230°C$; (j) truncated octahedra prepared at $250°C$. (From Ref. 43;
reprinted with permission.)

ligand concentrations yield a reduced rate of growth from the {111} faces compared to the {100} faces, which experience enhanced *relative* growth rates. Further, alkylthiols are more effective in controlling relative growth rates compared to alkylamines. The latter, a weaker Pb binder, consistently leads to large, thermodynamically stable cubic shapes. Finally, the reaction temperature plays a key role in determining particle morphology. The lowest temperatures (140°C) yield the metastable rod-based morphologies, with intermediate star shapes generated at moderate temperatures (180–230°C) and truncated octahedra isolated at the highest temperatures (250°C). Interestingly, the rod structures appear to form by preferential growth of {100} faces from truncated octahedra seed particles. For example, the "tadpole"-shaped monopods are shown by HR TEM studies to comprise truncated octahedra "heads" and [100]-axis "tails," resulting from growth from a (100) face. The star-shaped particles that form at 180°C are characterized by six triangular corners, comprising each of the six {100} faces. The {100} faces have "shrunk" into these six corners as a result of their rapid growth, similar to the replacement of the $(00\bar{1})$ face by slower growing faces during the formation of arrow-shaped CdSe particles (discussed earlier). The isolation of star-shaped particles at intermediate temperatures suggests that the relative growth rates of the {100} faces remain enhanced compared to the {111} faces at these temperatures. Further, the overall growth rate is enhanced as a result of the higher temperatures. The star-shaped particles that form at 230°C are rounded and represent a decrease in the differences in relative growth rates between the {100} faces and the {111} faces, the latter, higher-activation-barrier surface benefiting from the increase in temperature. A definitive shift from kinetic growth to thermodynamic growth is observed at 250°C (or at long growth times). Here, the differences in reactivity between the {100} and the {111} faces are negligible given the high-thermal-energy input that surmounts either face's activation barrier. The thermodynamic cube shape is, therefore, approximated by the shapes obtained under these growth conditions. In all temperature studies, the alkylthiol:precursor ratio was ~80:1 and monomer concentrations were kept high, conditions supporting controlled and kinetic growth, respectively.

The III–V semiconductors have proven amenable to solution-phase control of particle shape using an unusual synthetic route. Specifically, the method involves the solution-based catalyzed growth of III–V nanowhiskers [44]. In this method, referred to as the "solution–liquid–solid mechanism," a dispersion of nanometer-sized indium droplets in an organometallic reaction mixture serves as the catalytic sites for precursor decomposition and nanowhisker growth. As initially described, the method afforded no control over nanowhisker diameters, producing very broad diameter distributions and mean diameters far in excess of the strong-confinement regime for III–V semiconductors. Additionally, the nanowhiskers were insoluble, aggregating and precipitating upon growth. However, recent studies have demonstrated that the nanowhisker mean diameters and diameter distributions are controlled through the use of near-monodisperse metallic-catalyst nanoparticles [45,46]. The metallic nanoparticles are prepared over a range of sizes by heterogeneous seeded growth [47]. The solution–liquid–solid mechanism in conjunction with the use of these near-monodisperse catalyst nanoparticles and polymer stabilizers affords soluble InP and GaAs nanowires having diameters in the range 3.5–20 nm and diameter distributions of ± 15–20% (Fig. 20). The absorption spectra of the InP quantum wires contain discernible excitonic features from which the size dependence of the bandgap has

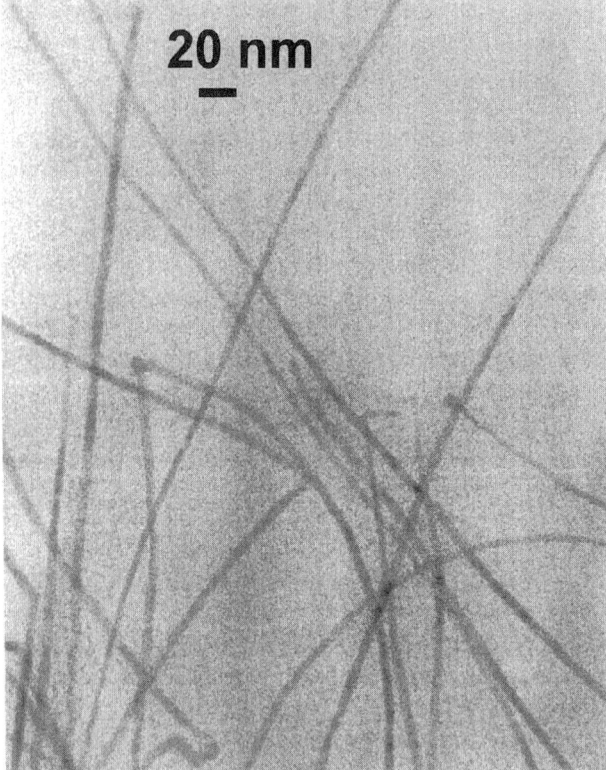

Figure 20 Transmission electron micrograph of InP quantum wires of diameter 4.49 ± 0.75 (±17%), grown from 9.88 ± 0.795 (±8.0%) In-catalyst nanoparticles. The values following the "±" symbols represent one standard deviation in the corresponding diameter distribution. (From Ref. 46, reprinted with permission.)

been determined and quantitatively compared to that in InP quantum dots [46]. Preliminary results indicate that II–VI quantum wires can be grown similarly [48]. A similar approach was applied to growth of insoluble, but size-monodispersed in diameter (4–5 nm), silicon nanowires. Here, reactions were conducted at elevated temperature and pressure (500°C and 200–270 bar, respectively) using alkanethiol-coated gold nanoclusters (2.5 ± 0.5 nm diameter) as the nucleation and growth "seeds" [49].

V. PHASE TRANSITIONS AND PHASE CONTROL

Nanocrystal quantum dots have been used as model systems to study solid–solid phase transitions [50–53]. The transitions, studied in CdSe, CdS, InP and Si nanocrystals [52], were induced by pressure applied to the nanoparticles in a diamond anvil cell by way of a pressure-transmitting solvent medium, ethyl-cyclohexane. Such transitions in bulk solids are typically complex and dominated by multiple nucleation events, the kinetics of which are controlled by crystalline defects that lower the barrier height to nucleation [50,53]. In nearly defect-free nanoparticles, the transitions can exhibit single-structural-domain behavior and are characterized by large kinetic barriers (Fig. 21). In contrast to original interpretations which described the phase transition in nanocrystals as "coherent" over the entire nanocrystal [50], the nucleation of the phase transition process was recently shown to be localized to specific crystallographic planes [53]. The simple unimolecular kinetics of the transition still support a single nucleation process; however, the transition is now thought to result from plane sliding as opposed to a coherent deformation process. Specifically, the sliding-plane mechanism involves shearing motion along the (001) crystallographic planes, as supported by detailed analyses of transformation times as a function of pressure and temperature.

Because of the large kinetic barriers in nanocrystal systems, their phase transformations are characterized by hysteresis loops (Fig. 21) [50,51,53]. The presence of a strong hysteresis signifies that the phase transition does not occur at the thermodynamic transition pressure and that time is required for the system to reach an equilibrium state. This delay is fortunate in that it permits detailed analysis of the transition kinetics even though the system is characterized by single-domain (finite-size) behavior. As alluded to, these analyses were used to determine the structural mechanism for transformation. Specifically, kinetics studies of transformation times as a function of temperature and pressure were used to determine relaxation times, or average times to overcome the kinetic barrier, and, thereby, rate constants. The temperature dependence of the rate constants led to the determination of activation energies for the forward and reverse transitions, and the pressure dependence

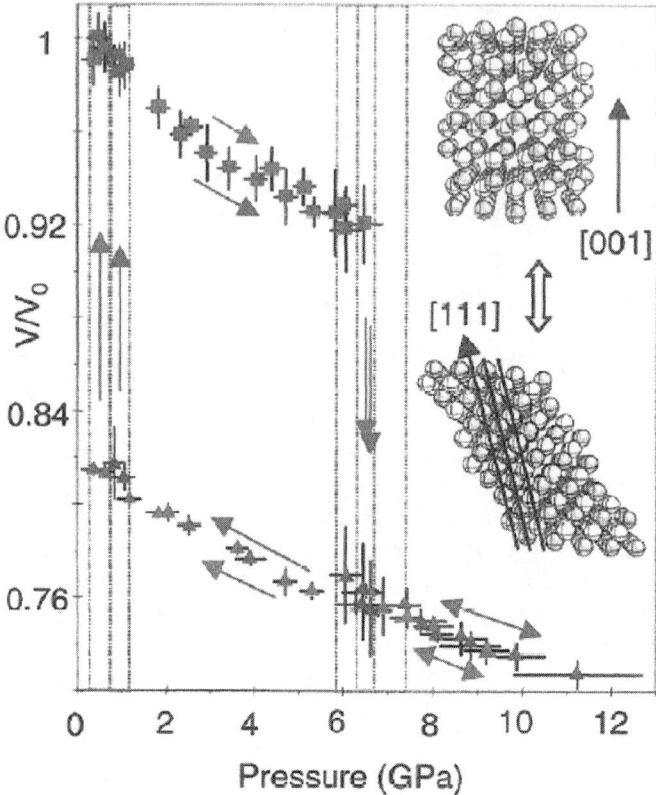

Figure 21 Two complete hysteresis cycles for 4.5-nm CdSe NQDs presented as unit cell volumes for the wurtzite sixfold-coordinated phase (triangles) and the rock-salt fourfold-coordinated phase (squares) versus pressure. Solid arrows indicate the direction of pressure change, and dotted boxes indicate the mixed-phase regions. Unlike bulk-phase transitions, the wurtzite to rock-salt transformation in nano-crystals is reversible and occurs without the formation of new high-energy defects, as indicated by overlapping hysteresis loops. The shape change that a sliding-plane transformation mechanism (see text) would induce is shown schematically on the right. (From Ref. 51; reprinted with permission.)

of the rate constants led to the determination of activation volumes for the process [53]. The latter represents the volume change between the starting structure and an intermediate transitional structure. Activation volumes for the two directions, wurtzite to rock salt and rock salt to wurtzite, respectively, were unequal and smaller for the latter, implying that the intermediate structure more closely resembles the four-coordinate structure. The activation volumes were also shown to be of opposite sign, indicating that the mechanism by which the phase transformation takes place involves a structure whose volume is in between that of the two end phases. Most significantly, the magnitude of the activation volume is small compared to the total volume change that is characteristic of the system (~0.2% versus 18%). The activation volume is equal to the critical nucleus size responsible for initiating the phase transformation—*defining the volume change associated with the nucleation event*.

The small size of the activation volume suggests that the structural mechanism for transformation cannot be a coherent one involving the entire nanocrystal [53]. Spread out over the entire volume of the nanocrystal, the activation volume would amount to a volume change smaller than that induced by thermal vibrations in the lattice. Therefore, a mechanism involving some fraction of a nanocrystal was considered. The nucleus was determined not to be three dimensional, as a sphere the size of the activation volume would be less than a single unit cell. Also, activation volumes were observed to increase with increasing particle size (in the direction of increasing pressure). There is no obvious mechanistic reason for a spherically shaped nucleus to increase in size with an increase in particle size. Further, additional observations have been made: (1) particle shape changes from cylindrical or elliptical to slablike upon transformation from the four-coordinate phase to the six-coordinate phase [51], (2) the stacking-fault density increases following a full pressure cycle from the four-coordinate through the six-coordinate and back to the four-coordinate structure [51], and (3) the entropic contribution to the free-energy barrier to transformation increases with increasing size (indicating the nucleation event can initiate from multiple sites) [53]. Together, the various experimental observations suggest that the mechanism involves a directionally dependent nucleation process that is not coherent over the whole nanocrystal. The specific proposed mechanism entails shearing of the (001) planes, with precedent found in martensitic phase transitions (Fig. 22) [51,53]. Further, the early observation that activation energy increases with size [50] likely results from the increased number of chemical bonds that must be broken for plane sliding to occur in large nanocrystals, compared to that in small nanocrystals. Such a mechanistic-level understanding of the phase transformation processes in nanocrystals is important because nanocrystal-based studies, due to their simple kinetics, may ultimately provide a better understanding of the hard-to-study, complex transformations that occur in bulk materials and geologic solids [53].

Phase control, much like shape control (Sect. IV), can be achieved in nanocrystal systems by operating in kinetic growth regimes. Materials synthesis strategies have typically relied upon the use of reaction conditions far from standard temperature and pressure (STP) to obtain nonmolecular

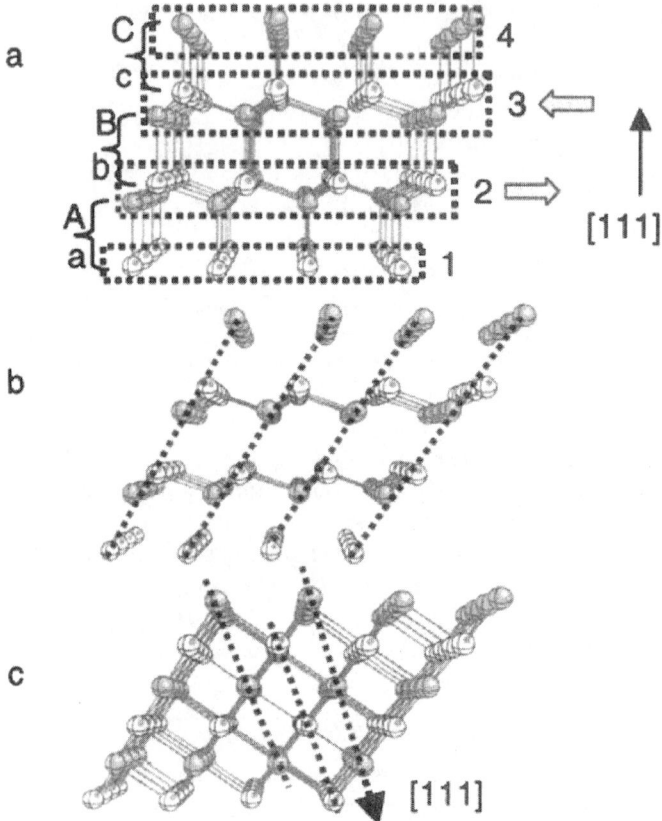

Figure 22 Schematic illustrating the sliding-plane transformation mechanism: (a) zinc-blende structure, where brackets denote (111) planes, dashed boxes show planes that slide together, and arrows indicate the directions of movement; (b) structure of (a) after successive sliding has occurred; (c) rock-salt structure, where dashed lines denote (111) planes. Structures are oriented the same in (parts a–c). (From Ref. 51, reprinted with permission.)

materials such as ceramics and semiconductors. The crystal-growth barriers to covalent nonmolecular solids are high and have historically been surmounted by employing relatively extreme conditions, comprising a direct assault on the thermodynamic barriers to solid-state growth. The interfacial processes of adsorption–desorption and surface migration permit atoms initially located at nonlattice sites on the surface of a growing crystal to relocate to a regular crystal lattice position. When these processes are inefficient or not functioning, amorphous material can result. Commonly, synthesis temperatures of $\geq 400\,^{\circ}\text{C}$ are required to promote these processes leading to crystalline growth [54,55]. Such conditions can preclude the formation of kinetic, or higher-energy, materials and can limit the selection of accessible materials to those formed under thermodynamic control—the lowest-energy structures [56,57]. In contrast, biological and organic–chemical synthetic strategies, often relying on catalyzed growth to surmount or lower-energy barriers, permit access to both lowest-energy and higher-energy products [56], as well as access to a greater variety of structural isomers compared to traditional, solid-state synthetic methods. The relatively low-temperature, surfactant-supported, solution-based reactions employed in the synthesis of NQDs provide for the possibility of forming kinetic phases (i.e., those phases that form the fastest under conditions that prevent equilibrium to the lowest-energy structures). Formation of the CdSe zinc-blende phase, as opposed to the wurzite structure, is likely a kinetic product of low-temperature growth. In general, however, examples are relatively limited. More examples are to be found in the preparation of nonmolecular *solid* thin films: electrodeposition onto single-crystal templating substrates [58], chemical vapor deposition using single-source precursors having both the target elements and the target structure built in [59], and reaction of nano-thin-film, multilayer reactants to grow metastable, superlattice compounds [60–62]. One clear example from the solution phase is that of the formation of the metastable, previously unknown, rhombohedral InS (R-InS) phase [63]. The organometallic precursor t-Bu$_3$In was reacted with H$_2$S(g) at ~200 °C in the presence of a protic reagent, benzenethiol. This reagent provided the apparent dual function of catalyzing efficient alkyl elimination and supplying some degree of surfactant stabilization. Although the starting materials were soluble, the final product was not. Nevertheless, characterization by TEM and powder X-ray diffraction (XRD) revealed that the solid-phase product was a new layered InS phase, structurally distinct from the thermodynamic network structure—orthorhombic β-InS. Further, the new phase was 10.6% less dense compared to β-InS and was, therefore, predicted to be a low-temperature kinetic structure. To confirm the relative kinetic-thermodynamic relationship between R-InS and β-InS, the new phase was placed back into an organic solvent (reflux temperature ~200 °C) in the presence of a molten indium metal flux. The metal flux (molten nanodroplets) provided a convenient recrystallization medium, effecting *equilibration* of the layered and network structures allowing conversion to the more stable, thermodynamic network β-InS. The same phase transition can occur by simple solid-state annealing; however, significantly higher temperatures (>400 °C) are required. That the flux-mediated process involves true, direct conversion of one phase to the other (rather than dissolution into the flux followed by nucleation and crystallization) was demonstrated by subjecting a sample powder containing significant amorphous content to the metal flux. The time required for complete phase transformation was several times that of the simple R-InS to β-InS conversion.

VI. NANOCRYSTAL DOPING

Incorporation of dopant ions into the crystal lattice structure of an NQD by direct substitution of constituent anions or cations involves synthetic challenges unique to the nanosize regime. Doping in nanoparticles entails synthetic constraints not present when considering doping at the macroscale. The requirements for relatively low growth temperatures (for solvent/ligand compatibility and controlled growth) and for low posttreatment temperatures (to prevent sintering), as well as the tendency of nanoparticles to efficiently exclude defects from their cores to their surfaces (see Sect. V) are nanoscale phenomena. Dopant ions in such systems can end up in the external sample matrix, bound to surface ligands, adhered directly to nanoparticle surfaces, doped into near-surface lattice sites, or doped into core lattice sites [64].

Given their dominance in the literature thus far, this section will focus on a class of doped materials known, in the bulk phase, as dilute magnetic semiconductors (DMSs). Semiconductors, bulk or nanoparticles, that are doped with magnetic ions are characterized by a *sp–d* exchange interaction between the host and the dopant, respectively. This interaction provides magnetic and magneto-optical properties that are unique to the doped material. In the case of doped nanoparticles, the presence of a dopant paramagnetic ion can essentially mimic the effects of a large external magnetic field. Magneto-optical experiments are, therefore, made possible merely by doping. For example, fluorescence line-narrowing studies on Mn-doped CdSe NQDs are consistent with previous studies on *undoped* NQDs in an external magnetic field [64]. Further, *nanosized* DMS materials provide the possibility for additional control over material properties as a result of enhanced carrier spatial confinement. Specifically, unusually strong interactions between electron and hole spins and the magnetic ion dopant should exist [65,66]. Such carrier spin interactions were observed in Cd(Mn)S [65] and in Zn(Mn)Se [66] as giant splitting of electron and hole spin sublevels using magnetic circular dichroism (MCD). Under ideal dopant conditions—a single Mn ion at the center of an NQD—it is predicted that significantly enhanced spin-level splitting, compared to that in bulk semimagnetic semiconductors, would result [65]. Host–dopant interactions are also apparent in simple PL experiments. Emission from a dopant ion such as Mn^{2+} can occur by way of energy transfer from NQD host to dopant and can be highly efficient (e.g., QY = 22% at 295 K and 75% below 50 K [66]) (Fig. 23). The

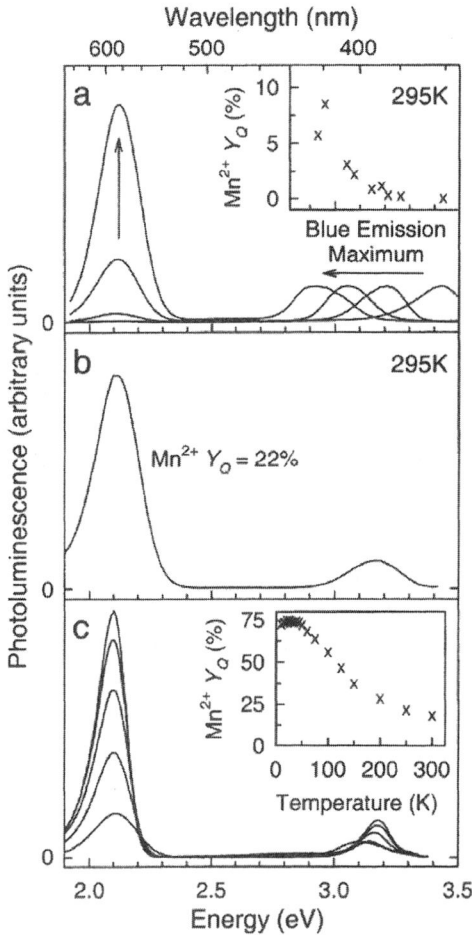

Figure 23 (a) Photoluminescence spectra taken at 295 K for a size series of Mn-doped ZnSe NQDs. As ZnSe emission ("blue emission") shifts to lower energies with increasing particle size, Mn emission increases in intensity. The particle diameters represented are <2.7, 2.8, 4.0, and 6.3 nm. The reaction Mn concentration (C_I) for each is 2.5%, whereas the actual (doped) concentration is ~1–5% of C_I. The inset shows Mn emission QY (Y_Q) versus the ZnSe emission maximum. (b) PL spectrum and Mn^{2+} QY (22%) for 2.85 nm Mn-doped ZnSe NQDs prepared using a C_I of 6.3%. (c) Temperature-dependent PL spectra for doped NQDs from (b); QY reaches a maximum of 75% below 50 K. (From Ref. 66, reprinted with permission.)

dopant emission signal occurs to the red of the NQD emission signal, or it overlaps NQD PL if the latter is dominated by deep-trap emission. Its presence has been cited as evidence for successful doping; however, the required electronic coupling can exist even when the "dopant" is located outside of the NQD [64]. Therefore, other methods are now preferred in determining the success or failure of a doping procedure.

Successful "core" doping was first achieved using low-temperature growth methods, such as room-temperature condensation from organometallic precursors in the presence of a coordinating surfactant [67] or room-temperature inverse-micelle methods [68–70]. Unfortunately, due to relatively poor NQD crystallinity and/or surface passivation, photoluminescence from undoped semiconductor nanoparticles prepared by such methods is generally characterized by weak and broad deep-trap emission. Thus, NQD quality is not optimized in such systems. Other low-temperature methods commonly used to prepare "doped" nanocrystals have been shown to yield only "dopant-associated" nanocrystals. For example, the common condensation reaction involving completely uncontrolled growth performed at room temperature by simple aqueous-based coprecipitation from inorganic salts (e.g., Na_2S and $CdSO_4$, with $MnSO_4$ as the dopant source), in the absence of organic ligand stabilizers, yields agglomerates of nano-sized domains and *unincorporated* dopant. [113]Cd- and ^{1}H-NMR (nuclear magnetic resonance) were used to demonstrate that Mn^{2+} remained outside of the NQD in these systems [71]. Doping into the crystalline lattice, therefore, appears to require some degree of control over particle growth when performed *at room temperature* (i.e., excluding higher-temperature, solid-state pyrolysis reactions that can yield well-doped nanocrystalline, although *not* quantum confined (>20 nm), material in the absence of any type of ligand control or influence [72]). The ability to distinguish between surface-associated and truly incorporated dopant ions is critical. Both can provide the necessary electronic coupling to achieve energy transfer and the resultant dopant emission signal, for example. Various additional characterization methods have been employed, such as NMR spectroscopy [64,71], electron paramagnetic resonance (EPR) [42,64–66,68,73], powder XRD, x-ray absorption fine-structure spectroscopy (XAFS) [73], chemical treatments (e.g., surface-exchange reactions and chemical etching–see below) [42,64], and ligand-field electronic absorption spectroscopy (see below) [74].

More recently, doped NQDs have been prepared by high-temperature pyrolysis of organometallic precursors in the presence of highly coordinating ligands: Zn(Mn)Se at an injection temperature of 310°C [66] and Cd(Mn)Se at an injection temperature of 350°C [64]. The undoped NQDs prepared by such methods are very well size selected (~4–7%), highly crystalline, and well passivated [64,66]. However, the dopant is incorporated into the NQD at low levels (~ ≤1 Mn per NQD). Despite input concentrations of ~0.5–5% dopant precursor, dopant incorporation is only ~0.025–0.125%. In the Cd(Mn)Se system, for example, Mn^{2+} incorporation into CdSe was limited to near-surface lattice sites, and the remaining Mn^{2+} was present merely as "surface-associated" ions. The location of the dopant ions following DMS NQD preparation was elucidated using a combination of chemical surface treatments and EPR measurements. Surface-exchange reactions, involving thorough replacement of TOPO/TOP capping ligands for pyridine, revealed that much of the dopant cations were only loosely associated with the NQD surface. EPR spectra following surface exchange showed a dramatic decrease in intensity following surface exchange (Fig. 24), indicating that the majority of dopant ions were not successfully incorporated into the CdSe crystal lattice. Further, even limited incorporation of Mn into the CdSe lattice required the use of a single-source Mn–Se precursor $[Mn_2(\mu\text{-SeMe})_2(CO)_8]$, rather than a simple Mn-only precursor [e.g., $MnMe_2$, $Mn(CO)_5Me$, $(MnTe (CO)_3(PEt_3)_2)_2$]. In the absence of the single-source dopant precursor, which

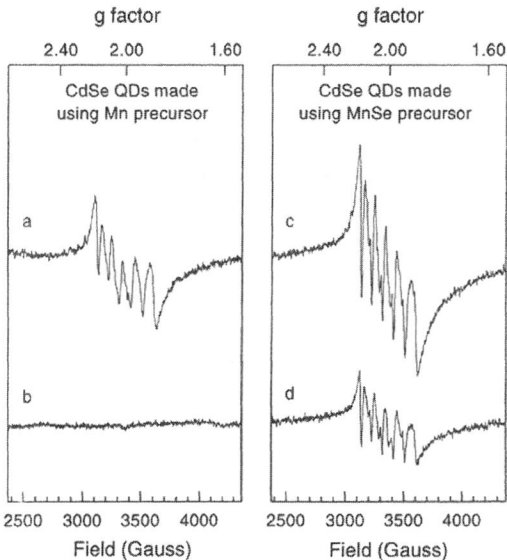

Figure 24 Low-temperature (5 K) EPR spectra of 4.0-nm-diameter Mn-doped CdSe NQDs prepared using (a,b) a Mn-only precursor and (c,d) the $Mn_2(\mu\text{-SeMe})_2(CO)_8$ single-source precursor. Before purification (a,c), both samples display the six-line pattern characteristic of Mn. After pyridine cap exchange (b,d), only the sample prepared with the single-source precursor retains the Mn signal. (From Ref. 64, reprinted with permission.)

apparently facilitated Mn^{2+} incorporation by supplying preformed Mn—Se bonds, EPR spectra following surface exchange with pyridine were structureless (Fig. 24a and 24b). Further, in the case of Cd(Mn)Se, chemical etching experiments were conducted to remove surface layers of the NQDs and, with them, any dopant that resided in these outer lattice layers. Etching revealed that the distribution of Mn in the CdSe was *not random*. Most Mn^{2+} dopant ions resided in the near-surface layers, and only a small fraction resided near the core. These results are suggestive of a "zone-refining" process [64] and consistent with the previously discussed tendency of NQDs to exclude defects. In the case of Zn(Mn)S DMSs, EPR and MCD experiments demonstrated that the majority of Mn^{2+} dopant resided well inside the NQD in high-symmetry, cubic Zn lattice sites. The dominant EPR signal comprised six-line spectra that exhibited hyperfine splitting of 60.4×10^{-4} cm^{-1}, similar to the splitting observed for Mn in bulk ZnSe (61.7×10^{-4} cm^{-1}) [66]. Also, the presence of giant spin sublevel splitting at zero applied field, as demonstrated by MCD, provided additional evidence that the Mn^{2+} dopant resided inside the NQD. The dopant-induced sublevel splitting occurs only when there is wave function overlap between the dopant and the confined electron–hole pair (i.e., only when Mn^{2+} resides inside the NQD) [65,66].

Sufficient experimental work has been conducted to begin to make a few general statements regarding solution-based preparation of DMS NQDs and the synthetic parameters that most strongly influence the success of the doping process. First, dopant ions can be excluded from the interior of the NQD to near-surface lattice sites when high-temperature nucleation and growth is employed [64] and/or limited (apparently) to approximately less than or equal to one dopant ion per NQD under such high-temperature conditions [64,66]. Lower-temperature approaches appear to provide higher doping levels. For example, a "moderate-temperature" organometallic-precursor approach was recently used to successfully internally dope CdS with Mn at very high levels: 2–12% as indicated by changes in x-ray diffraction patterns with increasing Mn concentrations (Fig. 25) (EPR hyperfine structure consistent with high-symmetry coordination of the dopant cations was most convincing for dopant concentrations ≤4%; Fig. 26 [42]). Exclusion from the lattice was only observed at dopant levels >15%. Single-source precursors were used for both core and dopant ions [$Cd(S_2CNEt_2)_2$ and $Mn(S_2CNEt_2)_2$, respectively], and the reaction temperature was 120°C. The particles were rather large and rod-shaped (Fig. 27). Significantly, if repeated at 300°C, dopant levels were limited to <2% Mn [42]. Additional temperature effects have been observed in annealing studies. Specifically, in $Cd_{0.95}Mn_{0.05}S$ NQDs, it has been demonstrated that postpreparative heat treatment can force Mn^{2+} to the surface. The Mn^{2+} PL signal decreases to zero following anneal treatments above 200°C, and no Mn is detected by EDS following

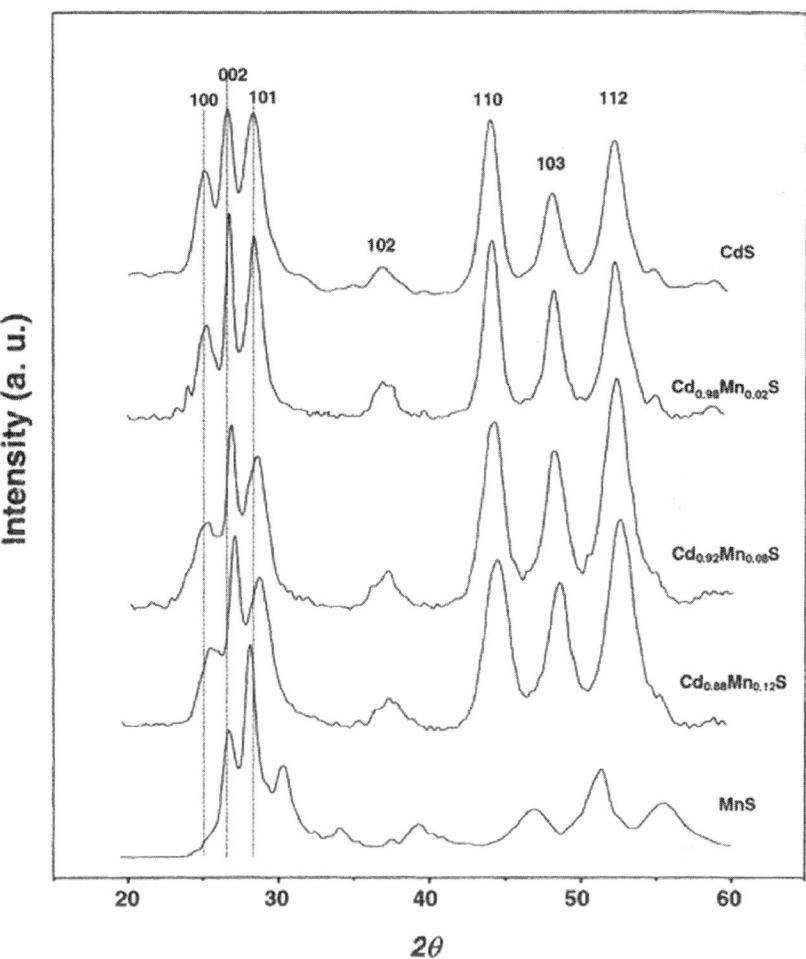

Figure 25 X-ray diffraction patterns for $Cd_{1-x}Mn_xS$ nanorods, where x in the doped samples varies from 2% to 8% to 12%. Peaks shift to higher 2θ with increasing Mn concentration. The magnitudes of the observed shifts are consistent with that predicted by Vegard's law, indicating a homogeneous distribution of Mn within the CdS matrix. (From Ref. 42; reprinted with permission.)

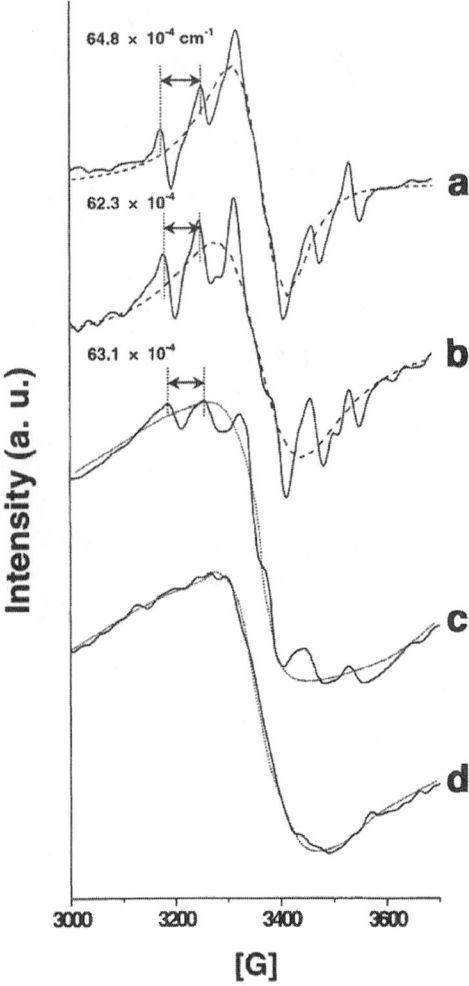

Figure 26 Electron paramagnetic resonance spectra for $Cd_{1-x}Mn_xS$ nanorods: (a) $x = 0.02$; (b) $x = 0.04$; (c) $x = 0.08$; (d) $x = 0.12$. Hyperfine splitting due to Mn ($I = 5/2$) is evident in (a–c). The background Lorentzian-curve pattern is attributed to Mn–Mn interactions. (From Ref. 42, reprinted with permission.)

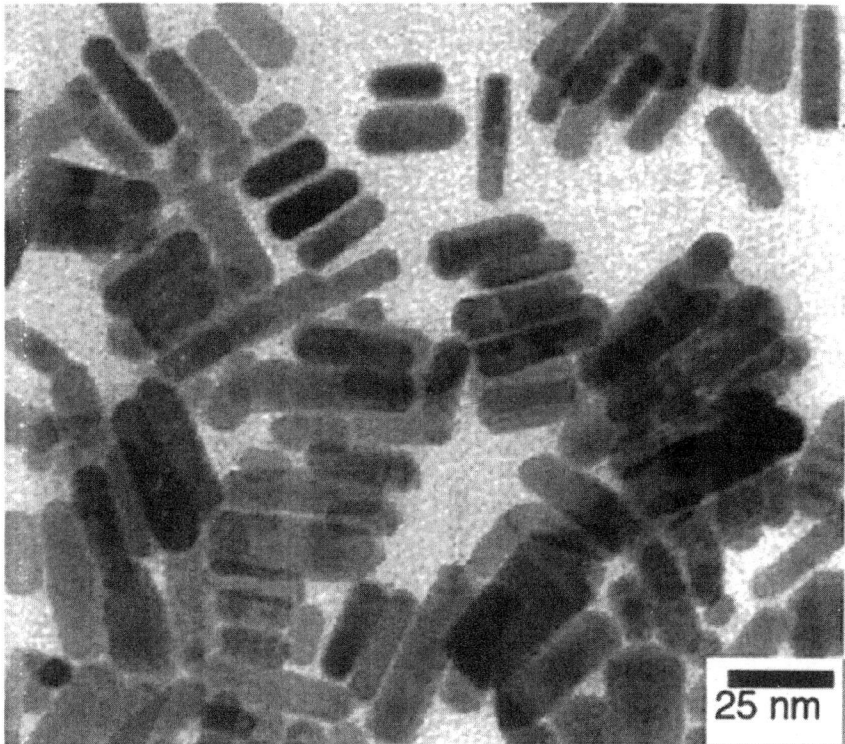

Figure 27 Transmission electron micrograph of $Cd_{0.88}Mn_{0.12}S$ rods, 7 nm wide with an aspect ratio of ~4. (From Ref. 42, reprinted with permission.)

anneal treatments above 120°C [75]. Finally, low-temperature inverse-micelle reactions appear to benefit from an "aging" process. The micelle solutions are allowed to "age" prior to ligand stabilization and workup. This process is thought to entail particle growth by Ostwald ripening and concurrent loss of dopant leading to 40–60% less Mn^{2+} [69] or Co^{2+} [74] in CdS particles. The resultant DMS NQDs, however, are of superior quality compared to unaged particles. For example, electronic absorption spectroscopy has been used to demonstrate a change in local environment for Co^{2+} in CdS NQDs dissolved in pyridine (a coordinating solvent) for unaged and aged samples. Unaged samples are dominated by surface-bound Co^{2+}, $(Co(\mu_4\text{-}S)_2(N(py))_2$ and $Co(\mu_4\text{-}S)_3(N(py))_1$, where $(\mu_4\text{-}S)$ refers to "lattice sulfides" and N(py) to pyridine coordination), and entirely lattice-bound Co^{2+} $[Co(\mu_4\text{-}S)_4]$ prevails in aged samples [74]. Interestingly, the aging process, which requires up to several days, can be substituted by an isocrystalline shell-growth process that requires only minutes to complete. In other words, following particle formation but prior to ligand stabilization, additional Cd and S (or Zn and S) precursor can be added, prompting "shell" growth that effectively encapsulates the isocrystalline core and the dopants. Structurally and optically, the shell–core particles behave like aged particles (better resolved hyperfine EPR signals and improved Mn^{2+} photoluminescence). Ligand-field electronic absorption spectroscopy was also used to distinguish between surface-bound and lattice-bound dopant ions, regardless of whether incorporation of ions into the host matrix occurred as a result of aging or isocrystalline shell growth (Fig. 28) [74].

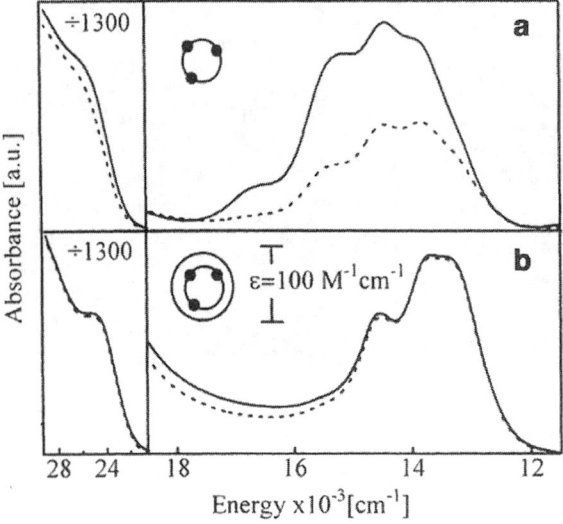

Figure 28 (a) Ligand-field absorption spectra (300 K) for 3.0-nm-diameter $Cd_{1-x}Co_xS$ (x = 0.023)-doped NQDs showing the CdS band-gap transitions (left panels) and the Co^{2+} v_3 ligand-field transitions (right panels). The NQDs were prepared by the standard coprecipitation method (see text) and subsequently suspended in pyridine. The spectra were taken 2 h (solid line) and 23 h (dashed line) after synthesis. The loss in Co^{2+} signal over time suggests that Co^{2+} is dissolving into the pyridine solvent. (b) Ligand-field absorption spectra (300 K) for 3.7-nm-diameter $Cd_{1-x}Co_xS$ (x = 0.009)-doped NQDs prepared by the isocrystalline (core)shell method, 2 h (solid line) and 28 h (dashed line) after synthesis. The reproducibility over time suggests that the (core)shell doped NQDs are stable to dissolution of Co^{2+} into the pyridine solvent. The absorption band shape of (a) is characteristic of Co^{2+} that resides on the CdS NQD surface, whereas that for (b) is characteristic of Co^{2+} that is incorporated into the crystal lattice structure of the host CdS. (From Ref. 74, reprinted with permission.)

An additional factor that strongly influences NQD doping is ionic radii mismatch. The apparent relative ease with which a high-temperature solution-based synthesis method was used to prepare internally doped Zn(Mn)Se DMSs [66] may, in large part, be attributable to the size matching of the core metal and dopant ionic radii: Zn^{2+} (0.80 Å) and Mn^{2+} (0.74 Å). In contrast, the relative difficulty of achieving even near-surface doping of Cd(Mn)Se by similar methods [64] may relate to the rather large size mismatch of the substituting cations: Cd^{2+} (0.97 Å) and Mn^{2+} (0.74 Å). Evidence from inverse-micelle preparations support this conclusion. Comparison of unaged Cd(Co)S and Zn(Co)S DMSs, where the ionic radius of Co^{2+} is 0.74 Å, revealed that the dopant is well distributed throughout the core lattice in the latter case, benefiting only minimally from an isocrystalline shell-growth step.

VII. NANOCRYSTAL ASSEMBLY AND ENCAPSULATION

Because of their chemical, size, shape, and properties tunability, NQDs have long been considered ideal building blocks for novel functional materials. Many conceived device applications require that NQDs be controllably assembled into organized structures, at a variety of length scales, and that these assemblies be macroscopically addressable. Even applications that make use of the optical properties of *individual* nanoparticles (e.g., fluorescent biolabeling) require controlled assembly of bio–nano conjugates. For these reasons, small- and large-scale assembly and encapsulation methods have been developed in which NQDs are manipulated as artificial atoms or molecules. Encapsulation has typically involved incorporating NQDs into organic polymers [76–80] or inorganic glasses [81–83]. Either may simply provide structural rigidity to an NQD ensemble, as well as protection from environmental degradation; or, the matrix material may be electronically, optically, or magnetically "active," where the encapsulant then provides for added functionality and/or device addressibility. Assembly approaches are as diverse as the targeted applications and have emerged from each of the traditional disciplines: physics/physical chemistry (e.g., self-assembly approaches), chemistry (chemical patterning of surfaces, e.g., dip-pen nanolithography [84,85], electric-field-directed assembly [86], etc.), biology (bio-inspired mineralization [87], DNA-directed assembly [85,88], etc.), and materials science (lithography-defined templating [89], etc.). The diversity of approaches suggests that the subject warrants a book of its own. Therefore, only a small subset of this field is reviewed here, namely self-assembly at the nanoscale.

Particles uniform in size, shape, composition, and surface chemistry can self-assemble from solution into highly ordered 2D and 3D solids (Fig. 29).

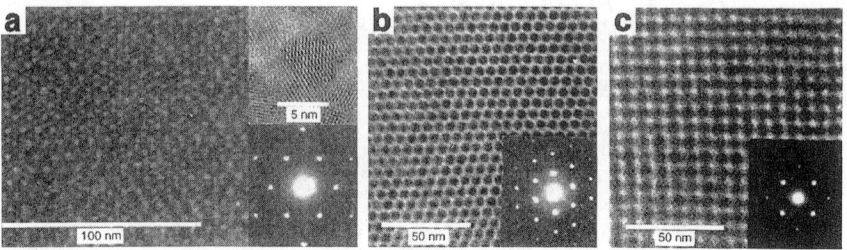

Figure 29 Transmission electron micrographs and electron diffraction (ED) patterns for CdSe NQD superlattices (SLs) of different orientations: (a) Main: $\langle 111 \rangle_{SL}$-oriented array of 6.4-nm-diameter NQDs five layers thick; upper right: HR-TEM of a single NQD with its $\langle 110 \rangle$ axis parallel to the electron beam and its $\langle 002 \rangle$ axis in the plane of the SL; lower right: small-angle ED pattern. (b) Main: face-centered-cubic (fcc) array of 4.8-nm-diameter NQDs ($\langle 101 \rangle_{SL}$ projection); lower right: small-angle ED pattern. (c) Main: fcc array of 4.8-nm-diameter NQDs ($\langle 100 \rangle_{SL}$ oriented); lower right: small-angle ED pattern. (From Ref. 90, reprinted with permission.)

The process is similar for particles ranging in size from cluster molecules to micron-sized colloidal particles [8]. It entails controlled destabilization and precipitation from a slowly evaporating solvent (Fig. 30). As the solution concentrates, interactions between particles become mildly attractive. Particle association is sufficiently slow, however, to prevent disordered aggregation. Instead, ordered assembly dominates by a reversible process of particle addition to the growing superlattice [10]. Fully formed ordered solids are commonly called colloidal crystals (Fig. 31), where the cluster, NQD, or colloid serves as the "artificial atom" building block. In the case of NQDs, the process is controlled by manipulating the polarity and the boiling point of the solvent [10]. The solvent polarity is chosen to ensure that mild attractive forces develop between the nanoparticles as the solvent evaporates. The boiling point is chosen to ensure that the evaporation process is sufficiently slow [10]. At the other extreme of very fast destabilization, by fast solvent evaporation or nonsolvent addition, a rapid increase in the "sticking coefficient" (particle attraction) and in the rate at which particles are added to the growing surface yields loosely associated fractal aggregates. Moderate destabilization rates, implemented by using "moderate-boiling" solvents, produce close-packed glassy solids having local order but lacking long-range order (Fig. 30a) [8]. For assembly of well-ordered NQD superlattices, the chosen solvent is typically a mixed solvent, involving both a lower-boiling alkane and a higher-boiling alcohol. The alkane evaporates more quickly than the alcohol, yielding relatively higher concentrations of the "destabilizing"

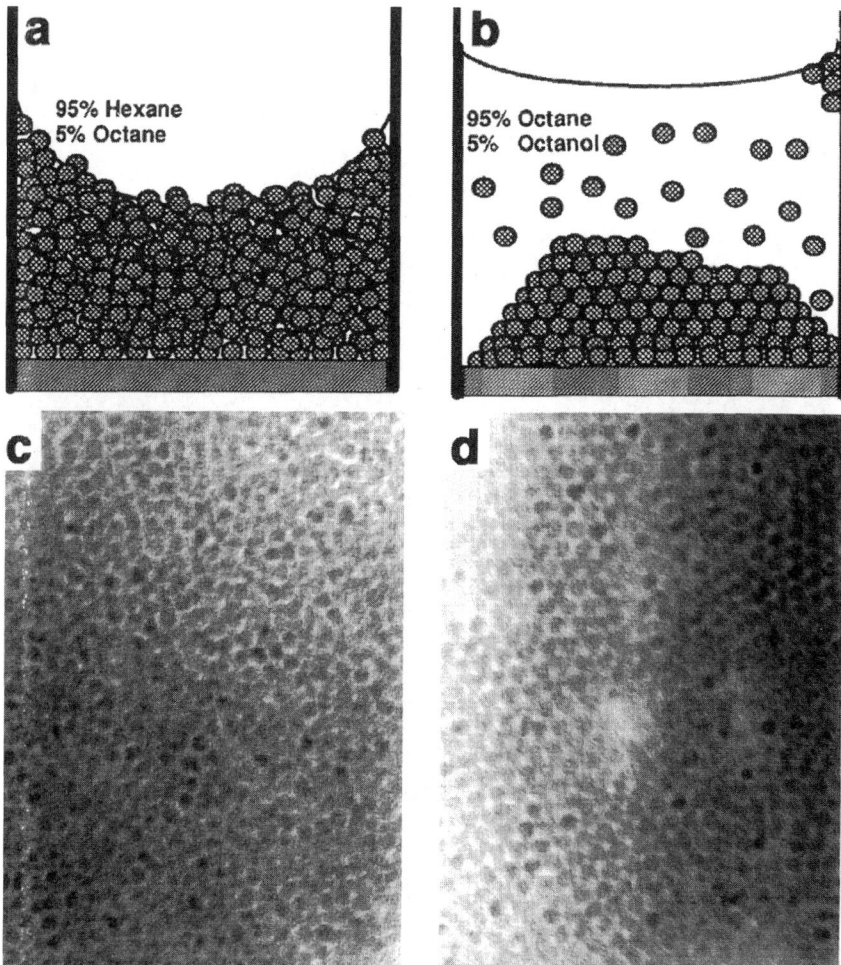

Figure 30 (a,b) Schematics illustrating the deposition conditions (solvent dependent) that yield three-dimensional, close-packed NQD solids (films nucleated heterogeneously at a surface or colloidal crystals nucleated homogeneously from solution) as (a) disordered glasses and (b) ordered superlattices. (c,d) TEMs revealing the long-range disorder characteristic of the glassy solids (c) and the long-range order characteristic of the crystalline solids (d). (From Ref. 8, reprinted with permission.)

Figure 31 Dark-field optical micrograph of faceted CdSe colloidal crystals. The crystals were prepared by slow self-assembly of 2.0-nm-diameter NQDs that nucleated homogeneously from solution. A mixed nonpolar/polar solvent system similar to that depicted in Fig. 30b was used in combination with careful regulation of solution temperature and pressure to provide the necessary conditions for controlled destabilization of the NQD starting solution. The colloidal crystals are stacked and range in size from 5 to 50 μm. (From Ref. 90, reprinted with permission.)

alcohol (assuming the NQDs are surface terminated with long-chain alkyls) over time (Fig. 30b) [8]. The process can be controlled by applying heat and/or vacuum to the system. Free-standing and surface-bound colloidal [face-centered-cubic (fcc)] crystals exhibiting long-range order over hundreds of microns have been prepared in this way. In addition, the individual NQDs can crystallographically orient as demonstrated by small-angle and wide-angle x-ray scattering for a CdSe NQD thin film in which the wurtzite c axis of the nanocrystals are aligned in the plane of the substrate [8,90].

Slow, controlled precipitation of highly ordered colloidal crystals has also been achieved by the method of "controlled oversaturation" [91]. Here, gentle destabilization is induced by slow diffusion of a nonsolvent into a NQD solution directly or through a "buffer" solution. The nonsolvent (e.g., methanol), buffer (e.g., propan-2-ol, if present), and the NQD solution (toluene solvent) are carefully layered in reaction tubes (Fig. 32a). Colloidal–crystal nucleation occurs on the tube walls and in the bulk solution. Although both methods, not buffered and buffered, yield well-ordered colloidal crystals from 100 to 200 μm in size, the latter process is relatively more controlled and produces faceted, hexagonal platelets (Fig. 32c), rather than ragged, irregularly shaped colloidal crystals (Fig. 32b). The CdSe NQD building blocks again yield (fcc) superlattices [91].

In the self-assembly of large colloidal crystals from nano-sized crystals, the nanocrystals behave as artificial atoms. The self-assembly process is largely driven by the relative favorability of the interaction between the NQD surface *ligands* and the solvent. Without a favorable interaction, particle cores begin to attract via van der Waals forces. Nanoparticles can also assemble into large crystals by a self-assembly process called "oriented attachment." In contrast with self-assembly by ligand-stabilized colloids, oriented attachment entails direct interaction between ligand-free (or almost ligand-free) nanocrystals [92]. The driving force for the assembly is the *lack* of surface-passivating ligands. The bare nanoparticles assemble in order to satisfy surface dangling bonds, and the assembly process is sufficiently reversible that *oriented* attachment dominates, leading to highly crystalline

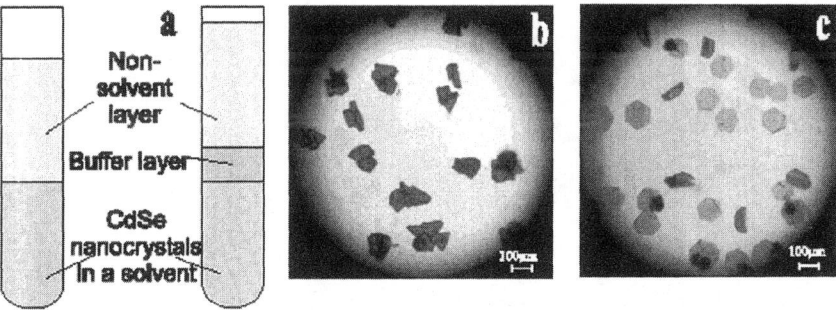

Figure 32 (a) Schematic illustrating the method of "controlled oversaturation" for the preparation of CdSe NQD colloidal crystals. (b) Optical micrograph of irregularly shaped colloidal crystals prepared without a "buffer–solvent" layer (i.e., faster nucleation). (c) Optical micrograph of faceted hexagonal colloidal crystals prepared with a "buffer–solvent" layer (i.e., slower nucleation). (From Ref. 91, reprinted with permission.)

macrostructures. Oriented attachment is known in a variety of systems, including natural mineral systems where chains and extended sheets are formed [93], epitaxial attachment of metal nanoparticles to metal substrates through a process of dislocation formation/movement and particle rotation in response to interfacial strain [94], chain formation from "artificial" nanoscale TiO_2 building blocks (Fig. 33) [95], ZnO rod formation from ZnO nanodots [92], and sheet formation from "artificial" nanoscale rhombohedral InS and InSe building blocks [63]. This crystal-growth mechanism may even provide an advantage compared to traditional atom-by-atom growth, as nanoscale inorganic building blocks are typically characterized by nearly perfect or perfect crystal structure, devoid of internal defects. Constructing large crystals from these preformed perfect crystallites may permit growth of extended solids having unusually low defect densities [96]. Alternatively, impurities (natural and intentional dopants) can perhaps be more easily incorporated into large crystals by oriented attachment of nanocrystals decorated with surface impurities [93,96]. Various types of dislocations, from edge to screw, can form as a result of interfacial distortions generated to accommodate coherency between surfaces that are not atomically flat (Fig. 34). Thus, the forces driving particles to attach (i.e., the drive to eliminate unsaturated surface bonds) can induce dislocation formation (Fig. 35) [97]. Predictably, attachment occurs on high-surface-energy faces [97]. For example, the rhombohedral InS (R-InS) and InSe (R-InSe) crystal structures feature planar, covalently bonded sheets that are four atomic layers

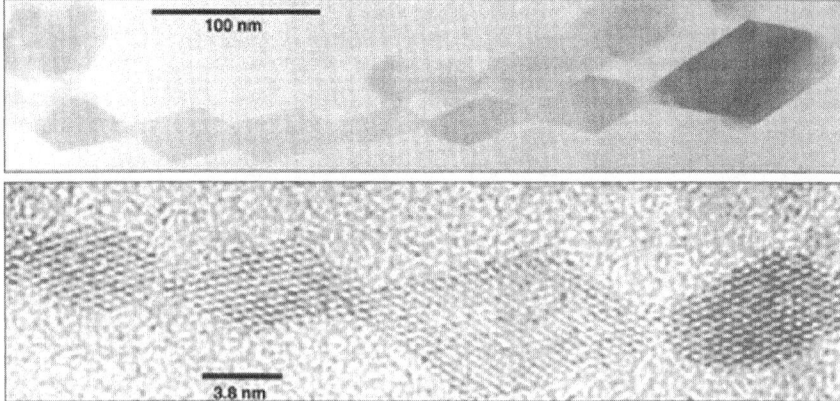

Figure 33   Transmission electron micrograph of TiO_2 nanocrystal aggregates that have assembled by "oriented attachment." The chains are oriented by a process of particle docking, aligning, and fusing. (From Ref. 95, reprinted with permission.)

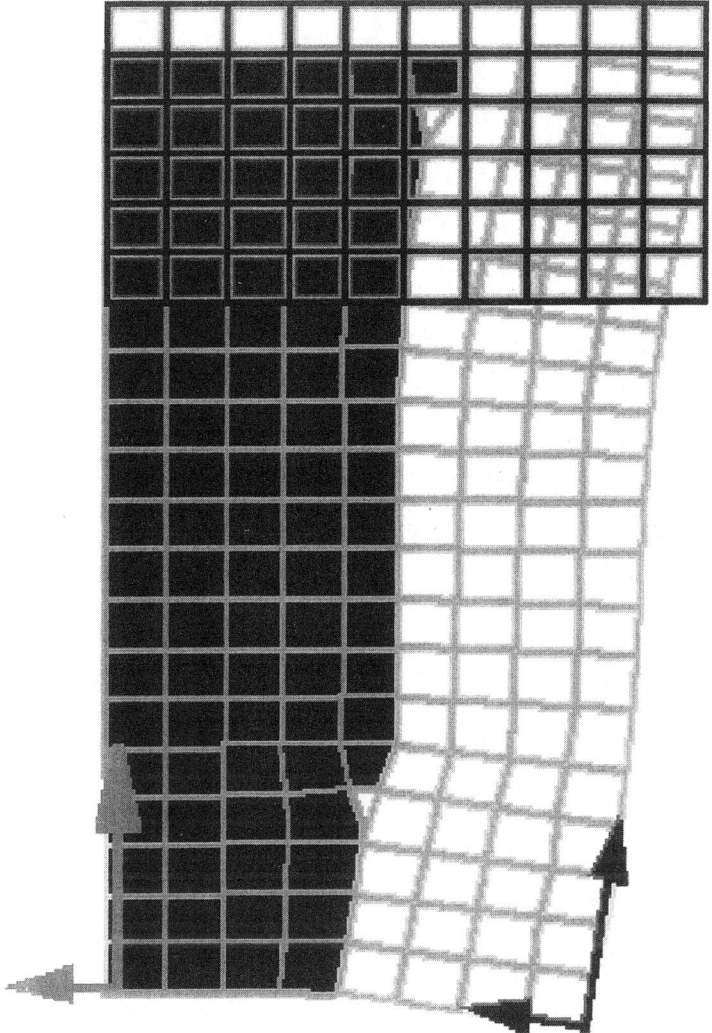

Figure 34 Schematic illustrating the process of dislocation generation in imperfect "oriented attachment." Three particles are shown. First, the two lower particles join with a small rotation being incorporated as a result of surface steps on the particle to the left. The third particle joins in an oriented fashion to the left-side crystal, but with a rotational misorientation relative to the right-side crystal. The diagram demonstrates the formation of two types of dislocation: edge (dislocation line normal to the page) and screw (dislocation line horizontal). (From Ref. 97, reprinted with permission.)

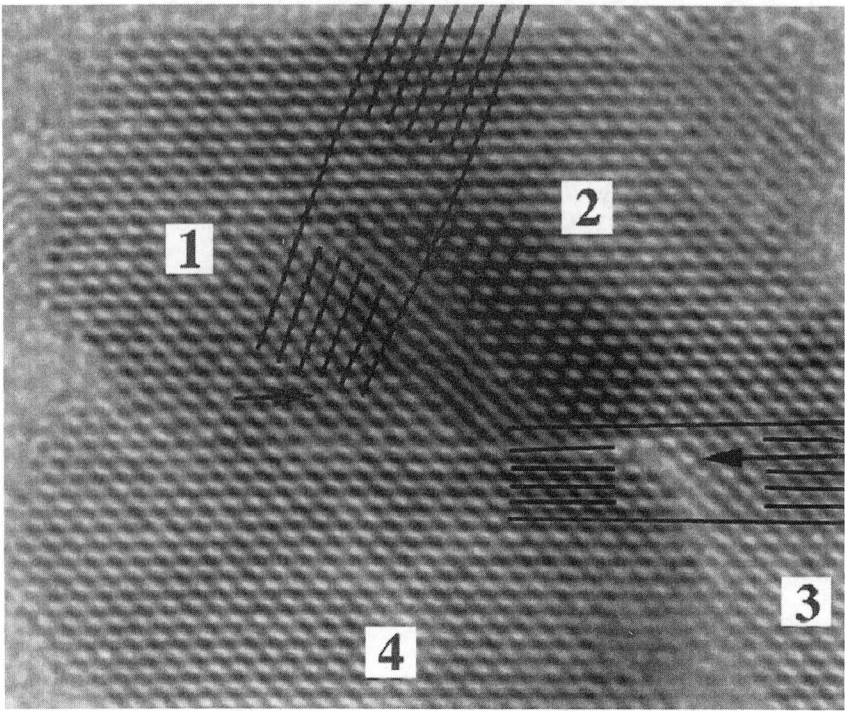

Figure 35   High-resolution TEM (along [100] TiO$_2$ anatase) of a crystal segment formed by oriented attachment of at least four particles (numbered 1–4). Arrowheads and lines (0.48 nm apart) show edge dislocations. (From Ref. 97; reprinted with permission.)

thick and are separated by van der Waals gaps. Solution-phase growth at low temperature (~200°C) generates nanocrystal platelets (5–20 nm) that self-assemble, or "self-attach," to form large micron-scale sheets. The underlying psuedographitic layered structure supports growth of nanocrystals in the form of 2D platelets. The van der Waals surfaces of the nanocrystal *ab* plane (the large-area plane in the platelets) are low-energy, coordinatively saturated "faces," whereas the edges of the nanocrystallites are characterized by higher-energy unsaturated sites. Therefore, it is at the edges that the nanoplatelets likely attach, generating the larger sheets. Electron-diffraction patterns collected perpendicular to the large-area sheet surfaces and over large areas (collection radii ≈ 150 nm) demonstrate not only that the crystallographic *c* axes are indeed perpendicular to the sheet surfaces but that the sheets diffract coherently, as single crystals would (Fig. 36). Therefore, assembly and

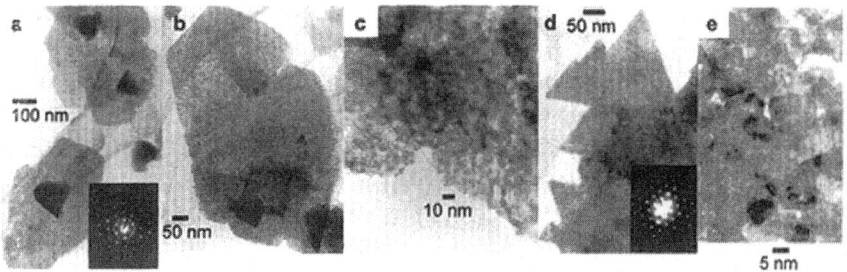

Figure 36 Transmission electron micrographs of InSe and InS colloid-crystal platelets formed by oriented assembly of nanosized building blocks: (a) InSe platelets approximating hexagonal shapes; inset, electron-diffraction pattern collected along the [0001] zone axis of a platelet; (b) InSe platelet revealing texture; (c) InSe platelet edge revealing constituent hexagonal nanocrystallites; (d) InS platelets with triangular features; inset, electron-diffraction pattern collected along the [0001] zone axis of a platelet; (e) InS platelet center revealing internal structure. (From Ref. 63, reprinted with permission.)

attachment of platelets into sheets proceeds in a crystallographically coherent fashion [63].

Nanocrystal self-assembly can also proceed by way of *electrostatic* or *covalent interactions*. In a recent example, oppositely charged CdS nanocrystals were mixed in different ratios and under controlled ionic strength. The positively charged CdS nanocrystals were prepared by surface modification with 2-(dimethylamino)ethanethiol, whereas the negatively charged nanocrystals were surface passivated with 3-mercaptopropionic acid. At high ionic strength, the particles repelled one another, regardless of relative particle concentrations, whereas at low ionic strength, oppositely charged particles developed an attractive potential. When present in equal proportions, the negatively and positively charged particles aggregated and precipitated from solution. Despite imperfect size dispersions and some differences in size between the oppositely charged particles, ordering was observed by low-angle x-ray diffraction. Consistent with self-assembly from nonpolar solvent systems, the degree of ordering in these aqueous-based systems was enhanced by slowing the precipitation rate. Also, when the positively and negatively charged particles were present in unequal amounts (e.g., ratios of 1 : 10), soluble moleculelike clusters were formed. These comprised a central particle of one charge surrounded by several particles of opposite charge. CdS–CdS clusters were demonstrated as were CdS-coated Au particles [98].

In a recent example of covalently driven assembly, disordered but densely packed CdSe nanocrystal monolayers were deposited onto *p*- and *n*-doped GaAs substrates. The GaAs substrates were either bare or pretreated with dithiol self-assembled monolayers (SAMs). In the absence of the 1,6-hexanedithiol layer, the NQDs were only physically absorbed to the substrate and did not withstand a thorough toluene wash. In contrast, NQDs self-assembled onto the dithiol SAMs were very robust due to covalent linkages between the NQDs and the exposed thiols [99]. Soluble, covalently linked NQD clusters have also been prepared. Dimers and larger-order NQD "molecules" were prepared from dilute solutions in which bifunctional linkers provided covalent attachment between two or several CdSe NQDs [100,101]. Improving yields and enhancing control over "molecule" size (dimer versus trimer, etc.) remain as future challenges.

Both electrostatic and covalent linkage strategies have been applied to biological labeling applications where fluorescent NQDs provide specific tags for cellular constituents [102]. The NQDs are negatively charged using, for example, either dihydrolipoic acid (DHLA) [103] or octylamine-modified polyacrylic acid [104] as a surface-capping agent (Fig. 37). In the former case, positively charged proteins are coupled *electrostatically* directly to the negatively charged NQD or indirectly through a positively charged leucine zipper peptide bridge [103]. In the latter case, coupling to antibodies, streptavidin, or other proteins is *covalent* to the polyacrylate cap via traditional carbodiimide chemistry [104]. Both the DHLA and the polyacrylate methods provide relatively simple NQD surface-modification procedures for

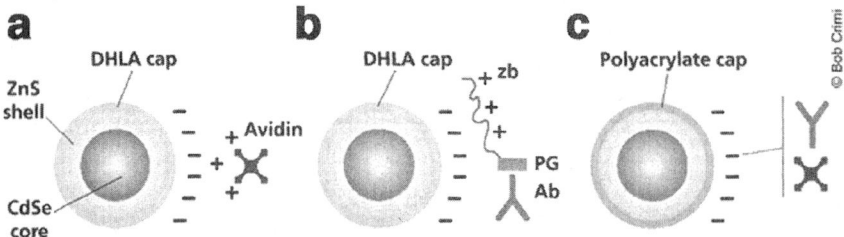

Figure 37 Three strategies for bioconjugation to NQDs as reviewed in Ref. 102. Highly luminescent (CdSe)ZnS (core)shell NQDs comprise the fluorescent probe. The NQD surfaces are negatively charged using the carboxylate groups of either dihydrolipoic acid (DHLA) (a,b) or an amphiphilic polymer (40% octylamine modified polyacrylic acid) (c). In (a,b), proteins are conjugated electrostatically to the DHLA–NQDs either (a) directly or (b) indirectly via a bridge comprising a positively charged leucine zipper peptide, zb, fused to a recombinant protein, PG, that binds to a primary antibody, Ab, with target specificity. In (c), covalent binding by way of traditional carbodiimide chemistry is used to couple antibodies, streptavidin, or other proteins to the polyacrylate cap. (From Ref. 102 reprinted with permission.)

achieving biological compatibility (i.e., retention of high QYs in PL), pathways for specific binding to biological targets, and biological inertness [102].

The above methods for NQD self-assembly represent only a fraction of this growing field. As such, they are not meant to be inclusive; rather, they are intended to provide theoretical background for understanding the chemical and physical issues involved in the assembly process and to highlight possible future directions for research. Further, we have intentionally avoided discussion of templated, directed, and active assembly of nanoscale building blocks, as this area is perhaps even more diverse and remains a subject of intense and active investigation.

ACKNOWLEDGMENTS

We would like to acknowledge Jennifer's husband, Howard Coe, for editing the text, transcribing and organizing references, and making graphical improvements to the figures. Jennifer would also like to thank Howard for his patience, support, and sense of humor throughout this project.

REFERENCES

1. Petroff P.M.; Medeiros-Riberio G. *MRS Bull.* 1996, *21*, 54.
2. Bimberg D.; Grundmann M.; Ledentsov N.N. *MRS Bull.* 1998, *23*, 31.
3. Yoffe A.D. *Adv. Phys.* 2001, *50*, 1.
4. Hu J.; Odom T.W.; Lieber C.M. *Acc. Chem. Res.* 1999, *32*, 435.
5. Lauhon L.J.; Gudiksen M.S.; Wang D.; Lieber C.M. *Nature* 2002, *420*, 57.
6. Borelli N.F.; Smith D.W. *J. Non-Crystal. Solids* 1994, *180*, 25.
7. Lipovskii A.; Kolobkova E.; Petrikov V.; Kang I.; Olkhovets A.; Krauss T.; Thomas M.; Silcox J.; Wise F.; Shen Q.; Kycia S. *Appl. Phys. Lett.* 1997, *71*, 3406.
8. Murray C.B.; Kagan C.R.; Bawendi M.G. *Annu. Rev. Mater. Sci.* 2000, *30*, 545.
9. La Mer V.K.; Dinegar R.H. *J. Am. Chem. Soc.* 1950, *72*, 4847.
10. Murray C.B.; Sun S.; Gaschler W.; Doyle H.; Betley T.A.; Kagan C.R. *IBM J. Res. Dev.* 2001, *45*, 47.
11. Nozik A.J.; Micic O.I. *MRS Bull.* 1998, *23*, 24.
12. Peng Z.A.; Peng X. *J. Am Chem. Soc.* 2001, *123*, 1389.
13. Higginson K.A.; Kuno M.; Bonevich J.; Qadri S.B.; Yousuf M.; Mattoussi H. *J. Phys. Chem.* 2002, *106*, 9982.
14. Peng X.; Wickham J.; Alivisatos A.P. *J. Am. Chem. Soc.* 1998, *120*, 5343.
15. Murray C.B.; Norris D.J.; Bawendi M.G. *J. Am. Chem. Soc.* 1993, *115*, 8706.
16. Qu L.; Peng Z.A.; Peng X. *Nano Lett.* 2001, *1*, 333.
17. Peng X.; Manna L.; Yang W.; Wickham J.; Scher E.; Kadavanich A.; Alivisatos A.P. *Nature* 2000, *404*, 59.

18. Qu L.; Peng X.G. *J. Am. Chem. Soc.* 2002, *124*, 2049.
19. Hines M.; Guyot-Sionnest P. *J. Phys. Chem. B* 1998, *102*, 3655.
20. Peng X.; Schlamp M.C.; Kadavanich A.V.; Alivisatos A.P. *J. Am. Chem. Soc.* 1997, *119*, 7019.
21. Wehrenberg B.L.; Wang C.J.; Guyot-Sionnest P. *J. Phys. Chem. B* 2002, *106*, 10,634.
22. Kershaw S.V.; Harrison M.; Rogach A.L.; Kornowski A. *J. Select. Topics Quantum Electron.* 2000, *6*, 534.
23. Gaponik N.; Talapin D.V.; Rogach A.L.; Hoppe K.; Shevchenko E.V.; Kornowski A.; Eychmüller A.; Weller H. *J. Phys. Chem. B* 2002, *106*, 7177.
24. Vossmeyer T.; Katsikis L.; Giersig M.; Popovic I.G.; Diesner K.; Chemseddine A.; Eychmüller A.; Weller H. *J. Phys. Chem.* 1994, *98*, 7665.
25. Rogach A.; Kershaw S.; Burt M.; Harrison M.; Kornowski A.; Eychmüller A.; Weller H. *Adv. Mater.* 1999, *11*, 552.
26. Rockenberger J.; Troger L.; Rogach A.L.; Tischer M.; Grundmann M.; Eychmüller A.; Weller H. *Ber. Bunsenges Phys. Chem.* 1998, *102*, 1.
27. Steigerwald M.L.; Alivisatos A.P.; Gibson J.M.; Harris T.D.; Kortan R.; Muller A.J.; Thayer A.M.; Duncan T.M.; Douglas D.C.; Brus L.E. *J. Am. Chem. Soc.* 1988, *110*, 3046.
28. O'Brien S.; Brus L.; Murray C.B. *J. Am. Chem. Soc.* 2001, *123*, 12085.
29. Hines M.A.; Guyot-Sionnest P. *J. Phys. Chem.* *100*, 468.
30. Dabbousi B.O.; Rodriguez-Viejo J.; Mikulec F.V.; Heine J.R.; Mattoussi H.; Ober R.; Jensen K.F.; Bawendi M.G. *J. Phys. Chem. B* 1997, *101*, 9463.
31. Micic O.I.; Smith B.B.; Nozik A.J. *J. Phys. Chem. B* 2000, *104*, 12,149.
32. Cao Y.W.; Banin U. *Agnew Chem. Int. Ed.* 1999, *38*, 3692.
33. Cao Y.W.; Banin U. *J. Am. Chem. Soc.* 2000, *122*, 9692.
34. Eychmüller A.; Mews A.; Weller H. *Chem. Phys. Lett.* 1993, *208*, 59.
35. Mews A.; Eychmüller A.; Giersig M.; Schoos D.; Weller H. *J. Phys. Chem.* 1994, *98*, 934.
36. Mews A.; Kadavanich A.V.; Banin U.; Alivisatos A.P. *Phys. Rev. B* 1996, *53*, R13,242.
37. Little R.B.; El-Sayed M.A.; Bryant G.W.; Burke S. *J. Chem. Phys.* 2001, *114*, 1813.
38. Manna L.; Scher E.C.; Alivisatos A.P. *J. Am. Chem. Soc.* 2000, *122*, 12,700.
39. Kazes M.; Lewis D.Y.; Ebenstein Y.; Mokari T.; Banin U. *Adv. Mater.* 2002, *14*, 317.
40. Manna L.; Scher E.C.; Li L.S.; Alivisatos A.P. *J. Am. Chem. Soc.* 2002, *124*, 7136.
41. Jun Y.; Lee S.M.; Kang N.M.; Cheon J. *J. Am. Chem. Soc.* 2001, *123*, 5150.
42. Jun Y.; Jung Y.; Cheon J. *J. Am. Chem. Soc* 2002, *124*, 615.
43. Lee S.M.; Jun Y.; Cho S.N.; Cheon J. *J. Am. Chem. Soc.* 2002, *124*, 11,244.
44. Trentler T.J.; Hickman K.M.; Goel S.C.; Viano A.M.; Gibbons P.C.; Buhro W. E. *Science* 1995, *270*, 1791.
45. Yu H.; Buhro W.E. *Adv. Mater.* 2003, *15*, 416–419.
46. Yu, H.; Loomis, R.A.; Buhro, W. E. submitted.

47. Yu H.; Gibbons P.C.; Kelton K.F.; Buhro W.E. *J. Am. Chem. Soc.* 2001, *123*, 9198.
48. Yu, H.; Buhro, W.E. unpublished results.
49. Holmes J.D.; Johnston K.P.; Doty R.C.; Korgel B.A. *Science* 2000, *287*, 1471.
50. Chen C.C.; Herhold A.B.; Johnson C.S.; Alivisatos A.P. *Science* 1997, *276*, 398.
51. Wickham J.N.; Herhold A.B.; Alivisatos A.P. *Phys. Rev. Lett.* 2000, *84*, 923.
52. Herhold A.B.; Chen C.C.; Johnson C.S.; Tolbert S.H.; Alivisatos A.P. *Phase Trans.* 1999, *68*, 1.
53. Jacobs K.; Zaziski D.; Scher E.C.; Herhold A.B.; Alivisatos A.P. *Science* 2001, *293*, 1803.
54. Stein A.; Keller S.W.; Mallouk T.E. *Science* 1993, *259*, 1558.
55. Buhro W.E.; Hickman K.M.; Trentler T.J. *Adv. Mater.* 1996, *8*, 685.
56. Brus L. *Science* 1997, *276*, 373.
57. Parkinson B. *Science* 1995, *270*, 1157.
58. Switzer J.A.; Shumsky M.G.; Bohannan E.W. *Science* 1999, *284*, 293.
59. Gillan E.G.; Barron A.R. *Chem. Mater.* 1997, *9*, 3037.
60. Sellinschegg H.; Stuckmeyer S.L.; Hornbostel M.D.; Johnson D.C. *Chem. Mater.* 1998, *10*, 1096.
61. Noh M.; Johnson D.C. *J. Am. Chem. Soc.* 1996, *118*, 9117.
62. Noh M.; Thiel J.; Johnson D.C. *Science* 1995, *270*, 1181.
63. Hollingsworth J.A.; Poorjay D.M.; Clearfield A.; Buhro W.E. *J. Am. Chem. Soc.* 2000, *122*, 3562.
64. Mikulec F.V.; Kuno M.; Bennati M.; Hall D.A.; Griffin R.G.; Bawendi M.G. *J. Am. Chem. Soc.* 2000, *122*, 2532.
65. Hoffman D.M.; Meyer B.K.; Ekimov A.I.; Merkulov I.A.; Efros A.L.; Rosen M.; Couino G.; Gacoin T.; Boilot J.P. *Solid State Commun.* 2000, *114*, 547.
66. Norris D.J.; Yao N.; Charnock F.T.; Kennedy T.A. *Nano Lett.* 2001, *1*, 3.
67. Bhargava R.N.; Gallagher D.; Hong X.; Nurmiko A. *Phys. Rev. Lett.* 1994, *72*, 416.
68. Levy L.; Hochepied J.F.; Pileni M.P. *J. Phys. Chem.* 1996, *100*, 18,322.
69. Levy L.; Feltin N.; Ingert D.; Pileni M.P. *Langmuir* 1999, *15*, 3386.
70. Couino G.; Esnouf S.; Gacoin T.; Boilot J.P. *J. Phys. Chem.* 1996, *100*, 20,021.
71. Ladizhansky V.; Hodes G.; Vega S. *J. Phys. Chem. B* 1998, *102*, 8505.
72. Kim K.W.; Cowen J.A.; Dhingfa S.; Kanatzidis M.G. *Mater. Res. Soc. Symp.* 1992, *272*, 27.
73. Bhargava R.N. *Jo. Lumin.* 1996, *70*, 85.
74. Radovanovic P.V.; Gamelin D.R. *J. Am. Chem. Soc.* 2001, *123*, 12,207.
75. Levy L.; Ingert D.; Feltin N.; Pileni M.P. *Adv. Mater.* 1998, *10*, 53.
76. Fogg D.E.; Radzilowski L.H.; Dabbousi B.O.; Schrock R.R.; Thomas E.L.; Bawendi M.G. *Macromolecules* 1997, *30*, 8433.
77. Greenham N.C.; Peng X.; Alivisatos A.P. *Synthetic Metals* 1997, *84*, 545.
78. Mattoussi H.; Radzilowski L.H.; Dabbousi B.O.; Fogg D.E.; Schrock R.R.; Thomas E.L.; Rubner M.F.; Bawendi M.G. *J. Appl. Phys.* 1999, *86*, 4390.
79. Huynh W.U.; Peng X.; Alivisatos A.P. *Adv. Mater.* 1999, *11*, 923.
80. Huynh W.U.; Dittmer J.J.; Alivisatos A.P. *Science* 2002, *295*, 2425.

81. Sundar V.C.; Eisler H.J.; Bawendi M.G. *Adv. Mater* 2002, *14*, 739.
82. Eisler H.J.; Sundar V.C.; Bawendi M.G.; Walsh M.; Smith H.I.; Klimov V.I. *Appl. Phys. Lett.* 2002, *80*, 4614.
83. Petruska M.A.; Malko A.V.; Klimov V.I. *Adv. Mater.* 2003, *15*, 610–613.
84. Piner R.D.; Zhu J.; Xu F.; Hong S.; Mirkin C.A. *Science* 1999, *283*, 661.
85. Mirkin C.A. *Inorg. Chem.* 2000, *39*, 2258.
86. Gao M.; Sun J.; Dulkeith E.; Gaponik N.; Lemmer U.; Feldmann J. *Langmuir* 2002, *18*, 4098.
87. Sone E.D.; Zubarev E.R.; Stupp S.I. *Angew. Chem. Int. Ed.* 2002, *41*, 1705.
88. Mitchell G.P.; Mirkin C.A.; Letsinger R.L. *J. Am. Chem Soc.* 1999, *121*, 8122.
89. Hua F.; Shi J.; Lvov Y.; Cui T. *Nano Lett.* 2002, *2*, 1219.
90. Murray C.B.; Kagan C.R.; Bawendi M.G. *Science* 1995, *270*, 1335.
91. Talapin D.V.; Shevchenko E.V.; Kornowski A.; Gaponik N.; Haase M.; Rogach A.L; Weller H. *Adv. Mater.* 2001, *13*, 1868.
92. Pacholski C.; Kornowski A.; Weller H. *Angew. Chem. Int. Ed.* 2002, *41*, 1188.
93. Banfield J.F.; Welch S.A.; Zang H.; Ebert T.T.; Penn R.L. *Science* 2000, *289*, 751.
94. Zhu H.L.; Averback R.S. *Phil. Mag.* 1996, *73*, 27.
95. Penn R.L.; Banfield J.F. *Geochim. Cosmochim. Acta* 1999, *63*, 1549.
96. Alivisatos A.P. *Science* 2000, *289*, 736.
97. Penn R.L.; Banfield J.F. *Science* 1998, *281*, 969.
98. Kolny J.; Kornowski A.; Weller H. *Nano Lett.* 2002, *2*, 361.
99. Marx E.; Ginger D.S.; Walzer K.; Stokbro K.; Greenham N.C. *Nano Lett.* 2002, *2*, 911.
100. Peng X.; Wilson T.E.; Alivisatos A.P.; Schultz P.G. *Angew. Chem. Int. Ed. Engl.* 1997, *36*, 145.
101. Petruska, M.A.; Klimov, V.I.; unpublished results.
102. Jovin T.M. *Nature Biotechnol.* 2003, *21*, 32.
103. Jaiswal J.K.; Mattoussi H.; Mauro J.M.; Simon S.M. *Nature Biotechnol.* 2003, *21*, 47.
104. Wu X.; Liu H.; Liu J.; Haley K.N.; Treadway J.A.; Larson P.; Ge N.; Peale F.; Bruchez M.P. *Nature Biotechnol.* 2003, *21*, 41.

2

Electronic Structure in Semiconductor Nanocrystals[*]

David J. Norris
University of Minnesota, Minneapolis, Minnesota, U.S.A.

I. INTRODUCTION

One of the primary motivations for studying nanometer-scale semiconductor crystallites, or nanocrystals, is to understand what happens when a semiconductor becomes small. This question has been studied not only for its fundamental importance but also for its practical significance. Because objects are rapidly shrinking in modern electronic and optoelectronic devices, we wish to understand their properties. To address this question, a variety of semiconductor nanocrystals have been investigated over the past two decades. Throughout these studies, one of the most important and versatile tools available to the experimentalist has been optical spectroscopy. In particular, it has allowed the description of how the electronic properties of these nanocrystals change with size. The purpose of this chapter is to review progress in this area.

The usefulness of spectroscopy stems from the inherent properties of bulk semiconductor crystals. Direct-gap semiconductors can absorb a photon when an electron is promoted directly from the valence band into the conduction band [1]. In this process, an electron–hole pair is created in the material, as depicted in Fig. 1a. However, if the size of the semiconductor structure becomes comparable to or smaller than the natural length scale of

---

[*] Portions of this chapter have been adapted (with permission) from D. J. Norris, Ph.D. thesis, MIT, 1995.

65

the electron–hole pair, the carriers are confined by the boundaries of the material. This phenomenon, which is known as the *quantum-size effect*, leads to atomic-like optical behavior in nanocrystals as the bulk bands become quantized (see Fig. 1b). Because, at the atomic level, the material remains structurally identical to the bulk crystal, this behavior arises solely due to its finite size. Therefore, by revealing this atomic-like behavior, simple optical data (e.g., absorption spectra) can give useful information about the nano-meter-size regime.

In many nanocrystal systems, this effect can be quite dramatic. Consequently, these materials, sometimes referred to as colloidal quantum dots, provide an easily realizable system for investigation of the nanometer-size regime. Once this was realized, early research endeavored to explain the underlying phenomenon [2–6]. It was shown (see below) that by modeling the quantum dot as a semiconductor inclusion embedded in an insulating matrix, as illustrated in Fig. 2a, the basic physics could be understood. Photoexcited carriers reside in a three-dimensional potential well, as shown in Fig. 2b. This causes the valence and conduction bands to be quantized into a ladder of hole and electron levels, respectively. Therefore, in contrast to the bulk absorption spectrum, which is a continuum above the bandgap of the semiconductor (E_g^s)

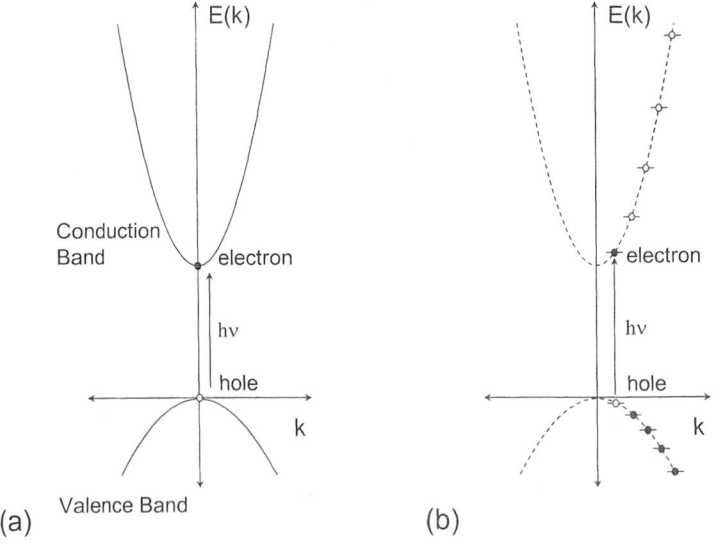

Figure 1 (a) Band diagram for a simple two-band model for direct-gap semiconductors. (b) Optical transitions in finite-size semiconductor nanocrystals are discrete due to the quantization of the bulk bands.

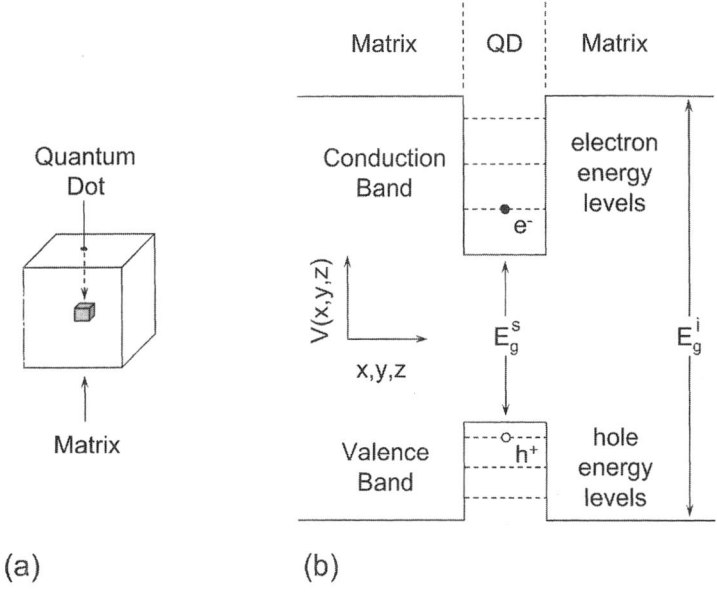

Figure 2 (a) Simple model of a nanocrystal (quantum dot) as a semiconductor inclusion embedded in an insulating matrix. (b) Potential well formed in any one dimension (x, y, or z) in the conduction and valence bands. The energy levels of the excited carriers (electrons and holes) become quantized due to the finite size of the semiconductor nanocrystal.

[1], spectra from semiconductor nanocrystals exhibit a series of discrete electronic transitions between these quantized levels. Accordingly, semiconductor nanocrystals are sometimes referred to as *artificial atoms*. Furthermore, because the energies of the electron and hole levels are quite sensitive to the amount of confinement, the optical spectra of nanocrystals are strongly dependent on the size of the crystallite.

To review recent progress in utilizing optical spectroscopy to understand this size dependence, we begin in Section II with a discussion of the basic theoretical concepts necessary to understand electronic structure in nanocrystals. Then, experimental data from the prototypical direct-gap semiconductor system, cadmium selenide (CdSe), is described in Section III. As the first system to be successfully prepared with extremely high quality [7], CdSe has been extensively studied. Indeed, it was the first system where the size dependence of the electronic structure was understood in detail [8,9]. Furthermore, this understanding led to the resolution of a long-standing mystery in semiconductor nanocrystals [10,11]. It was puzzling why these systems typically exhibit emission lifetimes that are two to three orders of magnitude longer than the bulk crystal. The origin of this phenomenon is explained in Section III. Afterward, Section IV discusses recent work which moves beyond CdSe. In particular, InAs nanocrystals are described [12]. As a narrow-gap, III–V semiconductor, this system presents significant differences from CdSe. Studies of InAs have both confirmed and prompted further refinements in the theoretical model of nanocrystals [13]. Finally, Section IV concludes by briefly outlining some of the remaining issues in the electronic structure of nanocrystals.

   Unfortunately, this chapter does not provide a comprehensive review of nanocrystal spectroscopy. Therefore, the reader is encouraged to look at the other chapters in this volume as well as the many excellent reviews and treatises that are now available on this topic [14–21].

II. THEORETICAL FRAMEWORK

A. Confinement Regimes

In Section I, we stated that the quantum size effect occurs when the size of the nanocrystal becomes comparable to or smaller than the *natural length scale* of the electron and hole. To be more precise, one can utilize the Bohr radius as a convenient length scale. In general, the Bohr radius of a particle is defined as

$$a_B = \varepsilon \frac{m}{m^*} a_0 \tag{1}$$

where ε is the dielectric constant of the material, m^* is the mass of the particle, m is the rest mass of the electron, and a_0 is the Bohr radius of the hydrogen atom [22]. (Note that throughout this chapter, the term *particle* refers to an atomic particle, such as an electron or hole, *not* the nanocrystal.) For the nanocrystal, it is convenient to consider three different Bohr radii: one for the electron (a_e), one for the hole (a_h), and one for the electron–hole pair or exciton (a_{exc}). The latter is a hydrogeniclike bound state that forms in bulk crystals due to the Coulombic attraction between an electron and hole. Using Eq. (1), each of these Bohr radii can be easily calculated. In the case of the exciton, the reduced mass of the electron–hole pair is used for m^*. With these values, three different limits can be considered [2]. First, when the nanocrystal radius, a, is much smaller than a_e, a_h, and a_{exc} (i.e., when $a < a_e, a_h, a_{exc}$), the electron and hole are each strongly confined by the nanocrystal boundary. This is referred to as the *strong confinement regime*. Second, when a is larger than both a_e and a_h, but is smaller than a_{exc} (i.e., when $a_e, a_h < a < a_{exc}$), only the center-of-mass motion of the exciton is confined. This limit is called the *weak confinement regime*. Finally, when a is between a_e and a_h (e.g., when $a_h <$ $a < a_e, a_{exc}$), one particle (e.g., the electron) is strongly confined and the other (e.g., the hole) is not. This is referred to as the *intermediate confinement regime*.

Of course, the confinement regime which is accessed in the experiment depends on the nanocrystal material and size. For example, because the exciton Bohr radius in InAs is 36 nm and nanocrystals are typically much smaller than this size, InAs nanocrystals are in the strong confinement regime. In contrast, CuCl has an exciton Bohr radius of 0.7 nm. Accordingly, CuCl nanocrystals are in the weak confinement regime. CdSe nanocrystals can be in either the strong confinement or the intermediate confinement regime, depending on the size of the nanocrystal, because a_{exc} is 6 nm.

B. The Particle-in-a-Sphere Model

Although the description of the different confinement regimes is useful, it does not provide a quantitative description of the size-dependent electronic properties. To move toward such a description, one can begin with a very simple model: the *particle-in-a-sphere* model [2,6]. In general, this model considers an arbitrary particle of mass m_0 inside a spherical potential well of radius a,

$$V(r) = \begin{cases} 0 & r < a \\ \infty & r > a \end{cases} \tag{2}$$

Following Flügge [23], the Schrödinger equation is solved yielding wave functions

$$\Phi_{n,\ell,m}(r, \theta, \phi) = C \frac{j_\ell(k_{n,\ell}r) \, Y_\ell^m(\theta, \phi)}{r} \tag{3}$$

where C is a normalization constant, $Y_\ell^m(\theta, \phi)$ is a spherical harmonic, $j_\ell(k_{n,\ell}r)$ is the ℓth-order spherical Bessel function, and

$$k_{n,\ell} = \frac{\alpha_{n,\ell}}{a} \tag{4}$$

with $\alpha_{n,\ell}$ the nth zero of j_ℓ. The energy of the particle is given by

$$E_{n,\ell} = \frac{\hbar^2 k_{n,\ell}^2}{2m_0} = \frac{\hbar^2 \alpha_{n,\ell}^2}{2m_0 a^2} \tag{5}$$

Due to the symmetry of the problem, the eigenfunctions [Eq. (3)] are simple atomic-like orbitals which can be labeled by the quantum numbers n (1, 2,

3...), $\ell(s, p, d, \ldots)$, and m. The energies [Eq. (5)] are identical to the kinetic energy of the free particle, except that the wave vector, $k_{n,\ell}$, is quantized by the spherical boundary condition. Note also that the energy is proportional to $1/a^2$ and, therefore, is strongly dependent on the size of the sphere.

At first glance, this model may not seem useful for the nanocrystal problem. The above particle is confined to an empty sphere, whereas the nanocrystal is filled with semiconductor atoms. However, by a series of approximations, the nanocrystal problem can be reduced to the particle-in-a-sphere form [Eq. (2)]. The photoexcited carriers (electrons and holes) may then be treated as particles inside a sphere of constant potential.

First, the bulk conduction and valence bands are approximated by simple isotropic bands within the *effective mass approximation*. According to Bloch's theorem, the electronic wave functions in a bulk crystal can be written as

$$\Psi_{nk}(\vec{r}) = u_{nk}(\vec{r}) \quad \exp(i\vec{k} \cdot \vec{r}) \tag{6}$$

where u_{nk} is a function with the periodicity of the crystal lattice and the wave functions are labeled by the band index n and wave vector k. The energy of these wave functions is typically described in a *band diagram*, a plot of E versus k. Although band diagrams are in general quite complex and difficult to calculate, in the effective mass approximation the bands are assumed to have simple parabolic forms near extrema in the band diagram. For example, because CdSe is a direct-gap semiconductor, both the valence-band maximum and conduction-band minimum occur at $k = 0$ (see Fig. 1a). In the effective mass approximation, the energy of the conduction ($n = c$) and valence ($n = v$) bands are approximated as

$$E_k^c = \frac{\hbar^2 k^2}{2m_{\text{eff}}^c} + E_g$$

$$\tag{7}$$

$$E_k^v = -\frac{\hbar^2 k^2}{2m_{\text{eff}}^v}$$

respectively, where E_g is the semiconductor bandgap and the energies are relative to the top of the valence band. In this approximation, the carriers behave as free particles with an *effective mass*, $m_{\text{eff}}^{c,v}$. Graphically, the effective mass accounts for the curvature of the conduction and valence bands at $k = 0$. Physically, the effective mass attempts to incorporate the complicated periodic potential felt by the carrier in the lattice. This approximation allows the semiconductor atoms in the lattice to be completely ignored and the electron and hole to be treated as if they were free particles, but with a different mass.

However, to utilize the effective mass approximation in the nanocrystal problem, the crystallites must be treated as a bulk sample. In other words, we assume that the single-particle (electron or hole) wave function can be written in terms of Bloch functions [Eq. (6)] and that the concept of an effective mass still has meaning in a small quantum dot. If this is reasonable, we can utilize the parabolic bands in Fig. 1a to determine the electron levels in the nanocrystal, as shown in Fig. 1b. This approximation, sometimes called the *envelope function approximation* [24,25], is valid when the nanocrystal diameter is much larger than the lattice constant of the material. In this case, the single-particle (sp) wave function can be written as a linear combination of Bloch functions

$$\Psi_{sp}(\vec{r}) = \sum_{k} C_{nk} u_{nk}(\vec{r}) \exp(\vec{k} \cdot \vec{r}) \tag{8}$$

where C_{nk} are expansion coefficients which ensure that the sum satisfies the spherical boundary condition of the nanocrystal. If we further assume that the functions u_{nk} have a weak k dependence, then Eq. (8) can be rewritten as

$$\Psi_{sp}(\vec{r}) = u_{n0}(\vec{r}) \sum_{k} C_{nk} \exp(i\vec{k} \cdot \vec{r}) = u_{n0}(\vec{r}) f_{sp}(\vec{r}) \tag{9}$$

where $f_{sp}(\vec{r})$ is the single-particle *envelope function*. Because the periodic functions u_{n0} can be determined within the *tight-binding approximation* [or linear combination of atomic orbitals (LCAOs) approximation] as a sum of atomic wave functions, $\varphi_{n,k}$

$$u_{n0}(\vec{r}) \approx \sum_{i} C_{nl} \varphi_n(\vec{r} - \vec{r}_i) \tag{10}$$

where the sum is over lattice sites and n represents the conduction band or valence band for the electron or hole, respectively, the nanocrystal problem is reduced to determining the envelope functions for the single-particle wave functions, f_{sp}. Fortunately, this is exactly the problem that is addressed by the particle-in-a-sphere model. For spherically shaped nanocrystals with a potential barrier that can be approximated as infinitely high, the envelope functions of the carriers are given by the particle-in-a-sphere solutions [Eq. (3)]. Therefore, each of the electron and hole levels depicted in Fig. 2b can be described by an atomic-like orbital that is confined within the nanocrystal (1S, 1P, 1D, 2S, etc.). The energy of these levels is described by Eq. (5), with the free particle mass m_0 replaced by $m_{eff}^{c,v}$.

So far, this treatment has completely ignored the Coulombic attraction between the electron and the hole, which leads to excitons in the bulk material. Of course, the Coulombic attraction still exists in the nanocrystal. However, how it is included depends on the confinement regime [2]. In the strong confinement regime, another approximation, the *strong confinement approximation*, is used to treat this term. According to Eq. (5), the confinement energy of each carrier scales as $1/a^2$. The Coulomb interaction scales as $1/a$. In sufficiently small crystallites, the quadratic confinement term dominates. Thus, in the strong confinement regime, the electron and hole can be treated independently and each is described as a particle in a sphere. The Coulomb term may then be added as a first-order energy correction, E_c. Therefore, using Eqs (3), (5), and (9), the electron–hole pair (ehp) states in nanocrystals are written as

$$\Psi_{\text{ehp}}(\vec{r}_e, \vec{r}_h) = \Psi_e(\vec{r})\Psi_h(\vec{r}_h)$$

$$= u_c f_e(\vec{r}_e) u_v f_h(\vec{r}_h)$$

$$= C\left(u_c \frac{j_{L_e}(k_{n_e,L_e} r_e) Y_{L_e}^{m_e}}{r_e}\right)\left(u_v \frac{j_{L_h}(k_{n_h,L_h} r_h) Y_{L_h}^{m_h}}{r_h}\right)$$

$$(11)$$

with energies

$$E_{\text{ehp}}(n_h L_h n_e L_e) = E_g + \frac{\hbar^2}{2a^2}\left\{\frac{\varphi_{n_h,L_h}^2}{m_{\text{eff}}^v} + \frac{\varphi_{n_e,L_e}^2}{m_{\text{eff}}^c}\right\} - E_c \qquad (12)$$

The states are labeled by the quantum numbers $n_h L_h n_e L_e$. For example, the lowest-pair state is written as $1S_h 1S_e$. For pair states with the electron in the $1S_e$ level, the first-order Coulomb correction, E_c, is $1.8e^2/\varepsilon a$, where ε is the dielectric constant of the semiconductor [4]. Equations (11) and (12) are usually referred to as the particle-in-a-sphere solutions to the nanocrystal spectrum.

C.   Optical Transition Probabilities

The probability to make an optical transition from the ground state, $|0\rangle$, to a particular electron–hole pair state is given by the dipole matrix element

$$P = \left|\langle\Psi_{\text{ehp}}|\vec{e}\cdot\hat{p}|0\rangle\right|^2 \qquad (13)$$

where $\vec{e}$ is the polarization vector of the light and $\hat{p}$ is the momentum operator. In the strong confinement regime, where the carriers are treated independently, Eq. (13) is commonly rewritten in terms of the single-particle states:

$$P = \left| \langle \Psi_e | \vec{e} \cdot \hat{p} | \Psi_h \rangle \right|^2 \tag{14}$$

Because the envelope functions are slowly varying in terms of $\vec{r}$, the operator $\hat{p}$ acts only on the unit cell portion (u_{nk}) of the wave function. Equation (14) is simplified to

$$P = \left| \langle u_c | \vec{e} \cdot \hat{p} | u_e \rangle \right|^2 \left| \langle f_e | f_h \rangle \right|^2 \tag{15}$$

In the particle-in-a-sphere model, this yields

$$P = \left| \langle u_c | \vec{e} \cdot \hat{p} | u_v \rangle \right|^2 \delta_{n_e, n_h} \delta_{L_e, L_h} \tag{16}$$

due to the orthonormality of the envelope functions. Therefore, simple selection rules ($\Delta n = 0$ and $\Delta L = 0$) are obtained.

D. A More Realistic Band Structure

In the above model, the bulk conduction and valence bands are approximated by simple parabolic bands (Fig. 1). However, the real band structure of II–VI and III–V semiconductors is typically more complicated. For example, whereas the conduction band in CdSe is fairly well described within the effective mass approximation, the valence band is not. The valence band arises from Se $4p$ atomic orbitals and is sixfold degenerate at $k = 0$, including spin. (In contrast, the conduction band arises from Cd $5s$ orbitals and is only twofold degenerate at $k = 0$.) This sixfold degeneracy leads to a valence-band substructure that modifies the results of the particle-in-a-sphere model [26].

To incorporate this structure in the most straightforward way, CdSe is often approximated as having an ideal diamondlike band structure, as illustrated in Fig. 3a. Although the bands are still assumed to be parabolic, due to strong spin-orbit coupling ($\Delta = 0.42$ eV in CdSe [27]) the valence band degeneracy at $k = 0$ is split into $p_{3/2}$ and $p_{1/2}$ subbands, where the subscript refers to the angular momentum $J = l + s$ ($l = 1, s = 1/2$), with l the orbital and s the spin contribution to the angular momentum. Away from $k = 0$, the $p_{3/2}$ band is further split into $J_m = \pm 3/2$ and $J_m = \pm 1/2$ subbands, where J_m is the projection of J. These three subbands are referred to as the heavy-hole (hh), light-hole (lh), and split-off-hole (so) subbands, as shown in Fig. 3a. Alternatively, they are sometimes referred to as the A, B, and C subbands, respectively.

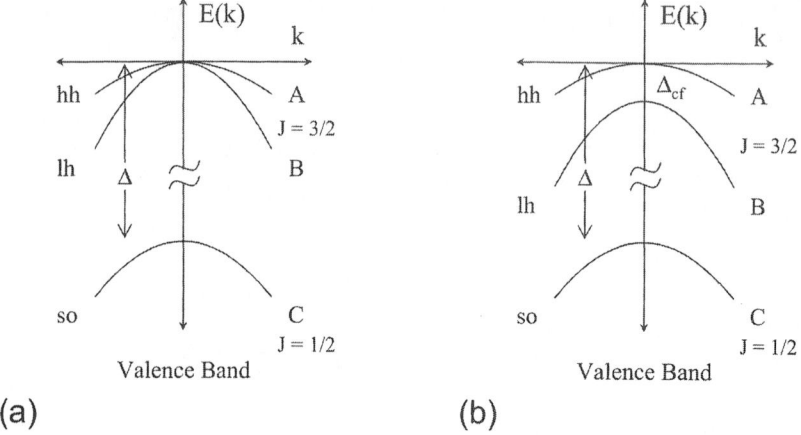

Figure 3 (a) Valence-band structure at $k = 0$ for diamondlike semiconductors. Due to spin-orbit coupling (Δ), the valence band is split into two bands ($J = 3/2$ and $J = 1/2$) at $k = 0$. Away from $k = 0$, the $J = 3/2$ band is further split into the $J_m = \pm 3/2$ heavy-hole (hh or A) and the $J_m = \pm 1/2$ light-hole (lh or B) subbands. The $J = 1/2$ band is referred to as the split-off (so or C) band. (b) The valence band structure for wurtzite CdSe near $k = 0$. Due to the crystal field of the hexagonal lattice the A and B bands are split by Δ_{cf} (25 meV) at $k = 0$.

For many semiconductors, the diamondlike band structure is a good approximation. In the particular case of CdSe, two additional complications arise. First, Fig. 3a ignores the crystal field splitting which occurs in materials with a wurtzite (or hexagonal) lattice. This lattice, with its unique c axis, has a crystal field which lifts the degeneracy of the A and B bands at $k = 0$, as shown in Fig. 3b. This *A–B splitting* is small in bulk CdSe ($\Delta_{cf} = 25$ meV [27]) and is often neglected in quantum-dot calculations. However, we discuss below how this term can cause additional splittings in the nanocrystal optical transitions.

The second complication is that, unlike the diamond structure, the hexagonal CdSe lattice does not have inversion symmetry. In detailed calculations, this lack of inversion symmetry leads to linear terms in k which further split the A and B subbands in Fig. 3b away from $k = 0$ [28]. Because these linear terms are extremely small, they are generally neglected and are ignored in the following.

E. The $k \cdot p$ Method

Because of the complexity in the band structure, the particle-in-a-sphere model [Eq. (11)] is insufficient for accurate nanocrystal calculations. Instead, a better description of the bulk bands must be incorporated into the theory. Although a variety of computational methods could be used, this route does not provide analytical expressions for the description of the bands. Thus, a more sophisticated effective mass approach, the $k \cdot p$ method, is typically used [29]. In this case, bulk bands are expanded analytically around a particular point in k-space, typically $k = 0$. Around this point, the band energies and wave functions are then expressed in terms of the periodic functions u_{nk} and their energies E_{nk}.

General expressions for u_{nk} and E_{nk} can be derived by considering the Bloch functions in Eq. (6). These functions are solutions of the Schrödinger equation for the single-particle Hamiltonian

$$H_0 = \frac{p^2}{2m_0} + V(x) \tag{17}$$

where $V(x)$ is the periodic potential of the crystal lattice. Using Eqs. (6) and (17), it is simple to show that the periodic functions, u_{nk}, satisfy the equation

$$\left[H_0 + \frac{1}{m_0}(k \cdot p) \right] u_{nk} = \lambda_{nk} u_{nk} \tag{18}$$

where

$$\lambda_{nk} = E_{nk} - \frac{k^2}{2m_0} \tag{19}$$

Because u_{n0} and E_{n0} are assumed known, Eq. (18) can be treated in perturbation theory around $k = 0$ with

$$H' = \frac{k \cdot p}{m_0} \tag{20}$$

Then, using nondegenerate perturbation theory to second order, one obtains the energies

$$E_{nk} = E_{n0} + \frac{k^2}{2m_0} + \frac{1}{m_0^2} \sum_{m \neq n} \frac{\left| \vec{k} \cdot \vec{p}_{nm} \right|^2}{E_{n0} - E_{m0}} \tag{21}$$

and functions

$$u_{nk} = u_{n0} + \frac{1}{m_0} \sum_{m \neq n} u_{m0} \frac{\vec{k} \cdot \vec{p}_{nm}}{E_{n0} - E_{m0}} \tag{22}$$

with

$$\vec{p}_{nm} = \langle u_{n0} | \vec{p} | u_{m0} \rangle \tag{23}$$

The summations in Eqs. (21) and (22) are over all bands $m \neq n$. As one might expect, the dispersion of band n is due to coupling with nearby bands. Also, note that inversion symmetry has been assumed. However, for hexagonal crystal lattices (i.e., wurtzite) like CdSe, the lack of inversion symmetry introduces linear terms in k into Eq. (21). Because these terms are typically small, they are generally neglected.

With the $k \cdot p$ approach, analytical expressions can be obtained which describe the bulk bands to second order in k. Whereas here the general method is outlined, the approach must be slightly modified for CdSe. First, for the CdSe valence band, degenerate perturbation theory must be used. In this case, the valence band must be diagonalized before coupling with other bands can be considered. Second, we have neglected spin-orbit coupling terms. However, these terms are easily added as can be seen in Ref. 29.

F.   The Luttinger Hamiltonian

For bulk diamondlike semiconductors, the six-fold degenerate valence band can be described by the Luttinger Hamiltonian [30,31]. This expression, a 6×6 matrix, is derived within the context of degenerate $k \cdot p$ perturbation theory [32]. The Hamiltonian is commonly simplified further using the *spherical approximation* [33–35]. Using this approach, only terms of spherical symmetry are considered. *Warping terms* of cubic symmetry are neglected and, if desired, treated as a perturbation. For nanocrystals, the Luttinger Hamiltonian (sometimes called the six-band model) is the initial starting point for including the valence-band degeneracies and obtaining the hole eigenstates and their energies. We note that because CdSe is wurtzite, as discussed earlier, use of the Luttinger Hamiltonian for CdSe quantum dots is an approximation. Most importantly, it does not include the crystal field splitting that is present in wurtzite lattices.

G.   The Kane Model

Although the Luttinger Hamiltonian is often suitable, particularly for describing the hole levels near k equal zero, for some situations it is necessary to go further. In particular, the Luttinger Hamiltonian does not include coupling between the valence and conduction bands, which can become significant e.g., in narrow-gap semiconductors (see Sect. IV. A). One approach would be to go to higher orders in $k \cdot p$ perturbation theory. However, because this can be quite cumbersome, Kane introduced an alternate procedure for bulk semiconductors, which is also widely used in nanoscale systems [36–38]. In the Kane model, a small subset of bands are treated exactly by explicit diagonalization of Eq. (18) (or the equivalent expression with the spin-orbit interaction included). This subset usually contains the bands of interest (e.g., the valence band and conduction band). Then, the influence of outlying bands is included within the second-order $k \cdot p$ approach. Due to the exact treatment of the important subset, the dispersion of each band is no longer strictly quadratic, as in Eq. (21). Therefore, the Kane model better describes band *nonparabolicities*. In particular, this approach is necessary for narrow-gap semiconductors, where significant coupling between the valence and conduction bands occurs.

For semiconductor quantum dots, a Kane-like treatment was first discussed by Sercel and Vahala [39,40]. More recently, such a description has been used to successfully describe experimental data on narrow-bandgap InAs nanocrystals [13]. Furthermore, even wide-bandgap semiconductor nanocrystals, such as CdSe, may require a more sophisticated Kane treatment of the coupling of the valence and conduction bands [41]. These issues will be discussed in Section IV.

III. CADMIUM SELENIDE NANOCRYSTALS

A. Samples

Although we focus here on nanocrystal spectroscopy, we cannot overemphasize the importance of sample quality in obtaining useful optical information. Indeed, a thorough understanding of the size dependence of the electronic structure in semiconductor quantum dots could not be achieved until sample preparation was well under control. Early spectroscopy (e.g., on II–VI semiconductor nanocrystals [5,42–51]) was constrained by distributions in the size and shape of the nanocrystals, which broaden all spectroscopic features, conceal optical transitions, and inhibit a complete investigation. Later, higher-quality samples became available in which many of the electronic states could be resolved [52–55]. However, the synthetic methods utilized to prepare these nanocrystals could not produce a complete series of such samples. Therefore, the optical studies were limited to one [52–54] or a few sizes [55].

Fortunately, this situation has dramatically changed since the introduction of the synthetic method of Murray et al. [7]. This procedure and subsequent variations [12,56–60] use a wet-chemical (organometallic) synthesis to fabricate high-quality nanocrystals. From the original synthesis, highly crystalline, nearly monodisperse [<4% root mean square (rms)] CdSe nanocrystals can be obtained with well-passivated surfaces. Furthermore, by controlling the growth conditions, such samples can be easily obtained from ~0.8 to ~6 nm in mean radius. Thus, a complete size series can be investigated. For optical experiments, such samples are ideal. In particular, the intensity of *deep-trap* emission, which dominates the luminescence behavior of dots prepared by many other methods, is very weak in these samples. Although the true origin of this emission is unknown, it is generally assumed to arise from surface defects which are deep in the bandgap. Instead of deep-trap emission, the newer nanocrystals exhibit strong band-edge luminescence with quantum yields measured as high as 90% at 10 K. At room temperature, the quantum yield is typically 10%. However, by encapsulating the CdSe nanocrystals in a higher-bandgap semiconductor, such as ZnS or CdS, the quantum yield can be further improved [61–63]. Emission efficiencies greater than 50% at room temperature have been reported.

B. Spectroscopic Methods

Samples obtained from these new synthetic procedures provided the first opportunity to study the size dependence of the electronic structure in detail. However, because even the best samples contain residual sample inhomogeneities which can broaden spectral features and conceal transitions, several optical techniques have been used to reduce these effects and maximize the information obtained. These techniques include *transient differential absorption* spectroscopy (TDA), *photoluminescence excitation* spectroscopy (PLE), and *fluorescence line-narrowing* spectroscopy (FLN), which are described further in this subsection. More recently, single-molecule spectroscopy [64], which can remove all inhomogeneities from the sample distribution, has been adapted to nanocrystals and many exciting results have been observed [65]. However, because *single quantum-dot* spectroscopy will be described elsewhere in this volume and these methods have mostly provided information about the emitting state (i.e., not the electronic-level structure), it will not be emphasized here.

From a historical perspective, the most common technique to obtain absorption information has been TDA, also called pump-probe or hole-burning spectroscopy [8,47,48,52–54,66–73]. This technique measures the absorption change induced by a spectrally narrow pump beam. TDA effectively increases the resolution of the spectrum by optically exciting a narrow subset of the quantum dots. By comparing the spectrum with and without this optical excitation, information about the absorption of the subset is revealed, with inhomogeneous broadening greatly reduced. Because the quantum dots within the subset are in an excited state, the TDA spectrum will reveal both the absence of a ground-state absorption (a bleach) and excited-state absorptions (also called pump-induced absorptions). Unfortunately, when pump-induced absorption features overlap with the bleach features of interest, the analysis becomes complicated and the usefulness of the technique diminishes.

To avoid this problem, many groups have utilized another optical technique, photoluminescence excitation (PLE) spectroscopy [8,9,54,74–76]. PLE is similar to TDA in that it selects a narrow subset of the sample distribution to obtain absorption information. However, in PLE experiments, one utilizes the emission of the nanocrystals. Thus, this technique is particularly suited to the efficient fluorescence observed in high-quality samples. PLE works by monitoring a spectrally narrow emission window within the inhomogenous emission feature while scanning the frequency of the excitation source. Because excited nanocrystals always relax to their first excited state before emission, the spectrum that is obtained reveals absorption information about the narrow subset of nanocrystals that emit.

An additional advantage of this technique is that emission information can be obtained during the same experiment. For example, fluorescence line-narrowing (FLN) spectroscopy can be used to measure the emission spectrum from a subset of the sample distribution. In particular, by exciting the nanocrystals on the low-energy side of the first absorption feature, only the largest dots in the distribution are excited.

Figure 4 demonstrates all of these techniques. In Fig. 4a, absorption and emission results are shown for a sample of CdSe nanocrystals with a mean radius of 1.9 nm. On this scale, only the lowest two excited electron–hole pair states are observed in the absorption spectrum (solid line). The emission spectrum (dashed line) is obtained by exciting the sample well above its first transition so that emission occurs from the entire sample distribution. This inhomogeneously broadened emission feature is referred to as the *full luminescence* spectrum. If, instead, a subset of the sample distribution is excited, a significantly narrowed and structured FLN spectrum is revealed. For example, when the sample in Fig. 4 is excited at the position of the downward arrow, a vibrational progression is clearly resolved (due to longitudinal optical phonons) in the emission spectrum. Similarly, by monitoring the emission at the position of the upward arrow, the PLE spectrum in Fig. 4b reveals absorption features with higher resolution than in Fig. 4a. Further, additional structure is observed within the lowest absorption feature. As we discuss later, these features (labeled α and β) represent fine structure present in the lowest electron–hole pair state and have important implications for quantum-dot emission. However, before discussing this fine structure, we first treat the size dependence of the electron structure in CdSe nanocrystals.

C. Size Dependence of the Electronic Structure

Although the absorption and PLE spectra in Fig. 4 show only the two lowest exciton features, high-quality samples reveal much more structure. For example, in Fig. 5, PLE results for a 2.8-nm-radius CdSe sample are shown along with its absorption and full luminescence spectra. These data cover a larger

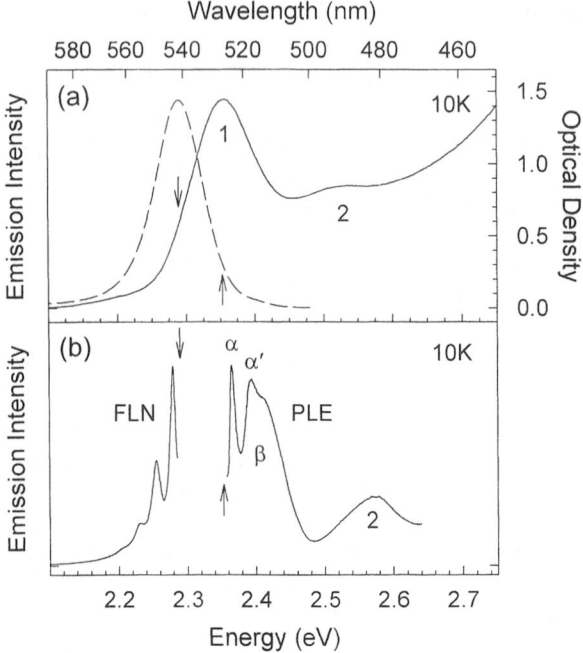

Figure 4 (a) Absorption (solid line) and full luminescence (dashed line) spectra for ~1.9-nm effective radius CdSe nanocrystals. (b) FLN and PLE spectra for the same sample. A longitudinal optical phonon progression is observed in FLN. Both narrow (α, α') and broad (β) absorption features are resolved in PLE. The downward (upward) arrows denote the excitation (emission) position used for FLN (PLE). (Adapted from Ref. 11.)

spectral range than Fig. 4 and show more of the spectrum. To determine how the electronic structure evolves with quantum-dot size, PLE data can be obtained for a large series of samples. Seven such spectra are shown in Fig. 6. The nanocrystals are arranged (top to bottom) in order of increasing radius from ~1.5 to ~4.3 nm. Quantum confinement clearly shifts the transitions to higher energy (> 0.5 eV) with decreasing size. The quality of these quantum dots also allows as many as eight absorption features to be resolved in a single spectrum.

By extracting peak positions from PLE data such as Fig. 6, the quantum-dot spectrum as a function of size is obtained. Figure 7 plots the result for a large dataset from CdSe nanocrystals. Although nanocrystal radius (or diameter) is not used as the x axis, Fig. 7 still represents a size-dependent

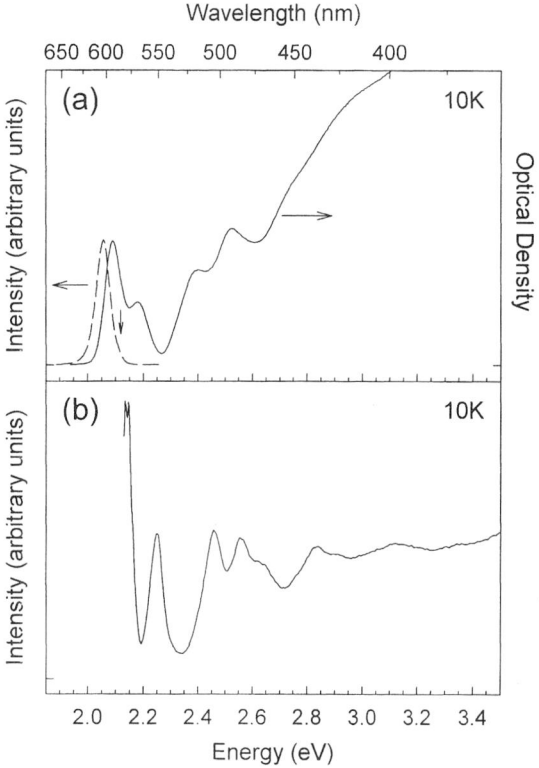

Figure 5 (a) Absorption (solid line) and full luminescence (dashed line) spectra for ~2.8-nm-radius CdSe nanocrystals. In luminescence, the sample was excited at 2.655 eV (467.0 nm). The downward arrow marks the emission position used in PLE. (b) PLE scan for the same sample. (Adapted from Ref. 9.)

plot. The x-axis label, the energy of the first excited state, is a strongly size-dependent parameter. It is also much easier to measure accurately than nanocrystal size. For the y axis, the energy relative to the first excited state is used. This is chosen, in part, to concentrate on the excited states. However, it is also chosen to eliminate the difficulty in comparing the data with theory. This point aside, Fig. 7 summarizes the size dependence of the first 10 transitions for CdSe quantum dots from ~ 1.2 to ~ 5.3 nm in radius. Because the exciton Bohr radius is 6 nm in CdSe, these data span the strong confinement regime in this material.

In order to understand this size dependence, one could begin with the simple particle-in-a-sphere model outlined in Section II.B. The complicated

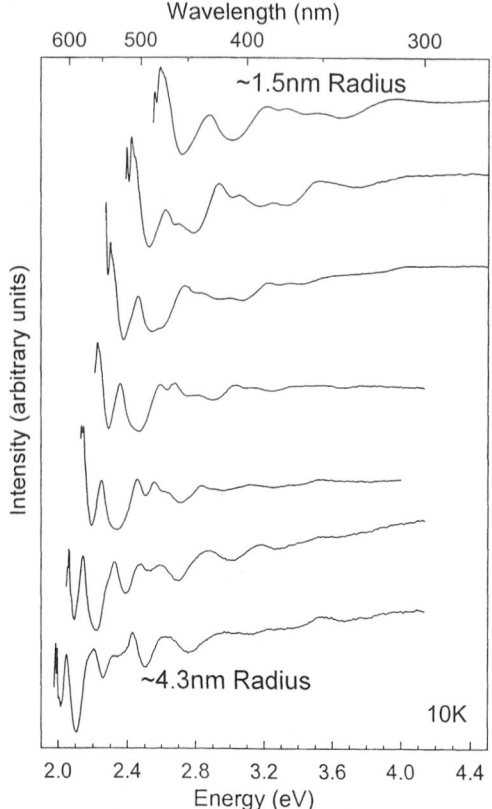

Figure 6 Normalized PLE scans for seven different-sized CdSe nanocrystal samples. Size increases from top to bottom and ranges from ~1.5 to ~4.3 nm in radius. (Adapted from Ref. 9.)

valence-band structure, shown in Fig. 3, could then be included by considering each subband (A, B, and C) as a simple parabolic band. In such a zeroth-order picture, each bulk subband would lead to a ladder of particle-in-a-sphere states for the hole, as shown in Fig. 8. Quantum-dot transitions would occur between these hole states and the electron levels arising from the bulk conduction band. However, this simplistic approach fails to describe the experimental absorption structure. In particular, two avoided crossings are present in Fig. 7 [between features (e) and (g) at ~2.0 eV and between features (e) and (c) above 2.2 eV] and these are not predicted by this particle-in-a-sphere model.

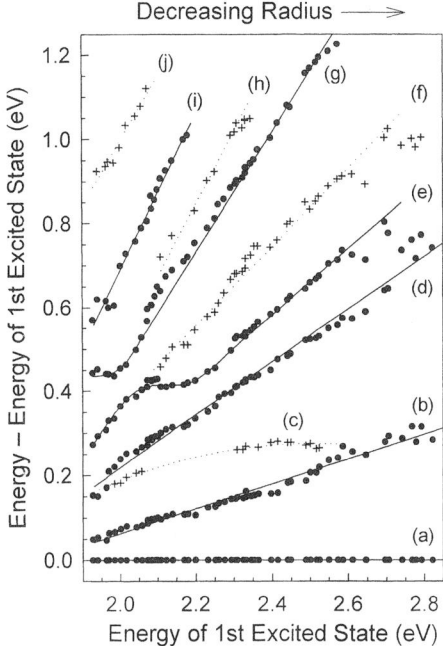

Figure 7 Size dependence of the electronic structure in CdSe nanocrystals. Peak positions are extracted from PLE data as in Fig. 6. Strong (weak) transitions are denoted by circles (crosses). The solid (dashed) lines are visual guides for the strong (weak) transitions to clarify their size evolution. (Adapted from Ref. 9.)

The problem lies in the assumption that each valence subband produces its own independent ladder of hole states. In reality, the hole states are mixed due to the underlying quantum mechanics. To help understand this effect, all of the relevant quantum numbers are summarized in Fig. 9. The total angular momentum of either the electron or hole (F_e or F_h) has two contributions: (a) a "unit-cell" contribution (J) due to the underlying atomic basis which forms the bulk bands and (b) an envelope function contribution (L) due to the particle-in-a-sphere orbital. To apply the zeroth-order illustration above (Fig. 8), we must assume that the quantum numbers describing each valence subband (J_h) and each envelope function (L_h) are conserved. However, when the Luttinger Hamiltonian is combined with a spherical potential, mixing between the bulk valence bands occurs. This effect, which was first shown for bulk impurity centers, [33–35] also mixes quantum-dot hole states [26,39, 40,55,77,78]. Only parity and the total hole angular momentum (F_h) are good quantum numbers. Neither L_h nor J_h are conserved. Therefore, each

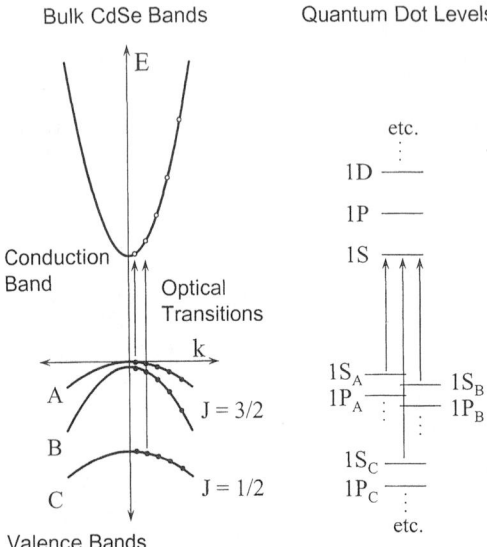

Figure 8 A simplistic model for describing the electronic structure in nanocrystals. Each valence band contributes a ladder of particle-in-a-sphere states for the hole. The optical transitions then occur between these hole states and the electron levels arising from the conduction band. This model fails to predict the observed structure due to mixing of the different hole ladders, as discussed in the text.

quantum-dot hole state is a mixture of the three valence subbands (*valence-band mixing*) as well as particle-in-a-sphere envelope functions with angular momentum L_h and $L_h + 2$ (*S-D mixing*). The three independent ladders of hole states, as shown in Fig. 8, are coupled. The electron levels, which originate in the simple conduction band that is largely unaffected by the valence-band complexities, can be assumed to be well described by the particle-in-a-sphere ladder. However, we will revisit this assumption below.

When theory includes these effects, the size dependence observed in Fig. 7 can be described. Using the approach of Efros et al. [55,77], in which the energies of the hole states are determined by solving the Luttinger Hamiltonian and the electron levels are calculated within the Kane model, strong agreement with the data is obtained, as shown in Figs. 10 and 11. Figure 10 compares theory with the lowest three transitions which exhibit simple size-dependent behavior (i.e., no avoided crossings). Figure 11 shows the avoided crossing regions. The transitions can be assigned and labeled by modified particle-in-a-sphere symbols which account for the valence-band mixing discussed earlier [9].

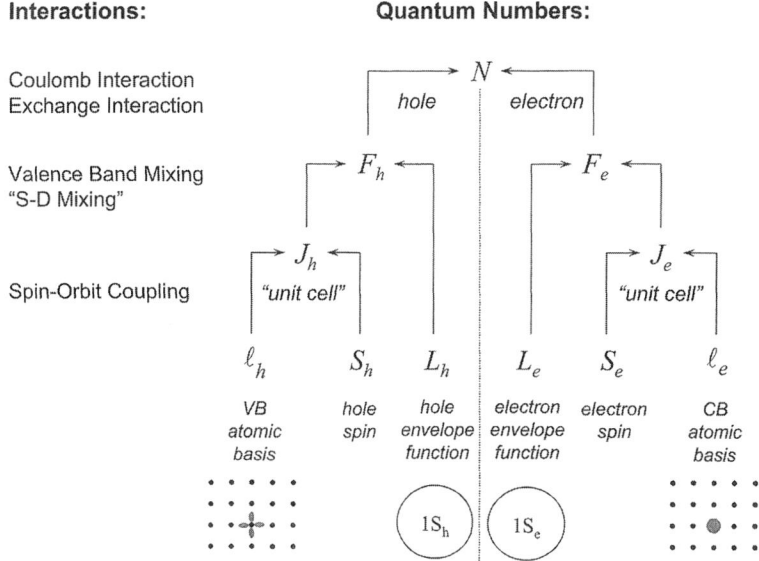

Figure 9 Summary of quantum numbers and important interactions in semi-conductor nanocrystals. The total electron–hole pair angular momentum (N) has contributions due to both the electron (F_e) and hole (F_h). Each carrier's angular momentum (F) may then be further broken down into a unit-cell component (J) due to the atomic basis (ℓ) and spin (s) of the particle and an envelope function component (L) due to the particle-in-a-sphere orbital.

Although theory clearly predicts the observed avoided crossings, Fig. 11 also demonstrates that theory underestimates the repulsion in both avoided crossing regions, causing theoretical deviation in the predictions of the $1S_{1/2}1S_e$ and $2S_{1/2}1S_e$ transitions. This discrepancy could be due to the Coulomb mixing of the electron–hole pair states, which is ignored by the model (via the strong confinement approximation). If included, this term would further couple the $nS_{1/2}1S_e$ transitions such that these states interact more strongly. In addition, the Coulomb term would cause the $1S_{1/2}1S_e$ and $2S_{1/2}1S_e$ states to avoid one another through their individual repulsion from the strongly allowed $1P_{3/2}1P_e$.

Despite these discrepancies, however, this theoretical approach is clearly on the right track. Therefore, this model can be used to understand the physics behind the avoided crossings. As discussed earlier, in the zeroth-order picture of Fig. 8, each valence subband contributes a ladder of hole states. Due to spin-orbit splitting (see Fig. 3) the C-band ladder is offset

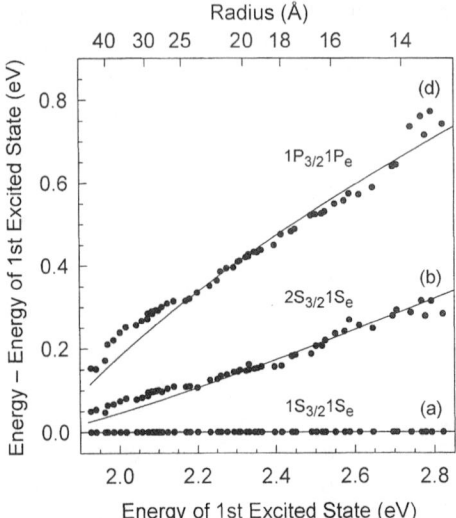

Figure 10 Theoretically predicted pair states (solid lines) assigned to features (a), (b), and (d) in Fig. 7. The experimental data is shown for comparison (circles). (Adapted from Ref. 9.)

0.42 eV below the A- and B-band ladders. This leads to possible resonances between hole levels from the A and B bands with C-band levels. Because the levels are spreading out with decreasing dot size, resonance conditions are satisfied only in certain special sizes. Figure 12 demonstrates the two resonances responsible for the observed avoided crossings. For simplicity, we treat the A and B bands together. In Fig. 12a (12b), the $2D$ ($1D$) level from the A and B bands is resonant with the $1S$ level from the C band. The size dependence of these levels is depicted in Fig. 12c. Due to both valence band mixing and S-D mixing, these resonant conditions lead to the observed avoided crossings. Although this description is based on the simple particle-in-a-sphere model of Fig. 8, the explanation has been shown to be consistent with a more detailed analysis [9].

D. Beyond the Spherical Approximation

The above success in describing the size dependence of the data was achieved within the *spherical approximation* (Sect. II.F). In this case, the bands are assumed to be spherically isotropic [i.e., warping terms in Eq. (21) are ignored]. Furthermore, the CdSe nanocrystals are assumed to have a spherical shape and a cubic crystal lattice (i.e., zinc blende). With these

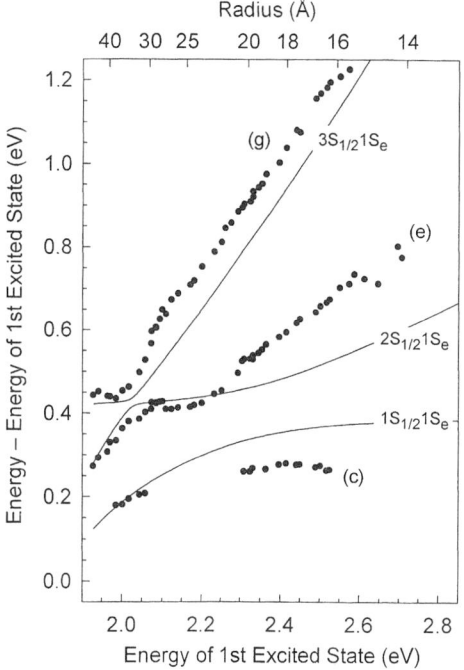

Figure 11 Theoretically predicted pair states (solid lines) assigned to features (c), (e), and (g) in Fig. 7. The experimental data is shown for comparison (circles). (Adapted from Ref. 9.)

assumptions, each of the electron–hole pair states is highly degenerate. For example, the first excited state ($1S_{3/2}1S_e$, which we refer to as the *band-edge exciton*) is eightfold degenerate. However, in reality, these degeneracies will be lifted by several second-order effects. First, as mentioned earlier, CdSe nanocrystals have a unixial crystal lattice (wurtzite) which leads to a splitting of the valence subbands (Fig. 3b) [79]. Second, electron micros-copy experiments show that CdSe nanocrystals are not spherical, but rather slightly prolate [7]. This shape anisotropy will split the electron–hole pair states [80]. Finally, the electron–hole exchange interaction, which is negli-gible in bulk CdSe, can lead to level splittings in nanocrystals due to enhanced overlap between the electron and hole [81–84]. Therefore, when all of these effects are considered, the initially eightfold degenerate band-edge exciton is split into five sublevels [10].

This *exciton fine structure* is depicted in the energy level diagram of Fig. 13. To describe the structure, two limits are considered. On the left of Fig.

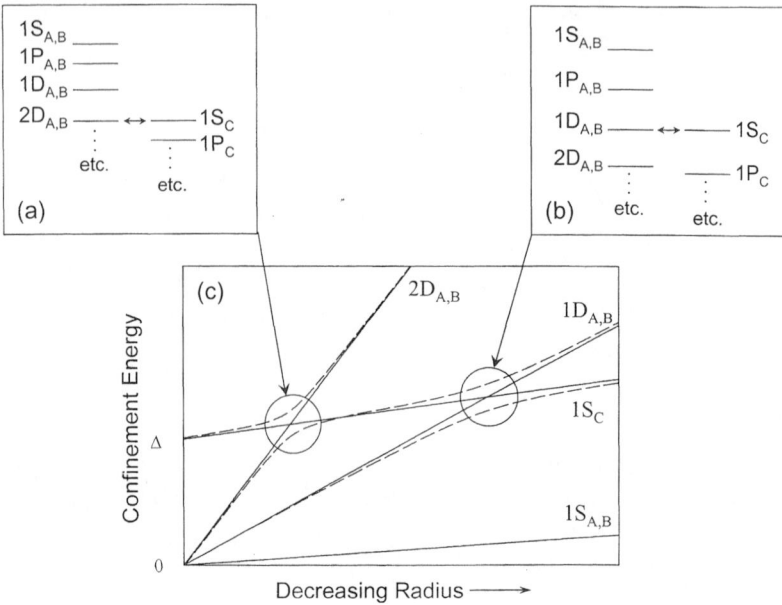

Figure 12 Schematics depicting the origin of the observed avoided crossings. For a particular nanocrystal size, a resonance occurs between a hole level from the A and B bands (combined for simplicity) and a hole level from the C band. Energy level diagrams for the hole states are shown in (a) and (b) for the two resonances responsible for the observed avoided crossings. (c) The energy of the hole states versus decreasing radius. The solid (dashed) lines represent the levels without (with) the valence band and *S-D* mixing.

13, the effect of the anisotropy of the crystal lattice and/or the nonspherical shape of the crystallite dominates. This corresponds to the bulk limit where the exchange interaction between the electron and hole is negligible (0.15 meV) [85]. The band-edge exciton is split into two fourfold degenerate states, analogous to the bulk A–B splitting (see Fig. 3b). The splitting occurs due to the reduction from spherical to uniaxial symmetry. However, because the exchange interaction is proportional to the overlap between the electron and hole, in small dots this term is strongly enhanced due to the confinement of the carriers [81–84]. Therefore, the right-hand side of Fig. 13 represents the small nanocrystal limit where the exchange interaction dominates. In this case, the important quantum number is the total angular momentum, N (see Fig. 9). Because $F_h = 3/2$ and $F_e = 1/2$, the band-edge exciton is split into a fivefold degenerate $N = 2$ state and a threefold degenerate $N = 1$ state. In the center

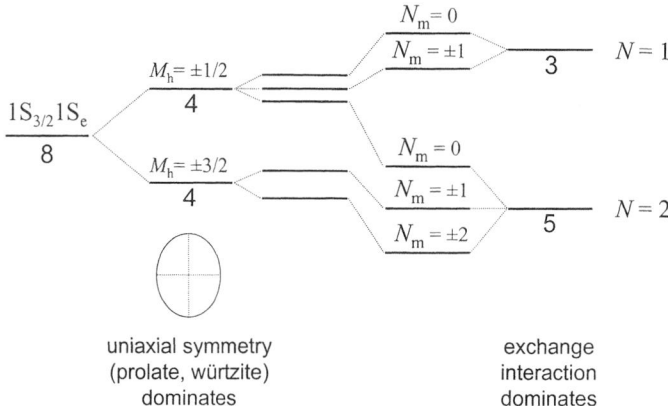

Figure 13 Energy level diagram describing the exciton fine structure. In the spherical model, the band-edge exciton ($1S_{3/2}1S_e$) is eightfold degenerate. This degeneracy is split by the nonspherical shape of the dots, their hexagonal (wurtzite) lattice, and the exchange interaction.

of Fig. 13, the correlation diagram between these two limits is shown. When both effects are included, the good quantum number is the projection of N along the unique crystal axis, N_m. The five sublevels are then labeled by $|N_m|$: one sublevel with $|N_m| = 2$, two with $|N_m| = 1$, and two with $|N_m| = 0$. Levels with $|N_m| > 0$ are twofold degenerate.

To include these effects into the theory, the anisotropy and exchange terms can be added as perturbations to the spherical model [10]. Figure 14 shows the calculated size dependence of the exciton fine structure. The five sublevels are labeled by $|N_m|$ with superscripts to distinguish upper (U) and lower (L) sublevels with the same $|N_m|$. Their energy, relative to the 1^L sublevel, is plotted versus *effective radius*, which is defined as

$$a_{\text{eff}} = \frac{1}{2}(b^2 c)^{1/3} \tag{24}$$

where b and c are the short and long axes of the nanocrystal, respectively. The enhancement of the exchange interaction with decreasing nanocrystal size is clearly evident in Fig. 14. Conversely, with increasing nanocrystal size, the sublevels converge upon the bulk A–B splitting, as expected.

E. The Dark Exciton

At first glance, one may feel that the exciton fine structure is only a small refinement to the theoretical model with no real impact on the properties of

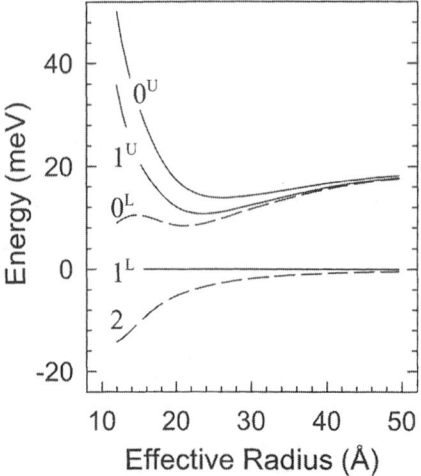

Figure 14 Calculated band-edge exciton structure versus effective radius. The sublevels are labeled by $|N_m|$ with superscripts to distinguish upper (U) and lower (L) sublevels with the same $|N_m|$. Positions are relative to 1^L. Optically active (passive) levels are shown as solid (dashed) lines. (Adapted from Ref. 11.)

nanocrystals. However, these splittings have helped explain a long-standing question in the emission behavior of CdSe nanocrystals. Although exciton recombination in bulk II–VI semiconductors occurs with a ~1-ns lifetime [86]. CdSe quantum dots can exhibit an ~1-μs radiative lifetime at 10 K [10,87–90]. This effect could perhaps be rationalized in early samples which were of poor quality and emitted weakly via deep-trap fluorescence. However, even high-quality samples, which emit strongly at the band edge, have long radiative lifetimes. To explain this behavior, the emission had been rationalized by many researchers as a *surface effect*. In this picture, the anomalous lifetime was explained by localization of the photoexcited electron and/or hole at the dot-matrix interface. Once the carriers are localized in surface traps, the decrease in carrier overlap increases the recombination time. The influence of the surface on emission was considered reasonable because these materials have such large surface-to-volume ratios (e.g., in an ~1.5-nm-radius nanocrystal, roughly one-third of the atoms are on the surface). This surface model could then explain the long radiative lifetimes, luminescence polarization results, and even the unexpectedly high longitudinal optical (LO) phonon coupling observed in emission.

However, as first proposed by Calcott et al, [81], the presence of exciton fine structure provides an alternative explanation for the anomalous emission behavior. Emission from the lowest band-edge state, $|N_m| = 2$, is optically forbidden in the electric dipole approximation. Relaxation of the electron–hole pair into this state, referred to as the *dark exciton*, can explain the long radiative lifetimes observed in CdSe quantum dots (QDs). Because two units of angular momentum are required to return to the ground state from the $|N_m| = 2$ sublevel, this transition is one-photon forbidden. However, less efficient, phonon-assisted transitions can occur, explaining the stronger LO phonon coupling of the emitting state. In addition, polarization effects observed in luminescence [89] can be rationalized by relaxation from the 1^L sublevel to the dark exciton [84].

F. Evidence for the Exciton Fine Structure

As we mentioned earlier, PLE spectra from high-quality samples often exhibit additional structure within the lowest electron–hole pair state. For example, in Fig. 4b, a narrow feature (α), its phonon replica (α'), and a broader feature (β) are observed. Although these data alone are not sufficient to prove the origin of these features, a careful analysis of a larger dataset has shown that they arise due to the exciton fine structure [11]. The analysis concludes that the spectra in Fig. 4b are consistent with the absorption and emission line shapes shown in Fig. 15. In this case, the emitting state is assigned to the dark exciton ($|N_m| = 2$), the narrow absorption feature (α) is assigned to the 1^L sublevel, and the broader feature (β) is assigned to a combination of the 1^U and the 0^U sublevels. Because it is optically passive, the 0^L sublevel remains unassigned.

This assignment is strongly supported by size-dependent studies. Figure 16 shows PLE and FLN data for a larger sample ($\sim$ 4.4-nm effective radius). Whereas in Fig. 4b, three band-edge states are resolved [a narrow emitting state, a narrow absorbing state (α), and a broad absorbing state (β)], in Fig. 16, four-band edge states are present: [a narrow emitting state and three narrow absorbing states (α, β_1, and β_2)]. Consequently, β_1 and β_2 can be assigned to the individual 1^U and 0^U sublevels. To be more quantitative, a whole series of sizes can be examined (see Fig. 17) to extract the experimental positions of the band-edge absorption and emission features as a function of size. In Fig. 18b, the positions of the absorbing (filled circles and squares) and emitting (open circles) features are plotted relative to the narrow absorption line, α (1^L). For larger samples, both the positions of β_1 and β_2 (pluses) and their weighted average (squares) are shown. Figure 17d shows the size dependence of the relative oscillator strengths of the optically allowed transitions. The strength of the upper states (1^U and 0^U) is combined because these states are not individually resolved in all of the data. Comparison of all of these data with theory (Figs. 17a and 17c) in-

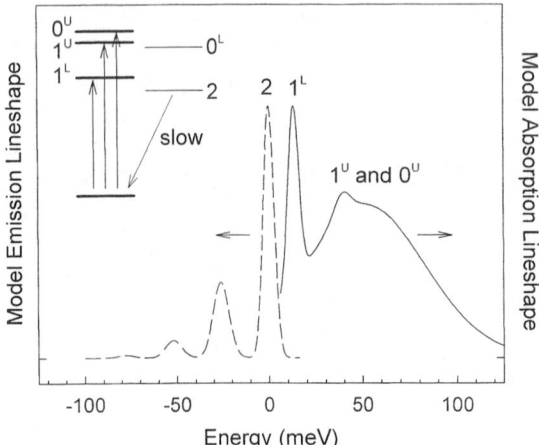

Figure 15 Absorption (solid line) and emission (dotted line) line shape extracted for the sample shown in Fig. 4 including LO phonon coupling. An energy level diagram illustrates the band-edge exciton structure. The sublevels are labeled as in Fig. 14. Optically active (passive) levels are shown as solid (dotted) lines. (Adapted from Ref. 11.)

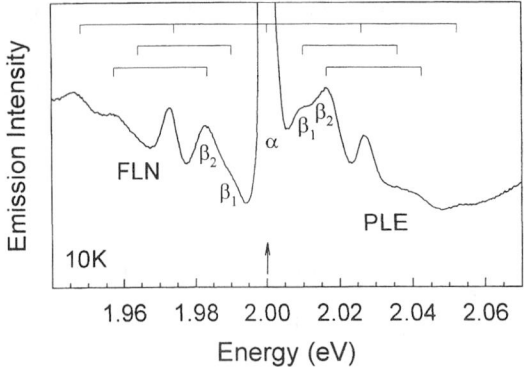

Figure 16 Normalized FLN and PLE data for an ~4.4-nm effective radius sample. The FLN excitation and PLE emission energies are the same and are designated by the arrow. Although emission arises from a single emitting state and its LO phonon replicas, three overlapping LO phonon progressions are observed in FLN due to the three band-edge absorption features (α, β_1, and β_2). Horizontal brackets connect the FLN and PLE and features with their LO phonon replicas. (Adapted from Ref. 11.)

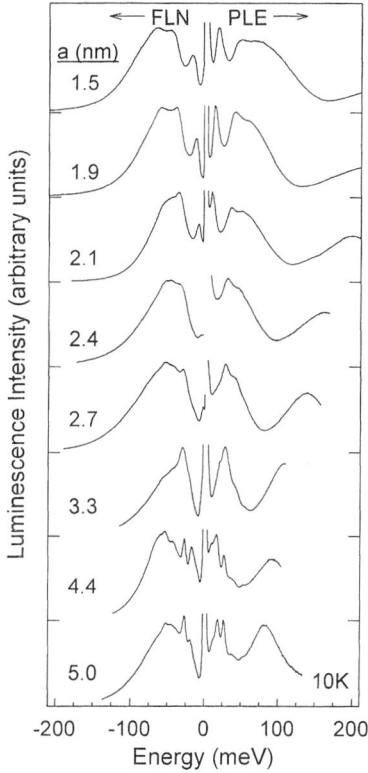

Figure 17   The size dependence of band-edge FLN/PLE spectra. (Adapted from Ref. 11.)

dicates that the model accurately reproduces many aspects of the data. Both the splitting between $|N_m| = 2$ and 1^L (the Stokes shift) and the splitting between 1^L and the upper states (1^U and 0^U) are described reasonably well. Also, the predicted trend in the oscillator strength is observed. These agreements are particularly significant because although the predicted structure strongly depends on the theoretical input parameters [10], only literature values were used in the theoretical calculation.

G.   Evidence for the "Dark Exciton"

In addition to the observation that CdSe nanocrystals exhibit long emission lifetimes, they also display other emission dynamics that point to the existence of the dark exciton. For example, Fig. 19a shows how the emission decay of a

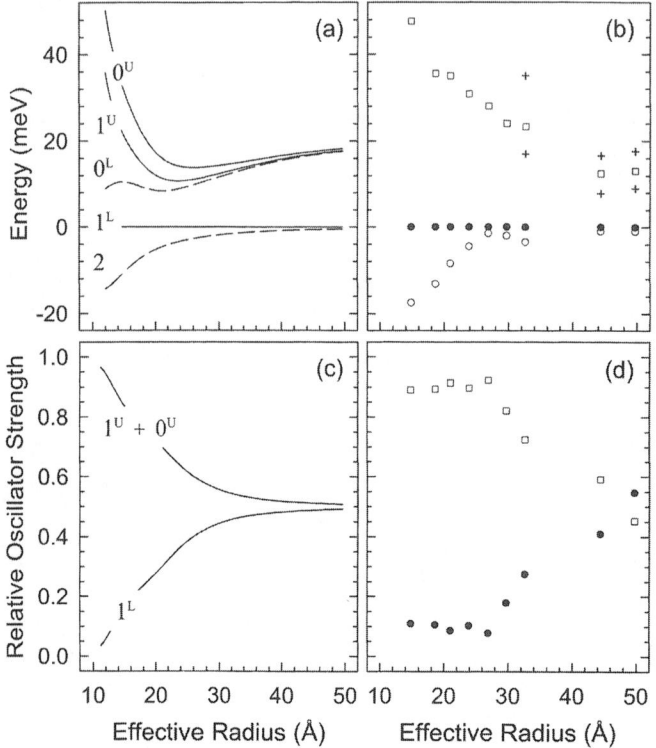

Figure 18 (a) Calculated band-edge exciton ($1S_{3/2}1S_e$) structure versus effective radius as in Fig. 14. (b) Position of the absorbing (filled circles and squares) and emitting (open circles) features extracted from Fig. 17. In samples where β_1 and β_2 are resolved, each position (shown as pluses) and their weighted average (squares) are shown. (c) Calculated relative oscillator strength of the optically allowed band-edge sublevels versus effective radius. The combined strength of 1^U and 0^U is shown. (d) Observed relative oscillator strength of the band-edge sublevels: 1^L (filled circles) and the combined strength of 1^U and 0^U (squares). (Adapted from Ref. 11.)

CdSe sample with a mean radius of 1.2 nm changes with an externally applied magnetic field. Obviously, the data indicate that the presence of a field strongly modifies the emission behavior. This fact, which is difficult to explain with other models (e.g., due to surface trapping), is easily explained by the dark-exciton model. Because thermalization processes are highly efficient, excited nanocrystals quickly relax into their lowest sublevel (the dark exciton). Furthermore, the separation between the dark exciton and the first optically allowed sublevel (1^L) is much larger than kT at cryogenic temperatures. Thus,

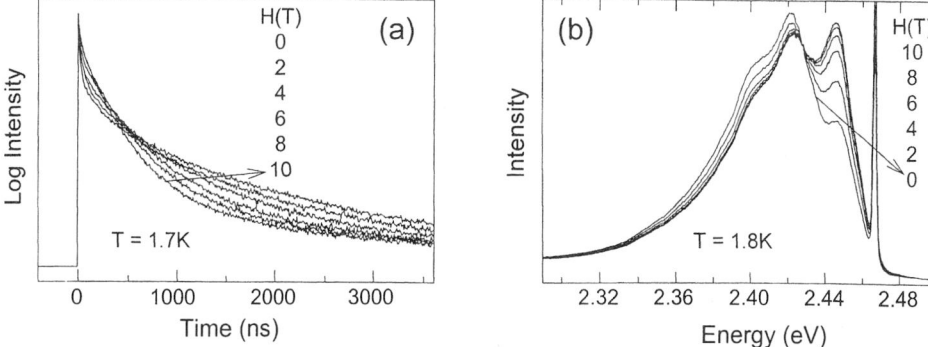

Figure 19   Magnetic field dependence of (a) emission decays recorded at the peak of the luminescence and (b) FLN spectra excited at the band edge (2.467 eV) for 1.2-nm-radius CdSe nanocrystals. The FLN spectra are normalized to their one phonon line. A small amount of the excitation laser is included to mark the pump position. Experiments were carried out in the Faraday geometry (magnetic field parallel to the light propagation vector). (Adapted from Ref. 10.)

the excited nanocrystal must return to the ground state from the dark exciton. The long (µs) emission is consistent with recombination from this weakly emitting state. However, because a strong magnetic field couples the dark exciton to the optically allowed sublevels, the emission lifetime should decrease in the presence of a magnetic field. Because the experimental fluorescence quantum yield remains essentially constant with field, this mixing leads to the decrease in the emission decay with increasing magnetic field [10].

Another peculiar effect which can easily be explained by the dark exciton is the influence of a magnetic field on the vibrational spectrum, which is demonstrated in Fig. 19b. A dramatic increase is observed in the relative strength of the zero-phonon line with increasing field. This behavior results from the dark exciton utilizing the phonons to relax to the ground state. In a simplistic picture, the dark exciton would have an infinite fluorescence lifetime in zero applied field because the photon cannot carry an angular momentum of 2. However, nature will always find some relaxation pathway, no matter how inefficient. In particular, the dark exciton can recombine via a LO-phonon-assisted, momentum-conserving transition [81]. In this case, the higher-phonon replicas are enhanced relative to the zero-phonon line. If an external magnetic field is applied, the dark exciton becomes partially allowed due to mixing with the optically allowed sublevels. Consequently, relaxation no longer relies on a phonon-assisted process and the strength of the zero-phonon line increases.

IV. BEYOND CdSe

A. Indium Arsenide Nanocrystals and the Pidgeon–Brown Model

The success in both the synthesis and the spectroscopy of CdSe has encouraged researchers to investigate other semiconductor systems. Although the synthetic methods used for CdSe can easily be extended to many of the II–VI semiconductors [7,58], much effort has been focused upon developing new classes of semiconductor nanocrystals, particularly those that may have high technological impact (e.g., silicon [91,92]). Among these, the system that is perhaps best to discuss here is InAs. As a zinc blende, direct-bandgap, III–V semiconductor, InAs is in many ways very similar to CdSe. Most importantly, InAs nanocrystals can be synthesized through a well-controlled organometallic route that can produce a series of different-sized colloidal samples [12]. These samples exhibit strong band-edge luminescence such that they are well suited to spectroscopic studies. On the other hand, InAs also has several important differences from CdSe. In particular, it has a narrow bandgap (0.41 eV). This implies that the coupling between the conduction and valence bands, which was largely ignored in our theoretical treatment of CdSe, will be important.

    To explore this issue, Banin et al. have performed detailed spectroscopic studies of high-quality Indium Arsenide nanocrystals [13,93,94]. Figure 6 in Chapter 8 shows size-dependent PLE data obtained from these samples. As in CdSe, the positions of all of the optical transitions can be extracted and plotted. The result is shown in Fig. 7 in Chapter 8. However, unlike CdSe, InAs nanocrystals are not well described by a six-band Luttinger Hamiltonian. Rather, the data require an eight-band Kane treatment (also called the Pidgeon–Brown model [37]), which explicitly includes coupling between the conduction and valence band [13,41]. With the eight-band model, the size dependence of the electronic structure can be well described, as shown in Fig. 7 of Chapter 8.

    Intuitively, one expects mixing between the conduction and valence band to become significant as the bandgap decreases. Quantitatively, this mixing has been shown to be related to

$$\sqrt{\frac{\Delta E_{e,h}}{E_g^s + \Delta E_{e,h}}} \tag{25}$$

where $\Delta E_{e,h}$ is the confinement energy of the electron or the hole [41]. As expected, the value of Eq. (25) becomes significant as ΔE approaches the width of the bandgap (i.e., in narrow-bandgap materials). However, unexpectedly, this equation also predicts that mixing can be significant in wide-gap semiconductors due to the square root dependence. Furthermore, because the electron is typically more strongly confined than the hole, the mixing should be more important for the electrons. Therefore, this analysis concludes that even in wide-gap semiconductors, an eight-band Pidgeon–Brown model may be necessary to accurately predict the size-dependent structure.

B. The Problem Swept Under the Rug

Although the effective mass models can quite successfully reproduce many aspects of the electronic structure, the reader may be troubled by a problem that was "swept under the rug" in Section III.c. In our discussion of the size-dependent data shown in Fig. 7, it was mentioned that the data were plotted relative to the energy of the first excited state, in part to avoid a difficulty with the theory. This difficulty is shown more clearly in Fig. 20, where the energy of the first excited state ($1S_{3/2}1S_e$) is plotted versus $1/a^2$. Surprisingly, the same theory that can quantitatively fit the data in Figs. 10 and 11 fails to predict the size dependence of the lowest transition (Fig. 20). Because the same problem arises in InAs nanocrystals, where a more sophisticated eight-band effective mass model was used, it is unlikely that this is caused by the inadequacies of

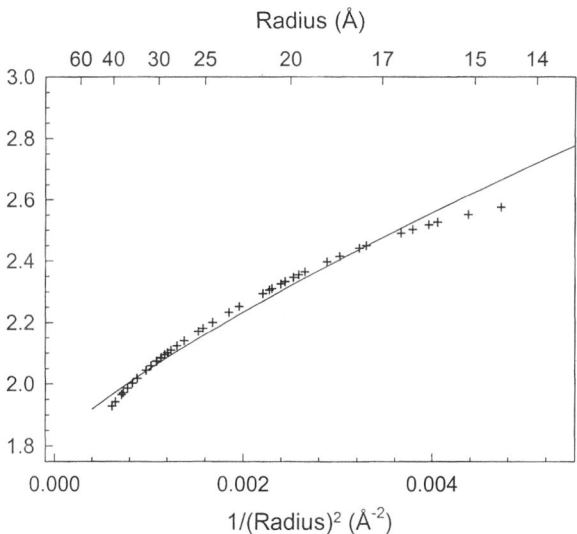

Figure 20 Energy of the first excited state ($1S_{3/2}1S_e$) in CdSe nanocrystals versus $1/radius^2$. The curve obtained from the same theory as in Figs. 10 and 11 (solid line) is compared with PLE data (crosses). (Adapted from Ref. 9.)

the six-band Luttinger Hamiltonian. Rather, the experiment suggests that an additional nonparabolicity is present in the bands that is not accounted for even by the eight-band model. However, the question remains: What the cause of this nonparabolicity?

The observation that the theory correctly predicts the transition energies when plotted relative to the first excited state is an important clue. Because most of the low-lying optical transitions share the same electron level $(1S_e)$, Fig. 20 implies that the theory is struggling to predict the size dependence of the strongly confined electrons. By plotting relative to the energy of the first excited state, Figs. 10 and 11 remove this troubling portion [8]. Then, the theory can accurately predict the transitions relative to this energy.

Because the underlying cause cannot be the mixing between the conduction and valence band, we must look for other explanations. Although the exact origin is still unknown, it is easy to speculate about several leading candidates. First, a general problem exists in how to theoretically treat the nanocrystal interface. In the simple particle-in-a-sphere model [Eq. (2)], the potential barrier at the surface was treated as infinitely high. This is theoretically desirable because it implies that the carrier wave function goes to zero at the interface. Of course, in reality, the barrier is finite and some penetration of the electron and hole into the surrounding medium must occur. This effect should be more dramatic for the electron, which is more strongly confined. To partially account for this effect, the models used to treat CdSe and InAs nanocrystals incorporated a finite "square well" potential barrier, V_e, for the electron. (The hole barrier was still assumed to be infinite.) However, in practice, V_e became simply a fitting parameter to better correct for deviations in Fig. 20. In addition, the use of a square potential barrier is not a rigorous treatment of the interface. In fact, how one should analytically approach such an interface is still an open theoretical problem. The resolution of this issue for the nanocrystal may require more sophisticated *general boundary condition* theories that have recently been developed [95].

A second candidate to explain Fig. 20 is the simplistic treatment of the Coulomb interaction, which is included only as a first-order perturbation. This approach not only misses additional couplings between levels but also, as recently pointed out by Efros et al. [41], ignores the expected size dependence in the dielectric constant. The effective dielectric constant of the nanocrystal should decrease with decreasing size. This implies that the perturbative approach underestimates the Coulomb interaction. Unfortunately, this effect has not yet been treated theoretically.

Finally, one could also worry, in general, about the breakdown of the effective mass and the envelope function approximations in extremely small nanocrystals. As discussed in Section II.B, we require the nanocrystals to be much larger than the lattice constant of the semiconductor. In extremely small nanocrystals, where the diameter may only be a few lattice constants, this is no longer the case. Therefore, it becomes an issue how small one can push the effective mass model before it breaks.

C. The Future

Clearly from the discussion in Section IV. B, important problems remain to be solved before a complete theoretical understanding about the electronic structure in nanocrystals is obtained. However, hopefully, this chapter has also demonstrated that we are clearly on the correct path. Further, theoretical issues are not the only area that needs attention. More experimental data are also necessary. Although much work has been done, it is surprising that after a decade of work on high-quality nanocrystal samples, detailed spectroscopic studies have only been performed on two compound semiconductors, CdSe and InAs. Hopefully, in the coming years, this list will be expanded. Our understanding will be truly tested only by applying it to new materials.

ACKNOWLEDGMENTS

The author gratefully acknowledges M. G. Bawendi, Al. L. Efros, C. B. Murray, and M. Nirmal, who have greatly contributed to the results and descriptions described in this chapter.

REFERENCES

1. Pankove, J.I. *Optical Processes in Semiconductors*; New York: Dover, 1971; 34 pp.
2. Efros, Al.L.; Efros, A.L. *Sov. Phys. Semicond.* 1982, *16*, 772.
3. Ekimov, A.I.; Onushchenko, A.A. *JETP Lett.* 1982, *34*, 345.
4. Brus, L.E. *J. Chem. Phys.* 1983, *79*, 5566.
5. Ekimov, A.I.; Efros, Al.L.; Onushchenko, A.A. *Solid State Commun.* 1985, *56*, 921.
6. Brus, L.E. *J. Chem. Phys.* 1984, *80*, 4403.
7. Murray, C.B.; Norris, D.J.; Bawendi, M.G. *J. Am. Chem. Soc.* 1993, *115*, 8706.
8. Norris, D.J.; Sacra, A.; Murray, C.B.; Bawendi, M.G. *Phys. Rev. Lett.* 1994, *72*, 2612.
9. Norris, D.J.; Bawendi, M.G. *Phys. Rev. B* 1996, *53*, 16,338.
10. Nirmal, M.; Norris, D.J.; Kuno, M.; Bawendi, M.G.; Efros, Al.L.; Rosen, M. *Phys. Rev. Lett.* 1995, *75*, 3728.

11. Norris, D.J.; Efros, Al.L.; Rosen, M.; Bawendi, M.G. *Phys. Rev. B* 1996, *53*, 16,347.
12. Guzelian, A.A.; Banin, U.; Kadavanich, A.V.; Peng, X.; Alivisatos, A.P. *Appl. Phys. Lett.* 1996, *69*, 1432.
13. Banin, U.; Lee, C.J.; Guzelian, A.A.; Kadavanich, A.V.; Alivisatos, A.P.; Jaskolski, W.; Bryant, G.W.; Efros, Al.L.; Rosen, M. *J. Chem. Phys.* 1998, *109*, 2306.
14. Brus, L. *Appl. Phys. A* 1991, *53*, 465.
15. Bányai, L.; Koch, S.W. *Semiconductor Quantum Dots.* Singapore: World Scientific, 1993.
16. Alivisatos, A.P. *J. Phys. Chem.* 1996, *100*, 13,226.
17. Alivisatos, A.P. *Science* 1996, *271*, 933.
18. Woggon, U. *Optical Properties of Semiconductor Quantum Dots.* Heidelberg: Springer-Verlag, 1997.
19. Nirmal, M.; Brus, L.E. *Acc. Chem. Res.* 1999, *32*, 407.
20. Gaponenko, S.V. *Optical Properties of Semiconductor Nanocrystals.* Cambridge: Cambridge University Press, 1999.
21. Eychmüller, A. *J. Phys. Chem. B* 2000, *104*, 6514.
22. Ashcroft, N.W.; Mermin, N.D. *Solid State Physics.* Orlando, FL: W.B. Saunders.
23. Flügge, S. *Practical Quantum Mechanics.* Berlin: Springer-Verlag, 1971; Vol 1, 155.
24. Bastard, G. *Wave Mechanics Applied to Semiconductor Heterostructures.* New York: Wiley, 1988.
25. Altarelli, M. In *Semiconductor Superlattices and Interfaces*; Stella A, ed. Amsterdam: North-Holland, 1993; 217.
26. Xia, J.B. *Phys. Rev. B* 1989, *40*, 8500.
27. Hellwege, K.H. *Landolt-Bornstein Numerical Data and Functional Relationships in Science and Technology, New Series.* Berlin: Springer-Verlag, 1982; Vol. 17b, Group III.
28. Aven, M.; Prener, J.S. *Physics and Chemistry of II–VI Compounds.* Amsterdam: North-Holland, 1967; 41 pp.
29. Kittel, C. *Quantum Theory of Solids.* New York: Wiley, 1987.
30. Luttinger, J.M. *Phys. Rev. B* 1956, *102*,1030.
31. Luttinger, J.M.; Kohn, W. *Phys. Rev.* 1955, *97*, 869.
32. Bir, G.L.; Pikus, G.E. *Symmetry and Strain-Induced Effects in Semiconductors.* New York: Wiley, 1974.
33. Lipari, N.O.; Baldereschi, A. *Phys. Rev. Lett.* 1973, *42*, 1660.
34. Baldereschi, A.; Lipari, N.O. *Phys. Rev. B* 1973, *8*, 2697.
35. Ge'lmont, B.L.; D'yakonov, M.I. *Sov. Phys. Semicond.* 1972, *5*, 1905.
36. Kane, E.O. *J. Phys. Chem. Solids* 1957, *1*, 249.
37. Pidgeon, C.R.; Brown, R.N. *Phys. Rev.* 1966, *146*, 575.
38. Kane, E.O. In *Narrow Band Semiconductors, Physics and Applications, Lecture Notes in Physics*; Zawadski W, ed. Berlin: Springer-Verlag, 1980; Vol. 133.
39. Vahala, K.J.; Sercel, P.C. *Phys. Rev. Lett.* 1990, *65*, 239.
40. Sercel, P.C.; Vahala, K.J. *Phys. Rev. B* 1990, *42*, 3690.

41. Efros, Al.L.; Rosen, M. *Phys. Rev. B* 1998, *58*, 7120.
42. Ekimov, A.I.; Onushchenko, A.A. *JETP Lett.* 1984, *40*.
43. Rossetti, R.; Hull, R.; Gibson, J.M.; Brus, L.E. *J. Chem. Phys.* 1985, *82*, 552.
44. Ekimov, A.I.; Onushchenko, A.A.; Efros, Al.L. *JETP Lett.* 1986, *43*, 376.
45. Chestnoy, N.; Hull, R.; Brus, L.E. *J. Chem. Phys.* 1986, *85*, 2237.
46. Borrelli, N.F.; Hall, D.W.; Holland, H.J.; Smith, D.W. *J. Appl. Phys.* 1987, *61*, 5399.
47. Alivisatos, A.P.; Harris, A.L.; Levinos, N.J.; Steigerwald, M.L.; Brus, L.E. *J. Chem. Phys.* 1988, *89*, 4001.
48. Roussignol, P.; Ricard, D.; Flytzanis, C.; Neuroth, N. *Phys. Rev. Lett.* 1989, *62*, 312.
49. Ekimov, A.I.; Efros, Al.L.; Ivanov, M.G.; Onushchenko, A.A.; Shumilov, S.K. *Solid State Commun.* 1989, *69*, 565.
50. Wang, Y.; Herron, N. *Phys. Rev. B* 1990, *42*, 7253.
51. Müller, M.P.A.; Lembke, U.; Woggon, U.; Rückmann, I. *J. Noncrystal. Solids* 1992, *144*, 240.
52. Peyghambarian, N.; Fluegel, B.; Hulin, D.; Migus, A.; Joffre, M.; Antonetti, A.; Koch, S.W.; Lindberg, M. *IEEE J. Quantum Electronic* 1989, *25*, 2516.
53. Esch, V.; Fluegel, B.; Khitrova, G.; Gibbs, H.M.; Jiajin, X.; Kang, K.; Koch, S.W.; Liu, L.C.; Risbud, S.H.; Peyghambarian, N. *Phys. Rev. B* 1990, *42*, 7450.
54. Bawendi, M.G.; Wilson, W.L.; Rothberg, L.; Carroll, P.J.; Jedju, T.M.; Steigerwald, M.L.; Brus, L.E. *Phys. Rev. Lett.* 1990, *65*, 1623.
55. Ekimov, A.I.; Hache, F.; Schanne-Klein, M.C.; Ricard, D.; Flytzanis, C.; Kudryavtsev, I.A.; Yazeva, T.V.; Rodina, A.V.; Efros, Al.L. *J. Opt. Soc. Am. B* 1993, *10*, 100.
56. Bowen Katari, J.E.; Colvin, V.L.; Alivisatos, A.P. *J. Phys. Chem.* 1994, *98*, 4109.
57. Micic, O.I.; Sprague, J.R.; Curtis, C.J.; Jones, K.M.; Machol, J.L.; Nozik, A.J.; Giessen, H.; Fluegel, B.; Mohs, G.; Peyhambarian, N. *J. Phys. Chem.* 1995, *99*, 7754.
58. Hines, M.A.; Guyot-Sionnest, P. *J. Phys. Chem. B* 1998, *102*, 3655.
59. Norris, D.J.; Yao, N.; Charnock, F.T.; Kennedy, T.A. *Nano Lett.* 2001, *1*, 3.
60. Peng, Z.A.; Peng, X. *J. Am. Chem. Soc.* 2001, *123*, 168.
61. Hines, M.A.; Guyot-Sionnest, P. *J. Phys. Chem.* 1996, *100*, 468.
62. Peng, X.; Schlamp, M.C.; Kadavanich, A.V.; Alivisatos, A.P. *J. Am. Chem. Soc.* 1997, *119*, 7019.
63. Dabbousi, B.O.; Rodriguez-Viejo, J.; Mikulec, F.V.; Heine, J.R.; Mattoussi, H.; Ober, R.; Jensen, K.F.; Bawendi, M.G. *J. Phys. Chem. B* 1997, *101*, 9463.
64. Moerner, W.E.; Orrit, M. *Science* 1999, *283*, 1670.
65. Empedocles, S.A.; Bawendi, M.G. *Acc. Chem. Res.* 1999, *32*, 389.
66. Hilinksi, E.F.; Lucas, P.A.; Wang, Y. *J. Chem. Phys.* 1988, *89*, 3435.
67. Park, S.H.; Morgan, R.A.; Hu, Y.Z.; Lindberg, M.; Koch, S.W.; Peyghambarian, N. *J. Opt. Soc. Am. B* 1990, *7*, 2097.
68. Norris, D.J.; Nirmal, M.; Murray, C.B.; Sacra, A.; Bawendi, M.G. *Z. Phys. D* 1993, *26*, 355.

69. Gaponenko, S.V.; Woggon, U.; Saleh, M.; Langbein, W.; Uhrig, A.; Müller, M.; Klingshirn, C. *J. Opt. Soc. Am. B* 1993, *10*, 1947.
70. Woggon, U.; Gaponenko, S.; Langbein, W.; Uhrig, A.; Klingshirn, C. *Phys. Rev. B* 1993, *47*, 3684.
71. Kang, K.I.; Kepner, A.D.; Gaponenko, S.V.; Koch, S.W.; Hu, Y.Z.; Peyghambarian, N. *Phys. Rev. B* 1993, *48*, 15,449.
72. Kang, K.; Kepner, A.D.; Hu, Y.Z.; Koch, S.W.; Peyghambarian, N.; Li, C.-Y.; Takada, T.; Kao, Y.; Mackenzie, J.D. *Appl. Phys. Lett.*, 1994, 64,1478.
73. Norris, D.J.; Bawendi, M.G. *J. Chem. Phys.* 1995, *103*, 5260.
74. Hoheisel, W.; Colvin, V.L.; Johnson, C.S.; Alivisatos, A.P. *J. Chem. Phys.* 1994, *101*, 8455.
75. de Oliveira, C.R.M.; Paula, A.M.d.; Filho, F.O.P.; Neto, J.A.M.; Barbosa, L.C.; Alves, O.L.; Menezes, E.A.; Rios, J.M.M.; Fragnito, H.L.; Cruz, C.H.B.; Cesar, C.L. *Appl. Phys. Lett.* 1995, *66*, 439.
76. Rodriguez, P.A.M.; Tamulaitis, G.; Yu, P.Y.; Risbud, S.H. *Solid State Commun.* 1995, *94*, 583.
77. Grigoryan, G.B.; Kazaryan, E.M.; Efros, A.L.; Yazeva, T.V. *Sov. Phys. Solid State* 1990, *32*, 1031.
78. Koch, S.W.; Hu, Y.Z.; Fluegel, B.; Peyghambarian, N. *J. Crystal Growth* 1992, *117*, 592.
79. Efros, Al.L. *Phys. Rev. B* 1992, *46*, 7448.
80. Efros, Al.L.; Rodina, A.V. *Phys. Rev. B* 1993, *47*, 10005.
81. Calcott, P.D.J.; Nash, K.J.; Canham, L.T.; Kane, M.J.; Brumhead, D. *J. Lumin* 1993, *57*, 257.
82. Takagahara, T. *Phys. Rev. B* 1993, *47*, 4569.
83. Nomura, S.; Segawa, Y.; Kobayashi, T. *Phys. Rev. B* 1994, *49*, 13571.
84. Chamarro, M.; Gourdon, C.; Lavallard, P.; Ekimov, A.I. *Jpn. J. Appl. Phys.* 1995, *34* (1), 12.
85. Kochereshko, V.P.; Mikhailov, G.V.; Ural'tsev, I.N. *Sov. Phys. Solid State* 1983, *25*, 439.
86. Henry, C.H.; Nassau, K. *Phys. Rev. B* 1970, *1*, 1628.
87. O'Neil, M.; Marohn, J.; McLendon, G. *J. Phys. Chem.* 1990, *94*, 4356.
88. Eychmüller, A.; Hasselbarth, A.; Katsikas, L.; Weller, H. *Ber. Bunsenges. Phys. Chem.* 1991, *95*, 79.
89. Bawendi, M.G.; Carroll, P.J.; Wilson, W.L.; Brus, L.E. *J. Chem. Phys.* 1992, *96*, 946.
90. Nirmal, M.; Murray, C.B.; Bawendi, M.G. *Phys. Rev. B* 1994, *50*, 2293.
91. Littau, K.A.; Szajowski, P.J.; Muller, A.J.; Kortan, A.R.; Brus, L.E. *J. Phys. Chem.* 1993, *97*, 1224.
92. Wilson, W.L.; Szajowski, P.F.; Brus, L.E. *Science* 1983, *262*, 1242.
93. Banin, U.; Lee, J.C.; Guzelian, A.A.; Kadavanich, A.V.; Alivisatos, A.P. *Superlattices Microstruct.* 1997, *22*, 559.
94. Cao, Y.-W.; Banin, U. *J. Am. Chem. Soc.* 2000, *122*, 9692.
95. Rodina, A.V.; Alekseev, A.Y.; Efros, Al.L.; Rosen, M.; Meyer, B.K. *Phys. Rev. B* 2002, *65*, 125,302.

3

Fine Structure and Polarization Properties of Band-Edge Excitons in Semiconductor Nanocrystals

Al. L. Efros
Naval Research Laboratory, Washington, D.C., U.S.A.

I. INTRODUCTION

It has been more than 10 years since the seminal paper of Bawendi et al. [1] on the resonantly excited photoluminescence (PL) in CdSe nanocrystals (NCs). This first investigation showed that in the case of resonant excitation, the PL from CdSe NCs shows a fine structure that is Stokes shifted relative to the lowest absorption band. The fine-structure PL consists of a zero-phonon line (ZPL) and longitudinal optical (LO) phonon satellites. The lifetime of PL at low temperatures is extremely long, on the order of 1000 ns. All of the major properties of the resonant PL in CdSe NCs have been described using the dark/bright-exciton model [2,3].

In this chapter, we review the dark/bright-exciton model and summarize the results of realistic multiband calculations of the band-edge exciton fine structure in quantum dots of semiconductors having a degenerate valence band. These calculations take into account the effect of the electron–hole-exchange interaction, nonsphericity of the crystal shape, and the intrinsic hexagonal lattice asymmetry. We describe also the effect of an external magnetic field on the fine structure, the transition oscillator strengths, and polarization properties of CdSe NCs. The results of these calculations are used to described unusual polarization properties of CdSe NCs, a size-dependent Stokes shift of the resonant PL, a fine structure in absorption, and the formation of a long-lived dark exciton. Particularly strong confirmation of our model is found in the magnetic field dependence of the dark-exciton decay time [2], magnetic circular dichroism of CdSe NCs, and the polarization of the CdSe NC PL in a strong magnetic field.

The chapter is organized as follows. In Section II, we calculate the energy structure of the band-edge exciton and obtain selection rules and transition oscillator strengths. In Section III, the effect of an external magnetic field on the fine-level structure and transition oscillator strengths is discussed. The polarization properties of the NC PL, Stokes shift of the resonant PL, field-induced shortening of the dark-exciton lifetime, magnetic circular dichroism, and PL polarization in strong magnetic fields are considered within the developed theoretical model in Section IV. The results are summarized and discussed in Section V.

II. FINE STRUCTURE OF THE BAND-EDGE EXCITON IN CdSe NANOCRYSTALS

A. Band-Edge Quantum-Size Levels

In semiconductor crystals that are smaller than the bulk exciton Bohr radius, the energy spectrum and the wave functions of electron–hole pairs can be approximated using the independent quantization of the electron and hole motions (the so-called strong confinement regime [4]). The electron and hole quantum confinement energies and their wave functions are found in the framework of the multiband effective mass approximation [5]. The formal procedure for deriving this method demands that the external potential be sufficiently smooth. In the case of nanosize semiconductor crystals, this requirement leads to the condition $2a \gg a_0$, where a is the crystals radius and a_0 is the lattice constant. In addition, the effective mass approximation holds only if the typical energies of electrons and holes are close to the bottom of the conduction band and to the top of the valence band, respectively. In practice, this means that the quantization energy must be much smaller than the energy distance to the next higher (lower)-energy extremum in the conduction (valence) band.

In the framework of the effective mass approximation, for spherically symmetric NCs having a cubic lattice structure, the first electron quantum-size level, $1S_e$, is doubly degenerate with respect to the spin projection. The first hole quantum-size level, $1S_{3/2}$, is fourfold degenerate with respect to the projection (M) of the total angular momentum, $\mathcal{K}$, ($M = 3/2, 1/2, -1/2$, and $-3/2$) [6,7]. The energies and wave functions of these quantum-size levels can be easily found in the parabolic approximation. The electrons energy levels and wave functions respectively are:

$$E_{1S} = \frac{\hbar^2 \pi^2}{2m_e a^2}$$

$$\psi_\alpha(\mathbf{r}) = \xi(\mathbf{r}) \mid S\alpha\rangle = \sqrt{\frac{2}{a}} \frac{\sin(\pi r/a)}{r} Y_{00}(\Omega) \mid S\alpha\rangle \tag{1}$$

where m_e is the electron effective mass, a is the NC radius, $Y_{lm}(\Omega)$ are spherical harmonic functions, $\mid S\alpha\rangle$ are the Bloch functions of the conduction band, and $\alpha = \uparrow (\downarrow)$ is the projection of the electron spin, $s_z = + (-) 1/2$. The energies and wave functions of holes in the fourfold degenerate valence band can be written respectively as

$$E_{3/2}(\beta) = \frac{\hbar^2 \varphi^2(\beta)}{2m_{hh} a^2} \tag{2}$$

$$\psi_M(\mathbf{r}) = 2 \sum_{l=0,2} R_l(r) (-1)^{M-3/2} \sum_{m+\mu=M} \begin{pmatrix} 3/2 & l & 3/2 \\ -\mu & m & -M \end{pmatrix} Y_{lm}(\Omega) u_\mu \tag{3}$$

where $\beta = m_{lh}/m_{hh}$ is the ratio of the light- to heavy-hole effective masses, $\varphi(\beta)$ is the first root of the equation [8–12]

$$j_0(\varphi) j_2\left(\sqrt{\beta}\varphi\right) + j_2(\varphi) j_0\left(\sqrt{\beta}\varphi\right) = 0 \tag{4}$$

where $j_n(x)$ are spherical Bessel functions, $\begin{pmatrix} i & k & l \\ m & n & p \end{pmatrix}$ are Wigner 3j-symbols, and u_μ ($\mu = \pm 1/2, \pm 3/2$) are the Bloch functions of the fourfold degenerate valence band Γ_8 [13]:

$$u_{3/2} = \frac{1}{\sqrt{2}}(X + iY) \uparrow, \qquad u_{-3/2} = \frac{i}{\sqrt{2}}(X - iY) \downarrow$$

$$u_{1/2} = \frac{i}{\sqrt{6}}[(X + iY) \downarrow - 2Z \uparrow] \tag{5}$$

$$u_{-1/2} = \frac{1}{\sqrt{6}}[(X - iY) \uparrow + 2Z \downarrow]$$

The radial functions $R_l(r)$ are [8,10–12]

$$R_2(r) = \frac{A}{a^{3/2}} \left(j_2(\varphi r/a) + \frac{j_0(\varphi)}{j_0(\varphi\sqrt{\beta})} j_2\left(\varphi\sqrt{\beta}r/a\right) \right)$$

$$R_0(r) = \frac{A}{a^{3/2}} \left(j_0(\varphi r/a) - \frac{j_0(\varphi)}{j_0(\varphi\sqrt{\beta})} j_0\left(\varphi\sqrt{\beta}r/a\right) \right) \tag{6}$$

where the constant A is determined by the normalization condition

$$\int dr r^2 [R_0^2(r) + R_2^2(r)] = 1 \tag{7}$$

The dependence of φ on β [12] is presented in Fig. 1a.

For spherical dots, the exciton ground state $(1S_{3/2}1S_e)$ is eightfold degenerate. However, the shape and the internal crystal structure anisotropy together with the electron–hole-exchange interaction lift this degeneracy. The

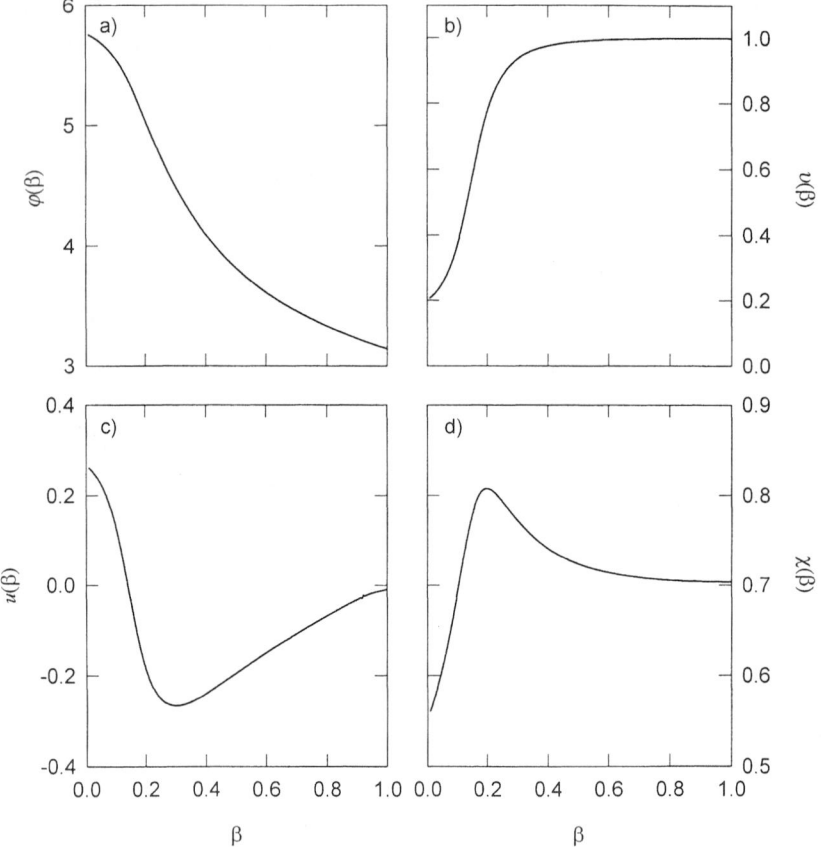

Figure 1 (a) The dependence of the hole ground-state function $\varphi(\beta)$ on the light- to heavy-hole effective mass ratio, β; (b) the dimensionless function $v(\beta)$ associated with hole-level splitting due to hexagonal lattice structure; (c) the dimensionless function $u(\beta)$ associated with hole-level splitting due to crystal shape asymmetry; (d) the dimensionless function $\chi(\beta)$ associated with exciton splitting due to the electron–hole-exchange interaction.

energy splitting and the transition oscillator strengths of the split-off states, as well as their order, are very sensitive to the NC size and shape, as shown in subsections B and C. We calculate this splitting neglecting the warping of the valence band and the nonparabolicity of the electron and light-hole energy spectra.

B. Energy Spectrum and Wave Functions

Nanocrystal asymmetry lifts the hole state degeneracy. The asymmetry has two sources: the intrinsic asymmetry of the hexagonal lattice structure of the crystal [12] and the non-spherical shape of the finite crystal [14]. Both split the fourfold degenerate hole state into two twofold degenerate states—a Kramer's doublet—having $|M| = 1/2$ and $3/2$, respectively.

The splitting due to the intrinsic hexagonal lattice structure, Δ_{int}, can be written [12]

$$\Delta_{int} = \Delta_{cr}v(\beta), \tag{8}$$

where Δ_{cr} is the crystal field splitting equal to the distance between the A and B valence subbands in bulk semiconductors having a hexagonal lattice structure (25 meV in CdSe). Equation (8) is obtained within the framework of the quasicubic model for the case when the crystal field splitting can be considered as a perturbation [12]. The Kramer's doublet splitting does not depend on the NC size but only on the ratio of the light- to heavy-hole effective masses. The dimensionless function $v(\beta)$ [12] that describes this dependence (shown in Fig. 1b) varies rapidly in the region $0 < \beta < 0.3$. The $|M| = 3/2$ state is the ground state.

We model the nonsphericity of a NC by assuming that it has an ellipsoidal shape. The deviation from the sphericity is quantitatively characterized by the ratio $c/b = 1 + \mu$ of the ellipsoid's major (c) to minor (b) axes, where μ is the NC ellipticity, which is positive for prolate particles and negative for oblate particles. The splitting arising from nonsphericity can be calculated in the first-order perturbation theory [14], which yields

$$\Delta_{sh} = 2\mu u(\beta)E_{3/2}(\beta) \tag{9}$$

where $E_{3/2}$ is the $1S_{3/2}$ ground-state hole energy for spherical NCs of radius $a = (b^2c)^{1/3}$. $E_{3/2}$ is inversely proportional to a^2 [see Eq. (2)] and the shape splitting is therefore a sensitive function of the NC size. The function $u(\beta)$ [14] is equal to $4/15$ at $\beta = 0$. It changes sign at $\beta = 0.14$, passes a minimum at $\beta \approx 0.3$, and, finally, becomes 0 at $\beta = 1$ (see Fig. 1c).

The net splitting of the hole state, $\Delta(a, \beta, \mu)$, is the sum of the crystal field and shape splitting:

$$\Delta(a, \beta, \mu) = \Delta_{sh} + \Delta_{int} \tag{10}$$

In crystals for which the function $u(\beta)$ is negative (this is, e.g., the case for CdSe for which $\beta = 0.28$ [15]), the net splitting decreases with size in prolate ($\mu > 0$) NCs. Even the order of the hole levels can change, with the $|M| = 1/2$ state becoming the hole ground level for sufficiently small crystals [16]. This can be qualitatively understood within a model of uncoupled A and B valence subbands. In prolate crystals, the energy of the lowest hole quantum-size level is determined by its motion in the plane perpendicular to the hexagonal axis. In this plane, the hole effective mass in the lowest subband A is smaller than that in the higher B subband [12]. Decreasing the size of the crystal causes a shift of the quantum-size level inversely proportional to both the effective mass and the square of the NC radius. The shift is therefore larger for the A subband than for the B subband and, as a result, it can change the order of the levels in small NCs. In oblate ($\mu < 0$) crystals, where the levels are determined by motion along the hexagonal axis, the B subband has the smaller mass. Hence, the net splitting increases with decreasing size and the states maintain their original order.

The eight-fold degeneracy of the spherical band-edge exciton is also broken by the electron–hole-exchange interaction which mixes different electron and hole spin states. This interaction can be described by the following expression [13,17]:

$$\hat{H}_{\text{exch}} = -\frac{2}{3}\varepsilon_{\text{exch}}(a_0)^3\delta(\mathbf{r}_\epsilon - \mathbf{r}_h)\boldsymbol{\sigma}\boldsymbol{J} \tag{11}$$

where $\boldsymbol{\sigma}$ is the electron Pauli spin-$1/2$ matrix, $\boldsymbol{J}$ is the hole spin-$3/2$ matrix, a_0 is the lattice constant, and $\varepsilon_{\text{exch}}$ is the exchange strength constant. In bulk crystals with cubic lattice structure, this term splits the eightfold degenerate ground exciton state into a fivefold degenerate optically passive state with total angular momentum 2 and a threefold degenerate optically active state with total angular momentum 1. This splitting can be expressed in terms of the bulk exciton Bohr radius, a_{ex}:

$$\hbar\omega_{\text{ST}} = \left(\frac{8}{3}\pi\right)\left(\frac{a_0}{a_{\text{ex}}}\right)^3\varepsilon_{\text{exch}} \tag{12}$$

In bulk crystals with hexagonal lattice structure, this term splits the exciton fourfold degenerate ground state into a triplet and a single state, separated by

$$\hbar\omega_{\text{ST}} = \left(\frac{2}{\pi}\right)\left(\frac{a_0}{a_{\text{ex}}}\right)^3\varepsilon_{\text{exch}} \tag{13}$$

Equations (12) and (13) allow one to evalute the exchange strength constant. In CdSe crystals, where $\hbar\omega_{\text{ST}} = 0.13$ meV [18], a value of $\varepsilon_{\text{exch}} = 450$ meV is obtained using $a_{\text{ex}} = 56$ Å.

Taken together, the hexagonal lattice structure, crystal shape asymmetry, and the electron–hole-exchange interaction split the original "spherical" eightfold degenerate exciton into five levels. The levels are labeled by the magnitude of the exciton total angular momentum projection, $F = M + s_z$: one level with $F = \pm2$, two with $F = \pm1$, and two with $F = 0$. The level energies, $\varepsilon_{|F|}$, are determined by solving the secular equation $\det(\hat{E} - \varepsilon_{|F|}) = 0$, where the matrix $\hat{E}$ consists of matrix elements of the asymmetry perturbations and the exchange interaction, $\hat{H}_{\text{exch}}$, taken between the exciton wave functions $\Psi_{\alpha,M}(\mathbf{r}_e, \mathbf{r}_h) = \psi_\alpha(\mathbf{r}_e)\psi_M(\mathbf{r}_h)$:

	$\uparrow,3/2$	$\uparrow,1/2$	$\uparrow,-1/2$	$\uparrow,-3/2$	$\downarrow,3/2$	$\downarrow,1/2$	$\downarrow,-1/2$	$\downarrow,-3/2$
$\uparrow,3/2$	$\dfrac{-3\eta}{2} - \dfrac{\Delta}{2}$	0	0	0	0	0	0	0
$\uparrow,1/2$	0	$\dfrac{-\eta}{2} - \dfrac{\Delta}{2}$	0	0	$-i\sqrt{3}\eta$	0	0	0
$\uparrow,-1/2$	0	0	$\dfrac{\eta}{2} + \dfrac{\Delta}{2}$	0	0	$-i2\eta$	0	0
$\uparrow,-3/2$	0	0	0	$\dfrac{3\eta}{2} - \dfrac{\Delta}{2}$	0	0	$-i\sqrt{3}\eta$	0
$\downarrow,3/2$	0	$i\sqrt{3}\eta$	0	0	$\dfrac{3\eta}{2} - \dfrac{\Delta}{2}$	0	0	0
$\downarrow,1/2$	0	0	$i2\eta$	0	0	$\dfrac{\eta}{2} + \dfrac{\Delta}{2}$	0	0
$\downarrow,-1/2$	0	0	0	$i\sqrt{3}\eta$	0	0	$\dfrac{-\eta}{2} + \dfrac{\Delta}{2}$	0
$\downarrow,-3/2$	0	0	0	0	0	0	0	$\dfrac{-3\eta}{2} - \dfrac{\Delta}{2}$

$$(14)$$

where $\eta = (a_{\text{ex}}/a)^3 \hbar\omega_{\text{ST}}\chi(\beta)$, and the dimensionless function $\chi(\beta)$ is written in terms of the electron and hole radial wave functions:

$$\chi(\beta) = \left(\frac{1}{6}\right)a^2 \int_0^a dr\, \sin^2(\pi r/a)[R_0^2(r) + 0.2R_2^2(r)] \qquad (15)$$

The dependence of χ on the parameter β is shown in Fig. 1d.

Solution of the secular equation yields five exciton levels. The energy of the exciton with total angular momentum projection $|F| = 2$ and its dependence on the crystal size is given by [3]

$$\varepsilon_2 = \frac{-3\eta}{2} - \frac{\Delta}{2} \qquad (16)$$

The respective wave functions are

$$\Psi_{-2}(\mathbf{r}_e, \mathbf{r}_h) = \Psi_{\downarrow, -3/2}(\mathbf{r}_e, \mathbf{r}_h) \tag{17}$$

$$\Psi_2(\mathbf{r}_e, \mathbf{r}_h) = \Psi_{\uparrow, 3/2}(\mathbf{r}_e, \mathbf{r}_h)$$

The energies and size dependences of the two levels, each with total momentum projection $|F| = 1$, are given by [3]

$$\varepsilon_1^{U,L} = \frac{\eta}{2} \pm \sqrt{\frac{(2\eta - \Delta)^2}{4} + 3\eta^2} \tag{18}$$

where U and L correspond to the upper ("+") or lower ("−") signs, respectively. We denote these states by $\pm 1^U$ and $\pm 1^L$, respectively (i.e., the upper and lower state with projection $F = \pm 1$). The corresponding wave functions for the states with $F = +1$ are*

$$\Psi_1^U(\mathbf{r}_e, \mathbf{r}_h) = -iC^+\Psi_{\uparrow, 1/2}(\mathbf{r}_e, \mathbf{r}_h) + C^-\Psi_{\downarrow, 3/2}(\mathbf{r}_e, \mathbf{r}_h)$$

$$\Psi_1^L(\mathbf{r}_e, \mathbf{r}_h) = +iC^-\Psi_{\uparrow, 1/2}(\mathbf{r}_e, \mathbf{r}_h) + C^+\Psi_{\downarrow, 3/2}(\mathbf{r}_e, \mathbf{r}_h) \tag{19}$$

whereas for the states with $F = -1$, the wave functions are

$$\Psi_{-1}^U(\mathbf{r}_e, \mathbf{r}_h) = -iC^-\Psi_{\uparrow, -3/2}(\mathbf{r}_e, \mathbf{r}_h) - C^+\Psi_{\downarrow, -1/2}(\mathbf{r}_e, \mathbf{r}_h)$$

$$\Psi_{-1}^L(\mathbf{r}_e, \mathbf{r}_h) = +iC^+\Psi_{\uparrow, -3/2}(\mathbf{r}_e, \mathbf{r}_h) + C^-\Psi_{\downarrow, -1/2}(\mathbf{r}_e, \mathbf{r}_h) \tag{20}$$

where

$$C^{\pm} = \left(\frac{\sqrt{f^2 - d} \pm f}{2\sqrt{f^2 - d}}\right)^{1/2} \tag{21}$$

$f = (-2\eta + \Delta)/2$, and $d = 3\eta^2$. The size-dependent energies of the two $F = 0$ exciton levels are given by

$$\varepsilon_0^{U,L} = \frac{\eta}{2} + \frac{\Delta}{2} \pm 2\eta \tag{22}$$

(we denote the two $F = 0$ states by 0^U and 0^L), with corresponding wave functions

$$\Psi_0^{U,L}(\mathbf{r}_e, \mathbf{r}_h) = \frac{1}{\sqrt{2}}[\mp i\Psi_{\uparrow, -1/2}(\mathbf{r}_e, \mathbf{r}_h) + \Psi_{\downarrow, 1/2}(\mathbf{r}_e, \mathbf{r}_h)] \tag{23}$$

---

* There are misprints in the function definitions of Eqs. (19) and (20) of Ref. 3.

In Eqs. (22) and (23), superscripts U and L correspond to the upper ("+") and the lower ("−") signs, respectively.

The size dependence of the band-edge exciton splitting calculated in Ref. 3 for hexagonal CdSe NCs of different shapes is shown in Fig. 2. The calculation were made using $\beta = 0.28$ [15]. In spherical nanocrystals (Fig. 2a), the $F = \pm 2$ state is the exciton ground state for all sizes and is optically

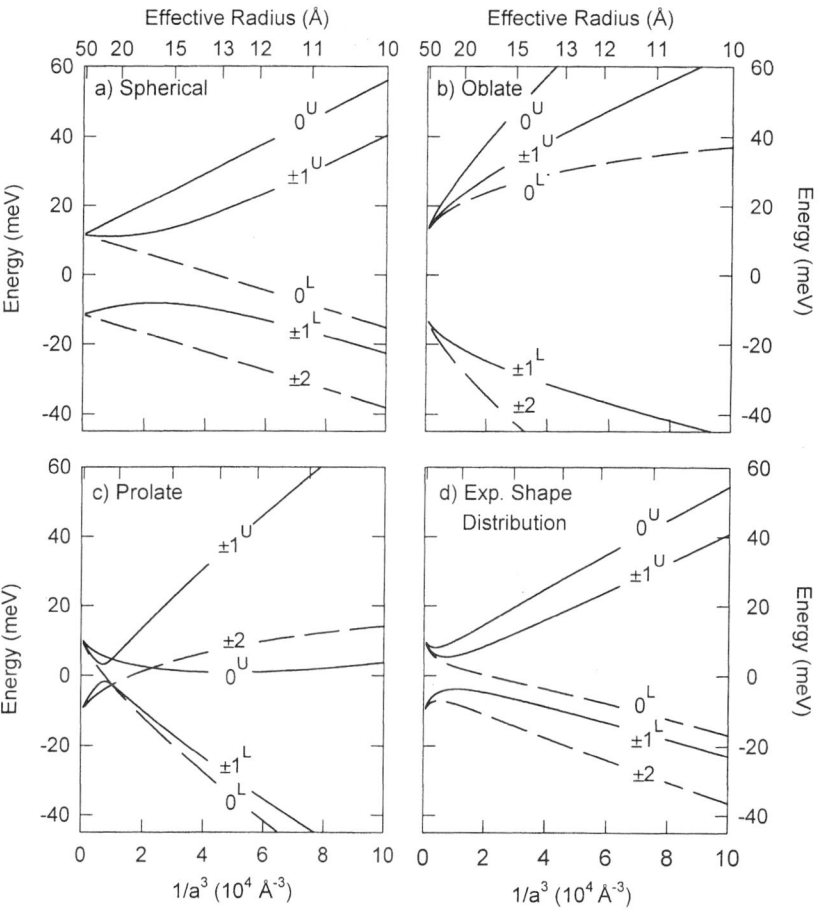

Figure 2 The size dependence of the exciton band-edge structure in ellipsoidal hexagonal CdSe quantum dots with ellipticity μ: (a) spherical dots ($\mu = 0$); (b) oblate dot ($\mu = -0.28$); (c) prolate dots ($\mu = 0.28$); (d) dots having a size-dependent ellipticity as determined from small-angle x-ray scattering and transmission electron microscopy measurements. Solid (dashed) lines indicate optically active (passive) levels.

passive, as was shown in Ref. 12. The separation between the ground state and the lower optically active $F = \pm 1$ state initially increases with decreasing size as $1/a^3$ but tends to $3\Delta/4$ for very small sizes. In oblate crystals (Fig. 2b), the order of the exciton levels is the same as in spherical ones. However, the splitting does not saturate, because in these crystals, Δ increases with decreasing NC size.

In prolate NCs, Δ becomes negative with decreasing size and this changes the order of the exciton levels at some value of the radius (Fig. 2c); in small NCs, the optically passive (as we show below) $F = 0$ state becomes the ground exciton state. The crossing occurs when Δ goes through 0. In NCs of this size, the shape asymmetry exactly compensates the asymmetry due to the hexagonal lattice structure [16]. The electronic structure of exciton levels have "spherical" symmetry although the NCs do not have spherical shape. As a result there is one fivefold degenerate exciton with total angular momentum 2 (which is reflected in the crossing of the 0^L, $\pm 1^L$, and ± 2 levels) and one threefold degenerate exciton state with total angular momentum 1 (reflected in the crossing of the 0^U and $\pm 1^U$ levels).

In Fig. 2d, the band-edge exciton fine structure is shown for the case for which the ellipticity varies with size.* This size-dependent ellipticity was experimentally observed in CdSe NCs using small-angle x-ray scattering (SAXS) and transmission electron microscopy (TEM) studies [19]. The level structure calculated for this case closely resembles that obtained for spherical crytals.

The size dependence of the band-edge exciton splitting in CdTe NCs with cubic lattice structure calculated for particles of different shapes is shown in Fig. 3. The calculation was done using the parameters $\beta = 0.086$ and $\hbar\omega_{ST} = 0.04$ meV. One can see that in the spherical NCs, the electron–hole-exchange interaction splits the eightfold degenerate band-edge exciton into a fivefold degenerate exciton with total angular momentum 2 and a threefold degenerate exciton with total angular momentum 1 (Fig. 3a). The NC shape asymmetry lifts the degeneracy of these states and completely determines the relative order of the exciton states (see Figs. 3b and 3c for comparison).

C. Selection Rules and Transition Oscillator Strengths

To describe the fine structure of the absorption and PL spectra, we calculate transition oscillator strengths for the lowest five exciton states. The mixing

---

* In accordance with SAXS and TEM measurements, the ellipticity was approximated by the polynomial $\mu(a) = 0.101 - 0.034a + 3.507 \times 10^{-3}a^2 - 1.177 \times 10^{-4}a^3 + 1.863 \times 10^{-6}a^4 - 1.418 \times 10^{-8}a^5 + 4.196 \times 10^{-11}a^6$.

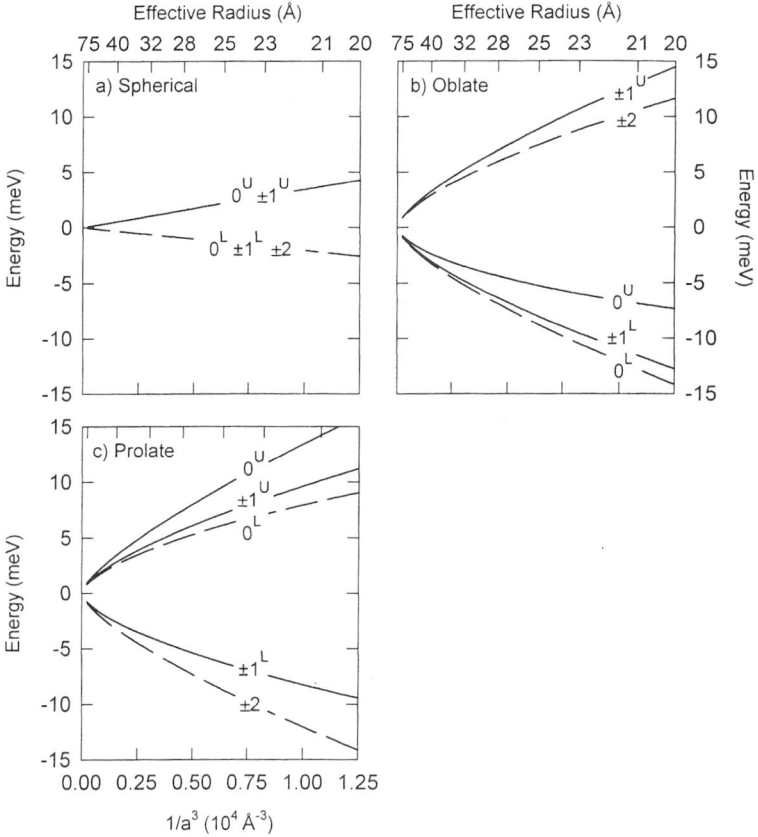

Figure 3 The size dependence of the exciton band-edge structure in ellipsoidal cubic CdTe quantum dots with ellipticity μ: (a) spherical dots ($\mu = 0$); (b) oblate dots ($\mu = -0.28$); (c) prolate dots ($\mu = 0.28$). Solid (dashed) lines indicate optically active (passive) levels.

between the electron and hole spin momentum states by the electron–hole-exchange interaction strongly affects the optical transition probabilities. The wave functions of the $|F| = 2$ exciton state, however, are unaffected by this interaction [see Eq. (17)]; it is optically passive in the dipole approximation because emitted or absorbed photons cannot have an angular momentum projection of ± 2. The probability of optical excitation or recombination of an exciton state with total angular momentum projection F is proportional to the square of the matrix element of the momentum operator $\mathbf{e}\hat{\mathbf{p}}$ between this state and the vacuum state

$$P_F = |\langle 0|\mathbf{e}\hat{\mathbf{p}}|\tilde{\Psi}_F\rangle|^2 \tag{24}$$

where $|0\rangle = \delta(\mathbf{r}_e - \mathbf{r}_h)$ and $\mathbf{e}$ is the polarization vector of the emitted or absorbed light. The momentum operator $\hat{\mathbf{p}}$ acts only on the valence-band Bloch functions [see Eq. (5)], and the exciton wave function, $\tilde{\Psi}_F$, is written in the electron–electron representation. Exciton wave functions in the electron–hole representation are transformed to the electron–electron representation by taking the complex conjugate of Eqs. (17), (19), and (23) and flipping the spin projections in the hole Bloch functions ($\uparrow$ and $\downarrow$ to $\downarrow$ and $\uparrow$, respectively).

To calculate the matrix element for a *linear polarized light*, we expand the scalar product $\mathbf{e}\hat{\mathbf{p}}$ as

$$\mathbf{e}\hat{\mathbf{p}} = e_z\hat{p}_z + \frac{1}{2}[e_-\hat{p}_+ + e_-\hat{p}_-] \tag{25}$$

where z is the direction of the hexagonal axis of the NC, $e_{\pm} = e_x \pm ie_y$, $\hat{p}_{\pm} = \hat{p}_x \pm i\hat{p}_y$, and $e_{x,y}$ and $p_{x,y}$ are the components of the polarization vector and the momentum operator, respectively, that are perpendicular to the NC hexagonal axis.

Using this expansion in Eq. (24), one can obtain for the exciton state with $F = 0$ [3]:

$$P_0^{U,L} = |\langle 0|\mathbf{e}\hat{\mathbf{p}}|\tilde{\Psi}_0^{U,L}\rangle|^2 = N_0^{U,L} \cos^2(\theta_{lp}) \tag{26}$$

where $N_0^L = 0$, $N_0^U = 4KP^2/3$, $P = \langle S|\hat{p}_z|Z\rangle$ is the Kane interband matrix element, θ_{lp} is the angle between the polarization vector of the emitted or absorbed light and the hexagonal axis of the crystal, and K is the square of the overlap integral [12]:

$$K = \frac{2}{a}\left|\int dr\, r \sin\left(\frac{\pi r}{a}\right) R_0(r)\right|^2 \tag{27}$$

The magnitude of K depends only on β and is independent of the NC size; hence, the excitation probability of the $F = 0$ state is also size independent. For the lower exciton state, 0^L, the transition probability is proportional to N_0^L and is identically zero. At the same time, the exchange interaction increases the transition probability for the upper 0^U exciton state (it is proportional to N_0^U) by a factor of 2. This result arises from the constructive and destructive interference of the wave functions of the two indistinguishable exciton states $|\uparrow, -1/2\rangle$ and $|\downarrow, 1/2\rangle$ [see Eq. (23)].

Using a similar procedure, one can obtain relative transition probabilities to/from the exciton state with $F = 1$:

$$P_1^{U,L} = N_1^{U,L} \sin^2(\theta_{lp}) \tag{28}$$

where

$$N_1^U = \left(\frac{2\sqrt{f^2 + d} - f + \sqrt{3d}}{6\sqrt{f^2 + d}} \right) KP^2,$$

$$N_1^L = \left(\frac{2\sqrt{f^2 + d} + f - \sqrt{3d}}{6\sqrt{f^2 + d}} \right) KP^2 \tag{29}$$

The excitation probability of the $F = -1$ state is equal to that of the $F = 1$ state. As a result, the total transition probability to the doubly-degenerate $|F| = 1$ exciton states is equal to $2P_1^{U,L}$.

Equations (26) and (28) show that the $F = 0$ and $|F| = 1$ state excitation probabilities for the linear polarized light differ in their dependence on the angle between the light polarization vector and the hexagonal axis of the crystal. If the crystal hexagonal axis is aligned perpendicular to the light direction, only the active $F = 0$ state can be excited. Alternatively, when the crystals are aligned along the light propagation direction, only the upper and lower $|F| = 1$ states will participate in the absorption. In the case of randomly oriented NCs, polarized excitation resonant with one of these exciton states selectively excites suitably oriented crystals, leading to polarized luminescence (polarization memory effect) [12]. This effect was experimentally observed in several studies [20,21]. Furthermore, large energy splitting between the $F = 0$ and $|F| = 1$ states can lead to different Stokes shifts in the polarized luminescence.

The selection rules and the relative transition probabilities for *circularly polarized light* are determined by the matrix element of the operator $e_\pm \hat{p}_\mp$, where the polarization vector, $e_\pm = e_x \pm ie_y$, and the momentum, $\hat{p}_\pm = \hat{p}_x \pm i\hat{p}_y$, $i\hat{p}_y$, lie in the plane perpendicular to the light propagation direction. In vector representation, this operator can be written as

$$e_\pm \hat{p}_\mp = \mathbf{e}\hat{\mathbf{p}} \pm ie'\hat{p} \tag{30}$$

where $\mathbf{e} \perp \mathbf{c}$, $\mathbf{c}$ is the unit vector parallel to the light propagation direction, and $\mathbf{e}' = (\mathbf{e} \times \mathbf{c})$; as a result of the $\mathbf{e}'$ definition, the scalar product $(\mathbf{ee}') = 0$. To calculate the matrix element in Eq. (24), we expand the operator of Eq. (30) in coordinates that are connected with the direction of the hexagonal axis of the NCs (z direction):

$$e_\pm \hat{p}_\mp = \boldsymbol{\varepsilon}^\pm \hat{\boldsymbol{p}} = \varepsilon_z^\pm \hat{p}_z + \frac{1}{2}[\varepsilon_+^\pm \hat{p}_- + \varepsilon_-^\pm \hat{p}_+] \tag{31}$$

where $\varepsilon^{\pm} = \mathbf{e} \pm i\mathbf{e}'$ and $\varepsilon_{\pm}^{\pm} = \varepsilon_x^{\pm} \pm i\varepsilon_y^{\pm}$. Substituting Eq. (31) into Eq. (24), we obtain the relative values of the optical transition probability to/from the exciton state having the total angular momentum projection F coursed by the absorption/emission of the $\sigma^{\pm}$ polarized light.

For the exciton state with $F = 0$, we obtain

$$P_0^{U,L}(\sigma^{\pm}) = |\langle 0|e_{\pm}\hat{p}_{\mp}|\tilde{\Psi}_0^{U,L}\rangle|^2 = |\varepsilon_{\mp}^{\pm}\langle 0|\hat{p}_z|\tilde{\Psi}_0^{U,L}\rangle|^2$$

$$= (e_z^2 + e_z'^2)N_0^{U,L} = N_0^{U,L}\sin^2(\theta) \qquad (32)$$

where θ is the angle between the crystal hexagonal axis and the light propagation direction. In deriving Eq. (32), we used the identity for three orthogonal vectors $(\mathbf{e}, \mathbf{e}', \text{and } \mathbf{c})$: $\cos^2(\theta_e) + \cos^2(\theta_e') + \cos^2(\theta) = 1$, where θ_e and θ_e' are the angles between the crystal hexagonal axis and the vectors $\mathbf{e}$ and $\mathbf{e}'$, respectively. One can see from Eq. (32) that the excitation probability of the upper $(+)$ $F = 0$ state does not depend on the NC size, and that for the lower state $(-)$, it is identically equal to zero. The lower $F = 0$ exciton state is always optically passive.

For the exciton states with $F = +1$, we obtain

$$P_{F=1}^{U,L}(\sigma^{\pm}) = |\langle 0|e_{\mp}\hat{p}_{\pm}|\tilde{\Psi}_1^{U,L}\rangle^2 = \frac{1}{4}|\varepsilon_-^{\mp}\langle 0|\hat{p}_+|\tilde{\Psi}_1^{U,L}\rangle|^2$$

$$= \frac{1}{4}|e_- \mp ie_-'|^2 N_1^{U,L} = N_1^{U,L}(1\pm\cos\theta)^2 \qquad (33)$$

Similar calculations yield the following expression for the excitation probability of the $F = -1$ state:

$$P_{F=-1}^{U,L}(\sigma^{\pm}) = N_1^{U,L}(1\mp\cos\theta)^2 \qquad (34)$$

Deriving Eqs. (33) and (34), we used the orthogonality condition $(\mathbf{ec}) = 0$. In a zero magnetic field, the exciton states $F = 1$ and $F = -1$ are degenerate and we cannot distinguish them in a system of randomly oriented crystals.

To find the probability of exciton excitation for a system of randomly oriented NCs, we average Eqs. (26) and (28) over all possible solid angles. The respective excitation probabilities are proportional to

$$\overline{P_0^L} = 0, \qquad \overline{P_0^U} = \frac{N_0^U}{3},$$

$$\overline{P_1^L} = \overline{P_{-1}^L} = \frac{2N_1^L}{3}, \qquad \overline{P_1^U} = \overline{P_{-1}^U} = \frac{2N_1^U}{3} \qquad (35)$$

There are three optically active states with relative oscillator strengths $\overline{P_0^U}$, $2\overline{P_1^U}$, and $2\overline{P_1^L}$. The size dependence of these strengths for different NC shapes is shown in Fig. 4 for hexagonal CdSe nanoparticles. It is seen that the NC shape strongly influences this dependence. For example, in prolate NCs (Fig. 4c), the $\pm 1^L$ state oscillator strength goes to zero if $\Delta = 0$; in this case, the crystal shape asymmetry exactly compensates the internal asymmetry due to the hexagonal lattice structure. For these NCs, the oscillator strength of all of the upper states (0^U, 1^U, and -1^U) are equal. Nevertheless, one can

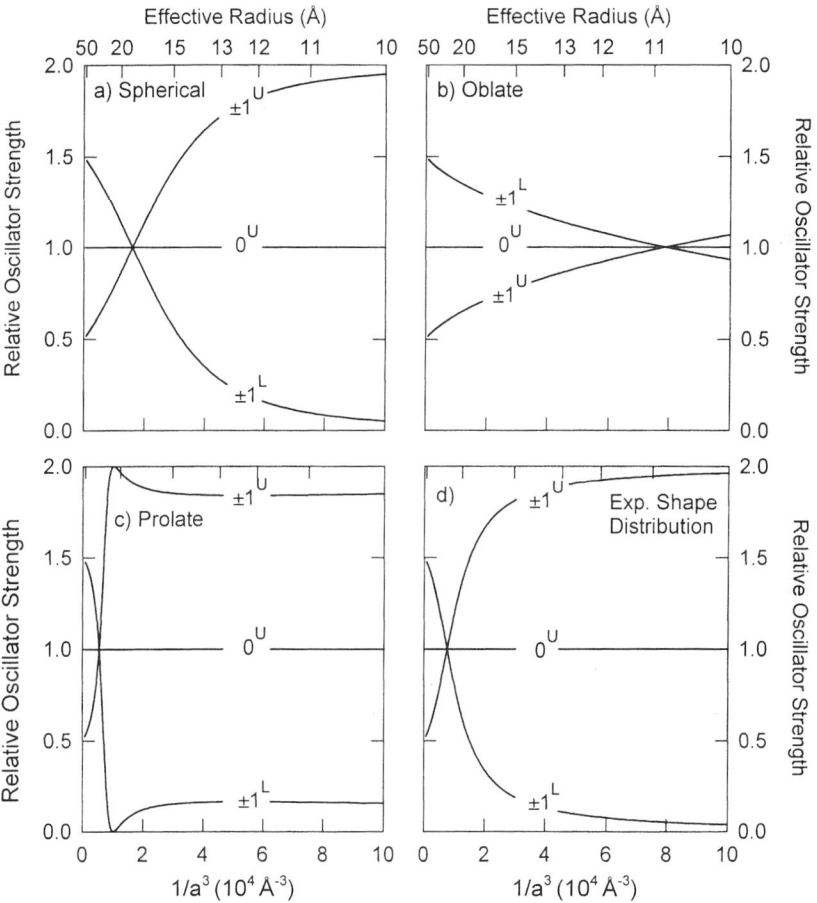

Figure 4   The size dependence of the oscillator strengths, relative to the that of the 0^U state, for the optically active states in hexagonal CdSe quantum dots with ellipticity μ: (a) spherical dots ($\mu = 0$); (b) oblate dots ($\mu = -0.28$); (c) prolate dots ($\mu = 0.28$); (d) dots having a size-dependent ellipticity as determined from SAXS and TEM measurements.

see that for all nanocrystal shapes, the excitation probability of the lower $|F| = 1$ ($\pm 1^L$) exciton state, $2\overline{P_1^L}$, decreases with size and that the upper $|F| = 1$ ($\pm 1^U$) gains its oscillator strength.

This behavior can be understood by examining the spherically symmetric limit. In spherical NCs, the exchange interaction leads to the formation of two exciton states—with total angular momenta 2 and 1. The ground state is the optically passive state with total angular momentum 2. This state is fivefold degenerate with respect to the total angular momentum projection. For small nanocrystals, the splitting of the exciton levels due to the nanocrystal asymmetry can be considered as a perturbation to the exchange interaction (the latter scales as $1/a^3$). In this situation, the wave functions of the $\pm 1^L$, 0^L, and ± 2 exciton states turn into the wave functions of the optically passive exciton with total angular momentum 2. The wave functions of the $\pm 1^U$ and 0^U exciton states become those of the optically active exciton states with total angular momentum 1. Therefore, these three states carry nearly all of the oscillator strength.

In large NCs, for all possible shapes, we can neglect the exchange interaction (which decreases as $1/a^3$), and thus there are only two fourfold degenerate exciton states (see Fig. 3). The splitting here is determined by the shape asymmetry and the intrinsic crystal field. In a system of randomly oriented crystals, the excitation probability of both of these states is the same: $\overline{P_0^U} + 2\overline{P_1^U} = 2\overline{P_1^L} = 2KP^2/3$ [12].

In Fig. 5, we show these dependences for variously shaped CdTe NCs with a cubic lattice structure. It is necessary to note here that despite the fact that the exchange interaction drastically changes the structure and the oscillator strengths of the band-edge exciton, the linear polarization properties of the nanocrystal (e.g., the linear polarization memory effect) are determined by the internal and crystal shape asymmetries. All linear polarization effects are proportional to the net splitting parameter Δ and become insignificant when $\Delta = 0$.

Our calculations show that the ground exciton state is always the optically passive dark exciton independent of the intrinsic lattice symmetry and the shape of the NCs. In spherical NCs with the cubic lattice structure, the ground exciton states has total angular momentum 2. It cannot be excited by the photon and cannot emit the photon directly in the electric-dipole approximation. This limitation holds also for the hexagonal CdSe NCs. They cannot emit or absorb photons directly, because the ground exciton state has the ± 2 angular momentum projections along the hexagonal axis. In small-size elongated NCs, the ground exciton state has a zero angular momentum projection; however, it was also shown to be the optically forbidden darkexciton state.

The radiative recombination of the dark exciton can only occur through some assisting processes that flip the electron-spin projection or change the

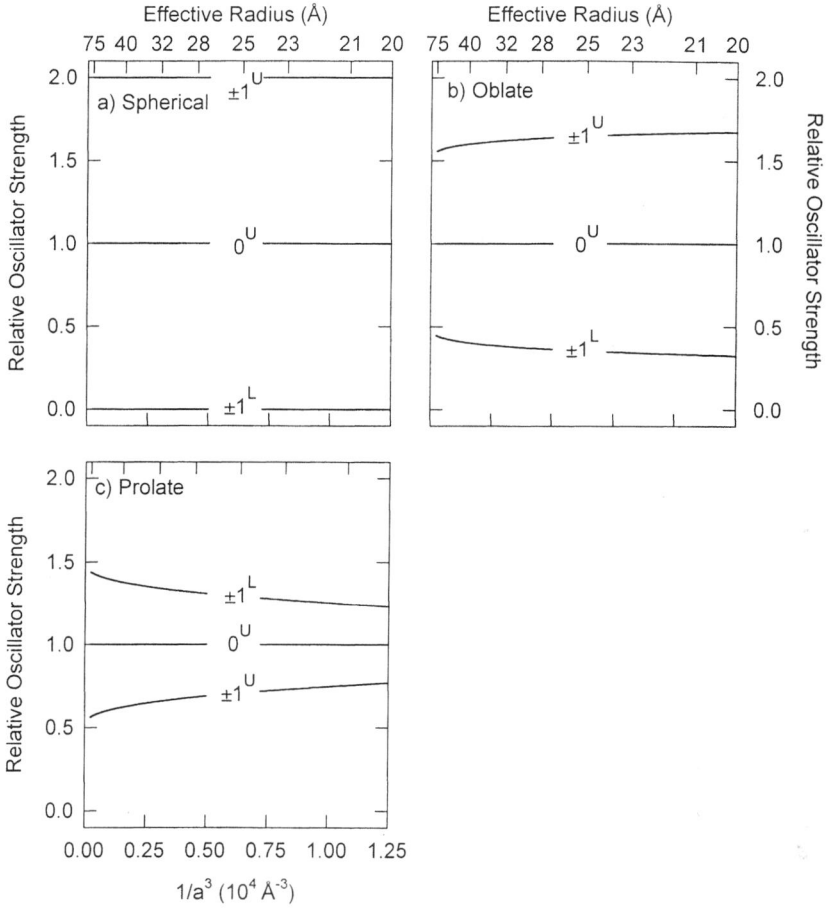

Figure 5 The size dependence of the oscillator strengths, relative to the that of the 0^U state, for the optically active states in cubic CdTe quantum dots with ellipticity μ: (a) spherical dots ($\mu = 0$); (b) oblate dots ($\mu = -0.28$); (c) prolate dots ($\mu = 0.28$).

hole angular momentum projection [12]. These can be optical phonon-assisted transitions, for example, and spherical phonons with the angular momentum 0 and 2 can participate in these transitions [22–24]. As a result, the polarization properties of the low-temperature PL are determined by the polarization properties of virtual optical transitions that are activated by the phonons. The external magnetic field can also activate the dark exciton if is not directed along the hexagonal axis of the NC. In this case, F is no longer a good quantum number and the ± 2 or 0 dark-exciton states are admixed with the optically active ± 1 bright-exciton states. This now allows the direct optical recombination of the exciton ground state. The polarization properties of this PL are determined by the symmetry of admixed states. We now consider the effect of an external magnetic field on the fine structure of the band-edge exciton.

III. FINE STRUCTURE OF THE BAND-EDGE EXCITONS IN MAGNETIC FIELDS

For nanosize quantum dots, the effect of an external magnetic field, H, on the band-edge exciton is well described as a molecular Zeeman effect:

$$\hat{H}_H = \frac{1}{2} g_e \mu_B \hat{\sigma} H - g_h \mu_B \hat{\mathscr{K}} H \tag{36}$$

where g_e is the g factor of the $1S$ electron state, g_h is the g factor of the hole $1S_{3/2}$ state, and μ_B is the Bohr magneton. For bulk CdSe, the electron g factor $g_e^b = 0.68$ [25]; however, due to the nonparabolicity of the conduction band, it depends strongly on the nanocrystal size [26,27]. The value of the hole g factor depends strongly on the structure of the valence band. The Appendix shows the expression for g_h that was derived in the Luttinger model [28] using the results of Ref. 29. In Eq. (29), we neglect the diamagnetic, H^2, terms because the dots are significantly smaller than the magnetic length [~115 Å at 10 T].

Treating the magnetic interaction as a perturbation, we can determine the influence of the magnetic field on the unperturbed exciton state using the perturbation matrix $\hat{E}_{-H}' = \langle \Psi_{\alpha,M} | \mu_B^{-1} \hat{H}_H | \Psi_{\alpha',M'} \rangle$:

	↑,3/2	↑,1/2	↑,-1/2	↑,-3/2	↓,3/2	↓,1/2	↓,-1/2	↓,-3/2
↑,3/2	$\frac{H_z(g_e - 3g_h)}{2}$	$\frac{-i\sqrt{3}g_h H_-}{2}$	0	0	$\frac{g_e H_-}{2}$	0	0	0
↑,1/2	$\frac{i\sqrt{3}g_h H_+}{2}$	$\frac{H_z(g_e - g_h)}{2}$	$-ig_h H_-$	0	0	$\frac{g_e H_-}{2}$	0	0
↑,-1/2	0	$ig_h H_+$	$\frac{H_z(g_e + g_h)}{2}$	$\frac{-i\sqrt{3}g_h H_-}{2}$	0	0	$\frac{g_e H_-}{2}$	0
↑,-3/2	0	0	$\frac{i\sqrt{3}g_h H_+}{2}$	$\frac{H_z(g_e + 3g_h)}{2}$	0	0	0	$\frac{g_e H_-}{2}$
↓,3/2	$\frac{g_e H_+}{2}$	0	0	0	$\frac{-H_z(g_e + 3g_h)}{2}$	$\frac{-i\sqrt{3}g_h H_-}{2}$	0	0
↓,1/2	0	$\frac{g_e H_+}{2}$	0	0	$\frac{i\sqrt{3}g_h H_+}{2}$	$\frac{-H_z(g_e + g_h)}{2}$	$-ig_h H_-$	0
↓,-1/2	0	0	$\frac{g_e H_+}{2}$	0	0	$ig_h H_+$	$\frac{-H_z(g_e - g_h)}{2}$	$\frac{-i\sqrt{3}g_h H_-}{2}$
↓,-3/2	0	0	0	$\frac{g_e H_+}{2}$	0	0	$\frac{i\sqrt{3}g_h H_+}{2}$	$\frac{-H_z(g_e - 3g_h)}{2}$

$$\tag{37}$$

where H_z is the magnetic field projection alon the crystal hexagonal axis and $H_\pm = H_x \pm iH_y$.

A. Zeeman Effect

One can see from Eq. (37) that the magnetic field leads to Zeeman splitting of the double-degenerate exciton states. For the ground dark-exciton state with angular momentum projection ± 2 this spliting, $\Delta\varepsilon_2 = \varepsilon_2 - \varepsilon_{-2}$, can be obtained directly from Eq. (37):

$$\Delta\varepsilon_2 = g_{\mathrm{cx},2}\mu_B H \cos\theta_H \tag{38}$$

where $g_{\mathrm{ex},2} = g_e - 3g_h$ and θ_H is the angle between the nanocrystal hexagonal axis and the magnetic field directions.

Considering the magnetic field terms in Eq. (37) as a perturbation, we determine the Zeeman splitting of the optically active $F = \pm 1$ state:

$$\begin{aligned}
\Delta\varepsilon_1^U &= \varepsilon_1^U - \varepsilon_{-1}^U = \mu_B H_z \{[(C^+)^2 - (C^-)^2]g_e - [(C^+)^2 + 3(C^-)^2]g_h\} \\
\Delta\varepsilon_1^L &= \varepsilon_1^L - \varepsilon_{-1}^L = \mu_B H_z \{[(C^-)^2 - (C^+)^2]g_e - [(C^-)^2 + 3(C^+)^2]g_h\}
\end{aligned} \tag{39}$$

This splitting (linear in the magnetic field) is proportional to the magnetic field projection H_z on the crystal hexagonal axis. Substituting Eq. (21) for $C^\pm$ into Eq. (39), one can get the dependence of this splitting on the NC radius:

$$\Delta\varepsilon_1^{U,L} = g_{\mathrm{ex},1}^{U,L}\mu_B H \cos\theta_H,$$

where

$$\begin{aligned}
g_{\mathrm{ex},1}^U &= g_e \frac{f}{\sqrt{f^2 + d}} - g_h \frac{2\sqrt{f^2 + d} - f}{\sqrt{f^2 + d}} \\
g_{\mathrm{ex},1}^L &= g_e \frac{-f}{\sqrt{f^2 + d}} - g_h \frac{2\sqrt{f^2 + d} + f}{\sqrt{f^2 + d}}
\end{aligned} \tag{40}$$

There is no splitting of the $F = 0$ optically active exciton state. In large NCs, for which one can neglect the exchange interaction ($\eta \ll \Delta$), $g_{\mathrm{ex},1}^U \approx g_e - g_h$ and $g_{\mathrm{ex},1}^L \approx -(g_e + 3g_h)$. In the opposite limit of $\eta \gg \Delta$, $g_{\mathrm{ex},1}^U \approx -(g_e + 5g_h)/2$ and $g_{\mathrm{ex},1}^L \approx (g_e - 3g_h)/2$. The average Zeeman splitting for a system of randomly oriented crystals is

$$\overline{\Delta\varepsilon_{\mathrm{ex},1}^{U,L}} = \frac{g_{\mathrm{ex},1}^{U,L}\mu_B H}{2}, \qquad \overline{\Delta\varepsilon_{\mathrm{ex},2}} = \frac{g_{\mathrm{ex},2}\mu_B H}{2} \tag{41}$$

for the $F = \pm 1^{U,L}$ and $F = \pm 2$ states, respectively.

B. Recombination of the Dark Exciton in Magnetic Fields

One can see from Eq. (37) that components of the magnetic field perpendicular to the hexagonal crystal axis mix the $F = \pm 2$ dark-exciton states with the respective optically active $F = \pm 1$ bright-exciton states. The dark-exciton state with $F = 0$ is also activated due to its admixture with the $F = \pm 1$ bright-exciton states. In small NCs, for which the level splittings are on the order of 10 meV, even the influence of strong magnetic fields can be considered as a perturbation. The case of large NCs for which η is of the same order as $\mu_B g_e H$ will be considered later. The admixture in the $F = 2$ state is given by

$$\Delta\Psi_2 = \frac{\mu_B H_-}{2} \times \left(\frac{g_e C^- - \sqrt{3}g_h C^+}{\varepsilon_2 - \varepsilon_1^+} \Psi_1^+ + \frac{\sqrt{3}g_h C^- + g_e C^+}{\varepsilon_2 - \varepsilon_1^-} \Psi_1^- \right), \quad (42)$$

where the constants $C^\pm$ are given in Eq. (21). The admixture in the $F = -2$ exciton state of the $F = -1$ exciton state is described similarly.

This admixture of the optically active bright-exciton states allows the optical recombination of the dark exciton. The radiative recombination rate of an exciton state F can be obtained by summing Eq. (24) over all light polarizations [30]:

$$\frac{1}{\tau_{|F|}} = \frac{4e^2\omega n_r}{3m_0^2 c^3 \hbar} |\langle 0|\hat{p}_\mu|\tilde{\Psi}_F\rangle|^2, \quad (43)$$

where ω and c are the light frequency and velocity, respectively, n_r is the refractive index, and m_0 is the free-electron mass. Using Eqs. (26) and (28), we obtain the radiative decay time for the upper exciton state with $F = 0$,

$$\frac{1}{\tau_0} = \frac{8\omega n_r P^2 K}{9 \times 137 m_0^2 c^2} \quad (44)$$

for the upper and lower exciton states with $|F| = 1$, correspondingly

$$\frac{1}{\tau_1^{U,L}} = \left(\frac{2\sqrt{f^2 + d} \mp f \pm \sqrt{3d}}{2\sqrt{f^2 + d}} \right) \frac{1}{\tau_0} \quad (45)$$

Using the admixture of the $|F| = 1$ states in the $|F| = 2$ exciton given by Eq. (42), we calculate the recombination rate of the $|F| = 2$ exciton in a magnetic field [3],

$$\frac{1}{\tau_2(H)} = \frac{3\mu_B^2 H^2 \sin^2(\theta_H)}{8\Delta^2} \left(2g_h - g_e \frac{2\eta + \Delta}{3\eta} \right)^2 \frac{1}{\tau_0} \quad (46)$$

The characteristic time τ_0 does not depend on the NC radius. For CdSe, calculations using $2P^2/m_0 = 19.0$ eV [31] give $\tau_0 = 1.5$ ns.

In large NCs, the magnetic field splitting $\mu_B g_e H$ is of the same order as the exchange interaction η and cannot be considered as a perturbation. At the same time, both of these energies are much smaller than the splitting due to the crystal asymmetry. We consider here the admixture in the $|F| = 2$ dark exciton of the lowest $|F| = 1$ exciton only. This problem can be calculated exactly. The magnetic field also lifts the degeneracy of the exciton states with respect to the sign of the total angular momentum projection F. The energies of the former $F = -2$ and $F = -1$ states are

$$\varepsilon^=_{-1,-2} = \frac{-\Delta + 3\mu_B g_h H_z}{2} \pm \frac{\sqrt{(3\eta + \mu_B g_e H_z)^2 + (\mu_B g_e)^2 H_\perp^2}}{2} \tag{47}$$

where $+ (-)$ refers to the $F = -1$ state with an $F = -2$ admixture ($F = -2$ state with an $F = -1$ admixture) and $H_\perp = \sqrt{H_x^2 + H_y^2}$. The corresponding wave functions are

$$\Psi^\pm_{-1,-2} = \left(\frac{\sqrt{p^2 + |n|^2} \pm p}{2\sqrt{p^2 + |n|^2}}\right)^{1/2} \Psi_{\uparrow,-3/2}$$

$$\mp \frac{n}{\left[2\sqrt{p^2 + |n|^2}\left(\sqrt{p^2 + |n|^2} \pm p\right)\right]^{1/2}} \Psi_{\downarrow,-3/2}, \tag{48}$$

where $n = \mu_B g_e H_+$ and $p = 3\eta + \mu_B g_e H_z$. The energies and wave functions of the former $F = 2, 1$ states are (using notation similar to that used just above)

$$\varepsilon^\pm_{1,2} = \frac{-\Delta + 3\mu_B g_h H_z}{2} \pm \frac{\sqrt{(3\eta - \mu_B g_e H_z)^2 + (\mu_B g_e)^2 H_\perp^2}}{2} \tag{49}$$

$$\Psi^\pm_{1,2} = \left(\frac{\sqrt{p'^2 + n'^2} \pm p'}{2\sqrt{p'^2 + n'^2}}\right)^{1/2} \Psi_{\downarrow,3/2}$$

$$\mp \frac{n'}{\left[2\sqrt{p'^2 + n'^2}\left(\sqrt{p'^2 + n'^2} \pm p'\right)\right]^{1/2}} \Psi_{\uparrow,3/2}, \tag{50}$$

where $n' = \mu_B g_e H_-$ and $p' = 3\eta - \mu_B g_e H_z$. As a result, the decay time of the dark exciton in an external magnetic field can be written [3]

$$\frac{1}{\tau_r(H, x)} = \frac{\sqrt{1 + \zeta^2 + 2\zeta x} - 1 - \zeta x}{2\sqrt{1 + \zeta^2 + 2\zeta x}} \frac{3}{2\tau_0} \tag{51}$$

where $x = \cos\theta$ and $\zeta = \mu_B g_e H / 3\eta$. The probability of exciton recombination increases in weak magnetic fields ($\zeta \ll 1$) as

$$\frac{(0.5\mu_B g_e H)^2}{(3\eta)^2} \frac{3\sin^2(\theta_H)}{2\tau_0}$$

and saturates in strong magnetic fields ($\zeta \gg 1$), reaching

$$\frac{3(1 - \cos\theta_H)}{4\tau_0} \left(1 - \frac{3\eta}{\mu_B g_e H}(1 + \cos\theta_H)\right).$$

One can see from Eqs. (46) and (51) that the recombination lifetime depends on the angle between the crystal hexagonal axis and the magnetic field. The recombination time is different for different crystal orientations, which leads to a nonexponetial time decay dependence for a system of randomly oriented crystals.

IV. EXPERIMENT

The fine structure of band-edge exciton spectra explains various unusual and unexpected properties of CdSe NC ensembles, including the linear polarization memory effect [20], circular polarized PL of a single CdSe nanocrystal [32], linear polarized PL of individual CdSe nanorods [33], Stokes shift of the resonant PL [2,3], fine structure of the resonant PL excitation spectra of CdSe NCs [34], shortening of the radiative decay time in magnetic field [2,3], magnetocircular dichroism [35], and polarization of the PL in a strong magnetic field [36]. Here, we will briefly discussed these results and provide their qualitative and quantitative explanations.

A. Polarization Properties of the Ground Dark-Exciton State

As we have already discussed, the ground exciton state in CdSe NCs does not have a dipole moment and it cannot emit light in electric-dipole approximation. The radiative recombination of the dark exciton can only occur through some assisting processes that flip the electron-spin projection or change the hole angular momentum projection [12]. As a result, the polarization properties of the low-temperature PL are determined by polarization properties of virtual optical transitions that are activated by phonons. The optical spherical phonons with the angular momenta 0 and 2 can participate in these transitions [22,23]. The phonons with the angular momentum 2, for example, mix the hole states with the angular momentum projections on the hexagonal axis $\pm 3/2$ and $\mp 1/2$ [24]. The phonons with angular momentum 1 allow one to flip the electron spin through the Rashba spin-orbital terms [12]. The polarization properties of the ground exciton state depend strongly on the angular momentum of phonons participating in the phonon-assisted optical transitions.

The calculations of the relative strength of the phonon-assisted transitions in NCs are unreliable because the strength of the exciton–optical phonon coupling in NCs is a still a controversial subject (see, e.g., Ref. 23). This makes it difficult to predict the polarization properties of the dark-exciton state. However, a single almost spherical CdSe NC [32] shows circularly polarized PL. The dark-exciton state in these crystals has total angular momentum projection $F = \pm 2$ (see Fig. 2d). This shows that the phonons with the angular momenta $l = 2$ and $l = 1$ can be responsible for the phonon-assisted recombination of the dark-exciton state.

The variation in the CdSe NC shape strongly affect their polarization properties. The individual CdSe NCs show a high degree of linear polarization (70%) when their aspect ratio changes from 1:1 to 1:2 [33]. This variation of the NC shape changes the order of the exciton levels and the exciton state with the angular momentum projection $F = 0$ becomes the ground state in elongated NCs, according to our calculations [3] (see Fig. 2c). However, this state is also the dark exciton state with a zero dipole transition matrix element. The linear PL polarization properties of the $F = 0$ state can be due to the $l = 0$ phonon-assisted transitions that mix dark- and bright-exciton states with $F = 0$.

The polarization properties of the individual nearly spherical CdSe nanocrystals and CdSe nanorods suggest that the interaction of the phonons with holes is the major mechanism of phonon-assisted recombination of the dark excitons in NCs.

B. Linear Polarization Memory Effect

The linear polarization memory effect in NC PL was observed by Bawendi et al. [20] for the case of the resonant excitation in the absorption band-edge tail. The sample was an ensemble of randomly oriented CdSe hexagonal NCs of 16 Å radius and the sample did not have any preferential axis. The polarization memory effect in such a sample is due to the selective excitation of some NCs which have a special orientation of their hexagonal axis relative to the polarization vector of the exciting light. The same NCs emit light with polarization completely determined by polarization properties of the ground exciton state and the NC orientation.

The theory of the polarization memory effect for an ensemble of randomly oriented CdSe NC was developed in Ref. 12. Here, we will consider only qualitative conclusions of Ref. 12 because it did not take the exchange electron–hole interaction into account.

The linear polarized light selectively excites NCs with the hexagonal axis predominantly parallel to the vector polarization of exciting light when the excitation frequency is in resonance with the $F = 0$ bright-exciton state. The emission of this nanocrystals is determined by the dark-exciton state and emitted light polarization vector is perpendicular to the hexagonal axis. This leads to the negative degree of the PL polarization as was observed in Ref. 20.

If the exciting light is in resonance with $F = \pm 1$ bright-exciton state, however, the degree of the linear polarization should be positive. This makes the experiments on the linear polarization memory effect very sensitive to the nanocrystal size distribution and the frequency of optical excitation.

C. Stokes Shift of the Resonant PL and Fine Structure of Bright-Exciton States

The strong evidence for the predicted band-edge fine structure has been found in fluorescence line-narrowing (FLN) experiments [2,3]. The resonant excitation of the samples in the red edge of the absorption spectrum selectively excites the largest dots from the ensemble. This selective excitation reduces the inhomogeneous broadening of the luminescence and results in spectrally narrow emission, which displays a well-resolved longitudinal optical (LO) phonon progression. In practice, the samples were excited at the spectral position for which the absorption was roughly one-third of the band-edge absorption peak. Figure 6 shows the FLN spectra for the size series considered in this chapter. The peak of the zero-LO-phonon line (ZPL) is observed to be shifted with respect to the excitation energy. This Stokes shift is size dependent and ranges from ~20 meV for small NCs to ~2 meV for large NCs. Changing the excitation wavelength does not noticeably affect the Stokes shift of the larger samples; however, it does make a difference for the smaller sizes. This difference was attributed to the excitation of different size dots within the size distribution of a sample, causing the observed Stokes shift to change. The effect is the largest in the case of small NCs because of the size dependence of the Stokes shift (see Fig. 7).

In terms of the proposed model, the excitation in the red edge of the absorption probes the lowest $|F| = 1$ bright-exciton state (see Fig. 2d). The transition to this state is followed by relaxation into the dark $|F| = 2$ state. The dark exciton finally recombines through phonon-assisted [2,12] or nuclear/paramagnetic spin-flip assisted transitions [2]. The observed Stokes

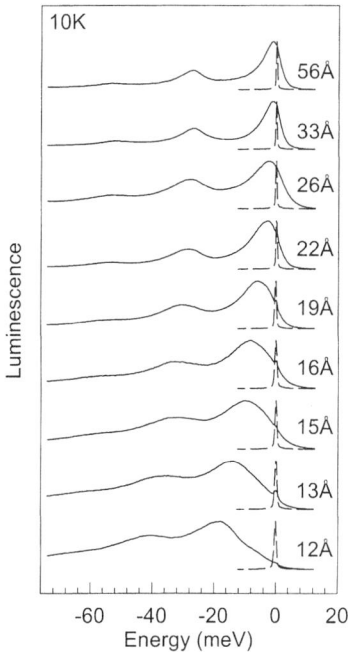

Figure 6 Normalized FLN spectra for CdSe NCs with radii between 12 and 56 Å. The mean radii of the dots are determined from SAXS and TEM measurements. A 10-Hz Q-switched Nd: YAG/dye laser system (~7 ns pulses) serves as the excitation source. Detection of the FLN signal is accomplished using a time-gated optical multichannel analyzer (OMA). The laser line is included in the figure (dotted line) for reference purposes. All FLN spectra are taken at 10 K.

shift is the difference in energy between the $\pm 1^L$ state and the dark ± 2 state; this difference increases with decreasing NC size.

The good agreement between the experimental data for the size-dependent Stokes shift and the values derived from the theory was found. Figure 7 compares experimental and theoretical results. The only parameters used in the theoretical calculation are taken from the literature: $a_{ex} = 56$ Å [7], $\hbar\omega_{ST} = 0.13$ meV [18], and $\beta = 0.28$ [15,37]. The comparison shows that there is good quantitative agreement between experiment and theory for large sizes. For small crystals, however, the theoretical splitting based on the size-dependent exchange interaction begins to underestimate the observed Stokes shift. This discrepancy may be explained, in part, by an additional contribution to the Stokes shift by phonons or dangling bonds that could

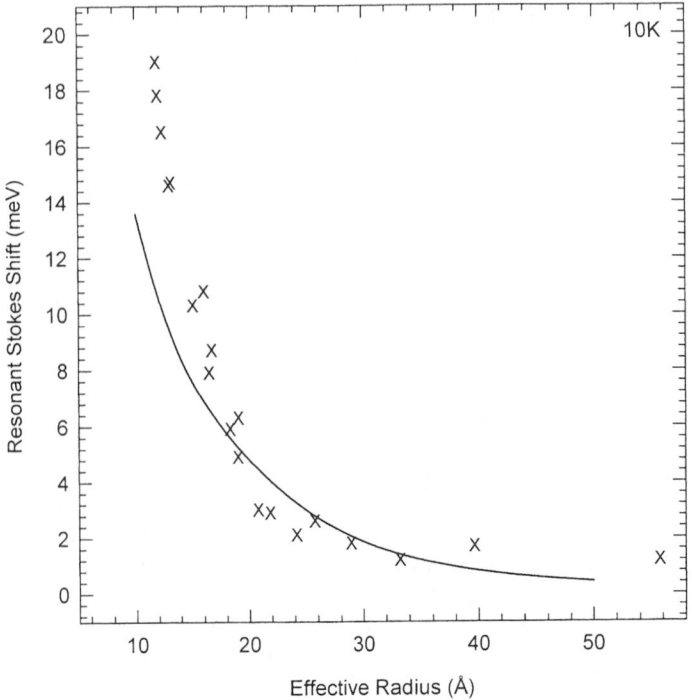

Figure 7 The size dependence of the resonant Stokes shift. This Stokes shift is the difference in energy between the pump energy and the peak of the ZPL in the FLN measurement. The points labeled X are the experimental values. The solid line is the theoretical size dependent splitting between the $\pm 1^L$ state and the ± 2 exciton ground state (see Fig. 2d).

form an exciton–phonon polaron [38] or an exciton–dangling bond magnetic polaron [36].

The resonant PL excitation (PLE) studies of CdSe NCs also confirms the predicted bright-dark exciton fine structure [39]. The resonant PLE experiment provides information on both the level splitting and the relative strength of optical transitions. Although there is a qualitative agreement between the experimental data and the theory, the theoretical model clearly fails to explain saturation of the relative oscillator transition strength in small NCs (see Fig. 9 in Chap. 2). This discrepancy can be due to the fact that the parabolic band approximation used to describe the conduction and valence bands overestimates the transition oscillator strength is small NCs. The detailed description of PLE experiments can be found in Chapter 2.

D. Dark-Exciton Lifetime in a Magnetic Field

Strong evidence for the dark-exciton state is provided by fluorescence line-narrowing (FLN) experiments as well as by studies of the luminescence decay in external magnetic fields. In Fig. 8a, we show the magnetic field dependence of the FLN spectra between 0 and 10 T for 12-Å-radius dots. Each spectrum is normalized to the zero-field, one-phonon line for clarity. In isolation, the ±2

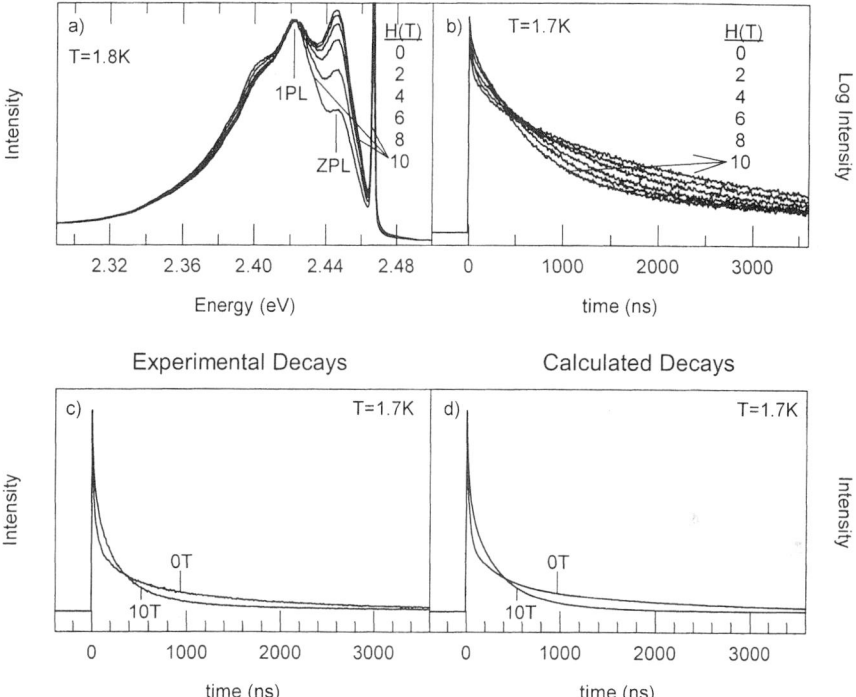

Figure 8 (a) The FLN spectra for 12-Å-radius dots as a function of an external magnetic field. The spectra are normalized to their one-phonon line (1PL). A small fraction of the excitation laser, which is included for reference, appears as the sharp feature at 2.467 eV to the blue of the ZPL; (b) luminescence decays for 12-Å-radius dots for magnetic fields between 0 and 10 T measured at the peak of the "full" luminescence (2.436 eV) and a pump energy of 2.736 eV. All experiments were done in the Faraday configuration ($\mathbf{H} \parallel \mathbf{k}$); (c) observed luminescence decays for 12-Å-radius dots at 0 and 10 T; (d) calculated decays based on the three-level model described in the text. Three weighted three-level systems were used to simulate the decay at zero field with different values of γ_2 (0.033, 0.0033, 0.00056 ns^{-1}) and weighting factors (1, 3.8, 15.3). γ_1 (0.1 ns^{-1}) and γ_{th} (0.026 ps^{-1}) were held fixed in all three systems.

state would have an infinite lifetime within the electric-dipole approximation, because the emitted photon cannot carry an angular momentum of 2. However, the dark exciton can recombine via LO-phonon-assisted, momentum-conserving transitions [39]. Spherical LO phonons with orbital angular momenta of 1 or 2 are expected to participate in these transitions; the selection rules are determined by the coupling mechanism [12,22]. Consequently, for zero field, the LO phonon replicas are strongly enhanced relative to the ZPL. With increasing magnetic field, however, the ± 2 level gains an optically active ± 1 character [Eq. (42)], diminishing the need for the LO-phonon-assisted recombination in dots for which the hexagonal axis is not parallel to the magnetic field. This explains the dramatic increase in the ZPL intensity relative to LO phonon replicas with increasing magnetic field.

The magnetic-field-induced admixture of the optically active ± 1 states also shortens the exciton radiative lifetime. Luminescence decays for 12-Å-radius NCs between 0 and 10 T at 1.7 K are shown in Fig. 8b. The sample was excited far to the blue of the first absorption maximum to avoid orientational selection in the excitation process. Excitons rapidly thermalize to the ground state through acoustic and optical phonon emission. The long microsecond luminescence at zero field is consistent with LO-phonon-assisted recombination from this ground state. Although the light emission occurs primarily from the ± 2 state, the long radiative lifetime of this state allows the thermally populated $\pm 1^L$ state to contribute to the luminescence also.

With increasing magnetic field, the luminescence lifetime decreases; because the quantum yield remains essentially constant, we intepret this result as due to an enhancement of the relative rate. The magnetic field dependence of the luminescence decays can be reproduced using three-level kinetics with $\pm 1^L$ and ± 2 emitting states [2]. The respective radiative rates from these states, $\Gamma_1(\theta_H, H)$ and $\Gamma_2(\theta_H, H)$, in a particular NC, depend on the angle θ_H between the magnetic field and the crystal hexagonal axis. The thermalization rate, Γ_{th} of the $\pm 1^L$ state to the ± 2 level is determined independently from picosecond time-resolved measurements. The population of the $\pm 1^L$ level is determined by microscopic reversibility. We assume that the magnetic field opens an additional channel for ground-state recombination via admixture in the ± 2 state of the ± 1 states: $\Gamma_2(\theta_H, H) = \Gamma_2(0, 0) + 1/\tau_2(\theta_H, H)$. This also causes a slight decrease in the recombination rate of the $\pm 1^L$ state.

The decay at zero field is multiexponential, presumably due to sample inhomogeneities (e.g., in shape and symmetry-breaking impurity contaminations). We describe the decay using three three-level systems, each having a different value of $\Gamma_2(0, 0)$ and each representing a class of dots within the inhomogeneous distribution. These three-level systems are then weighted to reproduce the zero field decay (Fig. 8c). We obtain average values of $1/\Gamma_2(0, 0)$ = 1.42 μs and $1/\Gamma_1(0, 0)$ = 10.0 ns. The latter value is in a good agreement with the theoretical value of the radiative lifetime for the $\pm 1^L$ state, $\tau_1^L = 13.3$ ns, calculated for a 12-Å nanocrystal using Eq. (45).

In a magnetic field, the angle-dependent decay rates $[\Gamma_1(\theta_H, H), \Gamma_2(\theta_H, H)]$ are determined from Eq. (46). The field-dependent decay was then calculated, averaging over all angles to account for the random orientation of the crystallite "c" axes. The calculation at 10 T (Fig. 8c) used a bulk value of $g_e = 0.68$ [25] and the calculated value for Δ (19.4 meV) and η (10.3 meV) for 12-Å-radius dots. The hole g factor was treated as a fitting parameter because its reliable value is not available. This procedure allowed an excellent agreement with the experiment for $g_h = -1.00$. However, the most recent measurements of the electron g factor yield $g_e \approx 1.4$ for NCs of 12 Å radius [26], implying that the hole g factors may require a reevaluation.

E. Magnetic Circular Dichroism of CdSe Nanocrystals

The splitting of the exciton levels in a magnetic field is usually much smaller than the inhomogeneous width of optical transitions and it cannot be seen directly in absorption spectra. However this splitting can be observed in magnetocircular dichroism (MCD) experiments in which the difference between the absorption coefficients, $\alpha^\pm$, for right and left circular polarized light ($\sigma^\pm$), respectively, in the presence of a magnetic field can be measured with high accuracy. Assuming that the inhomogeneous exciton line has a Gaussian shape, one finds for exciton states with identical Zeeman splitting (see e.g., Ref. 40):

$$\alpha_{MCD}(\varepsilon, \varepsilon_0) = \alpha^+(\varepsilon, H) - \alpha^-(\varepsilon, H)$$

$$= 2C\Delta\varepsilon \frac{\varepsilon - \varepsilon_0}{\sigma^2} \frac{1}{\sqrt{2\pi}\sigma} \exp\left(-\frac{(\varepsilon - \varepsilon_0)^2}{2\sigma^2}\right) \qquad (52)$$

where C is a constant related to the oscillator strength of the state, $\Delta\varepsilon$ is the field-dependent Zeeman splitting of the state, σ is the inhomogeneous linewidth of the state, and ε_0 is the position of the maximum of the transition at zero magnetic field. The MCD signal for a single line should have a typical derivative shape with extrema separated by 2σ. Its intensity is proportional to the Zeeman splitting of the levels $\Delta\varepsilon$ and grows linearly with magnetic field. To extract the absolute value of the splitting, one can normalize the MCD signal by $\alpha_{SUM} = \alpha^+ + \alpha^-$ (this procedure eliminates the unknown constant C).

Equation (52) can also be used for an ensemble of randomly oriented quantum dots. In this case, however, $\Delta\varepsilon$ characterizes the effective average splitting of the exciton states in a magnetic field because the Zeeman splitting in each CdSe NC depends on the angle (θ_H) between the magnetic field and the crystal axis [see Eq. (40)]. As a result of the random orientation of NC axes with respect to the light propagation direction, both polarizations ($\sigma^{\pm}$) can excite both states $F = \pm 1$ and their excitation probabilities depend on the angle (θ) between the light propagation direction and the NC axis [see Eqs. (33) and (34)]; in the MCD experiments, $\theta_H \equiv \theta$.

Let us assume that the inhomogeneous broadening of the exciton levels has a Gaussian shape. In this case, the absorption coefficient for the $\sigma^{\pm}$ polarized light due to the excitation of the $|F| = 1$ exciton states in NCs with a hexagonal oriented at the angle θ with respect to the light propagation direction has the form

$$\alpha_F^{\pm}(\varepsilon - \varepsilon_F^{U,L}) \sim \frac{N_1^{U,L}}{\sqrt{2\pi}\sigma_1}(1 \pm F\cos\theta)^2 \exp\left(-\frac{(\varepsilon - \varepsilon_F^{U,L})^2}{2\sigma_1^2}\right) \quad (53)$$

where $N_1^{U,L}$ is defined by Eq. (29) and σ_1 and ε_F are the linewidth and the average energy of the $|F| = 1$ exciton states for a given NC distribution, respectively. The splitting in the magnetic field leads to the MCD signal, α_{MCD}:

$$\alpha_{MCD}^{U,L} \sim \frac{1}{2}\sum_{F=\pm 1}\int[\alpha_F^+(\varepsilon - \varepsilon_F^{U,L}(H)) - \alpha_F^-(\varepsilon - \varepsilon_F^{U,L}(H))]d\cos\theta \quad (54)$$

This expression can be simplified because the intrinsic transition width is much larger than the Zeeman splitting. Substituting Eq. (53) into Eq. (54) and performing the integration, we find

$$\alpha_{MCD}^{U,L} \sim N_1^{U,L}(2\Delta\varepsilon)\frac{(\varepsilon - \overline{\varepsilon_1})}{\sigma_1^2}\frac{1}{\sqrt{2\pi}\sigma_1}\exp\left(-\frac{(\varepsilon - \overline{\varepsilon_1})^2}{2\sigma_1^2}\right) \quad (55)$$

where $\Delta\varepsilon = (\mu_B H g_{ex,1}^{U,L})/2$ [see Eqs. (40) and (41)] and $\overline{\varepsilon_1} = \overline{\varepsilon_1^{U,L}}$ is the average position of the exciton level in a zero magnetic field.

One can see that the magnitude of the MCD signal is proportional to $g_{ex,1}^{U,L}$. However, the absolute values of these g factors can be obtained only from the normalized MCD signal. Two cases must be considered, depending on whether the exciton line broadening is smaller or larger than the $F = 0$ and $|F| = 1$ exciton state splitting. In the former case, the sum of absorption coefficients for the σ^+ and σ^- polarized light can be obtained from Eq. (53) after the integration over angle θ:

$$\alpha_{SUM}^{U,L} \sim \frac{1}{2}\sum_{F=\pm}\int[\alpha_F^+(\varepsilon - \varepsilon_F^{U,L}(H)) + \alpha_F^-(\varepsilon - \varepsilon_F^{U,L}(H))]d\cos\theta$$

$$= \frac{2N_1^{U,L}}{\sqrt{2\pi}\sigma_1}\exp\left(-\frac{(\varepsilon - \overline{\varepsilon_1})^2}{2\sigma_1^2}\right) \quad (56)$$

The normalization of α_{MCD} by α_{SUM} allows one to extract the absolute value of a g factor.

In the case for which the inhomogeneous line broadening is larger than the fine-structure exciton splitting, both the upper and lower exciton states with $|F| = 1$ contribute to the MCD signal. In addition, the $F = 0$ exciton state contributes to the absorption and this contribution should be taken into account in normalization of the MCD signal. The absorption coefficient of the $F = 0$ exciton states for $\sigma^{\pm}$ polarized light depends also on the angle θ between the NC hexagonal axes and the light propagation direction:

$$\alpha_0^{\pm}(\varepsilon - \overline{\varepsilon_0}) \sim \frac{N_0^U}{\sqrt{2\pi}\sigma_0} \sin^2\theta \exp\left(-\frac{(\varepsilon - \overline{\varepsilon_0})^2}{2\sigma_0^2}\right) \tag{57}$$

where N_0^U is the constant defined by Eq. (26), σ_0 and $\overline{\varepsilon_0} = \overline{\varepsilon_0^U}$ are the linewidth and the average energy, respectively, of the optically active $F = 0$ exciton states for the given NC distribution.

Assuming that the inhomogeneous broadening for all exciton states is the same ($\sigma_1^U = \sigma_1^L = \sigma_0 = \sigma$) and is much larger than the exciton fine-structure splitting, we can obtain the following expression for the effective Zeeman splitting of the $S_{3/2}1S_e$ transition (after averaging over all solid angles):

$$\Delta\varepsilon = \mu_B H \frac{N_1^U g_{ex,1}^U + N_1^L g_{ex,1}^L}{N_1^U + N_1^L + N_0^U/4} = \mu_B H g_{eff} \tag{58}$$

The magnitude of the MCD signal is proportional to the magnetic field and its shape depends on the sign of the effective exciton g factor g_{eff}.

Experimental studies of CdSe NCs [35] show that the magnitude of the MCD signal for the $1S_{3/2}1S_e$ and $2S_{3/2}1S_e$ transitions increases linearly with magnetic field. At the same time, the measured shape of the MCD signal for these transitions was the reverse of each other and was described by the theoretical normalized MCD curve with the positive and negative size-dependent effective g factor, respectively ($g_{eff} > 0$ for the $1S_{3/2}1S_e$ transition and $g_{eff} < 0$ for the $2S_{3/2}1S_e$ transition). Equation (58) also gives opposite signs for the effective exciton g factor for these two transitions; however, it does not reproduce the experimental size dependence of g_{eff} [35]. The fact that the size dependence of the electron g factor was not taken into account in calculations might also be one of the possible reasons for this disagreement.

F. Polarization of the PL in Strong Magnetic Fields

An external magnetic field splits the ground dark-exciton state into two sublevels [see Eq. (38)]. The exciton sublevels are thermally populated if the time of the exciton momentum relaxation is faster then the exciton relaxation time. This unequal population of the exciton states with the angular momentum projection $F = +2$ and $F = -2$ on the hexagonal axis of the NCs leads to the circularly polarized PL. The effect can be observed in a strong magnetic field, H, or/and low temperatures, T (the ratio H/T controls the relative population of the exciton sublevels).

Figure 9a (Ref. 36) shows a characteristic PL spectrum from the 57-Å-diameter NCs at 1.45 K at both 0 and 60 T magnetic fields. The PL linewidth of ~60 meV is typical for NC samples and arises largely from the nonuniform size distribution of the NCs. The PL is unpolarized in the zero field and becomes circularly polarized if a magnetic field is applied. The σ^- (σ^+) polarized emission gains (loses) intensity with increasing field, as shown in Fig. 9b. The PL polarization degree, $P = (I_{\sigma^-} - I_{\sigma^+})/(I_{\sigma^-} + I_{\sigma^+})$, does not fully saturate even at 60 T (Fig. 9c). At 1.45 K, the polarization degree exhibits

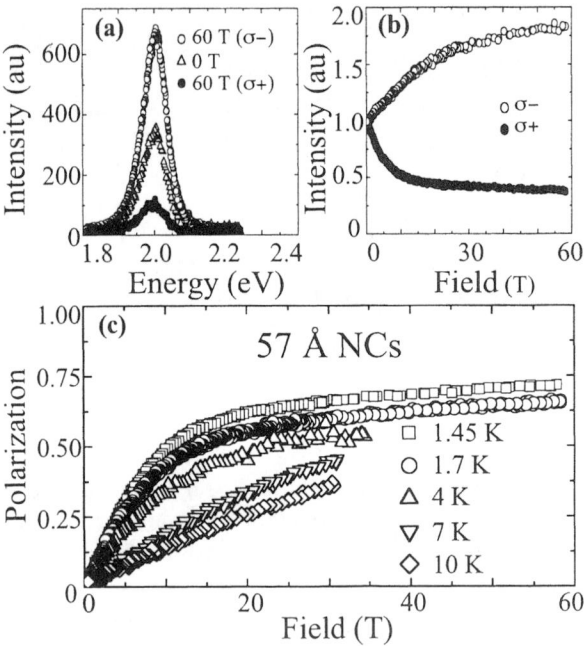

Figure 9 (a) Spectra of PL from 57-Å-diameter NCs at T = 1.45 K and at 0 and 60 T magnetic fields (σ^+ and σ^-). (b) The intensity of the σ^+ and σ^- PL versus magnetic field. (c) The degree of PL circular polarization at different temperatures. At 1.45 K, data show an initial saturation near 0.6 (~20 T) and subsequent slow growth to 0.73 (60 T), still well below complete polarization ($P = 1$).

a rapid growth at low fields, after which it rolls off at ~20 T at a value of ~0.6, well below complete saturation. At higher fields, the polarization degree does not remain constant, but rather continues to increase slowly, reaching ~0.73 at 60 T. Figure 1c also shows that the PL polarization degree for dark excitons drops quickly with increasing temperature, which is similar to the behavior for a thermal ensemble of optically active excitons distributed between two Zeeman-split sublevels.

The polarization data can be understood in terms of a fine structure of the band-edge exciton in above-considered magnetic field. The PL at low temperature is due to the radiative recombination of the dark exciton from the two $F = \pm 2$ sublevels that are activated by an external magnetic field [see Eq. (46)]. The polarization degree of PL depends on the relative population of these sublevels. The dark excitons with $F = \pm 2$ obtain the polarization properties of the bright excitons with $F = \pm 1$. In an ensemble of randomly oriented NCs, all characteristics (Zeeman splitting of the exciton sublevels, the degree of the dark-exciton activation, and the degree of the PL circular polarization) depend on the angle θ_H between the NC hexagonal axis and magnetic field that coincides in this case with the light propagation direction ($\theta = \theta_H$) [see Eqs. (33), (34), (38), and (46)].

Within the electric-dipole approximation, the relative probabilities of detecting $\sigma^\pm$ light from the $F = \pm 2$ excitons in NCs with axes oriented at angle θ with respect to the field are $P_{F=2}(\sigma^\pm) \sim (1 \pm \cos\theta)^2$ and $P_{F=-2}(\sigma^\pm)$ $\sim (1 \mp \cos\theta)^2$. The relative population of the $F = \pm 2$ exciton states is determined by the angular-dependent Zeeman splitting $\Delta = g_{ex,2}\mu_B H \cos\theta$ [see Eq. (38)]. Assuming the Boltzmann thermal distribution between these two exciton states, we obtain the following expression for the intensity of the detected PL with σ^+ and σ^- polarizations:

$$I_{\sigma^\pm}(x) = \frac{(1 \mp x)^2 e^{\Delta\beta/2} + (1 \pm x)^2 e^{-\Delta\beta/2}}{e^{\Delta\beta/2} + e^{-\Delta\beta/2}} \tag{59}$$

where $x = \cos\theta$ and $\beta = (k_B T)^{-1}$. Integrating over all orientations and computing the PL polarization degree, we obtain

$$P(H, T) = \frac{2 \int_0^1 dx\, x \tanh(0.5 g_{ex,2}\mu_B H\beta x)}{\int_0^1 dx\, (1 + x)^2} \tag{60}$$

In the limiting case of low temperatures (or high fields) when $g_{ex,2}\mu_B H\beta \gg 1$, this polarization $P(H, T) \to 0.75$. This is the maximum possible PL polarization that can be reached in a system of randomly oriented wurtzite NCs, and it should be noted that the data in Fig. 9c approach this limit at the lowest temperatures and highest fields.

One must also account for the influence of the magnetic field on the PL quantum efficiency of NCs. The PL quantum efficiency depends on the ratio of radiative (τ_r) and nonradiative (τ_{nr}) decay times. The magnetic field admixes the dark- and bright-exciton states and shortens the radiative decay time in NCs for which the hexagonal axis is not parallel to the magnetic field [see Eqs. (46) and (51)]. The PL quantum efficiency $q(H, x)$ increases in NCs with hexagonal axis predominantly oriented orthogonal to the field. As a result, the relative contribution of different NCs to the PL is also controlled by $q(H, x) = [1 + \tau_r(H, x)/\tau_{nr}]^{-1}$, where $\tau_r(H, x)$ in a strong magnetic field as defined by Eq. (51). Thus, the polarization degree becomes [36]

$$P(H, T) = \frac{2 \int_0^1 dx\, x \tanh(0.5 g_{ex,2} \mu_B H \beta x) q(H, x)}{\int_0^1 dx\, (1 + x)^2 q(H, x)} \tag{61}$$

which reduces to Eq. (60) in the limit $\tau_{nr} \gg \tau_r$, for which nonradiative transitions are negligible.

In the opposite limit of a low quantum efficiency ($\tau_{nr} \ll \tau_r$) the maximum degree of PL polarization is only 0.625 at low temperatures (this case can be realized, e.g., in low fields for which the state mixing is weak). The reduced degree of polarization is because NCs in which the hexagonal axis is perpendicular to the magnetic field are more emissive, but they have a zero g factor. With increasing magnetic field, however, the radiative decay time τ_r is

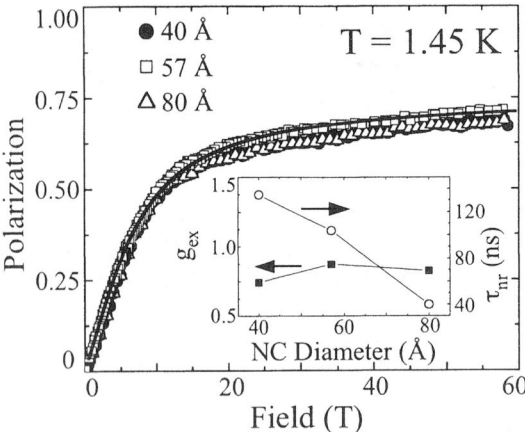

Figure 10 The nearly identical PL polarization in NCs of 40, 57, and 80 Å diameter for magnetic fields up to 60 T at low temperatures along with the calculated polarization given by Eq. (4). Inset: Extracted values of g_{ex} and τ_{nr} for these NCs.

decreased for the majority of NCs and a gradual growth of the maximum allowed polarization degree from 0.625 to 0.75 can occur. Precisely such behavior is observed in Fig. 9c, where the low-temperature polarization first saturates near ~0.6 and then grows slowly to ~0.73 at the highest fields.

This model [Eq. (61)] provides an excellent fit to the measured PL polarization data [36], as illustrated in Fig. 10, which displays experimental results of NCs of three different sizes ($T = 1.45$ K) along with results of the modeling for 57-Å NCs. Interestingly, all data are roughly equivalent; this stems from the balancing role played by τ_{nr} and the exchange interaction η [see Eq. (51)] both of which decrease with increasing NC size. The best-fit values of $g_{ex,2}$ and τ_{nr} are shown in the inset. The extracted exciton g factors are close to 0.9, which is much smaller than the value calculated for the dark-exciton state.

V. DISCUSSION AND CONCLUSIONS

We have shown that the dark/bright-exciton model presented in this chapter describes very well many important properties of the band-edge PL in CdSe NCs. The phenomena analyzed here include the fine structure of the PL excitation spectra, a Stokes shift of the resonant PL, the shortening of the radiative decay time in a magnetic field, the polarization memory effect, the transition from the circular polarized to the linear polarized PL induced by changing the NC shape, and PL polarization in strong magnetic fields. Some quantitative disagreements between experimental data and the theory are mainly due to some oversimplifications in the theoretical approach. For example, the theory considered here does not take into account the non-parabolicity of the conduction and the valence bands. In small NCs, the nonparabolicity can strongly modify the level structure, the electron–hole wave functions, the transition oscillator strength, the overlap integrals, the electron–hole exchange interactions, g factors, and so forth. All of these quantities, in principle, can be better described in the framework of a more rigorous approach.

The dark/bright-exciton model described in this chapter can also be applied to NCs with other than CdSe compositions. It was used, for example, to described the size-dependent Stokes shift in InAs NCs [41]. The penetration of the electron wave function under the barrier was important in InAs NCs for quantitative description of this dependence.

After a decade of studies and despite all the success of the dark/bright-exciton model, the PL in CdSe NCs still presents several unresolved puzzles. One of the unresolved issues is the temperature-dependent Stokes shift observed in one of the first studies of small-size CdSe NCs [42,43]. This effect may be due, for example, to the formation of an exciton–polaron that can be a source of an additional Stokes shift of the luminescence [38]. However, the "polaronic" model does not explain the presence of a zero LO phonon line in the absence of an external magnetic field. This line is strongly activated by the temperature. The effect of the temperature on the relative intensities of zero and phonon-assisted lines and PL decay times is very similar to the effect of an external magnetic field [42,43]. Interaction with paramagnetic defects in the lattice or surface dangling bonds can also provide an additional mechanism for the dark-exciton recombination. The spins of these defects can generate strong effective internal magnetic fields (potentially several tens of tesla) and induce spin-flip-assisted transitions of the ± 2 state, enabling the zero-phonon recombination. The electron interaction with these spins can also lead to the magnetic polaron formation, which may explain the temperature-dependent Stokes shift.

The electron-dangling bond spin interactions can also explain the unexpected temperature behavior of the dark-exciton g factor [36]. Similar to diluted magnetic semiconductors, the change in g_{ex} in CdSe NCs can be interpreted as arising from some exchange interaction. We therefore suggest that temperature-dependent g factors in NCs may be due to the interactions of uncompensated NC surface spins with photogenerated carriers within the NCs.

ACKNOWLEDGMENTS

I would like to thank my long-term collaborators A. I. Ekimov, A. Rodina, M. Rosen, M. G. Bawendi, D. Norris, M. Nirmal, K. Kuno, E. Johnston-Halperin, D. D. Awschalom, S. A. Crooker, P. Alivisatos, A. Nozik, and L. Brus for challenging and stimulating discussions which provided strong motivation for theoretical studies reviewed in this chapter. I also thank M. Nirmal for providing figures. This work was supported by the Office of Naval Research.

APPENDIX: CALCULATION OF THE HOLE g FACTOR

The expression for the g factor of a hole localized in a spherically symmetric potential was obtained by Gel'mont and D'yakonov [29]:

$$g_h = \frac{4}{5}\gamma_1 I_2 + \frac{8}{5}\gamma(I_1 - I_2) + 2\kappa\left(1 - \frac{4}{5}I_2\right) \tag{62}$$

where γ_1, γ, and κ are the Luttinger parameters [30] and $I_{1,2}$ are integrals of the hole radial wave functions [see Eq. (6)]:

$$I_1 = \int_0^a dr\, r^3 R_2 \frac{dR_0}{dr} : \qquad I_2 = \int_0^a dr\, r^2 R_2^2 \qquad (63)$$

These integrals depend only on the parameter β, and their variation with β is shown in Fig. 11. Using $\gamma_1 = 2.04$ and $\gamma = 0.58$ [15] and the relationship $\kappa = -2/3 + 5\gamma/3 - \gamma_1/3$ [44], one calculates that $g_h = -1.09$.

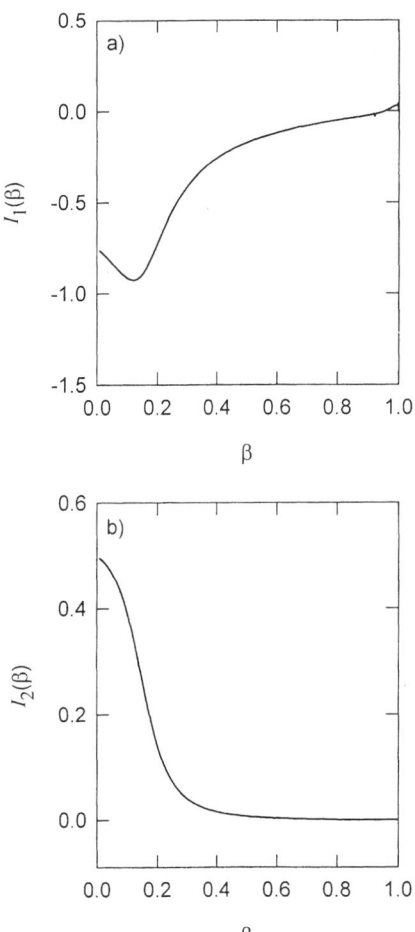

Figure 11 Dependence of the hole radial function integrals I_1 and I_2, which enter in the expression for the hole g factor, on the hole effective mass ratio β.

REFERENCES

1. Bawendi, M.G.; Wilson, W.L.; Rothberg, L.; Carroll, P.J.; Jedju, T.M.;
    Stegerwald, M.L.; Brus, L.E. Phys. Rev. Lett. 1990, 65, 1623.
2. Nirmal, M.; Norris, D.J.; Kuno, M.; Bawendi, M.G.; Efros, Al.L.; Rosen, M.
    Phys. Rev. Lett. 1995, 75, 3728.
3. Efros, Al.L.; Rosen, M.; Kuno, M.; Nirmal, M.; Norris, D.J.; Bawendi, M. Phys.
    Rev. B 1996, 54, 4843.
4. Efros, Al.L.; Efros, A.L. Fiz. Tekh. Poluprovodn. 1982, 16, (1209) Sov. Phys.
    Semicond. 1982, 16, 772.
5. Luttinger, J.M.; Kohn, W. Phys. Rev. 1955, 97, 869.
6. Grigoryan, G.B.; Kazaryan, E.M.; Efros, Al.L.; Yazeva, T.V. Fiz. Tverd. Tela
    1990, 32, 1772. Sov. Phys. Solid State 1990, 32, 1031.
7. Ekimov, A.I.; Hache, F.; Schanne-Klein, M.C.; Ricard, D.; Flytzanis, C.;
    Kudryavtsev, I.A.; Yazeva, T.V.; Rodina, A.V.; Efros, Al.L. J. Opt. Soc. Am. B
    1993, 10, 100.
8. Ekimov, A.I.; Onushchenko, A.A.; Plukhin, A.G.; Efros, Al.L. Zh. Eksp. Teor.
    Fiz. 1985, 88, 1490. Sov. Phys. JETP 1985, 61, 891.
9. Xia, J.B. Phys. Rev. B 1989, 40, 8500.
10. Vahala, K.J.; Sercel, P.C. Phys. Rev. Lett. 1990, 65, 239. Sercel, P.C.; Vahala,
    K.J. Phys. Rev. B 1990, 42, 3690.
11. Efros, Al.L.; Rodina, A.V. Solid State Commun. 1989, 72, 645.
12. Efros, Al.L. Phys. Rev. B 1992, 46, 7448.
13. Bir, G.L.; Pikus, G.E. Symmetry and Strain-Induced Effects in Semiconductors.
    Wiley: New York, 1975.
14. Efros, Al.L.; Rodina, A.V. Phys. Rev. B 1993, 47, 10005.
15. Norris, D.J.; Bawendi, M.G. Phys. Rev. B 1996, 53, 16338.
16. Brus, L.E.; unpublished.
17. Rashba, E.I. Zh. Eksp. Teor. Fiz. 1959, 36, 1703. Sov. Phys. JETP 1959, 9,
    1213.
18. Kochereshko, V.P.; Mikhailov, G.V.; Ural'tsev, I.N. Fiz. Tverd. Tela 1983, 25,
    759. Sov. Phys. Solid State 1983, 25, 439.
19. Murray, C.B.; Norris, D.J.; Bawendi, M.G. J. Am. Chem. Soc. 1993, 115, 8706.
20. Bawendi, M.G.; Carroll, P.J.; Wilson, W.L.; Brus, L.E. J. Chem. Phys. 1992, 96,
    946.
21. Chamarro, M.; Gourdon, C.; Lavallard, P.; Ekimov, A.I. Jpn. J. Appl. Phys.
    1995, 34 (Suppl. 1), 12.
22. Klein, M.C.; Hache, F.; Ricard, D.; Flytzanis, C. Phys. Rev. B 1990, 42, 11123.
23. Efros, Al.L. In: Leburton, J.-P., Pascual, J., Sotomayor-Torres, C. Eds. Phonons
    in Semiconductor Nanostructures. Boston: Kluwer Academic, 1993, 299.
24. Efros, Al.L.; Ekimov, A.I.; Kozlowski, F.; Petrova-Koch, V.; Schmidbaur, H.;
    Shumilov, S. Solid State Commun. 1991, 78, 853.
25. Piper, W.W. Proceedings of the International Conference 7th on II-VI Semi-
    conductor Compounds, W.A. Benjamin, New York, 1967, 839.
26. Gupta, J.A.; Awschalom, D.D.; Efros, Al.L.; Rodina, A.V. Phys. Rev. B 2002,
    66, 125307.

27. Rodina, A.V.; Efros, Al.L.; Yu., A. Alekseev, Phys. Rev. B, 2003, 67, 155312.
28. Luttinger, J.M. Phys. Rev. 1956, 102, 1030.
29. Gel'mont, B.L.; D'yakonov, M.I. Fiz. Tekh. Poluprovodn. 1973, 7, 2013. Sov. Phys. Semiconduct. 1973, 7, 1345.
30. Landau, L.D.; Lifshitz, E.M. Relativistic Quantum Theory, 2nd ed. Pergamon Press, Oxford, 1965.
31. Kapustina, A.V.; Petrov, B.V.; Rodina, A.V.; Seisyan, R.P. Fiz. Tverd. Tela 2000, 42, 1207. Phys. Solid State 2000, 42, 1242.
32. Embedocoles, S.A.; Neuhauser, R.; Bawendi, M.G. Nature 1999, 399, 126.
33. Hu, J.; Li, L.; Yang, W.; Manna, L.; Wang, L.; Alivisatos, A.P. Science 2001, 292, 2060.
34. Norris, D.J.; Efros, Al.L.; Rosen, M.; Bawendi, M.G. Phys. Rev. B 1996, 53, 16347.
35. Kuno, M.; Nirmal, N.; Bawendi, M.G.; Efros, Al.L.; Rosen, M. J. Chem. Phys. 1998, 108, 4242.
36. Johnston-Halperin, E.; Awschalom, D.D.; Crooker, S.A.; Efros, Al.L.; Rosen, M.; Peng, X.; Alivisatos, A.P. Phys. Rev. B 2001, 63, 205309.
37. Norris, D.J.; Sacra, A.; Murray, C.B.; Bawendi, M.G. Phys. Rev. Lett. 1994, 72, 2612.
38. Itoh, T.; Nishijima, M.; Ekimov, A.I.; Efros, Al.L.; Rosen, M. Phys. Rev. Lett. 1995, 74, 1645.
39. Calcott, P.D.J.; Nash, K.J.; Canham, L.T.; Kane, M.J.; Brumhead, D. J. Phys. Condens. Matter 1993, 5, L91.
40. Hoffman, D.; Ottinger, K.; Efros, Al.L.; Meyer, B.K. Phys. Rev. B 1997, 55, 9924.
41. Banin, U.; Lee, J.C.; Guzelian, A.A.; Kadavanich, A.V.; Alivisatos, A.P. Super-lattices Microstruct. 1997, 22, 559.
42. Nirmal, M.; Murray, C.B.; Norris, D.J.; Bawendi, M.G. Z. Phys. D 1993, 26, 361.
43. Nirmal, M.; Murray, C.B.; Bawendi, M.G. Phys. Rev. B 1994, 50, 2293.
44. Dresselhaus, G.; Kip, A.F.; Kittel, C. Phys. Rev. 1955, 98, 365.

4

Intraband Spectroscopy and Dynamics of Colloidal Semiconductor Quantum Dots

Philippe Guyot-Sionnest, Moonsub Shim, and Congjun Wang
James Franck Institute, Chicago, Illinois, U.S.A.

I. INTRODUCTION

Semiconductor nanocrystal colloids are most striking for the ease with which their color, determined by electronic absorption frequencies, can be controlled by size. Although most of the applications currently envisioned are based on the interband transitions, one should not overlook the intraband transitions. In the case of conduction-band states, these transitions are easily size-tunable through spectral regions of atmospheric transparencies (e.g., 3–5 μms and 8–10 μms). This tunability makes semiconductor nanocrystal colloids attractive subjects of study with potential applications in filters, detectors, lasers, and nonlinear optical elements.

The investigations of intraband (also called intersubband) transitions started in 1984 with studies of semiconductor quantum wells [1]. Progress in this field was rapid and led to the demonstration of photodetectors [2], nonlinear optical elements [3], and mid-infrared "quantum cascade lasers" [4].

Interest in the intraband transitions in quantum dots is more recent and is tied to the development of appropriate materials, but the essential motivation is to take advantage of the discrete transitions arising from three-dimensional (3D) confinement. As reviewed in Section II, because of strong intraband transitions, semiconductor nanocrystal colloids exhibit "infrared" (IR) optical properties that are unlikely to be achieved by organic molecular systems. Of potential practical interest is the fact that one single excess electron detectably affects the optical response of a small semiconductor nanocrystal, leading to nonlinear optical and large electro-optical responses in the infrared spectral range.

One motivation for studying quantum dots as building blocks for optical devices is that one can potentially control the energy and phase relaxation of excited electrons, issues that are essential, for example, to the operation of quantum-dots lasers. At present, unlike for quantum wells, there is a weak understanding of the role that phonons play in the relaxation mechanisms in quantum dots, mainly because the energy separation between the electronic states can be much larger than the phonon energy. Intraband spectroscopy is well suited for studies of relaxation processes because it allows one to decouple electron and hole dynamics.

II. BACKGROUND

As discussed in previous chapters of this book, the 3D confinement of electrons and holes leads to discrete energy states. Neglecting phonons and relaxation processes, these states are delta functionlike, and their energies are determined by the band structure of the semiconductor, the shape of the boundary, and the nature of the boundary conditions. For II-VI, III-V, and IV–VI semiconductors, the effective mass approximation based on the Luttinger model of the band structure has been used to reproduce interband spectra [5–8]. Given the uncertainties in the shape, the boundary conditions [9], the presence of surface charges [10] or dipoles [11], and so forth, one might expect the predictions to be rather crude, particularly for holes that are typically characterized by large effective masses. Nevertheless, the effective mass approximation, with its few adjustable parameters, provides a significant simplification over atomistic tight-binding or pseudopotential methods in the description of the delocalized states and optical transitions in quantum dots (see Chapter 3). The situation for the nondegenerate conduction band is the simplest. For a spherical box, the quantum-confined electronic wave functions are Bessel functions that satisfy the boundary conditions for, for example, a fixed finite external potential. The electronic states are described by the angular momentum (L) of the envelope function and denoted as $1S_e$, $1P_e$, $2S_e$, $1D_e$, based on standard convention. The optical selection rules are the same as for atomic spectra because the Bloch functions are derived from identical atomic wave functions, and, therefore, the allowed intraband transitions correspond to $\Delta L = \pm 1$. These selection rules are in contrast to interband transitions for which $\Delta L = 0, \pm 2$.

The solid line in Fig. 1 shows the energy separation between the first and second conduction-band states $1S_e$ and $1P_e$, respectively, plotted as a function of diameter for CdSe quantum dots. The $1S_e$ and $1P_e$ energies are calculated using the k·p approximation applied to a spherical quantum box and assuming an external potential of 8.9 eV, which is the value used by Norris and Bawendi to fit the interband absorption spectra [15]. With a delocalized hole in a $1S_{3/2}$ state that is initially created by band-edge photoexcitation, the electron–hole Coulomb contribution would increase the $1S_e$–$1P_e$ transition energy by about $0.05 - 0.1$ eV in the size range studied, as shown by the dotted line.

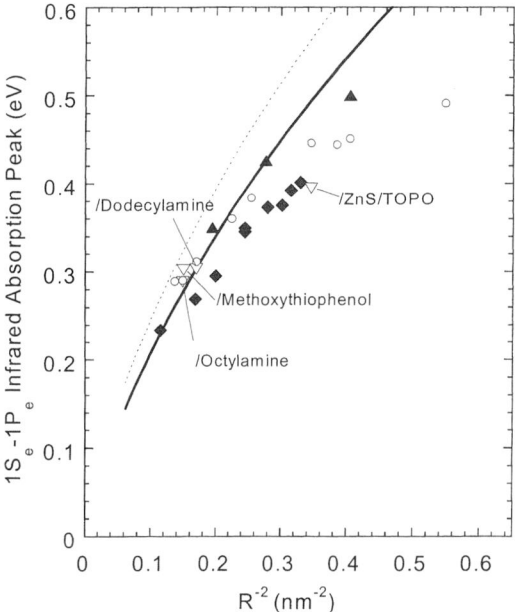

Figure 1 The $1S_e$–$1P_e$ transition energy as a function of the inverse radius squared for CdSe nanocrystals. The solid line is the result of calculations using the $k \cdot p$ model. The dashed line is the result of calculations that include Coulomb interactions with a delocalized hole [12]. The solid symbols are experimental results for photoexcited nanocrystals; solid triangles are measured with a laser (resolution 10 ps) [12] and solid diamonds are measured with a step-scan Fourier transform IR (FTIR) spectrometer (resolution ~40 ns) [13]. The open circles are results obtained for n-type, trioctyl phosphine oxide (TOPO)-capped nanocrystals, whereas the open triangles are results of measurements for n-type, surface-modified nanocrystals (the type of surface modification is indicated in the figure) [14].

As noted by Khurgin [16], the intraband transition has an oscillator strength that is the same order of magnitude as the value for the interband $1S_{3/2} - 1S_e$ transition. For a spherical box and using the parabolic approximation, the $1S_e-1P_e$ transition carries most of the oscillator strength of the electron in the $1S_e$ state (96%), with little left for transitions to higher P_e states. There should then be a large region of transparency between the IR intraband absorption and the band-edge absorption. In the parabolic approximation, the oscillator strength of the $1S_e-1P_e$ transition is proportional to $1/m^*$, where m^* is the effective electron mass. The magnitude of the oscillator strength is close to 10 for CdSe ($m^* \sim 0.12$), and it can be as high as 100 for other materials. In practice, as the quantum-dot size decreases, the $1S_e$ and $1P_e$ states occupy a higher-energy region of the conduction band and have a larger effective mass. Consequently, the $1S_e-1P_e$ oscillator strength becomes smaller, although it is still large by molecular standards.

III. EXPERIMENTAL OBSERVATIONS OF THE INTRABAND ABSORPTION IN COLLOIDAL QUANTUM DOTS

Intraband absorption in quantum dots was first observed in lithographically defined quantum dots in the far IR below the optical phonon frequency band [17]. For the more strongly confined epitaxial quantum dots and colloidal quantum dots, the intraband absorption lies in the mid-IR above the optical phonon bands. In epitaxial quantum dots, the intraband absorption was first detected by the infrared spectroscopy of n-doped materials, [18,19], whereas for the quantum dots grown using colloidal chemistry routes, the intraband spectra were initially recorded by IR probe spectroscopy after interband photoexcitation [12,20,21].

Figure 1 shows the peak position of the IR absorption for CdSe nanocrystals of various sizes measured under different experimental conditions. The experimental intraband spectra obtained using Fourier transform IR (FTIR) measurements on n-type colloidal CdSe nanocrystals are shown in Fig. 2 [14]. The peak position of the IR absorption is only weakly sensitive to the surface chemistry of the nanocrystals. This position is also not significantly dependent on whether the electrons are placed in the $1S_e$ by photoexcitation or by electron transfer. This result indicates that in the case of photoexcited nanocrystals, the electron–hole Coulomb interaction is weak or that on the fast timescale of these measurements (>6 ps), the hole has been localized. Time transients on the order of 1 ps in the mid-IR absorption region have been attributed to hole cooling dynamics [22,23]. It would be of interest to perform spectroscopic IR transient measurements of the intraband spectrum to monitor spectral shape changes on the subpicosecond timescale.

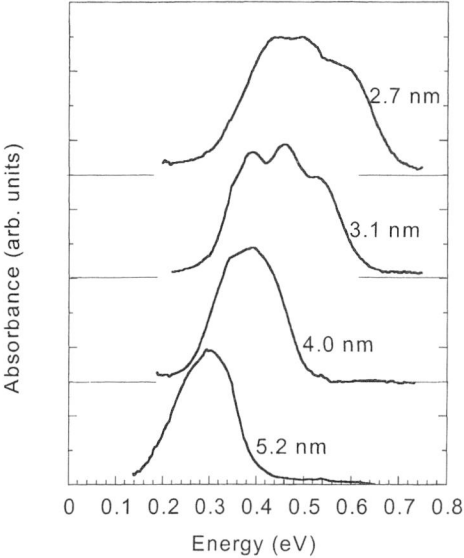

Figure 2 FTIR spectra of *n*-type CdSe nanocrystals with the indicated diameters. (From Ref. 14.)

Overall, the measured size dependence is in satisfactory agreement with the predictions of the $k \cdot p$ approximation, deviating more strongly at small sizes.

The large oscillator strength of the $1S_e$–$1P_e$ transition leads to strong optical changes upon photoexcitation of an electron–hole pair. Figure 3 shows the interband pump fluence dependence of the IR transmission for a sample of CdSe colloids. The intraband absorption cross sections derived from such plots agree within 30% with results of estimations [12]. In accord with expectations, the $1S_e$–$1P_e$ cross-section is similar to the interband cross-section at the band edge. This similarity is obvious in Figure 4, which shows the IR and visible spectral change upon electrochemical charge transfer in thin films (~0.5 µm) of CdSe nanocrystals [24]. In Fig. 4, the bleach of the first exciton peak at 2 eV is complete (ΔOD ~0.5), arising from the transfer of two electrons to each nanocrystal in the film, and the intraband absorbance at 0.27 eV is of the same magnitude (ΔOD ~0.8).

For CdSe samples synthesized by organometallic methods [25], the size dispersion, $\Delta R/R$, is typically 5–10%. Using a 10% size dispersion as a benchmark, noting that the $1S_e$–$1P_e$ transition energy scales at most as R^{-2}, and barring other broadening mechanisms, the overall IR inhomogenous linewidth [full width at half-maximum (FWHM)] should be less than ~23%

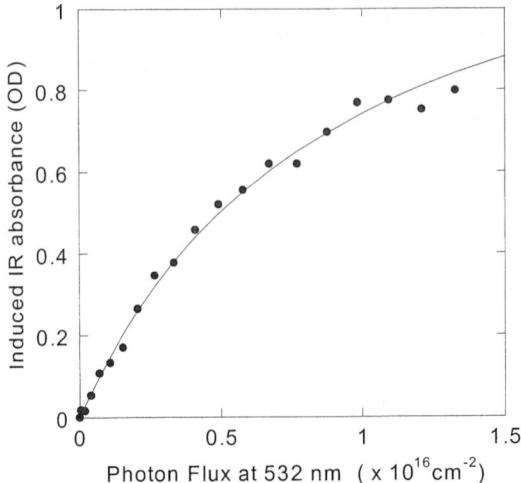

Figure 3 Induced IR absorbance of photoexcited CdSe nanocrystals of ~1.9 nm radius probed at 2.9 μm. The sample optical density (OD) at the 532-nm pump wavelength is 1.4. The temporal resolution is 10 ps.

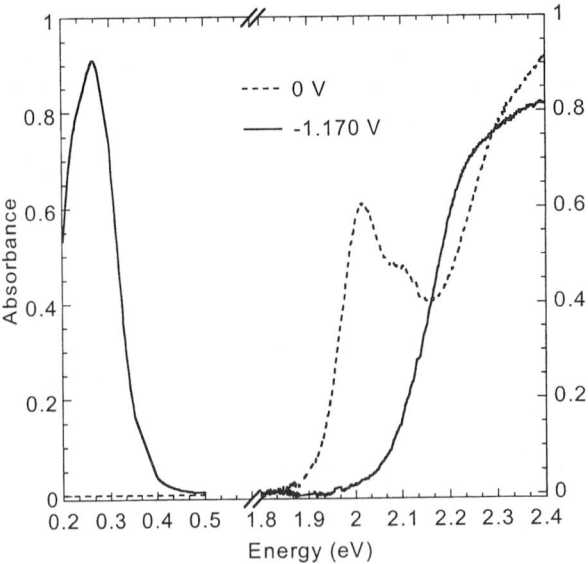

Figure 4 Electrochromic response of a thin (~0.5 μm) film of CdSe nanocrystals on a platinum electrode immersed in an electrolyte. The potential is referenced to Ag/AgCl. (From Ref. 24.)

of the center frequency. The experimental observations for CdSe nanocrystals are instead between 30% and 50% of the center frequency, and the linewidth increases as the particle size becomes smaller. In addition, as shown in Fig. 2, the spectra obtained for n-type nanocrystals show a multiple-peak structure. For future applications, it will be useful to identify and control the conditions necessary to achieve the narrowest intraband linewidths given a finite amount of size polydispersity. The most likely explanation for the large linewidth is the splitting of the $1P_e$ states. In one experimental observation, zinc-blende CdS nanocrystals exhibit a narrower linewidth than wurtzite CdSe or ZnO nanocrystals [14]. In order to narrow the linewidth, parameters such as nanocrystal shape and crystal symmetry can be investigated.

IV. INTRABAND ABSORPTION PROBING OF CARRIER DYNAMICS

Because the IR absorption is directly assigned to electrons with little contribution from holes, it is a convenient probe of the electron dynamics. This is an advantage of intraband spectroscopy over transient interband spectroscopy, because the latter yields signals that depend on both electron and hole dynamics. The combination of intraband spectroscopy with other techniques that probe the combined electron and hole response, such as interband transient spectroscopy, is a useful approach to analyaze the evolution of the exciton and a way to study which specific surface conditions affect the trapping processes and the fluorescence efficiency [26].

Figure 5 shows an example of the different time traces of the intraband absorption after the creation of an electron–hole pair in CdSe nanocrystals capped with various molecules. Whereas TOPO (trioctylphosphine oxide)-capped nanocrystals exhibit strong band-edge fluorescence, both thiophenol-capped and pyridine-capped nanocrystals have strongly reduced fluorescence. Intraband spectroscopy shows that the $1S$ electron relaxation dynamics differ dramatically depending on surface capping. For thiophenol, the electron in the $1S_e$ state is longer lived than for TOPO-capped samples, indicating that it is hole trapping that quenches fluorescence. The hole trap is presumably associated with a sulfur lone pair, which is stabilized by the conjugated ring. In contrast, for pyridine, which is also thought to be a hole trap because it is a strong electron donor, most of the excited electrons live only a short time in the $1S_e$ state. Therefore, it appears that pyridine strongly enhances electron trapping and/or fast (picosecond), nonradiative electron–hole recombination. Yet, there is a small percentage of nanocrystals (~5–10%) with a long-lived electron, and this must arise from a few of the nanocrystals undergoing complete charge separation. In fact, thiophenol-capped nano-

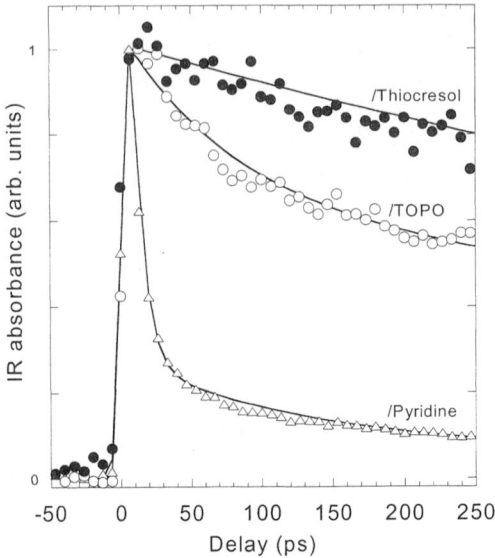

Figure 5 Transient IR absorbance at 2.9 μm for ~1.9-nm-radius CdSe nano-crystals with different surface passivations (indicated in the figure); the pump wavelength is 532 nm.

crystals [13] and pyridine-capped nanocrystals [27] exhibit remarkably long-lived electrons in the $1S_e$ state, in excess of 1 ms, for a small fraction of the photoexcited nanocrystals. It is certain that longer times are achievable with specifically designed charge-separating nanocrystals, which might then find applications, for example, in optical memory systems.

An important issue in quantum-dot research is the mechanism of linewidth broadening and energy relaxation. Indeed, electronic transitions are not purely delta functions because there is a coupling between the quantum-confined electronic states and other modes of excitations such as acoustic or optical phonons and surfaces states. The efficiency of these coupling processes affects the optical properties of the quantum dots. For band-edge laser action in quantum dots, it is beneficial to have a fast intraband relaxation down to the lasing states. For intraband lasers, we have the opposite situation, and one would benefit from slow intraband relaxation between the lasing states. Thus, intraband relaxation is a subject of great current interest.

Although lasers emitting at band-edge spectral energies have been made using quantum dots [28,29], the observed fast intraband relaxation is not well understood. In particular it is four to five orders of magnitude slower than the limit given by radiative relaxation (~100 ns). In quantum wells, one-optical-phonon relaxation processes provided an effective mechanism for electron energy relaxation because there is a continuum of electronic states. However, for strongly confined quantum dots in which the electronic energy separation is many times larger than the optical phonon energy, multiphonon process must be invoked, and these processes are expected to be too slow to explain the observed fast relaxation. This phenomenon is called the phonon bottle-neck [30,31]. Alternative explanations involve coupling to specific LO ± LA phonon combinations [32] or defect states [33] or to carrier-carrier scattering with electrons [34] or holes [35,36]. Intraband relaxation rates have been initially determined by fluorescence rise-time and transient interband spectroscopy. For the colloidal quantum dots, Klimov et al. were the first to observe the subpicosecond relaxation of transient bleach features attributed to the $1P_{3/2}$–$1P_e$ transition [37]. This very fast relaxation, still observed in the limit of a single electron–hole pair per nanocrystal, was explained by a fast Auger process involving scattering of the electron by the hole, as shown in the schematic in Fig. 6A. The electron–hole Auger relaxation is understood to be

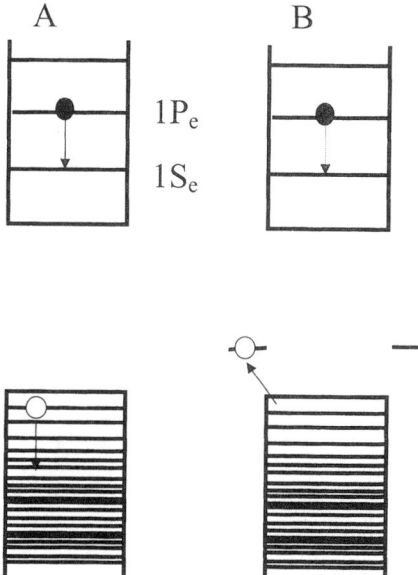

Figure 6 (A) Schematics of the electron–hole Auger-like relaxation process leading to fast electron intraband relaxation. (B) If the hole is captured by a surface trap, the electron–hole Auger coupling is reduced, leading to a slower intraband relaxation.

this fast because of the high density of hole states that are in resonance with the $1S_e–1P_e$ energy [36].

Two experiments using intraband spectroscopy have tested and strengthened this conclusion. In the first one, a photoexcited electron–hole pair in the nanocrystals is separated by providing surface hole traps that allow the degree of coupling of the electron and hole to be varied, according to Fig. 6B [38]. After a sufficient time delay, infrared pump-probe spectroscopy of the $1S_e–1P_e$ transition directly measured the intraband recovery rate, as shown in Fig. 7. CdSe–TOPO nanocrystals showed dominant fast (~1 ps) and weak long (~300 ps) components of the recovery rate. The fast component slowed down slightly to ~2 ps when the surface was capped by hole traps such as thiophenol or thiocresol and was much reduced in intensity for pyridine-capped nanocrystals. The interpretation of the data was that pyridine provides a hole trap that stabilizes the hole on the conjugated ring, strongly reducing the coupling to the electron, at least for the nanocrystals that escaped fast, nonradiative recombination. The slight reduction in the intra-band recovery rate for the thiol-capped nanocrystals is consistent with the thiol group localizing the hole on the surface via the sulfur lone pairs. The ubiquitous slow component is plausible evidence for the phonon bottleneck, but alternative interpretations of the bleach recovery data are possible (e.g., the slow component could be due to trapping from the $1P_e$ state rather than

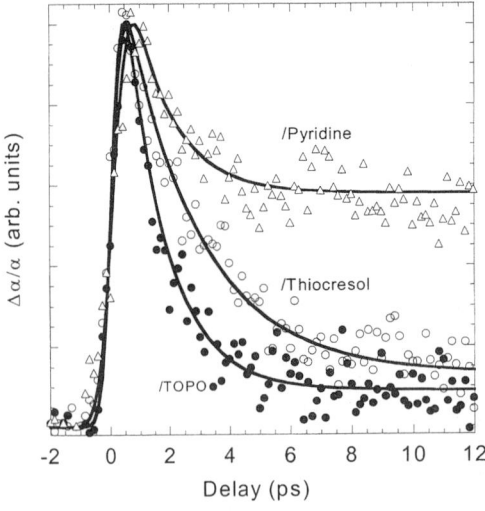

Figure 7 The recovery rate of the $1S_e–1P_e$ bleach after the intraband excitation of the photoexcited CdSe nanocrystals of 2.2-nm radius with different surface passivations (indicated in the figure). (From Ref. 38.)

from a long lifetime due to the phonon bottleneck). Further studies along these lines will require a better understanding of the electronic coupling of molecular surface ligands to the quantum states.

In another experiment by Klimov et al. [39], the strategy was similar except that the last pulse was an intraband probe of the $1S_{3/2}-1S_e$ and $1P_{3/2}-1P_e$ excitons, and the temporal resolution was ~0.3 ps. The samples compared were (CdSe)ZnS nanocrystals, which were expected to have no hole traps, and CdSe-pyridine, which were expected to have strong hole traps. The bleach time constants were 0.3 ps and 3 ps, respectively, which seems to confirm the role that the hole plays in the intraband relaxation. However, unlike the previous experiment, no slow component was observed. One possible explanation for the discrepancy is the smaller size of the nanocrystals in the second study, enhancing the role of the surface in the relaxation process.

There have been many more studies investigating the phonon bottleneck in epitaxial quantum dots. Although most reported a fast relaxation (<10 ps) [40,41] generally attributed to Auger-like processes, one study recently reported a slow (>100 ps) intraband relaxation for photoexcited InGaAs/GaAs quantum dots prepared with a single electron and no hole [42]. Furthermore, stimulated intraband emission has also been recently achieved [43,44]. Semiconductor colloidal quantum dots may also become efficient mid-IR "laser dyes" after one learns how to slow down the intraband relaxation, which will probably require better control of the surfaces states [33].

An attractive characteristic of quantum dots is their narrow spectral features. Narrow linewidths and long coherence times would be appealing in quantum logic operations using quantum dots [45]. On the other hand, at least for maximizing gain in laser applications, it is best if the overall linewidth is dominated by homogenous broadening [46,47]. Given that the methods to make quantum dots all lead to finite size dispersion, efforts to distinguish homogenous and inhomogeneous linewidths in samples have been pursued for more than a decade.

The first interband spectral hole burning [48–50] or photon-echo [51] measurements of colloidal quantum dots, performed at rather high powers and low repetition rates, yielded broad linewidths, typically >10 meV at low temperature. These results are now superseded by more recent low-intensity continuous-wave (CW) hole-burning [52] and accumulated photon-echo experiments, which have uncovered sub-meV homogeneous linewidths for the lowest interband absorption in various quantum-dot materials [53–55]. In parallel, single CdSe nanocrystal photoluminescence has yielded a linewidth of ~100 μeV for the emitting state [56]. For intraband optical applications, the linewidth of the $1S_e-1P_e$ state is of interest, but the broadening of the $1P_e$ state in the conduction band cannot be determined by interband spectroscopy because of the congestion of the hole states and hole dynamics that affect the linewidth broadening. Intraband hole burning or photon echo

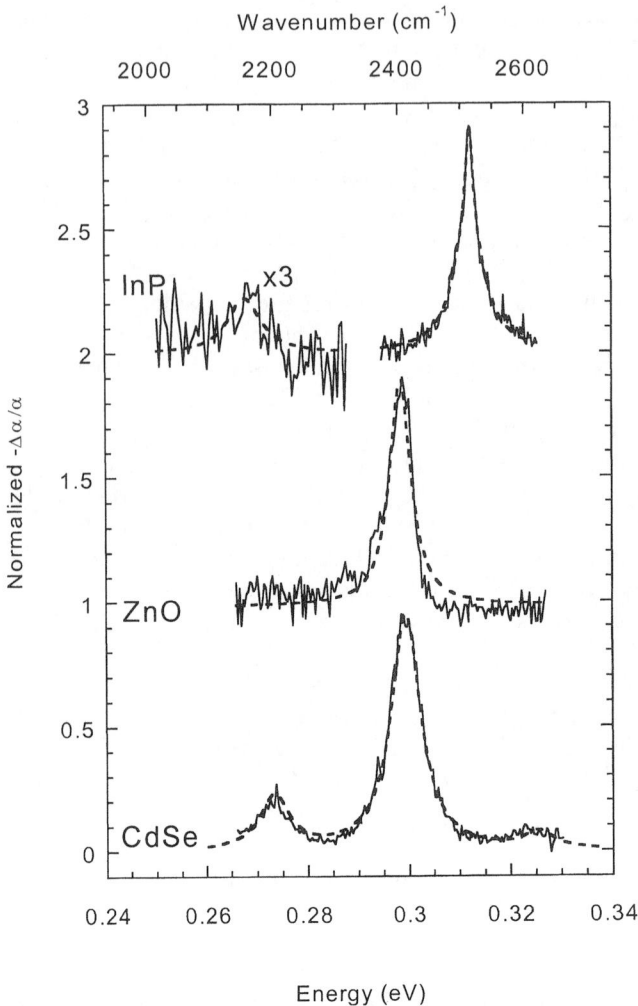

Figure 8 Spectral hole-burning results in the range of the $1S_e$–$1P_e$ transition for photoexcited CdSe, InP, and n-type ZnO nanocrystals at 10 K, demonstrating the narrow homogeneous linewidths (<2 meV) and weak electron–optical phonon coupling. (From Ref. 57.)

are two natural approaches to studying the broadening of intraband transitions. Figure 8 shows hole-burning spectra of CdSe, InP, and ZnO nanocrystal colloids at 10 K [57]. These spectra also exhibit the LO phonon replica. Their strength is in good agreement with the bulk electron–LO phonon coupling, with values of the Huang-Rhys factor of 0.2 for the spectrum of CdSe nanocrystals shown in Fig. 8. An interesting result is that for CdSe nanocrystals, the homogeneous width remains narrower than ~10 meV at 200 K for a transition at ~300 meV [57]. Compared to the overall inhomogeneous linewidth of ~ 100 meV, this value indicates that much progress is still possible in improving monodispersity of colloidal samples that should lead to a more precise control of intraband absorption features.

V. CONCLUSIONS

As emphasized throughout the other chapters of this book, semiconductor colloidal quantum dots hold an interesting place as chromophores. This chapter focuses on the mid-IR properties that arise due to their intraband transitions. Whereas organic molecules can be made with great control and complexity, the strong electron–electron interactions and large electron–vibration coupling of the conjugated carbon backbone might preclude the possibility of efficient organic chromophores in the mid-infrared. Compared to organic materials, inorganic semiconductor quantum dots have weaker electron–vibration coupling and weaker electron–electron interactions. As briefly reviewed in this chapter, the intraband transitions are spectrally well defined in the mid-infared, strong, size-tunable, and controllable by electron injection. The chemically synthesized quantum dots therefore have a unique appeal as mid-infrared "dyes" and have also a potential as nonlinear optical elements [58,59]. The improvement in the quality of nanocrystal materials, however, is still required to improve the control over the intraband transition energies and electronic relaxation pathways. Ultimately, the intraband response of the colloidal quantum dots may find widespread applications in mid-infrared technologies.

REFERENCES

1. West, L.C.; Eglash, S.J. Appl. Phys. Lett. 1985, *46*, 1156.
2. Levine, B.F.; Malik, R.J.; Walker, J.; Choi, K.K.; Bethea, C.G.; Kleinman, D.A.; Vandenberg, J.M. Appl. Phys. Lett. 1987, *50*, 273.
3. Rosencher, E.; Bois, P.; Nagle, J.; Costard, E.; Delaitre, S. Appl. Phys. Lett

1989, *55*, 1597. Fejer, M.M.; Yoo, S.J.B.; Byer, R.L.; Harwit, A.; Harris, J.S. Phys. Rev. Lett. 1989, *62*, 1041.

4. Faist, J., et al. Phys. Rev. Lett. 1996, *76*, 411. Appl. Phys. Lett. 1995, *67*, 3057.
5. Efros, A.L.; Rosen, M. Annu. Rev. Mater. Sci. 2000, *30*, 475.
6. Ekimov, A.I.; Hache, F.; Schanne-Klein, M.C.; Ricard, D.; Flytzanis, C.; Kudryatsev, I.A.; Yazeva, T.Y.; Rodina, A.V.; Efros, Al.L. J. Opt. Soc. Am. 1993, *10*, 100.
7. Banin, U.; Lee, C.J.; Guzelian, A.A.; Kadavanich, A.V.; Alivisatos, A.P.; Jaskolski, W.; Bryant, G.W.; Efros, A.L.; Rosen, M.J. Chem. Phys. 1998, *109*, 2306.
8. Kang, I.; Wise, F.W. J. Opt. Soc. Am. B 1997, *14*, 1632.
9. Sercel, P.; Efros, A.L.; Rosen, M. Phys. Rev. Lett. 1999, *83*, 2394.
10. Empedocles, S.A.; Norris, D.J.; Bawendi, M.G. Phys. Rev. Lett. 1996, *77*, 3873.
11. Blanton, S.A.; Leheny, R.L.; Hines, M.A.; Guyot-Sionnest, P. Phys. Rev. Lett. 1997, *79*, 865. Shim, M.; Guyot-Sionnest, P. J. Chem. Phys. 1999, *111*, 6855.
12. Guyot-Sionnest, P.; Hines, M.A. Appl. Phys. Lett. 1998, *72*, 686.
13. Shim, M.; Shilov, S.V.; Braiman, M.S.; Guyot-Sionnest, P. J. Chem. Phys. B 2000, *104*, 1494.
14. Shim, M.; Guyot-Sionnest, P. Nature 2000, *407*, 981.
15. Norris, D.J.; Bawendi, M.G. Phys. Rev. B 1996, *53*, 16338.
16. Khurgin, J. Appl. Phys. Lett. 1993, *62*, 1390.
17. Heitmann, D.; Kotthaus, J.P. Phys. Today 1993, *46*, 56.
18. Sauvage, S.; Boucaud, P.; Julien, F.H.; Gerard, J.M.; Thierry-Mieg, V. Appl. Phys. Lett. 1997, *71*, 2785.
19. Sauvage, S.; Boucaud, P.; Julien, F.H.; Gerard, J.M.; Marzin, J.Y. J. Appl. Phys. 1997, *82*, 3396.
20. Shum, K.; Wang, W.B.; Alfano, R.R.; Jones, K.J. Phys. Rev. Lett. 1992, *68*, 3904.
21. Mimura, Y.; Edamatsu, K.; Itoh, T. J. Luminescence 1996, *66–67*, 401.
22. Klimov, V.I.; Schwarz, Ch.J.; McBranch, D.W.; Leatherdale, C.A.; Bawendi, M.G. Phys. Rev. B 1999, *60*, R2177.
23. Klimov, V.I.; McBranch, D.W.; Laetherdale, C.A.; Bawendi, M.G. Phys. Rev. B 1999, *60*, 13740.
24. Wang, C.; Shim, M.; Guyot-Sionnest, P. submitted.
25. Murray, C.B.; Norris, D.J.; Bawendi, M.G. J. Am. Chem. Soc. 1993, *115*, 8706.
26. Klimov, V.I. J. Phys. Chem. B 2000, *104*, 6112.
27. Ginger, D.S.; Dhoot, A.S.; Finlayson, C.E.; Greenham, NC. Appl. Phys. Lett. 2000, *77*, 2816.
28. Kirstaeder, N.; Ledentsov, N.N.; Grundmann, M.; Bimberg, D.; Ustinov, V.M.; Ruminov, S.S.; Maximov, M.V.; Kop'ev, P.K.; Alferov, Zh.I.; Richter, U.; Werner, P.; Goesele, U.; Heydenreich, J. Electron. Lett. 1994, *30*, 1416.
29. Klimov, V.I.; Mikhailovsky, A.A.; Xu, S.; Malko, A.; Hollingsworth, J.A.; Laetherdale, C.A.; Eisler, H.J.; Bawendi, M.G. Science 2000, *290*, 314.
30. Bockelmann, U.; Bastard, G. Phys. Rev. B 1990, *42*, 8947.
31. Benisty, H.; Sotomayor-Torres, C.M.; Weisbuch, C. Phys. Rev. B 1991, *44*, 10,945.

32. Inoshita, T.; Sakaki, H. Physica B 1996, *227*, 373. Phys. Rev. B 1992, *46*, 7260.
33. Sercel, P.C. Phys. Rev. B 1995, *51*, 14,532. Schroeter, D.F.; Griffiths, D.J.; Sercel, P.C. Phys. Rev. B 1996, *54*, 1486.
34. Bockelmann, U.; Egeler, T. Phys. Rev. B 1992, *46*, 15,574.
35. Vurgaftman, I.; Singh, J. Appl. Phys. Lett. 1994, *64*, 232.
36. Efros, A.L.; Kharchenko, V.A.; Rosen, M. Solid State Commun. 1995, *93*, 281.
37. Klimov, V.I.; McBranch, D.W. Phys. Rev. Lett. 1998, *80*, 4028.
38. Guyot-Sionnest, P.; Shim, M.; Matranga, C.; Hines, M.A. Phys. Rev. B 1999, *60*, 2181.
39. Klimov, V.I.; Mikhailovsky, A.A.; McBranch, D.W.; Laetherdale, C.A.; Bawendi, M.G. Phys. Rev. B 2000, *61*, R13,349.
40. Sauvage, S.; Boucaud, P.; Glotin, F.; Prazers, R.; Ortega, J.M.; Lemaitre, A.; Gerard, J.M.; Thierry-Mieg, V. Appl. Phys. Lett. 1998, *73*, 3818.
41. Brasken, M.; Lindberg, M.; Sopanen, M.; Lipsanen, H.; Tulkki, L. Phys. Rev. B 1998, *58*, R15,993.
42. Urayama, J.; Nrris, T.B.; Singh, J.; Bhattacharya, P. Phys. Rev. Lett. 2001, *86*, 4930.
43. Krishna, S.; Bhattacharya, P.; McCann, P.J.; Namjou, K. Electron. Lett. 2000, *36*, 1550.
44. Krishna, S.; Battacharya, P.; Singh, J.; Norris, T.; Urayama, J.; McCann, P.J.; Namjou, K. IEEE J. Quantum Electron. 2001, *37*, 1066.
45. Chen, G.; Bonadeo, N.H.; Steel, D.G.; Gammon, D.; Katzer, D.S.; Park, D.; Sham, L.J. Science 1998, *289*, 1473.
46. Vahala, K.J. IEEE. J. Quantum Electron. 1991, *24*, 523.
47. Sugarawa, M.; Mukai, K.; Nakata, Y.; Ishikawa, H.; Sakamoto, A. Phys. Rev. B 2000, *61*, 7595.
48. Alivisatos, A.P.; Harris, A.L.; Levinos, N.J.; Steigerwald, M.L.; Brus, L.E. J. Chem. Phys. 1988, *89*, 4001.
49. Woggon, U.; Gaponenko, S.; Langbein, W.; Uhrig, A.; Klingshirn, C. Phys. Rev. B 1993, *47*, 3684.
50. Norris, D.J.; Sacra, A.; Murray, C.B.; Bawendi, M.G. Phys. Rev. Lett. 1994, *72*, 2612.
51. Mittleman, D.M.; Schoenlein, R.W.; Shiang, J.J.; Colvin, V.L.; Alivisatos, A.P.; Shank, C.V. Phys. Rev. B 1994, *49*, 14,435.
52. Palingis, P.; Wang, H. Appl. Phys. Lett. 2001, *78*, 1541.
53. Kuribayashi, R.; Inoue, K.; Sakoda, K.; Tsekihomskii, V.A.; Baranov, A.V. Phys. Rev. B 1998, *57*, R15,084.
54. Ikezawa, M.; Masumoto, Y. Phys. Rev. B 2000, *61*, 12,662.
55. Takemoto, K.; Hyun, B.-R.; Masumoto, Y. J. Lumin. 2000, *87–89*, 485.
56. Empedocles, S.A.; Norris, D.J.; Bawendi, M.G. Phys. Rev. Lett. 1996, *77*, 3873.
57. Shim, M. PhD thesis, The University of Chicago, 2001.
58. Sauvage, S.; Boucaud, P.; Glotin, F.; Prazers, R.; Ortega, M.; Lemaitre, A.; Gerard, M.; Thierry-Mieg, V. Phys. Rev. B 1999, *59*, 9830.
59. Brunhes, T.; Boucaud, P.; Sauvage, S.; Glotin, F.; Prazeres, R.; Ortega, J.M.; Lemaitre, A.; Gerard, J.M. Appl. Phys. Lett. 1999, *75*, 835.

5

Charge Carrier Dynamics and Optical Gain in Nanocrystal Quantum Dots: From Fundamental Photophysics to Quantum-Dot Lasing

Victor I. Klimov
Los Alamos National Laboratory, Los Alamos, New Mexico, U.S.A.

I. INTRODUCTION

Semiconductor materials are widely used in both optically and electrically pumped lasers. The use of semiconductor quantum-well (QW) structures as optical gain media has resulted in important advances in laser technology. QWs have a two-dimensional (2D), steplike density of electronic states that is nonzero at the band edge, enabling a higher concentration of carriers to contribute to the band-edge emission and leading to a reduced lasing threshold, improved temperature stability, and a narrower emission line. A further enhancement in the density of the band-edge states and an associated reduction in the lasing threshold is, in principle, possible using quantum wires and quantum dots (QDs) in which the confinement is in two and three dimensions, respectively. In very small dots, the spacing of the electronic states is much greater than the available thermal energy (strong confinement), inhibiting thermal depopulation of the lowest electronic states. This effect should result in a lasing threshold that is temperature insensitive at an excitation level of only one electron-hole (e–h) pair per dot on average [1,2]. Additionally, QDs in the strong confinement regime have an emission wavelength that is a pronounced function of size, adding the advantage of continuous spectral tunability over a wide energy range simply by changing the size of the dots.

Quantum-dot lasing was initially realized in an optically pumped device using relatively large (~10 nm) CdSe nanoparticles fabricated via high-temperature precipitation in glass matrices [3]. Later, this effect was observed for epitaxially grown nanoislands (epitaxial or self-assembled QDs) using both optical and electrical (injection) pumping [4,5]. As predicted, lasers based on epitaxial QDs show an enhanced performance in comparison with, for example, QW lasers and feature reduced thresholds, improved temperature stability, and high differential optical gain (important for achieving high modulation rates).

The success of the laser technology based on epitaxial QDs has been a strong motivation force for the development of laser devices based on ultrasmall, sub-10-nm chemically synthesized nanoparticles. Such nanoparticles, known also as colloidal or nanocrystal QDs (NQDs) can be routinely generated with narrow size dispersions (~5%) using organometallic reactions [6–8]. In the sub-10-nm size range, electronic interlevel spacings can exceed hundreds of millielectron volts and size-controlled spectral tunability over an energy range as wide as 1 eV can be achieved. Furthermore, improved schemes for surface passivation by, for example, overcoating NQDs with a shell of a wide-gap inorganic semiconductor [9] allow significant suppression of surface trapping and produce room-temperature photoluminescence (PL) quantum efficiencies greater than 50%. Additionally, due to their chemical flexibility, NQDs can be easily prepared as close-packed films [10] (NQD solids) or incorporated with high densities into polymers or sol–gel glasses. [11,12]. NQDs are, therefore, compatible with existing fiber-optic technologies and are useful as building blocks for the bottom-up assembly of various optical devices, including optical amplifiers and lasers.

Despite a decade of research that provided some indication of optical gain performance [13], strongly confined NQDs had failed to yield lasing in numerous efforts. Difficulties in achieving lasing have often been attributed to high nonradiative carrier losses due to trapping at surface defects, a direct consequence of the large surface-to-volume ratio characteristic of sub-10-nm particles. Another concern raised in several theoretical articles is the strongly reduced efficiency of electron–phonon interactions in the case of discrete, atomiclike energy structures, which are characteristic of small-size dots [14,15]. For discrete spectra, the availability of pairs of electronic states satisfying energy conservation in phonon-assisted processes is drastically reduced compared to the quasicontinuous spectra of bulk materials. This deficiency has been expected to significantly lower the efficiency of carrier cooling due to phonon emission (the effect known as a "phonon bottleneck"), leading to reduced carrier flows into the lowest "emitting" states and, hence, reduced PL efficiencies. However, the difficulties anticipated due to carrier surface trapping and the "phone bottleneck" turned out to be much less important compared to such largely unforeseen problems as nonradiative, multiparticle Auger recombination [16] and interference from photoinduced absorption (PA) due to carriers trapped at NQD interfaces [17].

In this chapter, we analyzed the underlying physics of processes relevant to optical amplification and lasing in strongly confined NQDs with a focus on sub-10-nm CdSe colloidal nanoparticles. Specifically, we discuss the issues of intraband carrier relaxation, multiparticle interactions, and photoinduced absorption in the context of NQD optical-gain properties. We also analyze the effect of NQD sample parameters, such as NQD size, surface passivation quality, matrix–solvent identity, and NQD densities on optical-gain performance. Furthermore, we illustrate lasing action in NQDs using several experimental examples in which different cavity configurations (e.g., microring and distributed feedback) were utilized.

II. ENERGY STRUCTURES AND INTRABAND RELAXATION IN NQDs

A. Band-Edge Optical Transitions

This chapter concentrates on sub-10-nm CdSe NQDs that can be fabricated, for example, by organometallic, colloidal methods described in Chapter 1. In bulk CdSe, the exciton Bohr radius is approximately 5 nm. Therefore, NQDs with sub-10-nm sizes (mean radii, $R < 5$ nm) correspond to the regime of strong quantum confinement for which electronic spectra are atomiclike and consist of well-separated energy states. Electronic structures and optical transitions in CdSe NQDs are analyzed in great detail in Chapter 2. Here, we only provide a brief description of the band-edge optical transitions, which is essential for understanding light-emitting and optical-gain properties of CdSe nanoparticles.

Within the effective mass approximation, NQD electron and hole wave functions can be presented as a product of periodic Bloch functions and envelope wave functions [18]. The Bloch function describes carrier motion in a periodic potential of a crystalline lattice, whereas the envelope function describes the motion in the NQD confinement potential. In the case of spherical NQDs, conduction-band states can be characterized by two quantum numbers: l, the angular momentum of the envelope wave function, and n, the number of the state within a series of states of the same symmetry. States with $l = 0, 1, 2 \ldots$ are labeled as $S, P, D, \ldots$ states. In CdSe NQDs, the three lowest electron states are $1S$, $1P$, and $1D$.

Because of the complexity of the valence-band structures, a description of hole states in CdSe NQDs requires the use of an additional quantum number, F, which is the total hole angular momentum (i.e., a sum of the Bloch function and envelope function momenta) [19]. In the notation for the hole states, the F number is usually shown as a subscript. For example, the notation for the lowest hole state is $1S_{3/2}$. The hole states are $(2F+1)$-fold degenerate. This degeneracy is lifted if the effects of the crystal field in a hexagonal lattice [20], NQD nonspherical shape [21], and e–h-exchange interactions [22,23] are taken into account, leading to a "fine-structure" splitting of hole levels [23].

Band-edge optical properties of NQDs are dominated by transitions involving the lowest electron $1S$ state and fine-structure states derived from the $1S_{3/2}$ hole level (Fig. 1a). The split-off $1S_{3/2}$ hole states form two groups of closely spaced levels separated by the gap that can be as large as tens of millielectron volts in small-size dots. The high-energy manifold of hole states is coupled to the $1S$ electron state by a strong optical transition observed in absorption spectra as the lowest $1S$ absorption maximum. The low-energy hole states are coupled to the $1S(e)$ state by a much weaker transition that gives rise to a band-edge PL band. The large, NQD size-dependent gap between the high- and low-energy manifolds of hole states (manifolds of "absorbing" and "emitting" states, respectively) is observed in optical spectra as a large Stokes shift between the lowest $1S$ absorption maximum and a band-edge PL detected under nonresonant excitation (Fig. 1b). This shift is often referred to as the "global" Stokes shift, in contrast to the "resonant" Stokes shift [24] observed under quasiresonant excitation in fluorescence line-narrowing experiments (see Chap. 2).

B. Electron Intraband Relaxation

In the case of both the optical and electrical pumping of semiconductor gain media, nonequilibrium charge carriers are usually injected with energies that are greater than the material's energy gap. Therefore, intraband energy relaxation, leading to the population buildup of the lowest "emitting" transition, is an important process in the sequence of events leading to light emission and, ultimately, to lasing. To ensure efficient pumping of electron/hole states involved in the "emitting" transition, energy relaxation in both conduction and valence bands has to be more efficient than carrier recombination due to both radiative and nonradiative processes.

In bulk II–VI semiconductors, carrier energy relaxation is dominated by the Fröhlich interactions with longitudinal optical (LO) phonons that lead to fast (typically subpicosecond) carrier cooling dynamics [25–27]. In QDs, even in the regime of weak confinement when the level spacing is only a few millielectron volts, the carrier relaxation mediated by interactions with phonons is hindered dramatically because of restrictions imposed by energy and momentum conservation, leading to a phenomenon called a "phonon

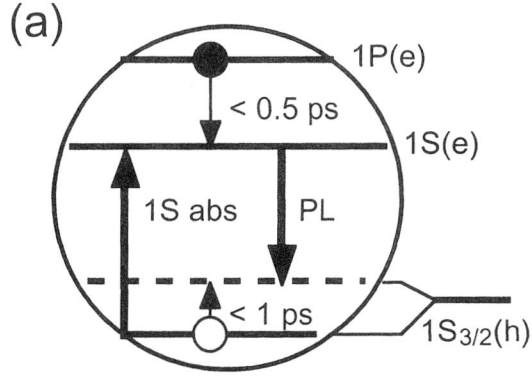

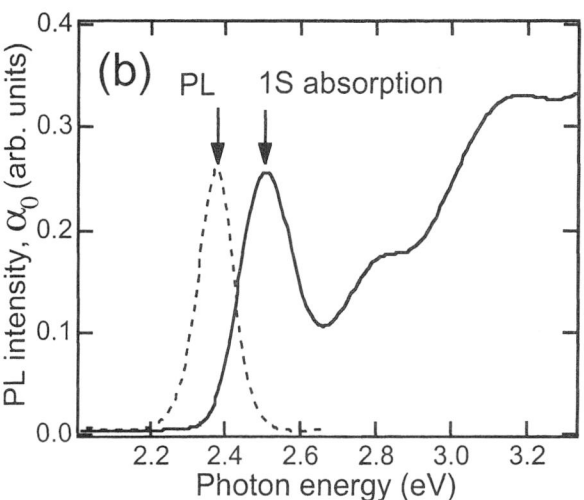

Figure 1 (a) Schematic of "absorbing" (labeled as "1S abs") and "emitting" (labeled as "PL") transitions in CdSe NQDs along with intraband relaxation processes leading to a population buildup of the "emitting" transition (these processes are discussed in Sects. II.A and II.B). (b) Absorption and PL spectra of CdSe NQDs with $R = 1.2$ nm ($T = 300$ K) illustrating a large, "global" Stokes shift between the $1S$ absorption peak and the PL maximum.

bottleneck" [14,28]. Further reduction in the energy loss rate is expected in the regime of strong confinement, for which the level spacing can be much greater than LO phonon energies and, hence, carrier–phonon scattering can only occur via weak multiphonon processes.

In II–VI NQDs, the energy separation between electron states is much greater than the separation between hole states; therefore, the "phonon bottleneck" was expected to affect electron intraband dynamics more strongly compared to hole intraband dynamics. Electron intraband dynamics in CdSe NQDs were extensively studied using an ultrafast transient absorption (TA), pump-probe technique [29–31]. TA is a nonlinear optical method which allows for the monitoring (in both time and spectral domains) of absorption changes caused by sample photoexcitation. In the ultrafast TA experiment, nonequilibrium charge carriers are rapidly injected into a material with a short, typically subpicosecond pump pulse. Absorption changes ($\Delta\alpha$) associated with photogenerated carriers are monitored with a second short probe pulse which can be derived from either a tunable, monochromatic (e.g., optical parametric amplifier), or broadband [e.g., femtosecond (fs) white-light continuum] light source.

One mechanism for carrier-induced absorption changes in NQDs is state filling that leads to the bleaching of optical transitions involving electronic states occupied by nonequilibrium carriers (a detailed analysis of mechanisms for resonant optical nonlinearities can be found in Refs. 32 and 33). Because of a high spectral density of valence-band states, TA bleaching signals are dominated by the filling of *electron* quantized levels. Therefore, by monitoring the dynamics of transitions associated, for example, with 1S and 1P electron states, one can directly evaluate the rate of the 1P-to-1S intraband relaxation.

An example of such dynamics recorded for CdSe NQDs with $R = 4.1$ nm at the positions of the $1S(e)$–$1S_{3/2}(h)$ and $1P(e)$–$1P_{3/2}(h)$ transitions (these transitions are referenced later as 1S and 1P, respectively) is displayed in Fig. 2a. These dynamics indicate a fast population decay (~500 fs time constant) of the 1P state, which is complementary to the growth of the 1S state population. The observed fast relaxation is despite the very large 1S–1P energy separation of about eight LO phonon energies. Interestingly, intraband relaxation becomes even faster with decreasing NQD size, as evident from a comparison of the 1S state population dynamics for NQDs of different radii shown in Fig. 2b. These data indicate a decrease in the 1S buildup time [$\tau_b(1S)$] with decreasing NQD radius: $\tau_b(1S)$ is 530 fs for $R = 4.1$ nm and shortens down to 120 fs for $R = 1.7$ nm, roughly following a linear size dependence.

Extremely fast electron relaxation, as well as a confinement-induced enhancement in the relaxation process, clearly indicate that energy relaxation

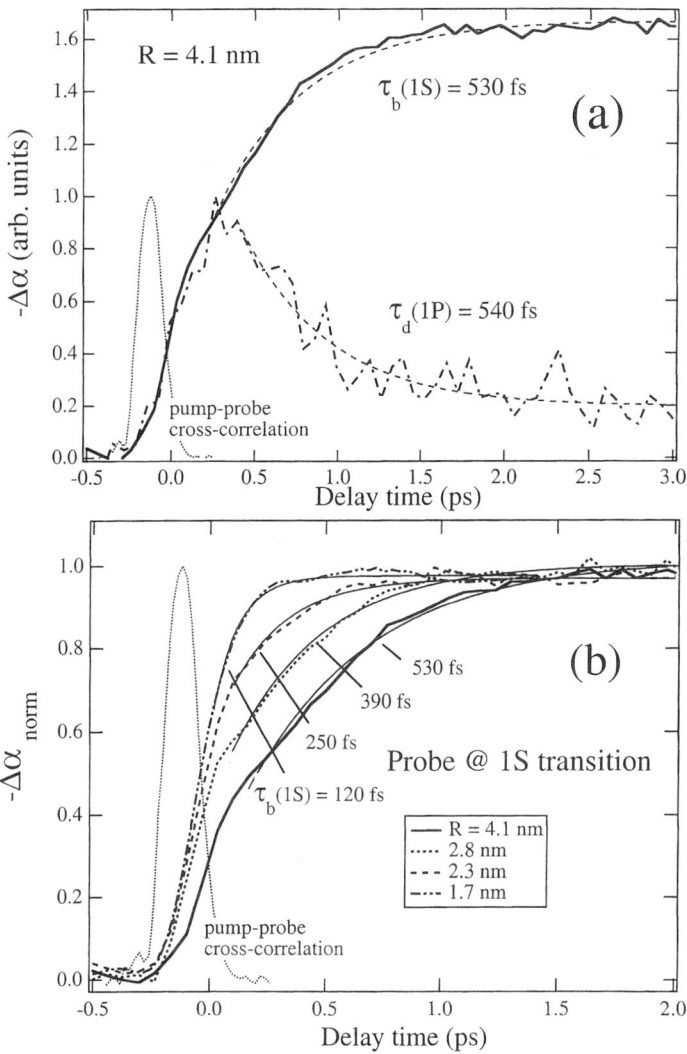

Figure 2 (a) Transient absorption dynamics recorded for CdSe NQDs (R = 4.1 nm) at the positions of the $1S$ (solid line) and $1P$ (dashed-dotted line) transitions which illustrate ultrafast, subpicosecond $1P$-to-$1S$ electron relaxation. Dashed lines are fits to a single-exponential growth/decay. The dotted line is the pump-probe cross-correlation. (b) The $1S$ bleaching buildup dynamics in CdSe NQDs with radii 1.7, 2.3, 2.8, and 4.1 nm (thick lines), fit to a single-exponential growth (thin solid lines). The thin dotted line is the pump-probe cross-correlation.

in NQDs is dominated by *nonphonon* energy-loss mechanisms. Recent works have suggested that coupling to defects [34], Auger interactions with carriers outside the NQD [35] or Auger-type e–h energy transfer [36] can lead to fast energy relaxation not limited by a "phonon bottleneck." The first two mechanisms are not intrinsic to NQDs and cannot explain relaxation data for colloidal samples [37]. These data indicate that in colloidal nanoparticles, energy relaxation does not show any significant dependence on NQD surface properties (i.e., the number of surface defects) and remains almost identical for different liquid- and solid-state matrices, including transparent optically passive glasses, polymers, and organic solvents, for which no carriers are generated outside the NQD.

The energy-loss mechanism proposed in Ref. 36 involves transfer of the excess energy of an electron to a hole, with subsequently fast hole relaxation through its dense spectrum of states. This mechanism is based on the intrinsic Auger-type e–h interactions and leads to significantly faster relaxation times than those for the multiphonon emission.

Most direct studies of the role of e–h interactions in intraband relaxation have used CdSe NQDs in which e–h coupling (separation) was controlled by the surface ligand [30,31]. In the presence of hole-accepting capping groups (pyridine in Refs. 30 and 31), the e–h coupling is strong immediately after photoexcitation (holes are inside the dot) but is reduced dramatically after hole transfer to pyridine (hole transfer time $\tau_T = 450$ fs). Therefore, the rate of electron relaxation should be strongly different before and after hole transfer to the capping group if this relaxation is indeed due to e–h coupling.

In order to monitor electron intraband dynamics at different stages of hole relaxation/transfer, one can use a three-pulse (pump–postpump–probe), femtosecond, TA experiment [38] schematically shown in Fig. 3. In this experiment, the sample is excited by a sequent of two ultrashort pulses [one in the visible and another in the infrared (IR) spectral range] and is probed by broadband pulses of a femtosecond white-light continuum. The visible interband pump is used to create an e–h pair in the NQD, whereas a time-delayed, intraband IR pump is used to resonantly reexcite an electron from the $1S$ to the $1P$ state. Performing reexcitation at different stages of hole transfer (e.g., before and after the transfer event) and monitoring its relaxation back to the ground state, it is possible to directly evaluate the role of e–h coupling on electron relaxation rates.

Applying the above experiment to a ZnS-overcoated control sample, one does not see any difference in electron dynamics for different reexcitation times (Fig. 4a), which is consistent with the fact that a confinement potential created by the ZnS layer inhibits e–h charge separation. In sharp contrast, electron dynamics in the pyridine-capped NQDs (Fig. 4b) show a strong dependence on the delay between visible and IR pulses (Δt_{IR}). At $\Delta t_{IR} = 70$ fs

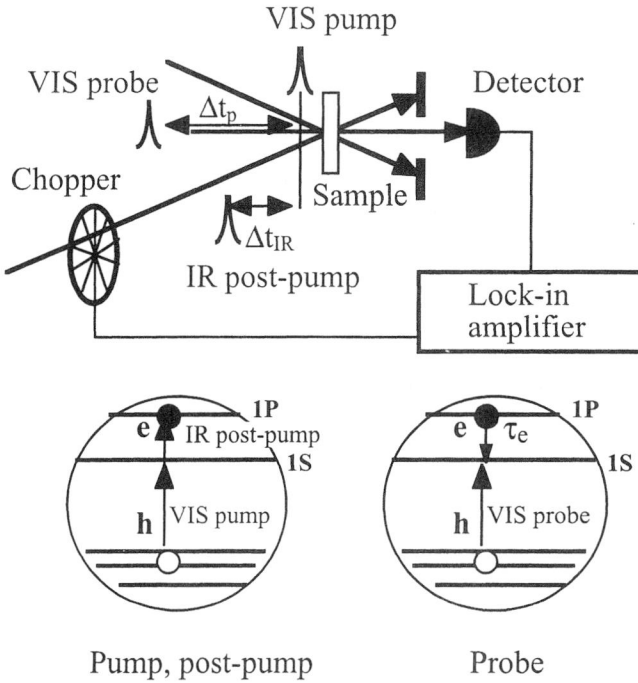

Figure 3 Schematic of a three-pulse (pump–postpump–probe) experiment and the excitation/relaxation processes monitored in it. An e–h pair is created via an interband excitation with a pump pulse in the visible spectral range. The electron is reexcited within the conduction band by an IR postpump pulse. The electron relaxation back to the ground state is monitored with a third pulse, probing interband absorption changes.

(diamonds in Fig. 4b), the relaxation constant is 250 fs, which is close to the value measured for ZnS-capped NQDs. With increasing reexcitation time, the electron relaxation time gradually increases up to 3 ps at $\Delta t_{IR} = 430$ fs (triangles in Fig. 4b). A further increase in Δt_{IR} does not lead to significant changes in electron intraband dynamics [compare traces taken at $\Delta t_{IR} = 430$ fs (triangles) and 600 fs (squares); Fig. 4b]. Importantly, the threshold delay of roughly 400 fs found in these experiments is very close to the hole-transfer time to a capping molecule inferred from two-pulse TA measurements [31]. This observation indicates that dramatic changes in electron relaxation result from hole transfer to the surface pyridine group, strongly suggesting that electron relaxation is dominated by e–h interactions (Auger-type e–h energy transfer).

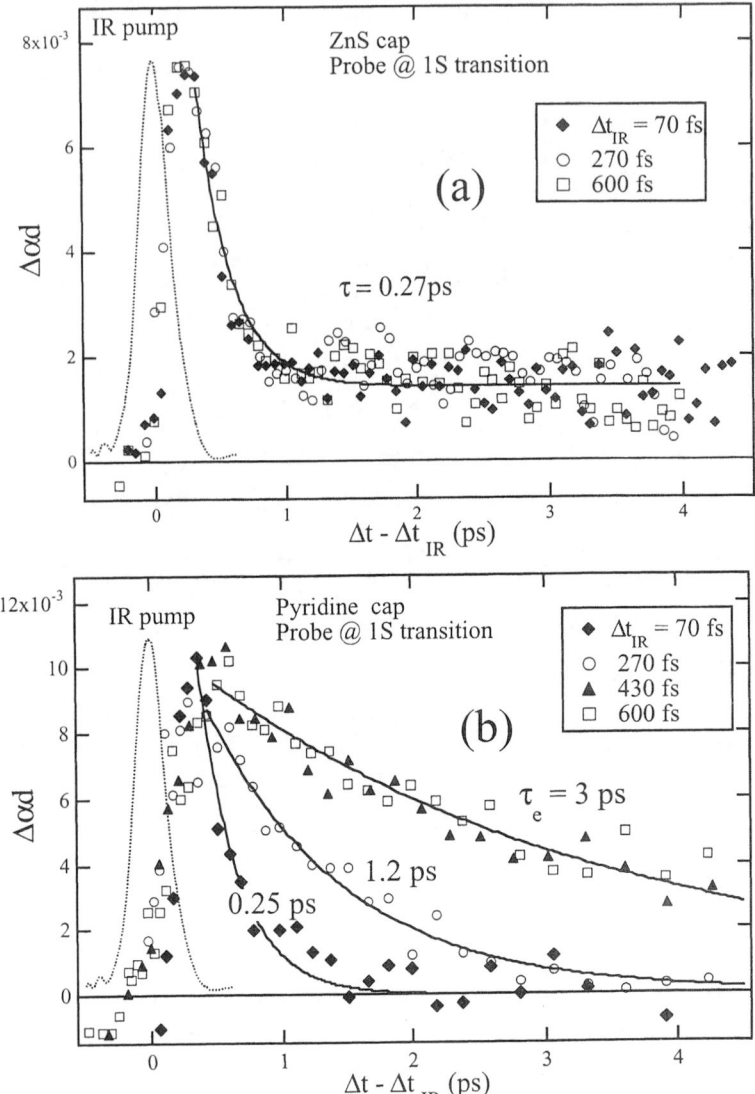

Figure 4 Dynamics of the IR postpump-induced 1S bleaching changes (electron intraband relaxation) detected at different delay times (Δt_{IR}) between the visible pump and IR postpump pulses for ZnS-capped (a) and pyridine-capped (b) CdSe NQDs (R = 1.15 nm) (symbols are experimental data points; lines are single-exponential fits).

C. Hole Intraband Relaxation

As analyzed in the previous subsection, energy transfer from electrons to holes dominates over other (e.g., phonon-related) mechanisms of intraband energy loses in the conduction band. However, because of a small interlevel spacing in the valence band (particularly, well above the energy gap), phonon emission by holes can provide an efficient mechanism for energy dissipation [39].

Experimental studies of hole intraband dynamics are complicated by difficulties in detecting valence-band populations. Most carrier intraband relaxation studies in NQDs have been performed using femtosecond TA experiments [29,30,40]. However, holes are only weakly pronounced in TA spectra because of a high spectral density of valence-band states resulting in a significant spread of hole populations over multiple levels. One of the approaches to detecting hole intraband relaxation is by monitoring the ultrafast dynamics of "hot" PL [39]. In contrast to state-filling-induced TA signals that are proportional to the sum of the electron and hole occupation numbers, PL is proportional to the product of these numbers. Therefore, electron and hole dynamics can, in principle, be decoupled by performing both PL and TA studies.

Experimental studies of initial PL relaxation require methods that can provide subpicosecond time resolution. Such resolution is possible using a femtosecond up-conversion experiment [41]. In this experiment, a sample emission is frequency mixed with an ultrashort (typically, femtosecond) gating pulse in a nonlinear optical crystal. The sum-frequency signal, spectrally selected with a monochromator, is proportional to an instant PL intensity at the moment defined by an arrival time of the gating pulse. Therefore, by scanning the gating pulse with respect to the PL pulse, one can obtain information about a PL temporal profile. Alternatively, by fixing the arrival time of the gating pulse and scanning a monochromator (and simulteneously adjusting a phase-matching angle in the nonlinear crystal), one can obtain time-resolved, "instant" emission spectra.

Figure 5(a) shows an example of "hot" PL spectra recorded for CdSe dots with a 1.8-nm mean radius. The pump photon energy used in these measurements (3 eV) is close to the energy of the $1S(e)$–$3S_{3/2}(h)$ transition (~3.2 eV). Therefore, for a significant number of dots in the sample, electrons are generated directly in the lowest $1S$ state, whereas photogenerated holes are very "hot" and carry a large excess energy of ~0.9 eV. "Hot" hole dynamics are well pronounced in time-resolved PL spectra (Fig. 5a) and indicate a gradual relaxation of holes from high-energy states (manifested, e.g., as a "hot" PL peak at ~2.8 eV) to the lowest "emitting" state giving rise to the band-edge PL band.

A descent of holes through the ladder of the valence-band states is easily seen in PL time transient recorded at different spectral energies (Fig. 5b).

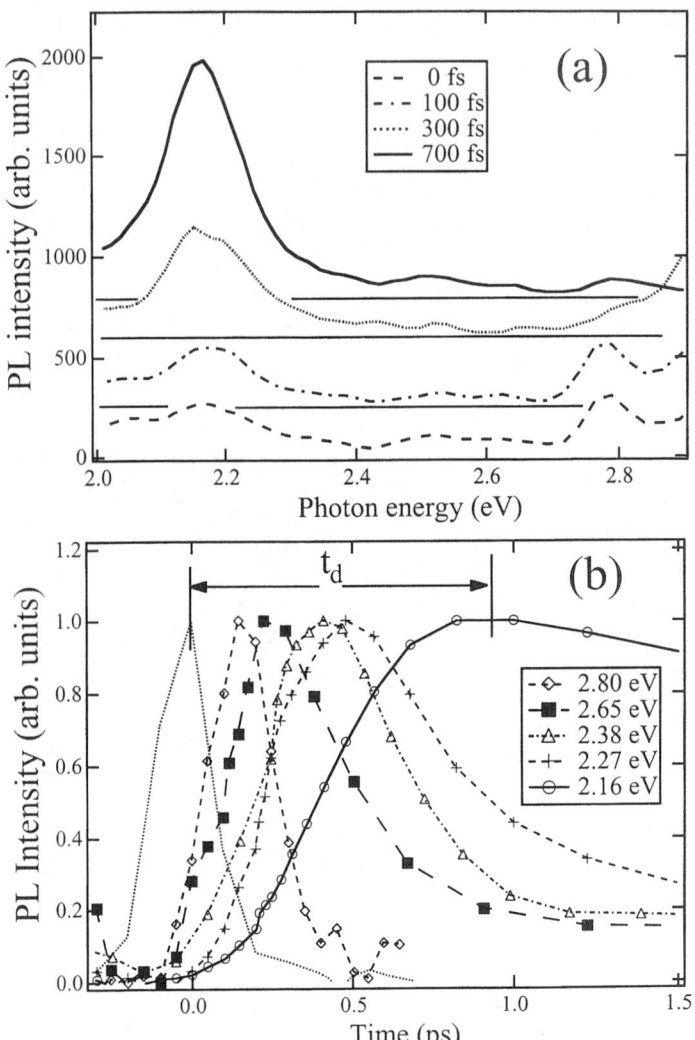

Figure 5 (a) Time-resolved PL spectra of CdSe NQDs ($R = 1.8$ nm) detected at 0, 100, 300, and 700 fs after excitation (spectra arbitrarily offset vertically for clarity). (b) Dynamics of "hot" PL detected at different spectral energies (symbols) along with a pump-pulse autocorrelation (dotted line).

These transients show a progressive increase in the delay (t_d) of the PL maximum with a reduction in the detection energy (E) as indicated by the plot in Fig. 6a (solid circles). The derivative, dE/dt_d, provides a measure of the hole energy-loss rate. For 1.8-nm dots, the energy relaxation rate is nearly constant (1.5 eV/ps) between 2.3 and 3.1 eV, but reduces dramatically (down to ~0.26 eV/ps) below the 1S absorption peak [i.e., at the stage of final hole relaxation between "absorbing" and "emitting" fine-structure states (see schematics of hole energy levels in Fig. 1a]. Two stages in hole relaxation are also observed for dots with radii 1.2 and 3 nm (Fig. 6a, open squares and open triangles, respectively). The spectral onset of the "slow" relaxation region is size dependent, closely following the position of the 1S absorption peak. The "fast" and "slow" relaxation rates were in the range 1.3–1.8 eV/ps and 0.19–0.3 eV/ps, respectively.

In Fig. 6b, we compare size-dependent relaxation rates observed for hole and electrons. In contrast to electron relaxation rates that increase with decreasing dot size, hole rates show an opposite trend. In this case, both the "fast" and "slow" rates decrease as the dot radius is decreased, indicating a relaxation mechanism that is different from the Auger-type energy transfer responsible for electron relaxation. The relaxation data obtained for samples with different surface passivations and in different solvents indicate that hole energy-loss rates are independent of NQD surface/interface properties, suggesting that hole relaxation does not result from coupling to surface defects or solvent molecules but rather is due to some intrinsic mechanisms such as coupling to NQD lattice vibrations (phonons).

The observed "fast" hole energy relaxation rates are close to those estimated for hole-LO phonon interactions in bulk CdSe (~1.4 eV/ps) [26], suggesting a "bulk"-like relaxation process or, more specifically, a cascade of single-phonon emission acts (one phonon per relaxation step). The fact that phonon emission by "hot" holes is apparently not hindered by the discrete character of the energy levels implies that the valence-band states form a very dense spectrum (*quasicontinuum*) at spectral energies above the 1S absorption peak. Large effective hole masses, the existence of three valence subbands strongly intermixed by quantum confinement [19], and the fine-structure splitting of valence band states [23] are all factors that can lead to a high density of hole states [the effect of fine-structure splitting is qualitatively illustrated in Fig. 7 taking into account S- and P-type hole states; an even denser spectrum is expected if the states of other symmetries (D, F, etc.) are also taken into account]. The hole energy structures are further smeared out by broadening due, for examples, to dephasing induced by elastic carrier–phonon scattering [42]. Another factor that simplifies the process of meeting energy conservation requirements in NQDs is the relaxation of momentum

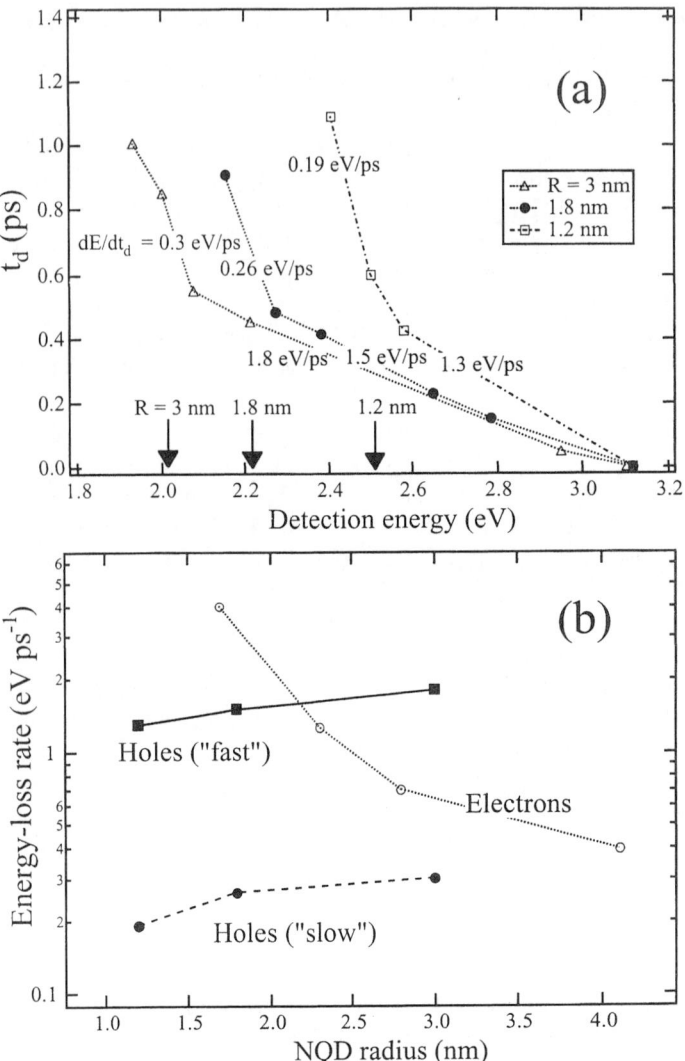

Figure 6 (a) The time delay of the "hot" PL maximum with respect to a pump pulse plotted as a function of the detection energy for CdSe NQDs of three different radii: 1.2, 1.8, and 3 nm. The energy-loss rates (dE/dt_d) derived from the plotted data at "fast" and "slow" stages of hole relaxation are indicated in the figure. The spectral onset of "slow" relaxation is compared with the position of the 1S absorption resonance (marked by arrows). (b) Energy-loss rates as a function of NQD radius for "fast" (solid squares) and "slow" (solid circles) stages of hole relaxation and for the 1P-to-1S electron relaxation (open circles).

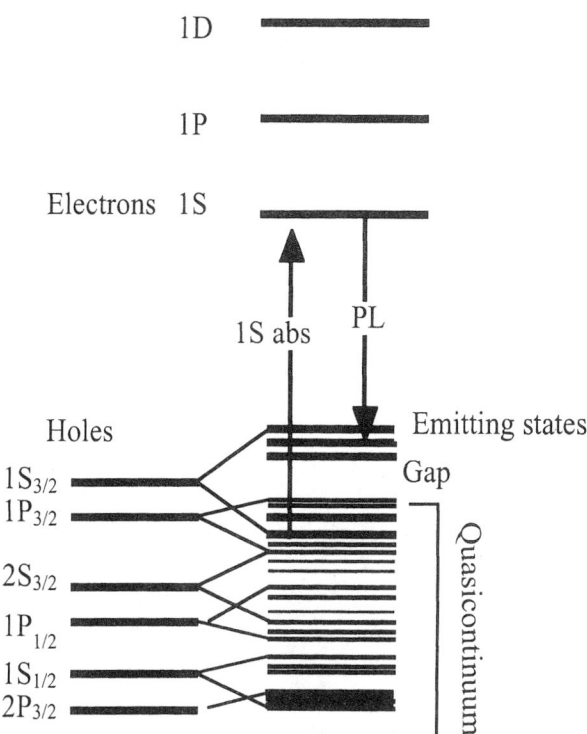

Figure 7 Schematic illustration of the formation of the valence-band quasicontinuum separated by an energy gap from the lowest "emitting" hole states. Arrows show the transitions observed in optical spectra as the $1S$ absorption peak and the band-edge PL.

conservation, which allows for a wider spread of k vectors (i.e., phonon energies) in phonon-assisted processes.

The reduction in the energy-loss rate in the final stage of hole relaxation (solid circles in Fig. 6b) can be explained by the fact that the hole spectrum becomes sparser at the band-edge energies (i.e., at energies close to the $1S$ absorption resonance). In the case of large interlevel separations, two or more phonons are required to satisfy the energy conservation during hole relaxation, leading to a reduced relaxation rate ("phonon bottleneck"). The fact that the onset for reduced rates occurs at energies close to the $1S$ absorption resonance implies that the "absorbing" and "emitting" valence-band manifolds are separated by a relatively wide energy gap, as discussed in Section II.A. Thus, the "phonon bottleneck," which is bypassed in the conduction band due to Auger-type e–h interactions, may still significantly affect hole dynamics, particularly in the final stage of hole relaxation between fine-structure, band-edge states.

However, despite being significantly slower than at high spectral energies, hole dynamics in the final stage of relaxation between "absorbing" and "emitting" states still remain on the sub-pisosecond timescale. Even for very small NQDs ($R = 1.2$ nm) for which the splitting of band-edge exciton states is particularly large, the "absorbing" transition shows a 700-fs decay that is complementary to the growth of the PL at the position of the "emitting" transition (Fig. 8).

Ultrafast population dynamics experimentally observed for both the lowest (i.e., "emitting") electron and hole quantized states indicate that despite the wide separation of electronic states, energy relaxation in strongly

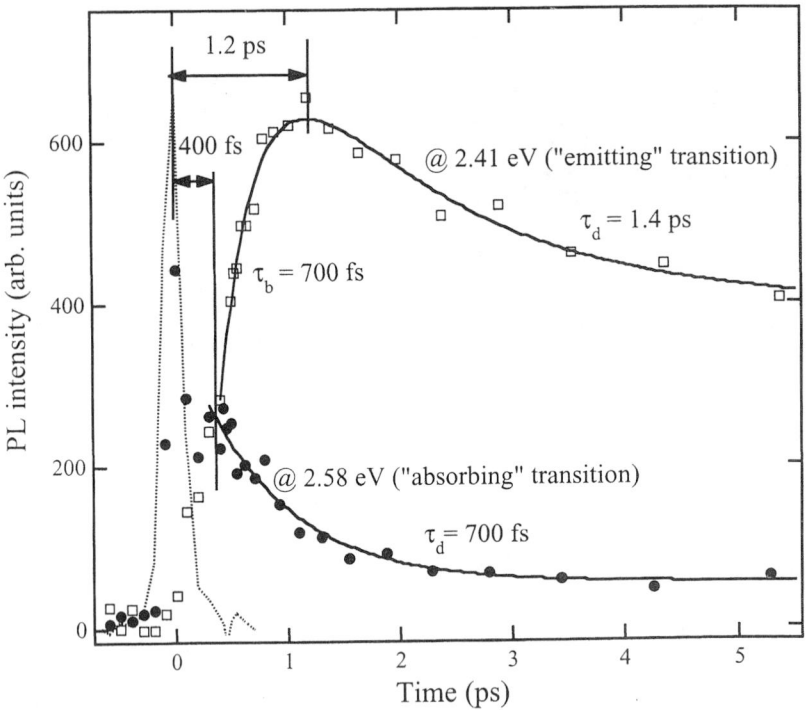

Figure 8 Complementary PL dynamics detected at the positions of the "absorbing" (solid circles) and "emitting" (open squares) transitions for 1.2-nm CdSe NQDs fit to a single- and a double-exponential function, respectively (dotted line is a pump-pulse autocorrelation).

confined NQDs is not significantly slower than in bulk materials. These data further imply that the optical-gain buildup in NQDs is not inhibited by the efficiency of intraband energy losses.

III. CARRIER TRAPPING AT INTERFACE STATES AND EXCITED-STATE ABSORPTION IN NQDs

A. Dynamics of Electron and Hole Surface Trapping

An important concern associated with light-emitting applications of NQDs is a high probability of carrier trapping at surface defects. For samples with incomplete surface passivation, surface trapping can lead to depopulation of NQD quantized states on picosecond timescales leading to significantly reduced band-edge PL efficiencies [37]. We briefly review some results of ultrafast studies on electron and hole surface trapping. These studies were performed in the low-intensity excitation regime (less than one e–h pair per dot on average was excited) for which the role of fast multiparticle Auger recombination (see Sec. IV. B) was insignificant.

Electron trapping dynamics have been studied using primarily a femtosecond TA experiment [29,37]. At the stage following energy relaxation, bleaching of the lowest ($1S$) absorption peak is almost entirely due to the population of the $1S$ electron state. The contribution of holes into this bleaching is not significant because the hole state involved in the $1S$ transition originates from a high-energy manifold of "absorbing" states that are not occupied after energy relaxation is finished.

The effect of the quality of surface passivation on electron dynamics is illustrated in Fig. 9, in which we compare the $1S$ bleaching decay for three samples with distinctly different surface properties [37]. The samples are freshly prepared NQDs passivated with organic molecules of trioctylphosphine oxide (TOPO) and trioctylphosphine (TOP) (TOPO-capped NQDs) (circles), the same NQDs but 8 months after preparation (squares), and freshly prepared NQDs with ZnS overcoating (core–shell structures) (crosses). As indicated by the PL quantum yield measurements, the quality of the "electronic" passivation of surface traps is progressively increased in going from the aged to the fresh TOPO-capped sample and then to the sample overcoated with ZnS. For all samples, the $1S$ bleaching decay shows two distinct regions: an initial region of fast sub-100-ps relaxation, which is followed by a slower nanosecond decay. The amplitude of the fast component is sensitive to the NQD surface passivation, indicating that it is likely due to surface-related electron relaxation. In freshly prepared TOPO-capped NQDs, the fast component is approximately 15% of the signal amplitude. Degradation in the surface passivation leads to an enhancement of this component up

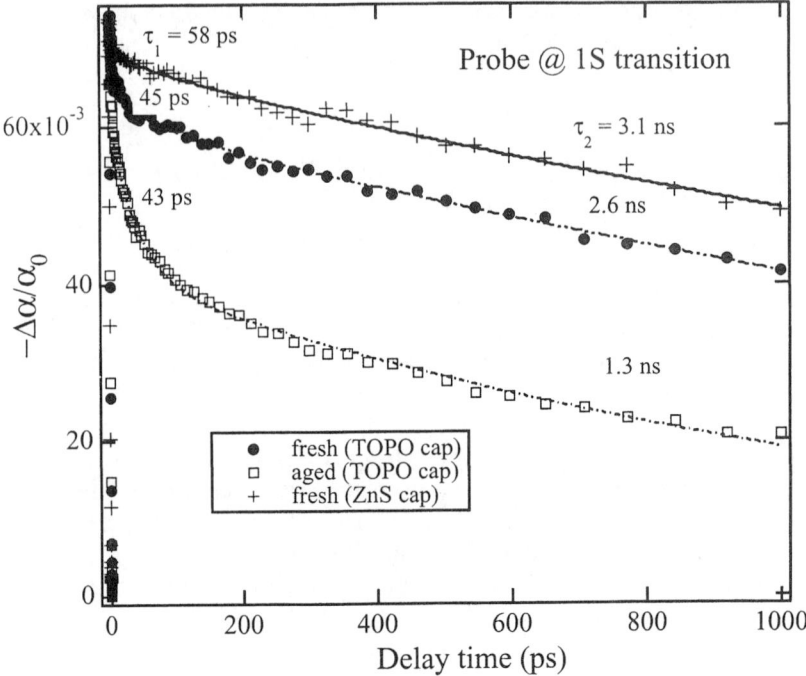

Figure 9 Dynamics of the 1S bleaching (2.26 eV) in 1.7-nm CdSe NQDs with different surface properties: fresh (solid circles) and aged (open squares) TOPO-capped NQDs, and NQDs overcoated with a ZnS layer (crosses). Lines are fits to a double-exponential decay.

to ~50% in the aged sample. On the other hand, overcoating with ZnS, which improves passivation of dangling bonds, leads to the reduction of the fast component down to ~7%.

Because valence-band populations are not well pronounced in TA signals detected using probe pulses resonant with interband (i.e., valence-to-conduction band) transitions, hole dynamics have been studied using either a TA experiment with intraband probe pulses [43] or time-resolved PL [39]. In the case of intraband TA measurements, one can emphasize a hole contribution to detected signals by tuning the energy of the probe photon in resonance with a selected intraband hole transition [38]. The sensitivity of PL dynamics to hole trapping is due to the fact that removal of a hole from the valence-band "emitting" state completely eliminates band-edge PL. However, because the removal of an electron from the 1S states has the same effect, extraction of hole dynamics from time-resolved PL traces may require independent measurements of electron relaxation using, for examples, a TA experiment [39].

In Fig. 10, we compare TA and PL dynamics for ZnS-overcoated CdSe NQDs with R = 1.2 nm. Both traces show initial fast picosecond decay followed by much slower nanosecond relaxation. The magnitude of the fast TA component, which is likely due to electron surface trapping (discussed earlier), ~25% of the signal amplitude, and its relaxation time is ~10 ps. The fast component in the PL trace is significantly greater in amplitude (~70%) and shows much faster initial decay (~2 ps). This difference in PL and TA temporal behavior is typical for CdSe NQDs and is even more pronounced in the case of TOPO-passivated samples, which show a greater ratio between the fast PL and TA components than ZnS-overcoated samples. The difference in TA and PL dynamics indicates that temporal evolution of PL signals is dominated not by electrons but by holes. The comparison of initial decay constants for TA and PL shows that hole trapping is significantly faster than electron trapping. Additionally, the fact that the fast PL component is

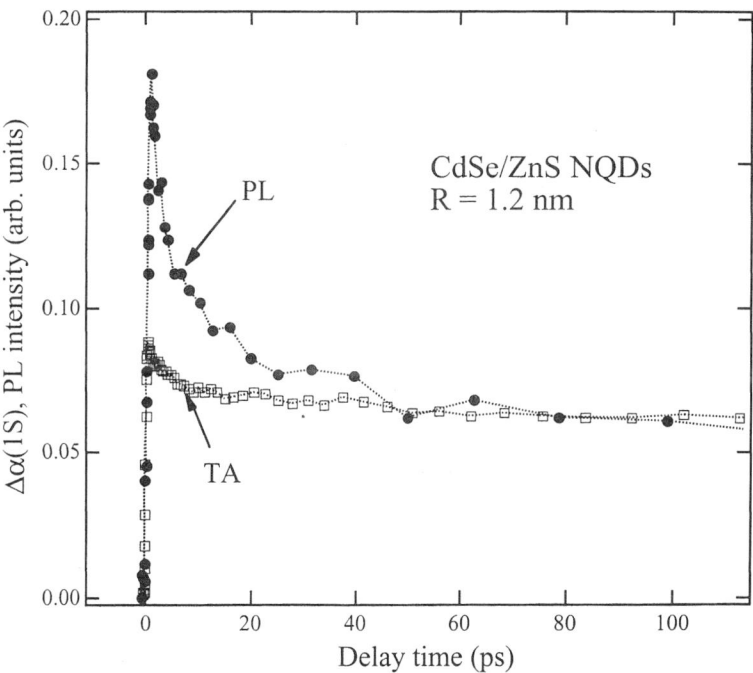

Figure 10 Comparison of $1S$ bleaching (squares) and PL (circles) dynamics recorded for ZnS-overcoated CdSe NQDs of 1.2-nm radius.

significantly greater than the fast TA component indicates that the dominant mechanism leading to reduced band-edge PL efficiencies in samples with incomplete surface passivation is hole trapping.

Carrier trapping at surface defects represents the most important mechanism for nonradiative carrier losses in the regime of low excitation densities. However, recently developed methods for surface treatment using both inorganic (e.g., core–shell structures [9]) and organic [8] capping allow significant suppression of this effect. These new methods provide routine fabrication of CdSe NQDs with PL quantum yields greater than ~80% at room temperature, strongly suggesting that yields near 100% are feasible.

B. Competition Between Optical Gain and Photoinduced Absorption

In addition to causing nonradiative carrier losses, surface trapping can deteriorate the lasing performance of NQDs because of excited-state absorption [i.e., photoinduced absorption (PA)] associated with carriers trapped at NQD interfaces. [17,44] As discussed next, PA is strongly sensitive to both NQD surface properties and the identity of the matrix material. Excited-state absorption, for example, develops in such commonly used solvents as hexane and toluene, in which it can completely suppress optical gain, the effect particularly pronounced in dots of small sizes.

The competition between optical gain and PA was studied in Ref. 44 using a femtosecond TA experiment. This experiment allows one to monitor the absorption of the sample with (α) and without (α_0) a pump. In absorption spectra, optical gain corresponds to $\alpha < 0$ [i.e., to pump-induced absorption bleaching ($\Delta \alpha = \alpha - \alpha_0 < 0$) that is greater than α_0 ($-\Delta\alpha/\alpha_0 > 1$)]. PA corresponds to $\Delta\alpha > 0$ (i.e., to the situation for which the absorption of the excited sample is greater than its ground-state absorption).

In ultrafast, pump-dependent studies of NQDs, it is convenient to characterize carrier densities in terms of NQD average populations, $\langle N \rangle$ (i.e., in terms of the number of e–h pairs per dot averaged over an ensemble). The initial NQD average population, $\langle N \rangle_0$, generated by a short pump pulse can be calculated using the expression $\langle N \rangle_0 = \sigma_a(\hbar\omega_p)J_p$, where J_p is the per-pulse pump fluence measured in photons/cm^2 and $\sigma_a(\hbar\omega_p)$ is the NQD absorption cross section at the pump spectral energy, $\hbar\omega_p$ [33]. For the case of optical excitation well above the NQD energy gap, the NQD absorption cross section can be estimated from

$$\sigma_a(\hbar\omega) = \frac{4\pi}{3} R^3 \frac{n_b}{n} |f(\hbar\omega)|^2 \alpha_b(\hbar\omega) \tag{1}$$

where α_b and n_b are the absorption coefficient and the refractive index of the bulk semiconductor, respectively, n is the index of the NQD sample, and f is the local-field factor that accounts for the difference in the field inside and outside the nanoparticle [45].

In this subsection, we briefly discuss the results of the pump-dependent TA studies of CdSe NQDs that focus on the effect of the NQD sizes and the solvent/matrix material on the strength of excited-state absorption [44]. Figure 11a displays absorption spectra of a hexane solution of small CdSe NQDs with $R = 1.2$ nm recorded at 2 ps after excitation for progressively higher pump intensities (the pump photon energy is 3.1 eV). These spectra do not show any evidence of gain ($\alpha < 0$), even at the highest pump density $\langle N_0 \rangle = 6$. Instead, the $1S$ absorption bleaching saturates slightly below a level $|\Delta\alpha|\alpha_0 \approx 1$ [i.e., right before a crossover to optical gain (Fig. 11b, squares)]. Such behavior indicates that the $1S$ transition is bleached by only one type of carrier (electrons), consistent with femtosecond PL data from Section II.C, indicating very fast hole relaxation from the "absorbing" (responsible for the $1S$ absorption) to the lower-energy "emitting" (involved in the PL transition) fine-structure state.

However, for this NQD size, optical gain is not detected at the position of the "emitting" transition either. In the region of this transition, the sample shows increased absorption (i.e., PA). In contrast to $1S$ bleaching, which saturates at high pump intensities, PA does not show saturation (circles in Fig. 11b) and, therefore, cannot be circumvented by simply increasing the excitation density.

Despite the fact that PA is a general feature of hexane solutions of CdSe NQDs, its relative contribution to band-edge TA signals is reduced for larger NQD sizes. This effect is illustrated in Fig. 12a, which displays pump-dependent TA spectra for TOPO-capped NQDs of 3.5-nm radius. These spectra clearly show the development of negative absorption (i.e., optical gain) at the position of the "emitting" transition. As indicated by the TA pump dependence in Fig. 12b (open squares), optical gain develops between one and two e–h pairs per dot on average and at $\langle N \rangle_0 \approx 17$, it reaches its maximum. Gain decreases at higher excitation densities, which is due to the increase in the relative contribution from excited-state absorption. At very high excitation powers ($\langle N \rangle_0 > 100$) the band-edge nonlinear optical response is dominated by PA.

Interestingly, the gain magnitude [$g = -(\alpha_0 + \Delta\alpha)$] at the maximum of the pump dependence exceeds the magnitude of the ground-state absorption ($g_{max} \approx 1.1\alpha_0$) [i.e., greater than the value expected for a complete population inversion (α_0)]. This discrepancy indicates a more complex gain mechanism in NQDs than simple state filling in a two-level system, as discussed in Sect. IV.A.

Figure 12b compares pump-dependent TA signals recorded at the position of the emitting transition for NQDs of different radii. These data clearly illustrate the competition between optical gain and PA as a function of

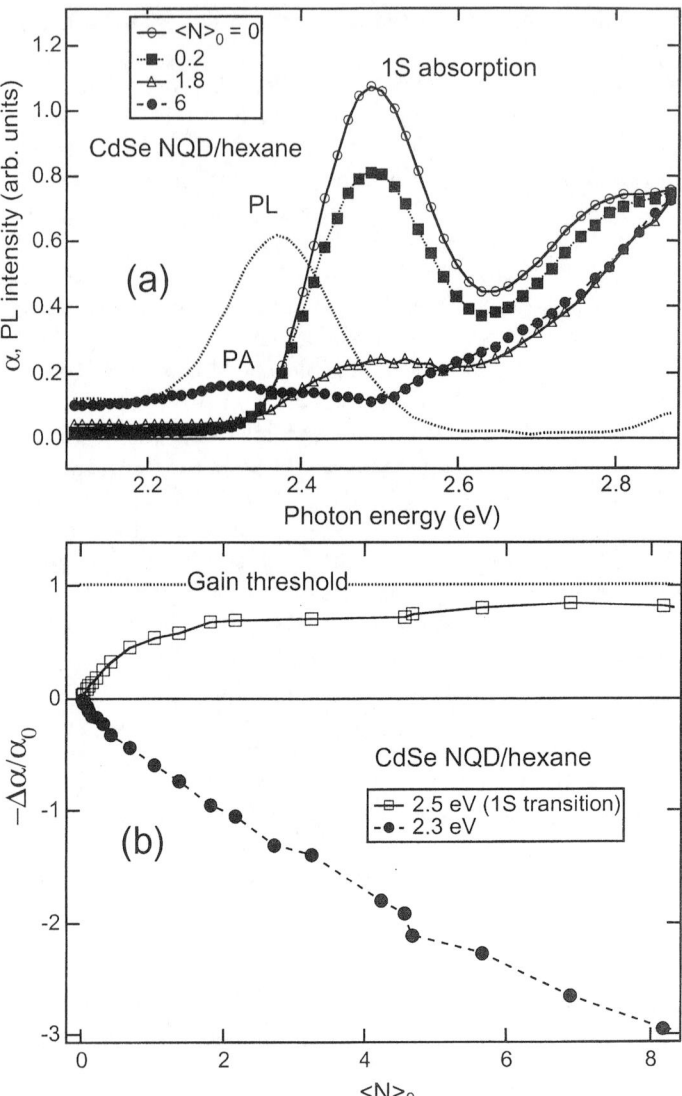

Figure 11 (a) Pump-intensity-dependent nonlinear absorption spectra (symbols) of 1.2-nm CdSe NQDs in hexane that show development of a strong PA band over-lapping with a PL spectrum (dotted line). (b) Pump-intensity dependence of nor-malized absorption changes at the positions of the 1S bleaching (squares) and the PL band (circles).

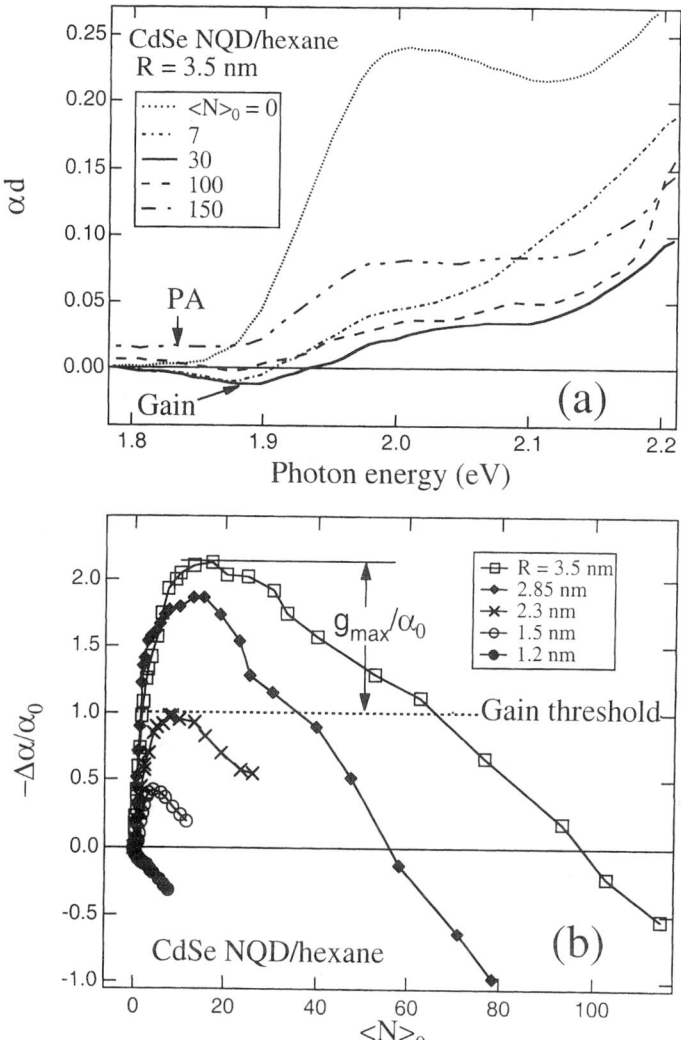

Figure 12 (a) Pump-intensity-dependent nonlinear absorption spectra of 3.5-nm CdSe NQDs in hexane that show the development of gain ($\alpha < 0$) followed by the transition to PA ($\alpha > \alpha_0$) with increasing pump level. (b) Pump dependence of normalized absorption changes in the spectral region of optical gain (or "potential" gain) as a function of NQD radii.

NQD size. For small NQD sizes (R = 1.2 and 1.5 nm), excited-state absorption overwhelms bleaching before the crossover to optical gain. The critical NQD radius above which optical gain is observed is ~2.3 nm. Excitation densities for which crossover to gain occurs (gain threshold) decrease as the dot size is increased. For dots of the "critical" 2.3-nm radius, the gain threshold is approached (but not crossed) at ~$\langle N \rangle_{th}$ = 8.5, whereas it is close to the theoretical limit of approximately one e–h pair per dot (see Sect. IV.A) for large 3.5-nm dots ($\langle N \rangle_{th}$ = 1.4 for R = 3.5 nm). This strong size dependence of the gain threshold for NQD hexane solutions is a direct result of the size-dependent interplay between gain and PA.

Analysis of TA data for NQDs with different surface passivations indicates that the PA is sensitive to the identity of the capping layer. For example, in the case of hexane solutions, the magnitude of excited-state absorption can be reduced by overcoating NQDs with a layer of a wide-gap semiconductor (e.g., using CdSe–ZnS core–shell structures). Excited-state absorption is also strongly dependent on the type of solvent/matrix material. Such commonly used solvents as toluene, chloroform, and heptamethylnonane (HTP) show a PA band comparable in intensity with that in hexane [46]. However, the PA is almost completely suppressed in TOP, one of the NQD growth solvents, and in such solid-state matrices as polyvinyl butyral or sol-gel titania [11], as well as in matrix-free, close-packed NQD films [46].

Figure 13 displays pump-dependent TA data for CdSe NQDs with R = 2.3 nm dissolved in TOP. In hexane solutions, the TA pump dependence for these dots only approaches the gain threshold but does not cross it (Fig. 12b, crosses). However, in the case of TOP, these NQDs clearly show a gain band at the position of the PL band. The transition from absorption to gain occurs at $\langle N \rangle_{th}$ = 1.8 (circles in Fig. 13b), which is much lower than the gain threshold observed in hexane (~8.5). This difference provides additional evidence of the size-dependent competition between gain and PA.

Figure 14 displays gain data for close-packed, matrix-free NQD films. Shown are normalized nonlinear absorption spectra (spectra detected in the presence of the optical pump) for three samples fabricated from NQDs with radii 2.1, 1.7, and 1.3 nm. For all samples, NQD radii are smaller than the "critical" radius of 2.3 nm, and in the case of hexane solutions, these NQDs do not show optical gain. In contrast to hexane-based samples, all solid-state samples show optical gain at the position of the emitting transition, clearly indicating suppression of PA. These data also provide an illustrative example of the size-controlled tunability of the gain band; it shifts by ~300 meV as the dot radius changes from 2.1 to 1.3 nm.

The strong effect of the identity of the solvent/matrix material on the strength of PA indicates that this feature is not intrinsic to NQDs but is rather a property of the NQD interface and/or matrix material [44]. One

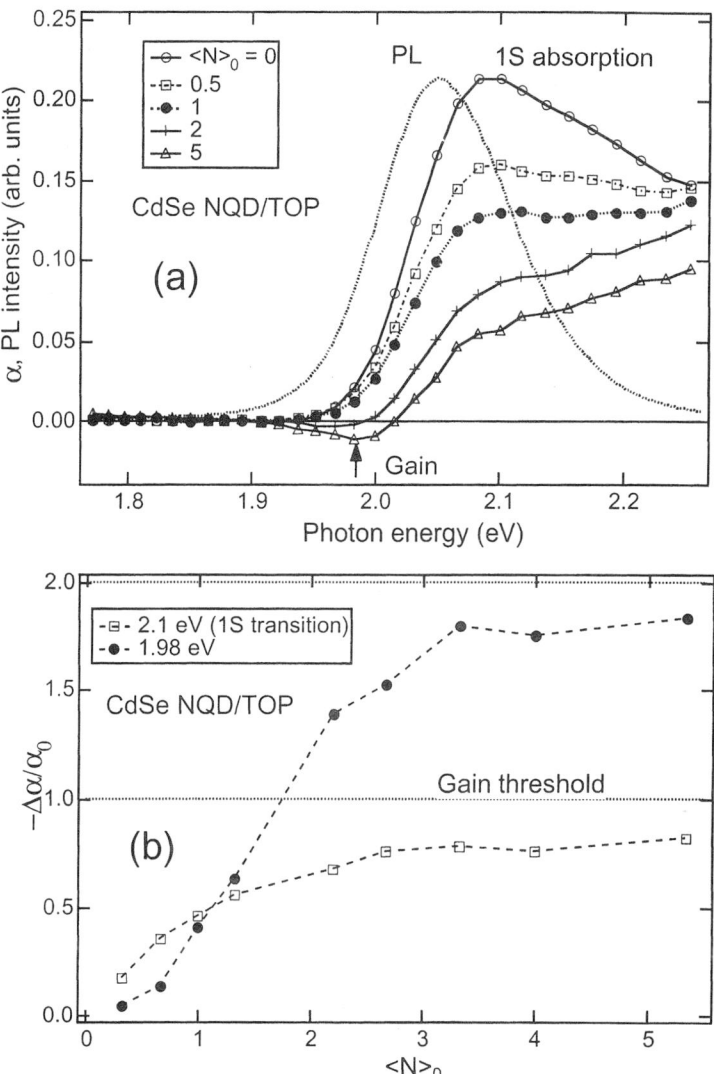

Figure 13 (a) Pump-intensity dependent TA spectra of CdSe NQDs (R = 2.3 nm) prepared in TOP which show the development of optical gain at the position of the "emitting" transition. (b) Pump-induced absorption changes at the position of the "emitting" (solid circles) and "absorbing" (open squares) transitions, plotted as a function of the average number of e–h pairs excited per dot.

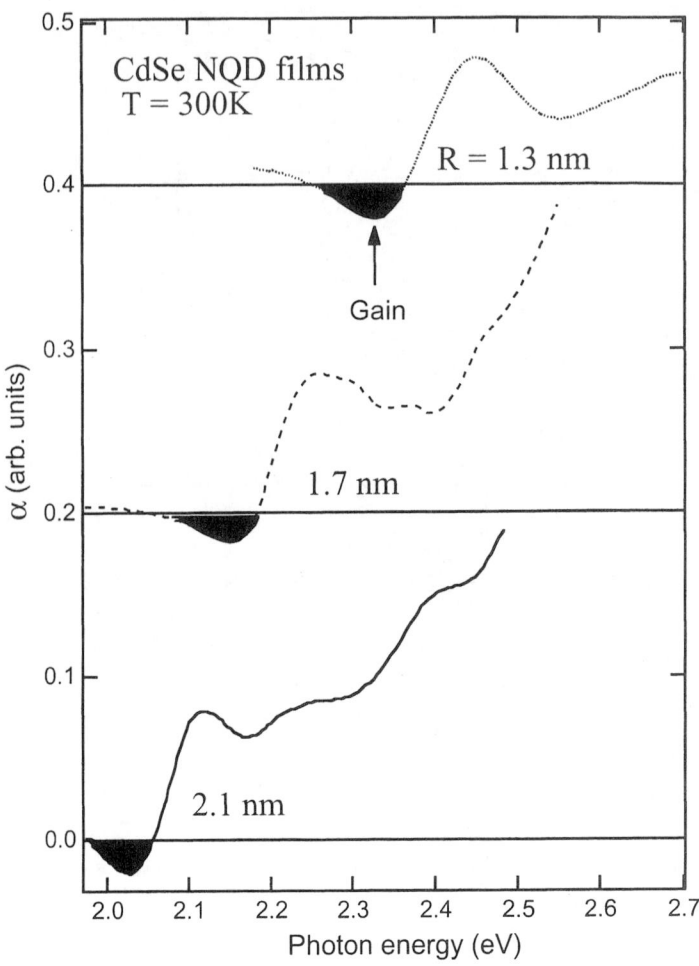

Figure 14 Nonlinear absorption/gain spectra of matrix-free CdSe NQD films fabricated from dots with R = 1.3, 1.7, and 2.1 nm (T = 300 K). The gain regions are shown in black.

possible explanation of PA is in terms of an excited-state absorption due to transitions that couple occupied NQD interface states to unoccupied states in the solvent/matrix. This explanation is consistent with both observations: the sensitivity of the PA strength to the quality of the surface passivation in hexane solutions and the complete suppression of PA in matrix-free films.

IV. MULTIPARTICLE EFFECTS AND OPTICAL GAIN IN NQDs

A. Band-Edge Transient Absorption and Optical Gain in NQDs

In terms of the TA spectroscopy, optical gain (i.e., negative absorption) corresponds to pump-induced absorption bleaching with a magnitude that is greater than sample absorption in the ground state. In strongly confined colloidal NQDs, pump-induced absorption changes are primarily due to state filling and Coulomb multiparticle interactions (the carrier-induced Stark effect) [32,33]. State filling arises from Pauli exclusion and leads to the bleaching of the interband optical transitions that involve populated quantized states. The Stark effect is associated with local fields generated by photoexcited carriers and leads to a shift of optical transitions and changes in transition oscillator strengths as a result of modifications in the selection rules. In contrast to state filling, which *selectively* affects only transitions involving populated states, the carrier-induced Stark effect does not have this selectivity and modifies NQD transitions that involve both occupied and unoccupied electronic states [47]. However, in the case of transitions that couple populated NQD states, the relative contribution of the Stark effect to TA signals is significantly smaller than that of state filling [33]. The state-filling effect, for example, is almost entirely responsible for the pump-dependent bleaching of the strong $1S$ absorption resonance [33]. On the other hand, involvement of Coulomb multiparticle interactions is essential for the explanation of optical-gain signals detected in the region of the lowest "emitting" transition (discussed later).

The absorption changes resulting from the state-filling effect are proportional to the sum of the electron and hole occupation numbers. If we present the linear absorption spectrum of NQDs as a superposition of separate absorption bands corresponding to different quantized optical transitions, the state-filling-induced absorption changes ($\Delta\alpha$) can be calculated using

$$\Delta\alpha(\hbar\omega) = -\sum_i a_i G_i(\hbar\omega - \hbar\omega_i)(n_i^e + n_i^h) \qquad (2)$$

in which $G_i(\hbar\omega - \hbar\omega_i)$ is the unit-area absorption profile of the $\hbar\omega_i$ transition, a_i is its area, and n_i^e and n_i^h are occupation numbers of the electron and hole states involved in the transition. Under thermal quasiequilibrium (after the intraband relaxation is finished), the occupation numbers can be found using the Fermi distribution function. Within this model, a normalized absorption change associated with a specific transition $(\hbar\omega_i)$ can be presented as follows:

$$-\frac{\Delta\alpha_i(\hbar\omega)}{\alpha_{0,i}(\hbar\omega)} = n_i^e + n_i^h \tag{3}$$

where $\alpha_{0,i}$ is the contribution of transition $\hbar\omega_i$ to the ground-state absorption spectrum.

As was discussed in Section II.A, the band-edge optical properties of NQDs are dominated by two transitions that couple the lowest $1S$ electron state to two manifolds of closely spaced hole states. These transitions are manifested in optical spectra as the $1S$ absorption (the "absorbing" transition) and the band-edge PL (the "emitting" transition) peaks (Fig. 1a and 1b). At moderate excitation levels for which the hole manifold of "absorbing" states is unoccupied, the pump-induced absorption change within the "absorbing" $1S$ transition $(\hbar\omega_{1S})$ is solely due to the $1S$ electron state occupation number:

$$-\frac{\Delta\alpha_{1S}(\hbar\omega)}{\alpha_0(\hbar\omega)} = n_{1S}^e \tag{4}$$

In order to use this expression for a comparison with an experiment, we need to average n_{1S}^e over different NQD populations existing in the NQD ensemble. We perform averaging assuming a Poisson distribution of NQD populations: $P(N) = \langle N \rangle^N e^{-\langle N \rangle}/N!$, where $P(N)$ is the probability of having N e–h pairs in a selected NQD when the average population of NQDs is $\langle N \rangle$. The Poisson distribution is applicable if the probability of generating the e–h pair in an NQD is independent of the number of e–h pairs already existing in it. This assumption is valid, for example, in the case of very low excitation intensities, $\langle N \rangle \ll 1$, or for the case where the pump–photon energy is much greater than the NQD energy gap.

Because of the twofold spin degeneracy of the $1S$ electron state, the $1S$ occupation number is 0.5 for $N = 1$ and 1 for $N \geq 2$. Therefore, the averaging over an ensemble can be performed as follows:

$$\langle n_{1S}^e \rangle = 0.5 P(1) + \sum_{i=2}^{\infty} P(i) = 1 - P(0) - 0.5 P(1)$$

$$= 1 - e^{-\langle N \rangle} \left(1 + \frac{\langle N \rangle}{2} \right) \tag{5}$$

Figure 15 displays $1S$ bleaching experimental data (symbols) for NQDs of three different radii (2.3, 1.7, 1.2 nm) in hexane. The state-filling model (solid line) describes well the whole dataset for the 2.3-nm sample. It also

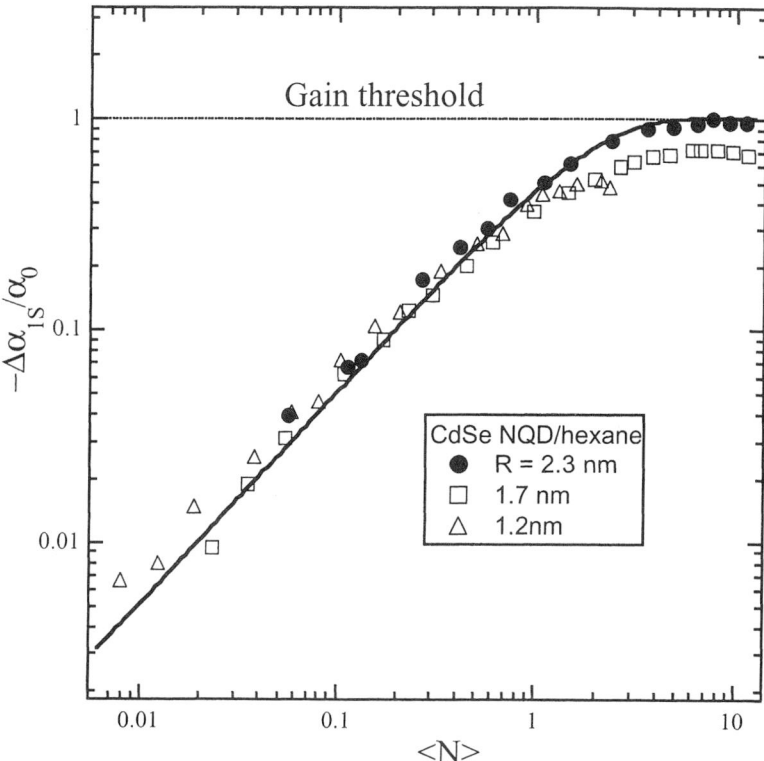

Figure 15 Pump dependence of normalized absorption changes at the position of the 1S transition (symbols) in CdSe NQDs with different mean radii (hexane solutions) plotted as a function of the average number of e–h pairs excited per dot. The line shows a result of calculations within the state-filling model assuming the Poison distribution of NQD populations.

reproduces well the low-intensity data ($\langle N \rangle \leq 1$) for dots of smaller sizes. However, for small NQD radii, experimental data points deviate from the model at high pump intensities, and this difference is likely due to a contribution from the excited-state absorption that becomes progressively more important with decreasing NQD radii (see Sect. III.B).

In order to describe pump-induced absorption changes within the lowest "emitting" transition ($\Delta\alpha_{em}$), we need to account for populations of the lowest electron and hole states:

$$-\frac{\Delta\alpha_{em}(\hbar\omega)}{\alpha_0(\hbar\omega)} = \langle n_{1S}^e + n_{em}^h \rangle \tag{6}$$

where n_{em}^h is the hole "emitting" state/manifold occupation number. The hole contribution to TA is strongly dependent on the degeneracy of the "emitting" hole state. In the case of the twofold degeneracy, the sum of average electron and hole occupation numbers can be presented as

$$\langle n_{1S}^e + n_{em}^h \rangle = 2\left[1 - e^{-\langle N \rangle}\left(1 + \frac{\langle N \rangle}{2}\right)\right] \qquad (7)$$

At the gain threshold, $-\Delta\alpha_{em}(\hbar\omega)/\alpha_0(\hbar\omega) = 1$ and, hence [see Eq. (3)],

$$\langle n_{1S}^e + n_{em}^h \rangle_{th} = 1 \qquad (8)$$

Using Eq. (8), we determine that the gain threshold corresponds to the condition $\langle N \rangle_{th} = 1.15$, which is close to the theoretical one e–h pair threshold obtained without statistical averaging of NQD occupation numbers [2].

The results of effective mass calculations [48], as well as measurements of temperature-dependent radiative lifetimes [49], suggest that at temperatures above ~50 K, the lowest "emitting" NQD state can be described in terms of a strong mixing of the lowest "dark" exciton (total momentum projection $J = \pm 2$) and the next bright-exciton state with $J = \pm 1$. This result suggests that the lowest "emitting" hole state can be characterized by a fourfold degeneracy, and its average population number can be calculated as $\langle n_{em}^h \rangle = 1 - 0.25 \sum_{i=1}^4 iP(i)$. Combining the latter expression with Eqs. (5) and (8), we obtain $\langle N \rangle_{th} = 1.55$. As expected, the gain threshold in the case of the fourfold degeneracy of the "emitting" hole state is higher than in the case of the twofold degeneracy.

The fact that optical gain develops at $\langle N \rangle > 1$ indicates that it arises from NQDs which contain multiple e–h pairs (i.e., NQD multiexciton states). Because of the Auger recombination, the intrinsic lifetime of multiexciton states in NQDs rapidly decreases as the number of e–h pairs per dot is increased (see Sect. IV.B), suggesting that optical gain is primarily due to two-pair states (biexcitons) which have the longest relaxation time among the states with $N \geq 2$ [50].

Optical-gain pump dependences and gain thresholds were studied experimentally in Ref. 50 using matrix-free films of CdSe NQDs. Figure 16 displays room-temperature emission and linear (α_0) and nonlinear (α) absorption spectra of TOPO-capped CdSe NQDs with $R = 2.5$ nm. The nonlinear absorption/gain and emission spectra were taken at a pump fluence $W = 1$ mJ/cm, which was above both the gain threshold and the threshold for the development of amplified spontaneous emission (ASE) observed as a

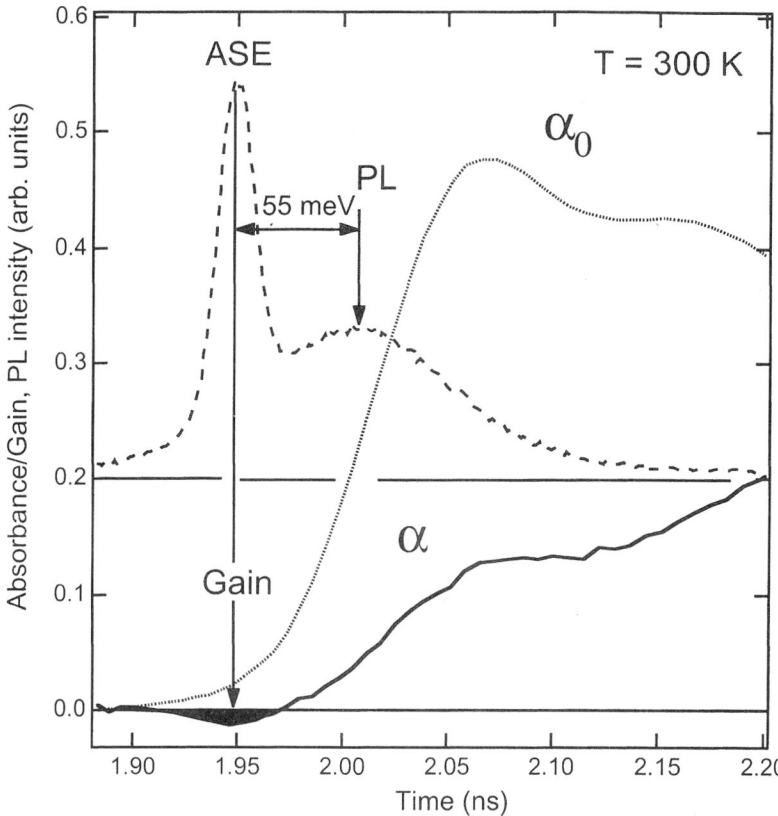

Figure 16 A room-temperature emission spectrum (dashed line) of a close-packed film of CdSe NQDs ($R = 2.5$ nm) in comparison with linear (dotted line) and nonlinear (solid line) absorption spectra ($W = 1$ mJ/cm); the gain region is shown in black and the emission spectrum is offset vertically for clarity.

sharp peak at 1.95 eV. The gain spectral maximum and the ASE peak were red-shifted with respect to the maximum of the "subthreshold" spontaneous PL by ~55 meV (the origin of this shift is discussed below). Depending on the dot sizes, NQD surface conditions, and the sample temperature, the gain threshold (W_{th}) varied from ~0.3 to ~10 mJ/cm, increasing with decreasing dot size and/or increasing sample temperature. The measured threshold fluences corresponded to initial numbers of photogenerated e–h pairs, $\langle N \rangle_0$, from ~1.5 to ~5 per dot on average. The increase in the gain threshold at room temperature relative to thresholds observed at cryogenic temperatures was attributed to temperature-activated hole trapping that could lead to ultrafast (subpicosecond to picosecond time constants) depopulation of quantized states on timescales comparable to those of the gain buildup.

Figure 17 displays normalized absorption changes measured for the solid-state CdSe NQD film ($T = 80$ K) as a function of the pump intensity within both the $1S$ bleaching (open circles) and the gain band (solid squares). We compare these data with results of the state filling model (dashed and dashed-dotted lines) obtained assuming a fourfold degeneracy of the lowest

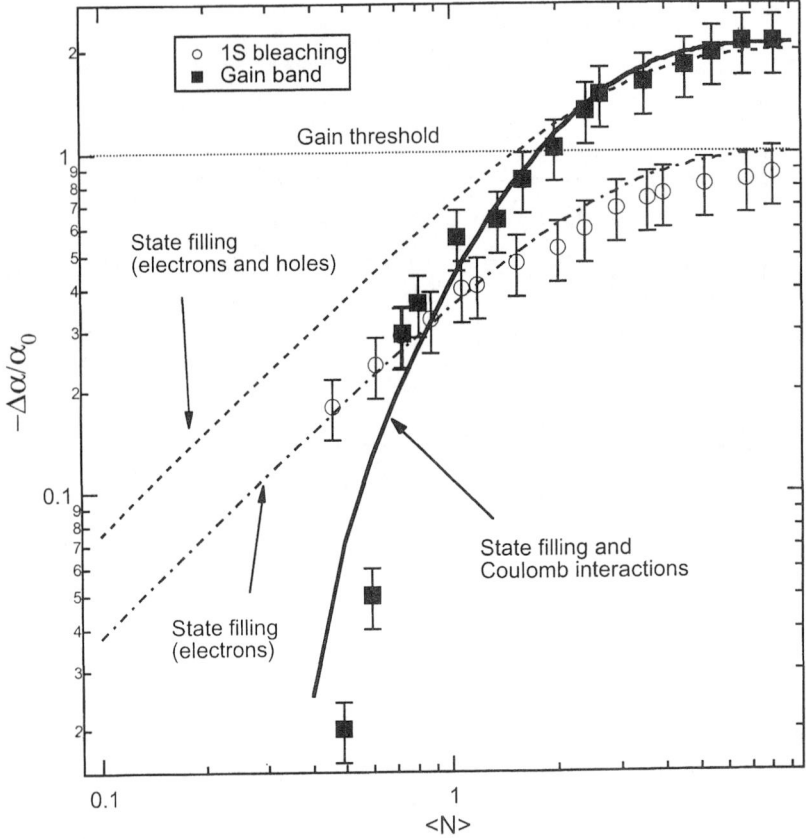

Figure 17 Pump-intensity dependence of normalized absorption changes for a CdSe NQD film (dot radius 2.5 nm) measured at the position of the $1S$ absorption maximum (open circles) and the maximum of the gain band (solid squares) (sample temperature is 80 K). Experimental data are compared to results of the modeling assuming either the pure state-filling mechanism (dashed and dashed-dotted line) or state filling together with Coulomb interactions (solid line).

hole state. Although state filling explains the saturation of the $1S$ transition (dashed-dotted line), it fails to describe the pump dependence of the TA signals within the gain band (dashed line). Experimental data clearly indicate a threshold in the development of $\Delta\alpha$, which is not predicated by state-filling reasoning.

The thresholdlike behavior of the TA signal within the gain band can be explained in terms of Coulomb multipartilce interactions [50] that lead to a shift of the emission band of multiexcitons (specifically, biexcitons) with respect to that of a single exciton [51,52]. The radiative decay of a biexction produces both a photon and an exciton. The energy of the photon resulting from this decay is determined by the expression $\hbar\omega_{bx} = 2E_x^i - \delta E_2 - E_x^f$, where E_x^i is the energy of the excitons comprising the biexcion, δE_2 is the exciton–exciton interaction energy (if two excitons attract to each other, this energy can be treated as a biexciton binding energy), and E_x^f is the energy of the exciton that is left as a product of the biexciton recombination. The biexciton decay can, in principle, produce several emission lines with positions that are determined by the energy of the exciton in the final state, E_x^f (Fig. 18). In the case of the biexciton formed by two ground-state excitons (E_x^0), the shift of the biexciton emission line with respect to the single-exciton line ($\hbar\omega_x$) is given by the expression $\delta E_{bx} = \delta E_2 + (E_x^f - E_x^0)$. This expression indicates that a significant red shift of the biexciton emission band can arise from both confinement-enhanced two-e–h-pair interactions [47,53] and/or the existence of decay channels that produce an excited-state exciton rather than a ground-state exciton [54]. Figure 18 schematically shows two possible biexciton decay channels resulting in emission lines at $\hbar\omega_{bx,0}$ and $\hbar\omega_{bx,1}$ that correspond to two different final exciton states (these states are associated with "emitting" and "absorbing" band-edge transitions). The red shift of the $\hbar\omega_{bx,0}$ band with respect to the single-exciton emission spectrum is due to the exciton–exciton interaction energy (δE_2), whereas the red shift of the $\hbar\omega_{bx,2}$ band is due to both δE_2 and the energy difference between ground-state and excited-state exciton.

Because biexcitons dominate optical gain (and hence stimulated emission), the "biexciton" red shift is at least one of the reasons for large shifts between the ASE peak and the center of the spontaneous PL observed experimentally (Fig. 16). However, other factors, such as interference from the strong $1S$ absorption, can also contribute to the ASE red shift observed in experimental spectra.

Multiparticle effects can be phenomenologically incorporated into the state-filling model by introducing biexciton shifts into both the absorption line of NQDs occupied with a single e–h pair and the emission line of NQDs that contain multiple e–h pairs [47]. The latter assumption presumes that multiexcitons contribute to gain primarily via a biexciton state that is formed

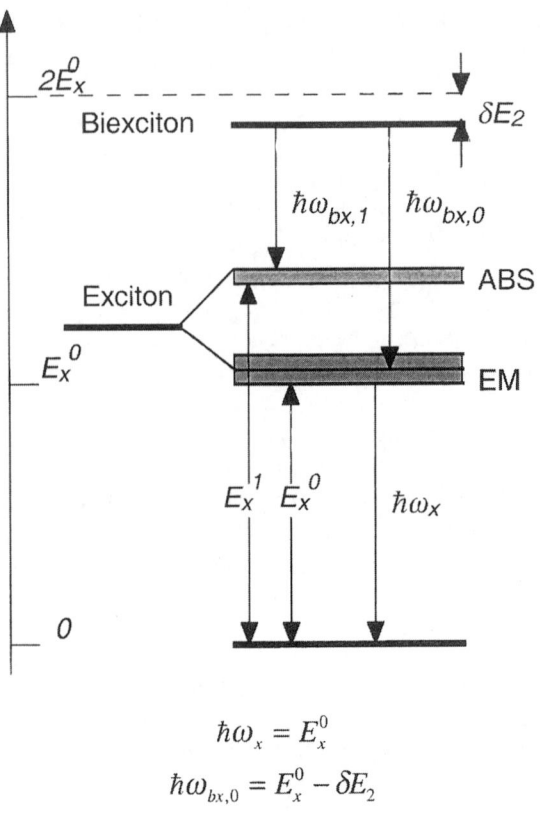

$$\hbar\omega_x = E_x^0$$

$$\hbar\omega_{bx,0} = E_x^0 - \delta E_2$$

$$\hbar\omega_{bx,1} = E_x^0 - \delta E_2 - (E_x^1 - E_x^0)$$

Figure 18 Schematic of transitions illustrating two possible decay channels for a biexciton composed of two ground-state excitons interacting with the energy δE_2. These two channels (denoted as $\hbar\omega_{bx,0}$ and $\hbar\omega_{bx,1}$) produce an exciton occupying either the lowest manifold of "emitting" states (denoted as EM) or a higher-lying manifold of "absorbing" states (denoted as ABS); exciton energies for these two manifolds are E_x^1 and E_x^1, respectively.

after a rapid Auger decay of states with $N \geq 3$. We used this model in order to describe pump-dependent TA/gain signals in Fig. 17 (solid line). The best fit to the experimental data was obtained for a biexciton shift of 0.4Γ, where Γ was the width of the emitting transition (a full width measured at a half-maximum). In contrast to a purely state-filling model (dashed line), which only provides a reasonable fit at high excitation densities ($\langle N \rangle > 2$), the "biexcitonic" model (solid line) describes well the experimental dataset across the entire intensity range $\langle N \rangle = 0.5$–8.

Within the state-filling model corrected for multiparticle interactions, the thresholdlike behavior in the development of optical gain is a direct consequence of a nonzero biexciton shift which "generates" a new "emitting" transition at a spectral energy that is spectrally offset with respect to the single-exciton "emitting" transition. Another important consequence of biexciton interactions is their effect on the magnitude of the optical gain. In the state-filling model, the maximum of the optical gain, corresponding to a complete population inversion, is limited by the magnitude of the ground-state absorption [$g_{max}(\hbar\omega) = \alpha_0(\hbar\omega)$]. In the case of multiparicle interactions, the gain amplitude can, in principle, exceed this level, which is due to a red shift of multiparticle transitions into the region of the reduced ground-state absorption. Gain with a magnitude that is greater than the ground-state absorption is observed for both NQD solutions (Fig. 12b) and solid-state samples (Fig. 17), providing an addition piece of evidence that Coulomb interactions play am important role in optical-gain properties of NQDs.

B. Multiparticle Auger Recombination and Optical Gain Dynamics

An important mechanism for nonradativie carrier losses in NQDs is multiparticle Auger recombination. The Auger recombination is a process in which the e–h recombination energy is not emitted as a photon but is transferred to a third particle (an electron or a hole) that is reexcited to a higher energy state within the dot or outside it (the latter effect is known as Auger ionization). The Auger recombination has a relatively low efficiency in bulk semiconductors, for which significant thermal energies are required to activate the effect [55,56]. However, Auger decay is greatly enhanced in quantum-confined systems, in which the relaxation in momentum conservation removes the activation barrier [57,58].

In well-passivated dots, Auger recombination dominates the multiparticle decay dynamics. Furthermore, because the Auger recombination has the same activation threshold as optical gain ($\langle N \rangle > 1$), it is unavoidable in the regime of optical amplification. The effect of Auger recombination on gain dynamics is clearly seen in Fig. 19, which shows TA time transients recorded in the region on the gain band for dots of 2.5-nm radius before (open circles) and after (solid squares and open diamonds) the gain threshold. Before the threshold ($\langle N \rangle_0 = 0.5$), the bleaching signal shows a slow decay that is in contrast to the fast signal relaxation observed above the threshold ($\langle N \rangle_0 = 2.3$ and 8) (i.e., at pump powers for which muliparticle Auger decay becomes active).

In bulk semiconductors, Auger decay is characterized by a cubic carrier decay rate $C_A n_{eh}^3$, where C_A is the Auger constant and n_{eh} is the carrier density. In this case, one can also introduce an instant time constant, $\tau_A = (C_A n_{eh}^2)^{-1}$,

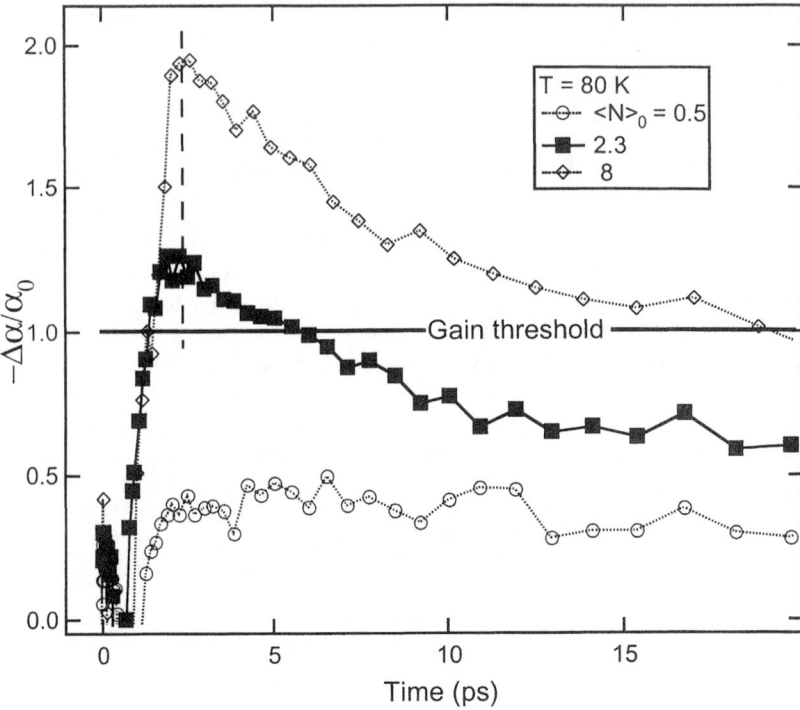

Figure 19 Dynamics of pump-induced absorption changes at the maximum of the gain band (sample temperature is 80 K). The initial numbers of e–h pairs per dot, $\langle N \rangle_0$, injected by a pump pulse are indicated in the figure.

that changes continuously as the carrier density is reduced during recombination. In contrast to this continuous decay, Auger recombination in NQDs occurs via a sequence of "quantized" steps from N to $N - 1$, $N - 2$, ..., and, finally, to the one-e–h-pair state, with each step characterized by a discrete exponential decay constant (Fig. 20a, inset). The decay of multiple-pair states in NQDs via a sequence of quantized steps can be described by the following set of coupled rate equations:

$$\frac{dn_N}{dt} = -\frac{n_N}{\tau_N}$$

$$\frac{dn_{N-1}}{dt} = \frac{n_N}{\tau_N} - \frac{n_{N-1}}{\tau_{N-1}}$$

$$\vdots$$

$$\frac{dn_1}{dt} = \frac{n_2}{\tau_2} - \frac{n_1}{\tau_1}$$

(9)

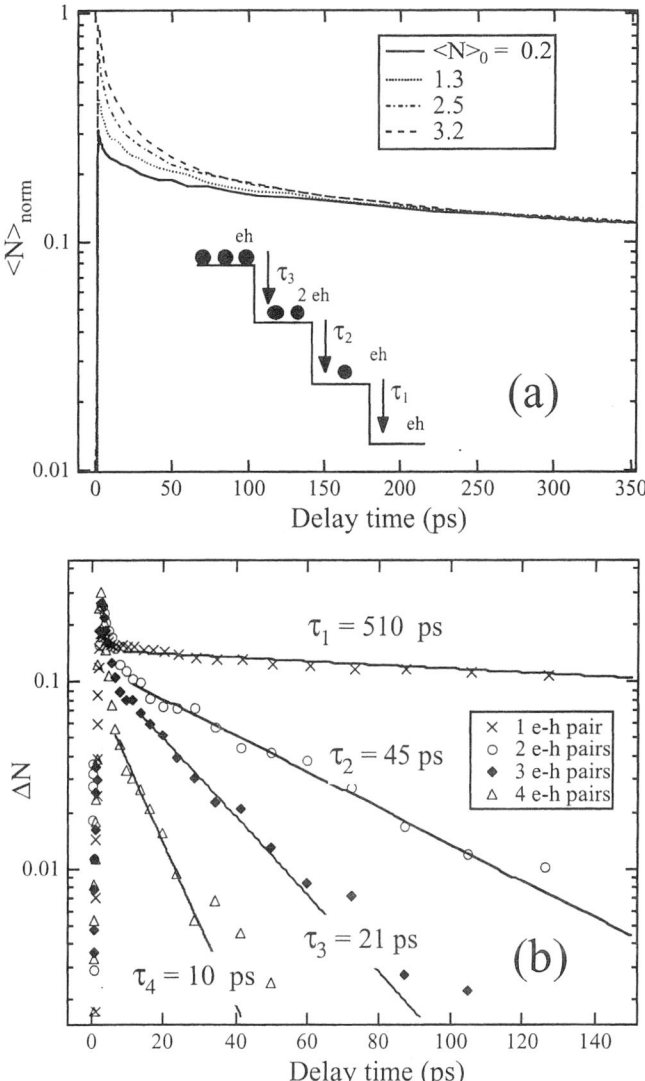

Figure 20 (a) Pump-dependent dynamics of NQD average populations in the 1.7-nm sample normalized to match their "tails." Inset: Quantized steps in quantum-confined Auger recombination in NQDs. (b) Dynamics of the one-, two-, three-, and four- e–h-pair states (symbols) extracted from the TA data for the 1.7-nm sample (see text), fit to a single exponential decay (solid lines).

where n_i $(i = 1, 2 \ldots, N)$ is the concentration of dots in the sample that contain i e–h pairs and τ_i is the lifetime of the i-pair NQD state. If one neglects recombination during the pump pulse, these equations have a straightforward solution for time-dependent NQD average populations: $\langle N(t) \rangle = \sum_{i=1}^{N} A_i e^{\frac{-t}{\tau_i}}$, where A_i are the coefficients determined by initial carrier densities. In the case for which absorption saturation at the pump wavelength is small, the initial conditions for Eq. (9) can be calculated using the Poisson distribution:

$$n_m(t = 0) = n_{\mathrm{NQD}} \frac{\langle N \rangle_0^m}{m!} e^{-\langle N \rangle_0}, \quad m = 1, 2, \ldots, N \tag{10}$$

where n_{NQD} is the concentration of NQDs in a sample.

Quantized steps during Auger decay in strongly confined CdSe NQDs were experimentally resolved in Ref. 16 using a femtosecond TA experiment. In this work, dynamics of NQD populations, $\langle N(t) \rangle$, were derived from time-resolved $1S$ bleaching signals. Figure 20a displays these dynamics measured for a 1.7-nm sample at carrier injection levels, $\langle N \rangle_0$, from 0.2 to 3.2. In Ref. 16, a simple subtractive procedure was used to extract *single* exponential dynamics that are characteristic of the decay of different multiple-pair NQD states from the measured $\langle N(t) \rangle$ time transients. This procedure is based on the fact that at long times after photoexcitation, the decay is governed by singly-excited NQDs independent of the initial carrier density. This observation allows one to normalize time transients taken at different pump levels to match their long-time-delay "tails," as shown in Fig. 20a. By subtracting the low-pump-intensity trace ($\langle N \rangle_0 < 1$; single e–h pair decay) from traces recorded at $\langle N \rangle_0 > 1$, it was possible to derive the dynamics, $\Delta N(t)$, due to relaxation of multiple-pair states with $N \geq 2$. For initial carrier densities $1 < \langle N \rangle_0 < 2$, this procedure yielded the two-pair state dynamics (open circles in Fig. 20b), with a faster initial component due to the contribution from states with a larger number of excited pairs. By further subtracting two-pair decay from the $\Delta N(t)$ transients detected at $\langle N \rangle_0 > 2$, one could derive dynamics of states with $N \geq 3$. This procedure could be repeated to extract dynamics of NQD states with $N = 4, 5$, and so forth.

The extracted dynamics of the two-, three-, four-pair states are shown in Fig. 20b. These dynamics indicate that the carrier decay becomes progressively faster with increasing number of e–h pairs per NQD, as expected for Auger recombination. In bulk semiconductor arguments, the effective decay time constant, τ_N, in the Auger regime is given by the expression $\tau_N^{-1} = C_A (N/V_0)^2$ $(N \geq 2)$ that predicts the following scaling of times for four-, three-, and two-pair relaxation: $\tau_4 : \tau_3 : \tau_2 = 0.25 : 0.44 : 1$. This scaling is very close to the one observed experimentally $(0.22 : 0.47 : 1)$, [16], indicating that the decay rates for quantum-confined Auger recombination are cubic with respect to the carrier density $(dn_{\mathrm{eh}}/dt \propto -n_{\mathrm{eh}}^3)$, just as in bulk materials.

Transient absorption data obtained for NQDs of different sizes (Fig. 21a) indicate that the τ_2 time constant rapidly decreases with decreasing NQD size following a cubic size dependence ($\tau_2 \propto R^3$) (see data shown by solid circles in Fig. 21b). The two-pair lifetime shortens from 363 ps to only 6 ps as the dot radius is decreased from 4.1 to 1.2 nm. Interestingly, the time constants measured for the three- and four-pair decay (open squares and solid triangles in Fig. 21b, respectively) follow the same cubic size dependence as that of the two-pair state such that the time-constant ratios predicted by the bulk-semiconductor model hold for all NQD sizes. Simple bulk-material reasoning would suggest that the enhancement in the Auger decay in smaller particles is caused by an effective increase in carrier concentrations resulting from the increased spatial confinement. This reasoning would predict the R^6 scaling for Auger times [$\tau_N \propto (N/V_0)^{-2} \propto R^6$]. However, the experimentally measured size dependence is R^3, indicating that in 3D-confined systems, the Auger "constant" depends on the particle size. For CdSe NQDs, C_A scales approximately as R^3 and decreases from $\sim 7 \times 10^{-29}$ to $\sim 2 \times 10^{-30}$ cm^6/s as the dot radius is reduced from 4.1 to 1.2 nm (Fig. 22). For all NQD sizes, the C_A values calculated by using lifetimes of two-, three-, and four-e–h pair states are close to each other (compare data shown by circles, squares, and triangles in Fig. 22), indicating that the cubic density dependence of the Auger rates holds for all sizes between 1 and 4 nm.

Because of their strong size dependence, Auger effects, which play a minor role in relatively large epitaxial dots, become significant in strongly confined colloidal nanoparticles. For moderately well-passivated dots, non-radiative Auger relaxation of doubly-excited dots (NQD biexcitons) is more efficient than the surface trapping that imposes an *intrinsic* limit on the lifetime of the optical gain in strongly confined NQDs. The dominance of the Auger effect in NQDs also implies that even relatively poorly passivated samples could behave as well as the best samples with respect to gain and stimulated emission because as long as trapping rates are lower than Auger rates, the two-e–h-pair lifetime is determined by the intrinsic Auger decay.

Very short optical-gain lifetimes imply that the regimes of light amplification and lasing in NQDs are easiest to achieve using a pulsed excitation with a pulse duration (τ_p) that is shorter than the Auger decay time of biexcitons (τ_2). When $\tau_p \ll \tau_2$, the gain threshold, W_{th}, expressed in terms of a pump fluence (J/CM^2) can be estimated from the condition ($W_{th}/\hbar\omega_p)\sigma_p \approx 1$. At energies well above the energy gap, σ_p is proportional to R^3 [see Eq. (1)], indicating that the threshold pump fluence rapidly increases ($W_{th} \propto R^{-3}$) with decreasing dot size. For continuous-wave (cw) excitation, the threshold pump intensity (I_{th}) can be estimated from the ratio of W_{th} and τ_2: $I_{th} \approx \hbar\omega_p(\sigma_p\tau_2)^{-1}$. Because τ_2 is proportional to R^3 (see Fig. 21b), the threshold cw intensity is proportional to R^{-6}. This dependence indicates that cw pumping becomes

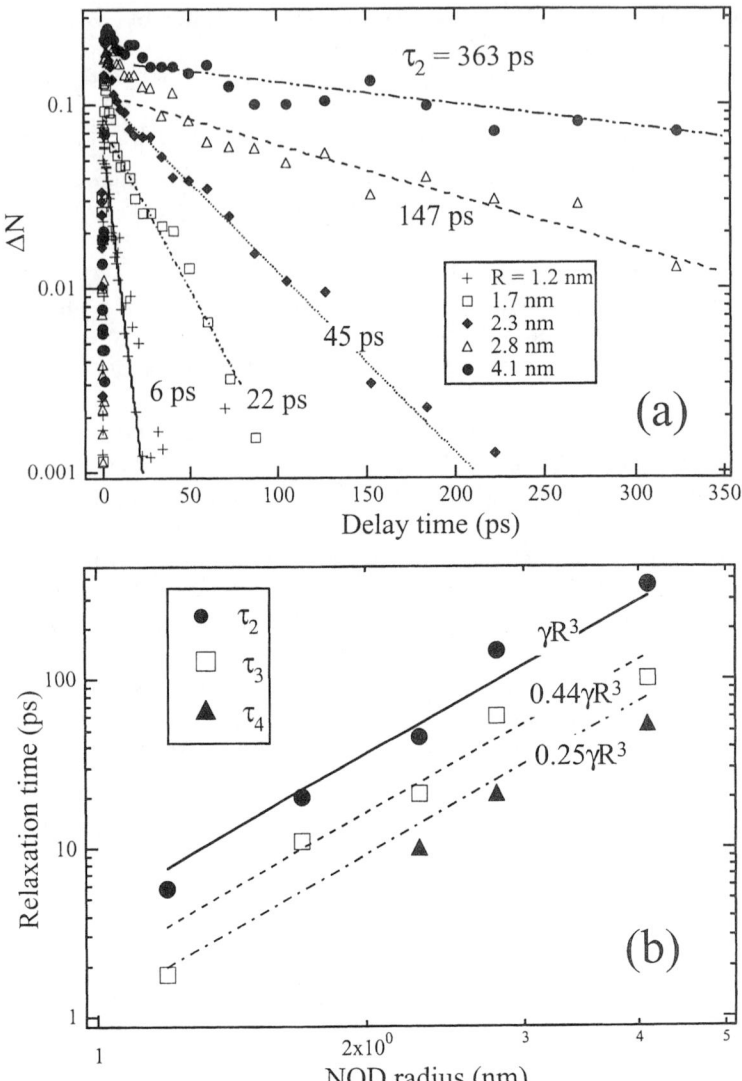

Figure 21 (a) Two-e–h-pair (biexciton) dynamics in NQDs with radii 1.2, 1.7, 2.3, 2.8, and 4.1 nm (symbols) fit to a single-exponential decay (lines): (b) Size dependence of relaxation time constants of two-, three-, and four-e–h pair states (symbols), fit to the dependence γR^3, $0.44\gamma R^3$, and $0.25\gamma R^3$, respectively (lines); γ is a size-independent constant.

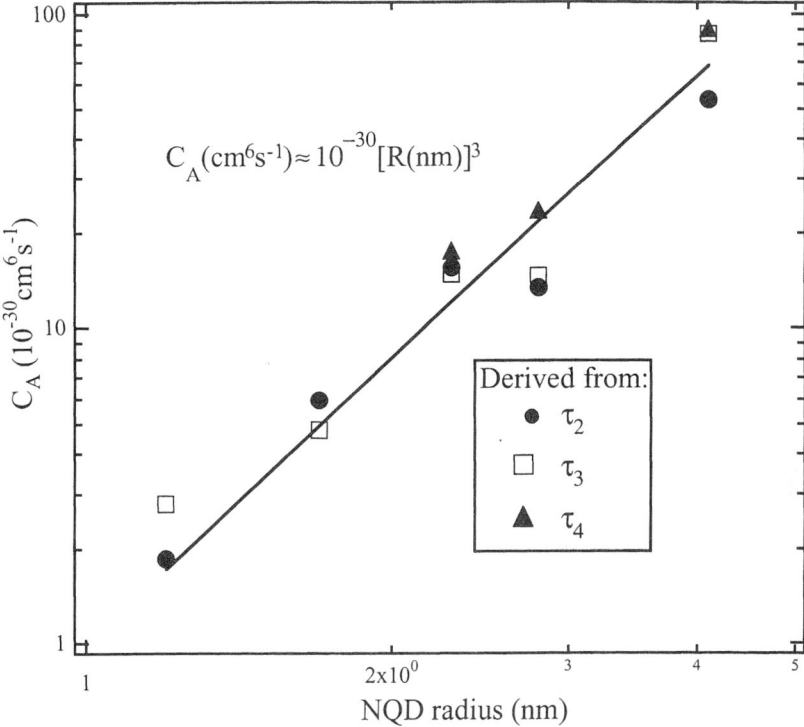

Figure 22 Size dependence of the Auger constant (symbols), fit to the dependence $C_A \propto R^3$ (line); solid circles, open squares, and solid triangles correspond to data derived from the lifetimes of the two-, three-, and four-pair states, respectively.

progressively less efficient, compared to pulsed pumping, with decreasing dot size.

V.   OPTICAL AMPLIFICATION AND LASING IN NQDs

A.   Observations of Amplified Spontaneous Emission

To overcome the problem of intrinsic ultrafast gain decay, development of amplified emission must occur on timescales that are shorter than those for the Auger decay. The ASE buildup time (τ_{ASE}) is inversely proportional to the gain magnitude: $\tau_{ASE} = n/gc$, where c is the light velocity. One can further rewrite this expression in terms of the gain cross section ($\sigma_g = g/n_{NQD}$) and

the volume fraction (filling factor) of semiconductor material in the sample
($\xi = 4\pi n_{NQD} R^3/3$):

$$\tau_{ASE} = \frac{4\pi R^3}{3} \frac{n}{\xi \sigma_g c} \tag{11}$$

In the case for which the gain decay is dominated by the intrinsic two-pair
Auger recombination (decay time τ_2), ASE can only be observed if $\tau_{ASE} < \tau_2$.
The Auger data discussed in the previous subsection indicate that τ_2 scales
with dot size as R^3 ($\tau_2 \approx \beta R^3$, where $\beta \approx 5$ ps/nm^3) which allows one to derive
the following critial condition for the development of ASE:

$$\xi > \frac{4\pi n}{3c\beta\sigma_g} \tag{12}$$

From this condition one can see that one approach to overcoming fast Auger
recombination is by using samples with high NQD densities. For $\sigma_0 = 10^{-17}$
cm^2 (a lower estimate for the gain cross section; Ref. 59) and $n = 2$ (an upper
estimate for the refractive index), Eq. (12) indicates that minimal filling
factors required for achieving the ASE regime are of the order of 0.005
(0.5%). The densities in excess of the critical value can be achieved in matrix-
free, close-packed films [59], glass [3], and sol-gel titania [11] matrices, and
concentrated solutions [60].

The highest dot densities are provided by close-packed NQD films
(NQD solids). For example, films made of 1.3-nm CdSe NQDs capped with
TOPO have filling factors as high as ~20% (1.1-nm length of a capping
molecule is assumed) for which the condition given by Eq. (12) is easily
satisfied. Solid-state films of close-packed CdSe NQDs were used in Ref. 59
to demonstrate ASE with a color that is controlled by NQD sizes. These films
were fabricated by drop casting from hexane/octane solutions using either
TOPO-capped or ZnS-overcoated dots. Transmission electron microscopy
(TEM) indicated random close-packing of dots in the samples (Fig. 23a). The
films' emission color was tunable over an entire visible range by changing the
dot size (Fig. 23a).

The effect of light amplification that leads to the development of a sharp
ASE band above a certain pump threshold is illustrated in Fig. 23b. This
figure shows pump-dependent emission spectra recorded for a CdSe NQD
film fabricated from TOPO-capped CdSe NQDs with $R = 2.1$ nm. The
spectra were taken using a frequency-doubled output of an amplified Ti-
sapphire laser as a pump (the photon energy was 3.1 eV, the pulse duration
was 100 fs); the ASE was detected at the edge of the film, which acted as an
optical waveguide. At pump levels of ~1.5 e–h per dot, a sharp ASE peak
develops on the low-energy side of the spontaneous emission band. The pump

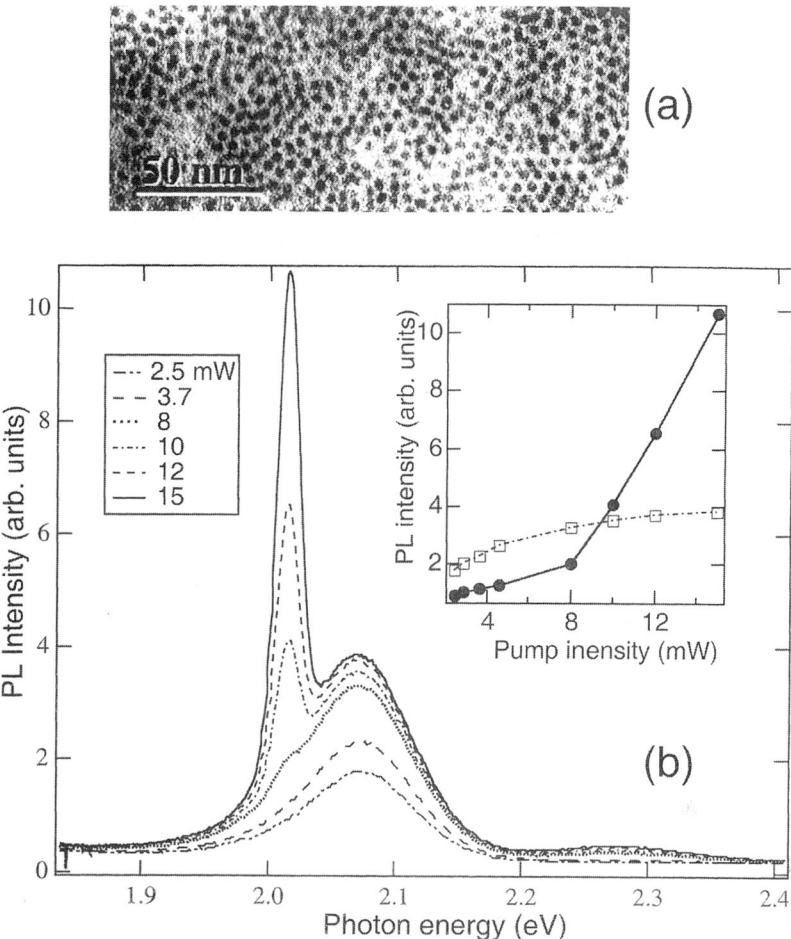

Figure 23 (a) A typical TEM image of a matrix-free NQD film indicating random close-packing. (b) Development of a sharp ASE band as a function of pump intensity in PL spectra of the film fabricated from CdSe NQDs with $R = 2.1$ nm ($T = 80$ K). Inset: Superlinear pump dependence of the ASE intensity (solid circles) showing a clear threshold compared to the sublinear dependence of the PL intensity outside the sharp ASE peak (open squares).

intensity dependence of this peak (Fig. 23b, inset) shows a threshold behavior that is a clear signature of optical amplification.

As expected from quantum-confinement effects, the ASE band is spectrally tunable by changing the dot size. The ASE spectra in Fig. 24 ($T = 80$ K) indicate that the emission color can be tuned from red (1.95 eV) to green (2.3 eV) using CdSe NQDs with radii from 2.5 to 1.3 nm [50]. On the red-orange side of the spectrum, the ASE regime was realized using as-prepared TOPO-capped samples. However, because of the enhanced hole surface trapping in small dots, the use of ZnS overcoating was essential to observe ASE in the yellow-green region. As a result of strong quantum confinement, the ASE band in the smallest dots is blue-shifted with respect to the ASE band in bulk CdSe by more than 0.5 eV.

The NQD films also show ASE at room temperature. Interestingly, the same pump fluences that are used to excite room temperature ASE in CdSe NQDs are not sufficient to produce light amplification in bulk CdSe samples. In bulk CdSe, optical gain can be due to both low-threshold excitonic and

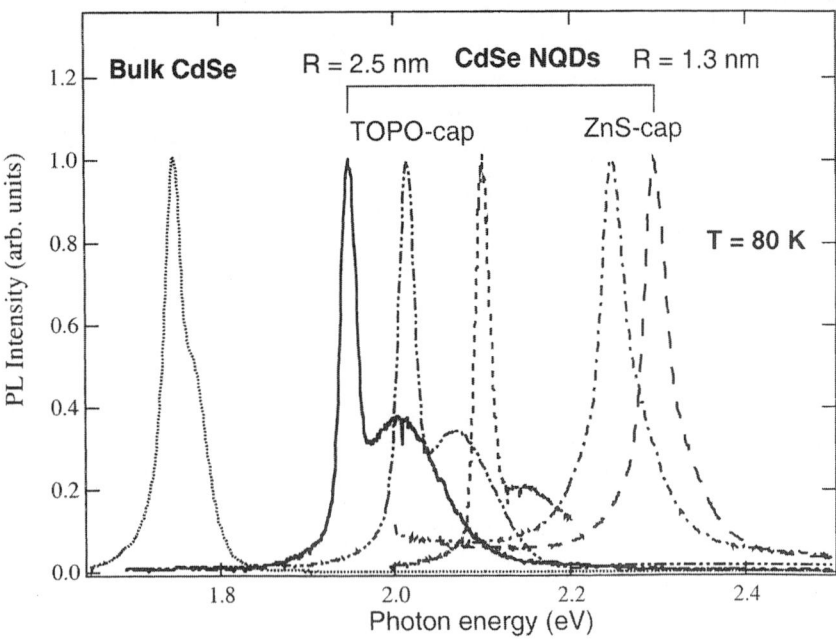

Figure 24 Size-controlled ASE spectra of close-packed films fabricated from either TOPO- or ZnS-capped CdSe NQDs with radii from 1.3 to 2.5 nm in comparison with an ASE spectrum of bulk CdSe ($T = 80$ K).

high-threshold e–h plasma mechanisms. Thermal dissociation of excitons at room temperature (the exciton binding energy in CdSe is 16 meV) results in a significantly increased threshold for light amplification. Because of a large interlevel spacing, "quantum-confined" excitons in NQDs are more robust than bulk excitons, allowing one to excite room-temperature ASE at pump levels comparable to those at cryogenic temperatures. This is an illustrative example of the enhanced temperature stability of optical gain expected for strongly confined NQDs.

B. Modal Gain Studies

One of the most direct ways of quantifying optical-gain properties of materials is by using "variable-stripe-length" measurements of ASE [61]. In these measurements, the excitation beam is focused onto the sample with a cylindrical lens into a narrow stripe of variable length l and the emission intensity is monitored as a function of the stripe length (see inset in Fig. 25a). This experiment allows one to determine net or modal gain (g_m), which represents a difference between material gain (g) and linear optical losses due, for example, to light scattering.

Figure 25a shows a set of room-temperature emission spectra for a CdSe NQD film (mean dot radius $R = 2.5$ nm) recorded using a "variable-stripe-length" configuration. The development of ASE (a narrow peak at ~635 nm) occurs at a threshold stripe length, l_{th}, of ~0.06 cm. The ASE threshold is clearly seen in the plot of the emission intensity (I) versus the stripe length as a sharp increase in the slope at $l > l_{th}$ (Fig. 25b). To derive the magnitude of modal gain, the I-versus-l dependence is usually analyzed using the expression $I = A(e^{g_m l} - 1)/g_m$, in which A is a constant proportional to the spontaneous emission power density [61]. This dependence should, in principle, describe both an initial, linear intensity growth below the ASE threshold (spontaneous emission regime) and a fast, exponential increase above it (stimulated emission regime). However, in the case of NQD samples, this simple expression does not allow one to simultaneously fit both regimes (see fit shown by dashed line in Fig. 25b). This discrepancy is fundamentally linked to the fact that in NQDs, spontaneous and stimulated emission arise from different types of excitation.

As indicated by the analysis in previous sections, optical gain and hence, stimulated emission in NQDs are dominated by quantum-confined biexcitons (doubly-excited dots). However, because of their short lifetimes, which are limited by nonradiative Auger recombination, biexcitons are not pronounced in spontaneous emission. Simple estimations based on the comparison of efficiencies of Auger decay and radiative recombination indicate that the contribution of biexcitons to time-integrated spontaneous emission is less

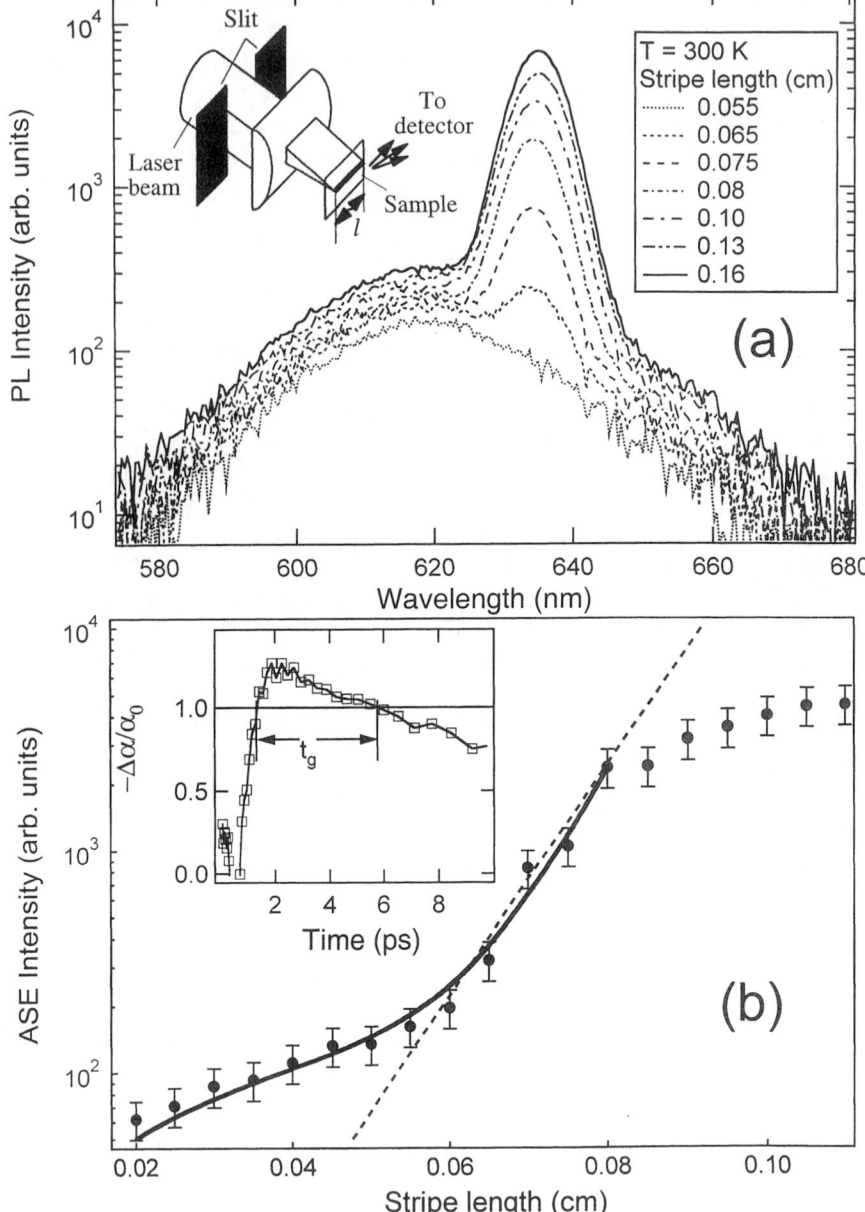

than 1%. Acceleration of radiative decay in the regime of stimulated emission as well as selective amplification of propagating waveguide modes make the contribution of biexcitons dominant over the contribution of excitons in the sample edge emission above the ASE threshold.

In order to take into account the existence of two types of emitter (excitons and biexcitons) and the fact that optical gain in NQDs is dominated by biexcitons and spontaneous emission is primarily due to single excitons, the ASE formula can be rewritten as $I = A_x l + A_{bx}(e^{g_{bx} l} - 1)/g_{bx}$, where A_x and A_{bx} are constants proportional to spontaneous emission power densities for excitons and biexcitons, respectively, and g_{bx} is the biexciton modal gain. Using this formula, one can simultaneously fit both below- and above-threshold regions of the I-versus-l dependence (solid line in Fig. 25b), which yields $g_{bx} = 155 \text{ cm}^{-1}$. This estimate represents effective modal gain per unit stripe length measurd for waveguide modes. To estimate the modal gain in the absence of waveguiding effects (i.e., gain per unit propagation length along the ray), we can scale g_{bx} by a correction factor $\beta = n_{gl}/n \approx 0.83$ (n_{gl} is the index of a glass substrate), which yields $g_m = \beta g_{bx} \approx 130 \text{ cm}^{-1}$ (this estimate assumes that the ASE is dominated by the mode that corresponds to the critical angle of the total internal reflection). Because of the transient nature of optical gain in the case of pulsed excitation, the above-derived quantities represent time-averaged characteristics of the NQD gain medium (i.e., gain average over the gain lifetime).

The inset in Fig. 25b displays dynamics of normalized, pump-induced absorption changes $(-\Delta \alpha/\alpha_0)$ in the same NQD film as the one used in the ASE studies discussed earlier, These dynamics indicate that optical gain $(-\Delta \alpha/\alpha_0 > 1)$ only exists during time $t_g \approx 5$ ps following photoexcitation. As discussed earlier, the fast gain decay is due to intrinsic, nonradiative Auger recombination and, possibly, to ultrafast hole trapping [50]. The short gain lifetimes are likely responsible for saturation of the ASE intensity observed in "variable-stripe-length" measurements at $l > l_s \approx 0.08$ cm (Fig. 25b). The light propagation time corresponding to the "saturation" length is ~6 ps, which is a value close to the gain lifetime (~5 ps). This observation indicates that under pulsed-excitation conditions, the transient nature of the NQD gain

---

Figure 25 Room-temperature emission spectra of a CdSe NQD film recorded using the "variable-stripe-length" configuration (shown schematically in the inset); pump fluence is 1 mJ/cm^2. (b) Solid circles describe the ASE intensity as a function of the stripe length. The dashed line is a fit to $I = A(e^{g l} - 1)/g$ ($g = 120 \text{ cm}^{-1}$), whereas the solid line is a fit to $I = A_x l + A_{bx} (e^{g_{bx} l} - 1)/g_{bx}$ ($g_{bx} = 155 \text{ cm}^{-1}$); note a difference in gain magnitudes produced by these two fits. Inset: Gain dynamics measured using a TA experiment (t_g is a gain lifetime).

can contribute to the saturation of ASE intensities along with a well-known mechanism of the gain medium depletion [61].

C. Lasing Experiments

The lasing regime in NQDs is possible by combining high-density NQD materials with optical cavities to provide efficient optical feedback. Several types of cavity have been utilized to demonstrate NQD lasing, including polystyrene microspheres [17], glass microcapillary tubes [46,50,62], and distributed feedback (DFB) resonators [63].

In Ref. 62, cylindrical NQD microcavities were fabricated by drawing concentrated hexane solutions of TOPO-capped CdSe NQDs ($R = 2.5$ nm) into microcapillary tubes (the inner tube diameter was 80 μm). The solvent was then allowed to evaporate. To increase the thickness of the NQD layer deposited on the inner wall of the tube, the above procedure was repeated several times. Cylindrical NQD microcavities formed in this way (inset in Fig. 26a) can support both planar-waveguide-like modes that develop along the tube length and whispering gallery (WG) modes developing around the inner circumference of the tube in the regime of total internal reflection. For the modes propagating along the tube, one can only achieve the ASE regime (no optical feedback is present), whereas WG modes can support a true lasing action (microring lasing) [64].

Depending on the sample, it is observed that as the pump level is increased, either microring lasing or ASE along the tube can develop first. This difference is likely due to uncontrolled variations in the optical quality of the NQD layer along and around the tube from one sample to another. For example, for the data shown in Fig. 26a (sample temperature is 80 K), light amplification along the tube (the ASE peak at ~613 nm) develops before the WG lasing. As the pump level is increased further, periodic oscillations due to lasing into WG modes develop on the top of the broader ASE peak.

The data in Fig. 26b, which were taken for a different sample, provide an example of pure microring lasing. Lasing into a single, sharp WG mode

---

Figure 26   Pump-dependent emission spectra of NQDs solids (sample temperature is 80 K) incorporated into microcapillary tubes (NQD cylindrical microcavities) demonstrating different emission regimes: (a) ASE along the tube length develops before microring lasing; (b) the regime of pure microring lasing (the emission spectra are shown by lines and the gain spectrum is shown by solid circles); the emission spectra are taken at progressively higher fluences of 1, 1.4, 1.6, and 2.8 mJ/cm². Insets: (a) schematics of a NDQ microcavity; (b) pump-fluence dependence of the intensity of the 612.0-nm mode demonstrating a sharp transition to the lasing regime.

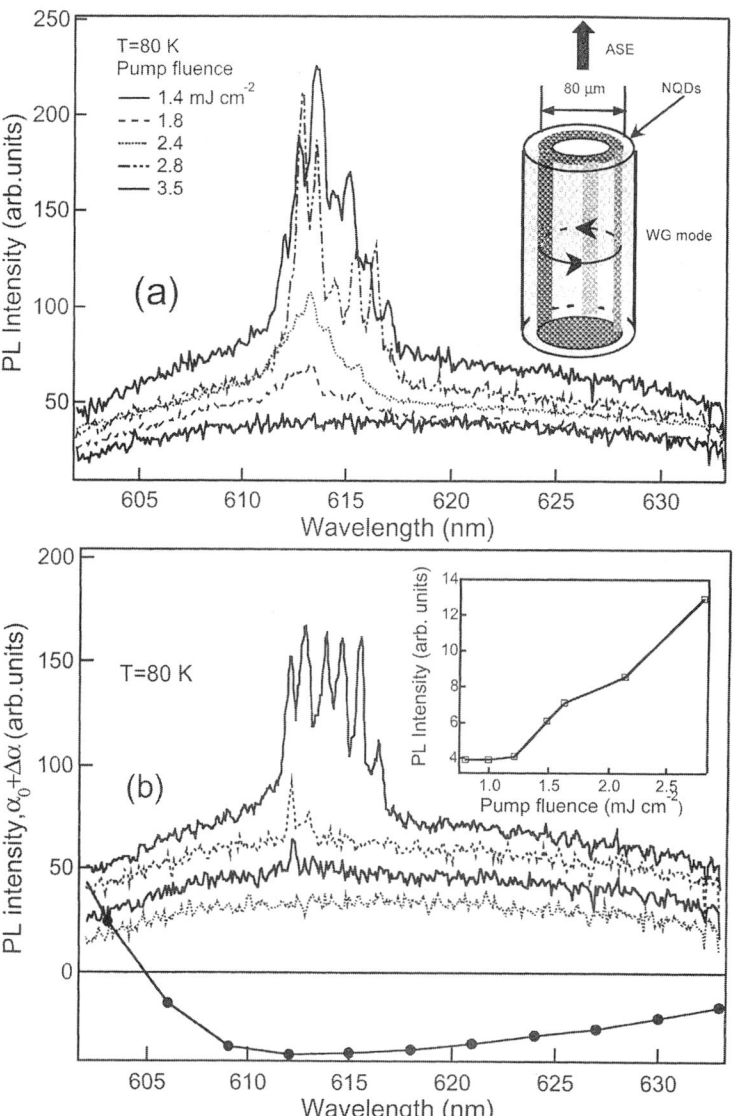

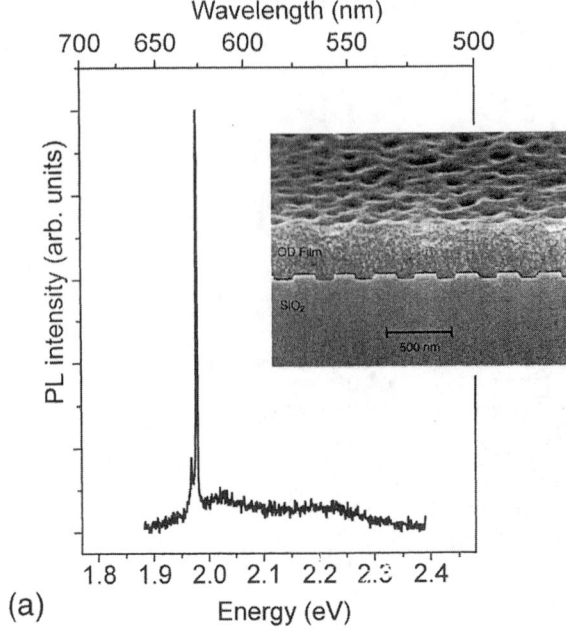

(a)

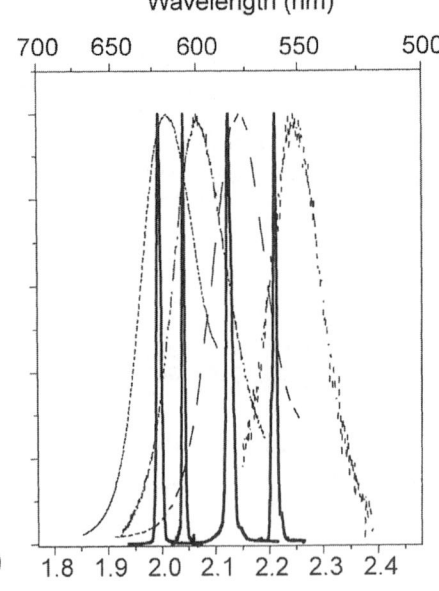

(b)

develops at ~1.25 mJ/cm^2 (the lasing threshold is clearly seen in the pump-dependent mode intensity in the inset in Fig. 26b). The position of this mode (612.0 nm) corresponds to the optical-gain maximum (solid circles in Fig. 26b). As the pump fluence is increased, additional WG modes develop on the low-energy side of the 612.0-nm mode, which is consistent with an asymmetric shape of the gain band. From the intermode spacing (~0.75 nm), one can derive an effective index of the gain medium of 1.8, indicating that WG modes are confined primarily within the NQD layer.

In Ref. 63, NQD lasing structures were fabricated by combining sol-gel titania waveguides [11] containing large NQD volume fractions with DFB gratings (Fig. 27a, inset). The gratings were fabricated by patterning thermal oxide layers on silicon substrates. The structures were completed by spin-coating thin NQD/titania films on the top of the gratings.

In the DFB/NQD structure, the refractive index is spatially modulated due to the thickness modulation of the nanocrystal/titania film caused by the underlying grating. This modulation leads to the formation of the stop band with a width of ~10 nm. Above the lasing threshold, a sharp mode at the edge of the stop band grows superlinearly as the pump power is increased (see spectrum in Fig. 27a). The emission from the front of the structure collapses into a visible laser beam, and the luminescence spot on the film tightens into a bright narrow line indicating a transition to the lasing regime. The DFB structures were used to demonstrate lasing with a tunable emission color using NQDs of different sizes, as illustrated in Fig. 27b. This figure shows sharp lasing lines (solid lines) that are compared to spontaneous emission spectra (dashed lines) recorded in the "subthreshold" regime. The lasing spectrum was tunable from 621 to 560 nm by changing the NQD radii from 2.7 to 1.7 nm.

VI. CONCLUSIONS AND OUTLOOK

Optical gain and lasing in strongly confined, colloidal quantum dots is a new, exciting area in NQD research. Due to very small sizes and the strong

---

Figure 27 (a) The room-temperature laser spectrum of a DFB device fabricated using core-shell CdSe/ZnS NQDs with a 2.5-nm CdSe core radius. Inset: the scanning electron microphotograph of a typical DFB device used in Ref. 63. (b) Spontaneous emission spectra (dashed lines) of CdSe NQDs with radii 2.7, 2.4, 2.1, and 1.7 nm (size decreases from left to right) in comparison with lasing spectra (solid lines) of the same samples incorporated into a DFB device. The lasing emission occurs at 621 nm ($R = 2.7$ nm), 607 nm ($R = 2.4$ nm), 583 nm ($R = 2.1$ nm), and 560 nm ($R = 1.7$ nm).

coupling between "volume" and interface states, optical-gain properties of NQDs are significantly different from those of well-studied semiconductor nanostructures fabricated by epitaxial techniques. The specificity of colloidal nanosystems results in unexpected complications for the development of optical gain and stimulated emission such as highly efficient, nonradiative Auger recombination, and strong photoinduced absorption associated with carriers trapped at NQD interfaces. On the other hand, NQDs offer numerous advantages for applications as optical-gain media. These advantages are due to the wide-range tunability of emission color provided by the facile manipulation of NQD sizes, large separations between energy states, and the flexibility of the chemical syntheses used for the NQD fabrication. Using appropriate materials, one can fabricate strongly confined NQDs emitting in spectral ranges from the far-infrared to ultraviolet. Furthermore, NQDs can be easily incorporated into transparent host matrices and different photonic structures including microcavities and photonic crystals.

The first demonstrations of NQD lasing devices operating with a wide range of colors indicate a high technological potential of NQD materials as new types of tunable lasing medium. On the other hand, this work also opens new and interesting avenues in fundamental research, specifically in the arena of light–matter interactions in photonic structures. Tunable electronic structures in NQDs, combined, for example, with a tunable density of photonic states in high finesse microcavities, can provide a wide range of control over electron–photon coupling and allow, in principle, a realization of a "strong-coupling" regime between NQDs and cavity modes. Such a regime is required for single-dot lasing and quantum information processing using, for example, NQD spins coupled through a microcavity mode.

There are currently several important challenges in NQD lasing research. Some of them are associated with materials quality issues and some are of conceptual character. On the materials' side, further improvements in monodispersity and surface passivation are required to improve NQD lasing performance. There is also a need for new, high-quality NQD materials for the extension of operational wavelengths into infrared and ultraviolet spectral ranges. Recent demonstration of high-quality PbSe NQDs [65] is promising for realizing lasing in the near-infrared spectral range, specifically in the range of communication wavelengths. Further work on reliable procedures for incorporating NQDs into solid-state matrices is an important step toward durable and stable NQD materials for lasing applications.

An important conceptual challenge focuses on carrier electrical injection. Previous approaches to electrical pumping of NQDs were based on combining NQDs with conducting polymers in hybrid organic–inorganic structures [66,67]. The performance of these devices was severely limited because of low carrier mobilities in both polymer and NQD device components and poor photostability of polymers. Recent work on InP NQD/ polymer structures [68] indicates improvement in the performance of hybrid systems. However, a many-orders improvement in the injection efficiency is still required for achieving the gain threshold.

One possible strategy to improve injection efficiencies can be through combining "soft" colloidal methods with traditional epitaxial techniques for incorporating dots into high-quality "injection" layers of wide-gap semiconductors. A possible technique that is "gentle" enough to be compatible with colloidal dots is energetic neutral atom beam epitaxy. This method utilizes a beam of neutral atoms carrying significant kinetic energy of several electron volts. The beam energy is sufficient for the activation of nonthermal surface chemical reactions, eliminating the need for substrate heating in order to grow high-quality films for NQD encapsulation.

Because of highly efficient nonradiative Auger decay, the realization of injection pumping of NQD lasing devices is significantly more difficult than pumping simple, "subthreshold" light emitters. Suppression of Auger recombination would simplify achieving lasing in the case of both optical and electrical pumping. One possible approach for reducing Auger rates is based on utilizing NQD shape control [69]. Previous work indicates that in spherical nanoparticles, Auger rates scale inversely with NQD volume [16]. If this scaling is also valid for nonspherical, elongated nanoparticles (quantum rods), the efficiency of Auger recombination should decrease with increasing nanoparticle aspect ratio. Because for elongated particles the energy gap is primarily determined by the length of the "short" axis [70], quantum rods can, in principle, provide independent (or weakly dependent) controls for both the position of the emission band (by varying the dimension along the "short" axis) and the efficiency of Auger recombination (by changing the "long" axis length).

Another interesting problem for future research is "single-exciton" versus "biexciton" mechanisms for optical gain. The "biexcitonic" gain origin is a consequence of the degeneracy of the lowest emitting transition. If the degeneracy is split using, for example, interactions with magnetic impurities [71], the gain can, in principle, be realized using NQD single-exciton states. In addition to reduced threshold, "single-exciton lasing" would eliminate the problem of ultrafast gain decay due to multiparticle Auger recombination.

ACKNOWLEDGMENTS

I would like to acknowledge contributions of A. A. Mikhailovsky, J. A. Hollingsworth, M. A. Petruska, A. V. Malko, H. Htoon, and S. Xu to the work reviewed here. I am also grateful to Al. L. Efros and M. G. Bawendi for numerous discussions on the photophysics of quantum dots. I would like to thank my wife, Tatiana, for her patience and continuing support throughout my research career and, specifically, during the work on this chapter.

This work was supported by Los Alamos Directed Research and Development Funds, and the U.S. Department of Energy, Office of Sciences, Division of Chemical Sciences.

REFERENCES

1. Arakawa, Y.; Sakaki, H. Appl. Phys. Lett. 1982, *40*, 939.
2. Asada, M.; Miyamoto, Y.; Suematsu, Y. IEEE J. Quantum Electron 1986, *22*, 1915.
3. Vandyshev, Y.V.; Dneprovskii, V.S.; Klimov, V.I.; Okorokov, D.K. JETP Lett. 1991, *54*, 442.
4. Ledentsov, N.N.; Ustinov, V.M.; Egorov, A.Y.; Zhukov, A.E.; Maksimov, M.V.; Tabatadze, I.G.; Kopev, P.S. Semiconductors 1994, *28*, 832.
5. Kistaedter, N.; Ledentsov, N.N.; Grundmann, M.; Bimber, D.; Ustinov, V.M.; Ruvimov, S.S.; Maximov, M.V.; Kopev, P.S.; Alferov, Z.I.; Richter, U. Electron. Lett. 1994, *30*, 1416.
6. Murray, C.B.; Norris, D.J.; Bawendi, M.G. J. Am. Chem. Soc. 1993, *115*, 8706.
7. Alivisatos, A.P. Science 1996, *271*, 933.
8. Qu, L.; Peng, X. J. Am. Chem. Soc. 2002, *124*, 2049.
9. Hines, M.; Guyot-Sionnest, P. J. Phys. Chem. 1996, *100*, 468.
10. Murray, C.B.; Kagan, C.B.; Bawendi, M.G. Science 1995, *270*, 1335.
11. Sundar, V.C.; Eisler, H.-J.; Bawendi, M.G. Adv. Mater. 2002, *14*, 739.
12. Petruska, M.A.; Malko, A.V.; Voyles, P.M.; Klimov, V.I. Adv. Mater. 2003, *15*, 610.
13. Gindele, F.; Westphaeling, R.; Woggon, U.; Spanhel, L.; Ptatschek, V. Appl. Phys. Lett. 1997, *71*, 2181.
14. Benisty, H.; Sotomayor-Torres, C.; Weisbuch, C. Phys. Rev. B 1991, *44*, 10945.
15. Inoshita, T.; Sakaki, H. Phys. Rev. B 1992, *46*, 7260.
16. Klimov, V.I.; Mikhailovsky, A.A.; McBranch, D.W.; Leatherdale, C.A.; Bawendi, M.G. Science 2000, *287*, 1011.
17. Klimov, V.I.; Bawendi, M.G. MRS Bull. 2001, *26*, 998.
18. Efros, Al.L.; Efros, A.L. Sov. Phys. Semicond. 1982, *16*, 772.
19. Ekimov, A.I.; Hache, F.; Schanne-Klein, M.C.; Ricard, D.; Flytzanis, C.; Kudryavtsev, I.A.; Yazeva, T.V.; Rodina, A.V.; Efros, Al.L. J. Opt. Soc. Am. B 1993, *10*, 100.
20. Efros, Al.L. Phys. Rev. B 1992, *46*, 7448.
21. Efros, Al.L.; Rodina, A.V. Phys. Rev. B 1993, *47*, 10,005.
22. Takagahara, T. Phys Rev. B 1993, *47*, 4569.
23. Nirmal, M.; Norris, D.J.; Kuno, M.; Bawendi, M.G.; Efros, Al.L.; Rosen, M. Phys. Rev. Lett. 1995, *75*, 3728.

24. Norris, D.J.; Efros, Al.L.; Rosen, M.; Bawendi, M.G. Phys Rev. B 1996, *53*, 16347.

25. Conwell, E. *High Field Transport in Semiconductors*; New York: Academic Press, 1967.

26. Prabhu, S.; Vengurlekar, A.; Shah, J. Phys. Rev. B 1995, *51*, 14,233.

27. Klimov, V.; Haring Bolivar, P.; Kurz, H. Phys. Rev. B 1995, *52*, 4728.

28. Bockelmann, U.; Bastard, G. Phys. Rev. B 1990, *42*, 8947.

29. Klimov, V.I.; McBranch, D.W. Phys. Rev. Lett. 1998, *80*, 4028.

30. Guyot-Sionnest, P.; Shim, M.; Matranga, C.; Hines, M. Phys. Rev. B 1999, *60*, 2181.

31. Klimov, V.I.; Mikhailovsky, A.A.; McBranch, D.W.; Leatherdale, C.A.; Bawendi, M.G. Phys. Rev. B 2000, *61*, 13,349.

32. Klimov, V.I. In *Handbook on Nanostructured Materials and Nanotechnology*; Nalwa, H.S. Ed.; San Diego, CA: Academic Press, 1999.

33. Klimov, V.I. J. Phys. Chem. B 2000, *104*, 6112.

34. Sercel, P. Phys. Rev. B 1995, *51*, 14,532.

35. Bockelmann, U.; Egler, T. Phys. Rev. B 1992, *46*, 15,574.

36. Efros, A.L.; Kharhenko, V.A.; Rosen, M. Solid State Commun. 1995, *93*, 281.

37. Klimov, V.I.; McBranch, D.W.; Leatherdale, C.A.; Bawendi, M.G. Phys. Rev. B 2000, *60*, 13,740.

38. Mikhailovsky, A.A.; Xu, S.; Klimov, V.I. Rev. Sci. Instrum. 2002, *73*, 136.

39. Xu, S.; Mikhailovsky, A.A.; Hollingsworth, J.A.; Klimov, V.I. Phys. Rev. B 2002, *65*, 53,191.

40. Woggon, U.; Giessen, H.; Gindele, F.; Wind, O.; Fluegel, B.; Peyghambarian, N. Phys. Rev. B 1996, *54*, 17,681.

41. Shah, J. IEEE J. Quantum Electron 1988, *24*, 276.

42. Uskov, A.V.; Jauho, A.P.; Tromborg, B.; Mork, J.; Lang, R. Phys. Rev. Lett. 2000, *85*, 1516.

43. Klimov, V.I.; Schwarz, C.J.; McBranch, D.W.; Letherdale, C.A.; Bawendi, M.G. Phys. Rev. B 1999, *60*, 2177.

44. Malko, A.V.; Hollingsworth, J.A.; Petruska, M.A.; Mikhailovsky, A.A.; Klimov, V.I., unpublished results.

45. Ricard, D.; Ghanassi, M.; Schanneklein, M. Opt. Commun. 1994, *108*, 311.

46. Klimov, V.I.; Mikhailovsky, A.A.; Hollingsworth, J.A.; Malko, A.V.; Leatherdale, C.A.; Bawendi, M.G. Proc. Electrochem. Soc. 2001, *19*, 321.

47. Klimov, V.; Hunsche, S.; Kurz, H. Phys. Rev. B 1994, *50*, 8110.

48. Efros, A.L.; Rosen, M. Phys. Rev. B 1996, *54*, 4843.

49. Crooker, S.A.; Barrick, T.; Hollingsworth, J.A.; Klimov, V.I. Appl. Phys. Lett. 2003, *82*, 2793.

50. Mikhailovsky, A.A.; Malko, A.V.; Hollingsworth, J.A.; Bawendi, M.G.; Klimov, V.I. Appl. Phys. Lett. 2002, *80*, 2380.

51. Hu, H.G.Y.; Peyghambarian, N.; Koch, S. Phys. Rev. B 1996, *53*, 4814.

52. Giessen, H.; Butty, J.; Woggon, U.; Fluegel, B.; Mohs, G.; Hu, Y.Z.; Koch, S.W.; Peyghambarian, N.; Koch, S. Phase Transitions 1999, *68*, 59.

53. Kang, K.; Kepner, A.; Gaponenko, S.; Koch, S.; Hu, Y.; Peyghambarian, N. Phys. Rev. B 1993, *48*, 15,449.
54. Shumway, J.; Franceschetti, A.; Zunger, A. Phys. Rev. B 2001, *63*, 53,161.
55. Chatterji, D. *The Theory of Auger Transitions*; London: Academic Press, 1976.
56. Landsberg, P. *Recombination in Semiconductors*; Cambridge: Cambridge University Press, 1991.
57. Chepic, D.; Efros, Al.L.; Ekimov, A.; Ivanov, M.; Kharchenko, V.A.; Kudriavtsev, I. J. Lumin 1990, *47*, 113.
58. Kharchenko, V.A.; Rosen, M. J. Lumin 1996, *70*, 158.
59. Klimov, V.I.; Mikhailovsky, A.A.; Xu, S.; Malko, A.; Hollingsworth, J.A.; Leatherdale, C.A.; Eisler, H.J.; Bawendi, M.G. Science 2000, *290*, 314.
60. Kazes, M.; Lewis, D.Y.; Evenstein, Y.; Mokari, T.; Banin, U. Adv. Mater. 2002, *14*, 317.
61. Shaklee, K.L.; Nahory, R.E.; Leheny, R.F. J. Lumin 1973, *7*, 284.
62. Malko, A.V.; Mikhailovsky, A.A.; Petruska, M.A.; Hollingsworth, J.A.; Htoon, H.; Bawendi, M.G.; Klimov, V.I. Appl. Phys. Lett. 2002, *81*, 1303.
63. Eisler, H.-J.; Sundar, V.C.; Bawendi, M.G.; Walsh, M.; Smith, H.I.; Klimov, V.I. Appl. Phys. Lett. 2002, *80*, 4614.
64. McCall, S.L.; Levi, A.F.; Slusher, R.E.; Pearton, S.J.; Logan, R.A. Appl. Phys. Lett. 1992, *60*, 289.
65. Murray, C.B.; Sun, S.; Gaschler, W.; Doyle, H.; Betley, T.A.; Kagan, C.R. IBM J. Res. Dev. 2001, *45*, 47.
66. Colvin, V.L.; Schlamp, M.C.; Alivisatos, A.P. Nature 1994, *370*, 354.
67. Dabbousi, B.O.; Bawendi, M.G.; Onitsuka, O.; Rubner, M.F. Appl. Phys. Lett. 1995, *66*, 1316.
68. Tessler, N.; Medvedev, V.; Kazes, M.; Kan, S.; Banin, U. Science 2002, *295*, 1506.
69. Htoon, H.; Hollingsworth, J.A.; Malko, A.V.; Dickerson, R.; Klimov, V.I. Appl. Phys. Lett. 2003, *82*, 4776.
70. Li, L.; Ho, J.; Yang, W.; Alivisatos, A.P. Nano Lett. 2001, *1*, 349.
71. Norris, D.J.; Yao, N.; Charnock, F.T.; Kennedy, T.A. Nano Lett. 2001, *1*, 3.

6

Optical Dynamics in Single Semiconductor Quantum Dots

K. T. Shimizu and M. G. Bawendi
*Massachusetts Institute of Technology,
Cambridge, Massachusetts, U.S.A.*

I. INTRODUCTION

In the pursuit to realize the potential of semiconductor nanocrystallite quantum dots (QDs) as nanomaterials for future biological and solid-state, electro-optical applications [1–4], the exact photophysics of these colloidal QDs becomes ever more significant. Although these QDs have been described as artificial atoms because of their proposed discrete energy levels [5,6], they differ from elemental atoms in their inherent size inhomogeneity. The size-dependent, quantum-confined properties that make this material novel also hinder the particles from being studied in detail. In other words, spectral and emission intensity features may vary for each QD depending on their size, shape, and degree of defect passivation. This disparity in the sample can lead to ensemble averaging, where the average value masks the individual's distinctive properties [7]. As the ultimate limit in achieving a narrow size distribution for physical study, we examine the QDs on an individual basis. Much work has been done in the field of single-molecule spectroscopy [8–11], investigating absorption, emission, lifetime, and polarization properties of these molecules. In this review, we focus on the mechanisms underlying the dynamic inhomogeneities—spectral diffusion and fluorescence intermittency—observed in the emission properties of single CdSe and CdTe colloidal QDs. Described frequently in a myriad of single chromophore studies [12–16], *spectral diffusion* refers to discrete and continuous changes in the emitting wavelength as a function of time, whereas *fluorescence intermittency* refers to the *"on–off"* emission intensity fluctuations that occur on the timescale of microseconds to minutes. Both of these QD phenomena, observed at cryogenic and room temperatures under continuous photoexcitation, give insight into a rich array of electrostatic dynamics intrinsically occurring in and around each individual QD.

II. SINGLE-QUANTUM-DOT SPECTROSCOPY

We studied many individual QDs simultaneously using a home-built, epifluorescence microscope coupled with fast data storage and data analysis. This setup is also referred to as a wide-field or far-field microscope due to the diffraction-limited, spatial resolution of the excitation light source. The basic components, shown in Fig. 1, are similar to most optical microscopes: a light source, microscope objective, stages for x-y manipulation and focusing of the objective relative to the sample, a spectrometer, and coupled-charge device (CCD) camera. To access the emission from individual chromophores, a 90% reflective silver mirror is placed as shown in Fig. 1 to allow for a small fraction of the excitation light to pass into the objective and the collection of 90% of the emitted light. Suitable optic filters are used to remove any residual excitation light. For all of the experiments discussed here, single-QD emission images and spectra were recorded with a bin size of 100 ms for durations of 1 h under continuous wave (cw), 514-nm, Ar ion laser excitation. However, for strictly emission intensity measurements, time resolution as fast as 500 µs is possible using an avalanche photodiode. The low-temperature studies were performed using a cold-finger, liquid-helium cryostat with a long-working-distance air objective (N.A. 0.7), whereas room-temperature studies were performed using an oil-immersion objective (N.A. 1.25). The raw data are collected in a series of consecutive images to form nearly continuous three-dimensional datasets as shown in Fig. 1b. The dark spots represent emission from individual QDs spaced ~1 µm apart. An advantage of a CCD camera over avalanche photodiode or photomultiplier tube detection is that spectral data of single QDs can be obtained in one frame using a monochromator. Moreover, all of the dots imaged on the entrance slit of the monochromator are observed in parallel. If only relative frequency changes need to be addressed, then the entrance slit can be removed entirely, allowing parallel tracking of emission frequencies and intensities of up to 50 QDs simultaneously. In cases where spectral information is not needed, such as in Fig. 1b, up to 200 QDs can be imaged simultaneously. The data analysis program then retrieves the time–frequency (or space)–intensity emission trajectories for all of these QDs. This highly

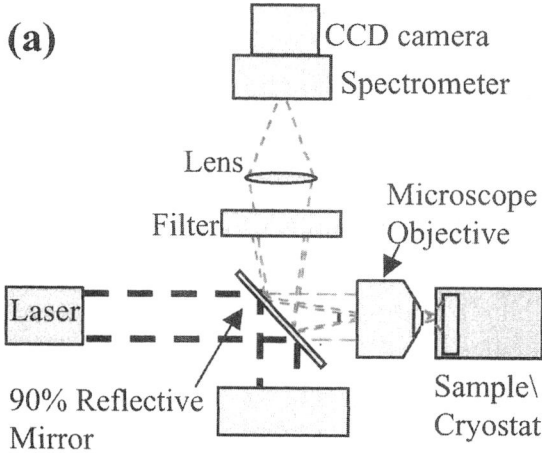

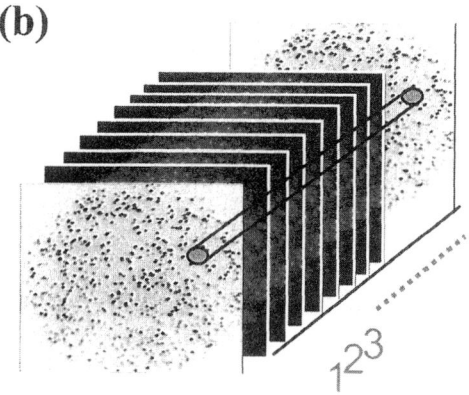

Figure 1 (a) Optical microscope schematic: 514 nm continuous-wave Ar ion laser excitation is used for most of the experiments described. The 90% reflective aluminum mirror angled at 45° allows for low-intensity excitation of the QD sample and highly efficient collection of the emitted light using the same objective. (b) Sample dataset of single QD images taken continuously on the intensified CCD camera. The arrow and highlighted circle indicate how the same QD can be traced throughout the entire set after the data have been collected.

parallel form of data acquisition is vital for a proper statistical sampling of the entire population.

The CdSe QDs are prepared following the method of Murray et al. [17] and protected with ZnS overcoating [18,19], whereas the CdTe samples were prepared following Ref. 20. All single QD samples are highly diluted and spin-cast in a 0.2–0.5-μm thin film of poly(methyl methacrylate) (PMMA) on a crystalline quartz substrate.

III. SPECTRAL DIFFUSION AND FLOURESCENCE INTERMITTENCY

Figures 2 and 3 showcase the typical, phenomenological behavior of fluorescence intermittency and spectral dynamics observed. An illustrative 3000-s time trace of fluorescence intermittency is shown for a CdSe/ZnS QD at 10 K and at 300 K in Figs. 2a and 2b, respectively. At first glance, there are clear differences between the blinking behaviors at these different temperatures. The QD appears to be emitting considerably more often at low temperature and the QD appears to turn *on* and *off* more frequently at room temperature (RT). However, by expanding a small section of the time trace, the similarities between these traces at different temperatures and the self-similarity of the traces on different timescales can be observed. The spectrally resolved time traces shown in Figs. 3b and 3c compare the spectral shifting for QDs at 10 K and at RT, respectively. At RT, the emission spectral peak widths range from 50 to 80 meV, whereas at 10 K, the characteristic phonon progression shown in Fig. 3a verifies the presence of CdSe QDs. Ultranarrow peak widths for the zero-phonon emission as small as 120 μeV have been previously observed at 10 K [21]. At either temperature, spectral shifts as large as 50 meV were observed in our experiments. Figure 4 shows the large variation in spectrally dynamic time traces from three QDs observed simultaneously at 10 K. The spectrum in Fig. 4a shows sharp emission lines with nearly constant frequency and intensity. The spectrum in Fig. 4b shows some pronounced spectral shifts and a few blinking events; the spectrum in Fig. 4c is fluctuating in frequency and shows a number of blinking events on a much faster timescale.

Early investigations of blinking and spectral diffusion have shed some light regarding these novel properties. The blinking of QDs showed a dependence on the surface overcoating, temperature, and excitation intensity. Individually or in any combination, increased thickness of ZnS overcoating, lower temperatures, and lower excitation intensity all decreased the blinking rate [12,22]. However, these earlier experiments were restricted to small numbers of QDs studied—one QD at a time—using confocal microscopy.

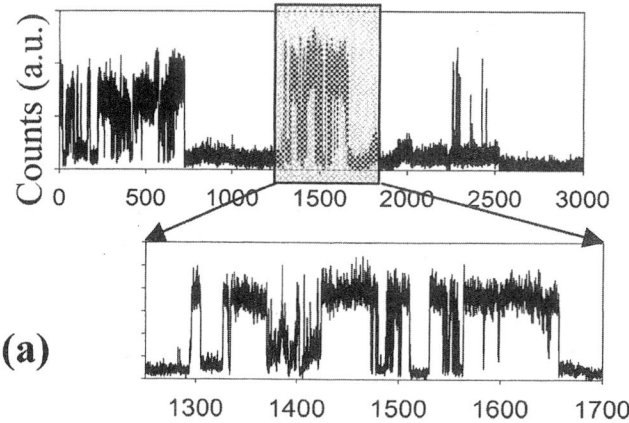

(a)

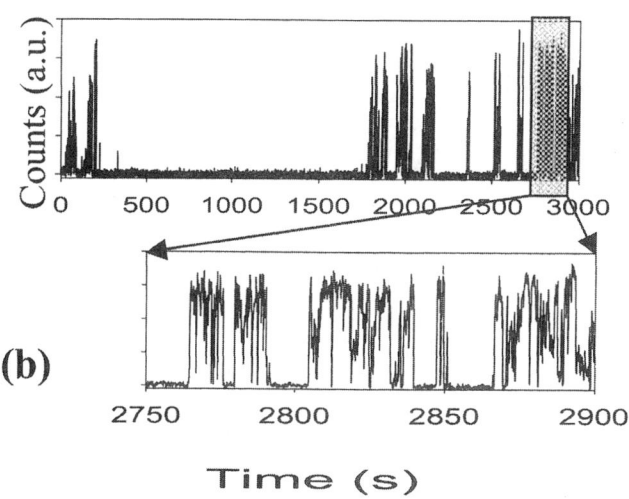

(b)

Time (s)

Figure 2 Representative intensity time traces for (a) 10 K and (b) room temperature. The magnified region exemplifies how similar the time traces look on different timescales and the similar nature between the time traces even at different temperatures (for short time regimes).

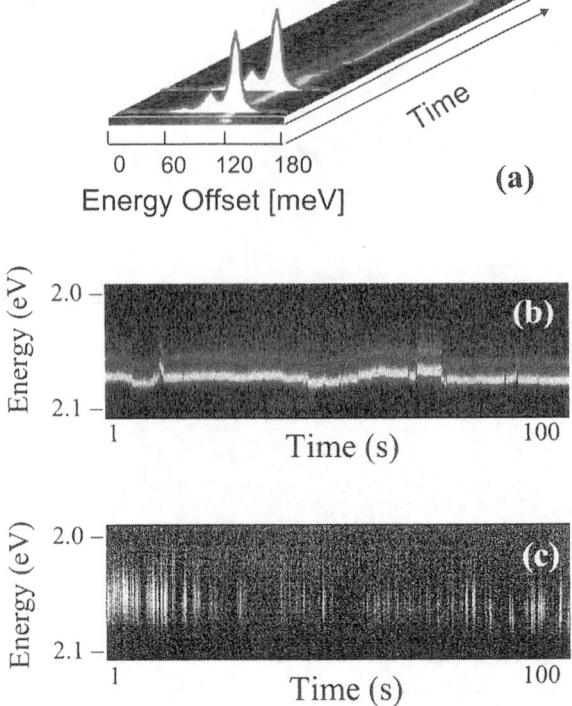

Figure 3 (a) Spectral time trace of a single CdSe/ZnS QD at 10 K. The phonon progression (~25 meV) can be seen to the red of the strongest zero-phonon peak. A comparison between spectral time traces for (b) 10 K and (c) room temperature shows that spectral diffusion is present at both temperatures.

In addition to the intensity data, spectral behavior similar to spectral diffusion was emulated by use of external dc electric fields [13]. The same external fields probed a changing, local electric dipole around each QD, indicating some changing local electric field around the QD. Despite these studies, uncertainty in the underlying physical mechanism remains.

IV. CORRELATION BETWEEN SPECTRAL DIFFUSION AND BLINKING

The spectral information in Figs. 4a–4c clearly shows that for a single quantum dot under the perturbations of its environment, there are many possible

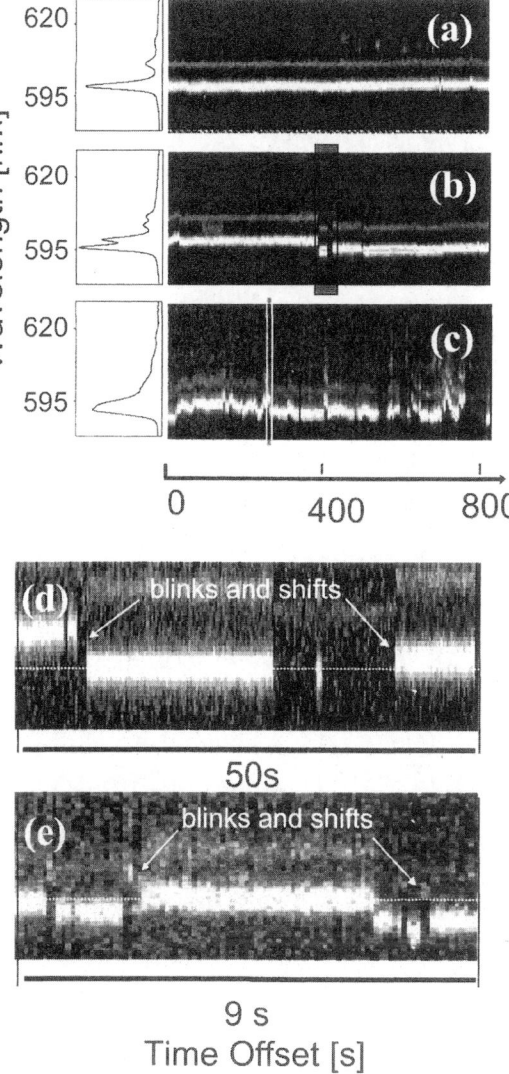

Figure 4 (a–c) Low-temperature spectral time trace of three CdSe/ZnS QDs demonstrating the different dynamics observed simultaneously on short timescales. The resulting averaged spectrum is plotted for each dot. The boxed areas in (b) and (c) are magnified and shown in (d) and (e). The blinking back *on* after a dark period is accompanied by a large spectral shift. The white dotted line is drawn in (d) and (e) as a guide to the eye.

transition energies. In fact, these emission dynamics suggest a QD intimately coupled to and reacting to a fluctuating environment. Through the concurrent measure of spectral diffusion and fluorescence intermittency, we examine the extent of this influence from the QD environment and observed an unexpected relationship between spectral diffusion and blinking. Zooming into the time traces of Figs. 4b and 4c reveals a surprising *correlation* between blinking and spectral shifting. As shown in Fig. 4d, magnifying the marked region in the time trace of Fig. 4b reveals a pronounced correlation between individual spectral jumps and blinking: following a blink-*off* period, the blink-*on* event is accompanied by a shift in the emission energy. Furthermore, as shown in Fig. 4e, zooming into the time trace of Fig. 4c reveals a similar correlation. As in Fig. 4d, the trace shows dark periods that are accompanied by discontinuous jumps in the emission frequency. The periods between shifts in Figs. 4d and 4e, however, differ by nearly an order of magnitude in timescales. Due to our limited time resolution, no blinking events shorter than 100 ms can be detected. Any fluorescence change that is faster than the 'blink-and-shift' event shown in Fig. 4e is not resolved by our apparatus and appears in a statistical analysis as a large frequency shift during an apparent *on*-time period. This limitation weakens the experimentally observed correlation between blinking and frequency shifts. Nevertheless, a statistically measurable difference between shifts following *on* and *off* times can be extracted from our results.

Because changes in the emitting state cannot be observed when the QD is *off*, we compare the net shifts in the spectral positions between the initial and final emission frequency of each *on* and *off* event. The histogram of net spectral shifts during the *on* times, shown in Fig. 5a, reveals a nearly Gaussian distribution (dark line) with a 3.8-meV full width at half-maximum. However, the histogram for the *off*-time spectral shifts in Fig. 5b shows a distribution better described as a sum of two distributions: a Gaussian distribution of small shifts and a distribution of large spectral shifts located in the tails of the Gaussian profile. To illustrate the difference between the distributions of *on*- and *off*-time spectral shifts, the *on*-time spectral distribution is subtracted from the *off*-time spectral distribution shown in Fig. 5c. Even with our limited time resolution, this difference histogram shows that large spectral shifts occur significantly more often during *off* times (longer than 100 ms) than during *on* times; hence, large spectral shifts are more likely to accompany a blink-*off* event than during the time the QD is *on*. This statistical treatment does not try to assess the distribution of QDs that show this correlation but rather confirms the strong correlation between spectral shifting and blinking events in the QDs observed. The *off*-time histogram, plotted on a logarithmic scale in Fig. 5d, shows that a single Gaussian

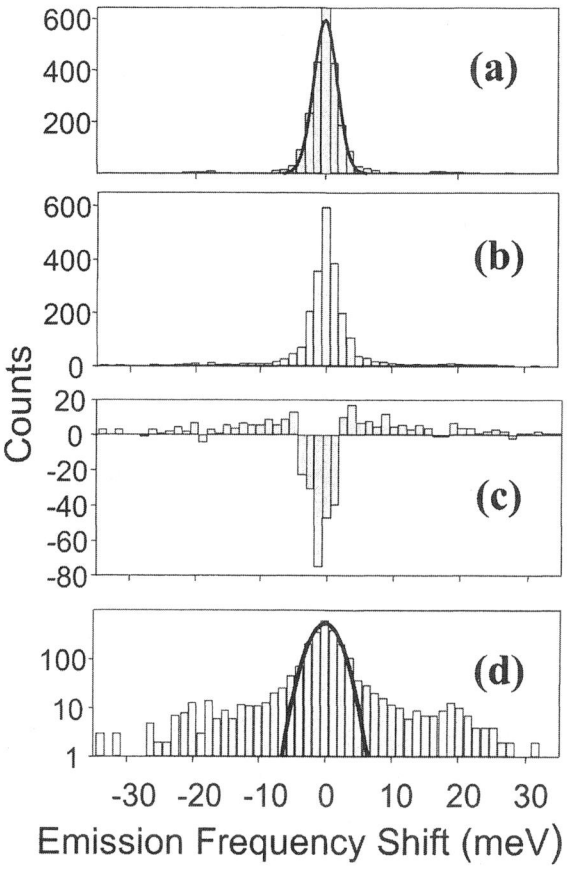

Figure 5 Distribution of net spectral shifts between the initial and final emission frequency for 2400 *on* and *off* periods of CdSe/ZnS QDs at 10 K. (a) Histogram of net spectral shifts for the on period shows a Gaussian distribution of shifts. The dark line is a best fit to Gaussian profile. (b) Histogram for *off* periods displays large counts in the wings of a similar Gaussian distribution. (c) Subtracting the *on*-period distribution from the *off*-period distribution magnifies the large counts in the wings of the Gaussian distribution. This quantifies the correlation that the large spectral shifts accompany an *off* event (longer than 100 ms) more than an *on* event. (d) A logarithmic plot of the histogram shows a clearer indication of the non-Gaussian distribution in the net spectral shifts during the *off* times. The dark line is a best fit to a Gaussian profile.

distribution (dark line) does not describe the distribution of *off*-time spectral shifts.*

Moreover, this correlation differs from blinking *caused* by spectral shifting observed in single molecules such as pentacene in a *p*-terphenyl matrix [23,24]. In single-molecule experiments, the chromophore is resonantly excited into a single absorbing state, and a spectral shift of the absorbing state results in a dark period because the excitation is no longer in resonance. In our experiments, we excite nonresonantly into a large density of states above the band edge [25].

The initial model for CdSe QD fluorescence intermittency [12,26] adapted a theoretical model for photodarkening observed in CdSe-QD-doped glasses [27] with the blinking phenomenon under the high excitation intensity used for single-QD spectroscopy. In the photodarkening experiments, Chepic et al. [28] described a QD with a single delocalized charge carrier (hole or electron) as a dark QD. When a charged QD absorbs a photon and creates an exciton, it becomes a quasi-three-particle system. The energy transfer from the exciton to the lone charge carrier and nonradiative relaxation of the charge carrier ($\sim$100 ps) [29] is predicted to be faster than the radiative recombination rate of the exciton (100 ns to 1 μs). Therefore, within this model, a charged QD is a dark QD. The transition from a bright to a dark QD occurs through the trapping of an electron or hole leaving a single delocalized hole or electron in the QD core. The switch from a dark to a bright QD then occurs through recapture of the initially localized electron (hole) back into the QD core or through capture of another electron (hole) from nearby traps. When the electron–hole pair recombines, the QD core is no longer a site for exciton-electron (exciton–hole) energy transfer. Concomitantly, Empedocles and Bawendi [13] showed evidence that spectral diffusion shifts were caused by a changing local electric field around the QD, where the magnitude of this changing local electric field was consistent with a single electron and hole trapped near the surface of the QD.

We can now combine both models to explain the correlation shown in Fig. 4. Using the assumption that a charged QD is a dark QD [30], there are four possible mechanisms, illustrated in Figs. 6a–6d, for the transition back to a bright QD. Electrostatic force microscopy (EFM) studies on single CdSe QDs recently showed positive charges present on some of the QDs,[†] even after

---

* The *on*-time histogram also has weak tails on top of the Gaussia distribution because of apparent *on*-time spectral shifts that may have occurred during an *off* time faster than the time resolution (100 ms) allowed by our present setup.

[†] Note we make a distinction between a charged QD where the charge is delocalized in the core of the QD (a dark QD) and a charged QD where the charge refers to trapped charges localized on the surface or in the organic shell surrounding the QD (not necessarily a dark QD).

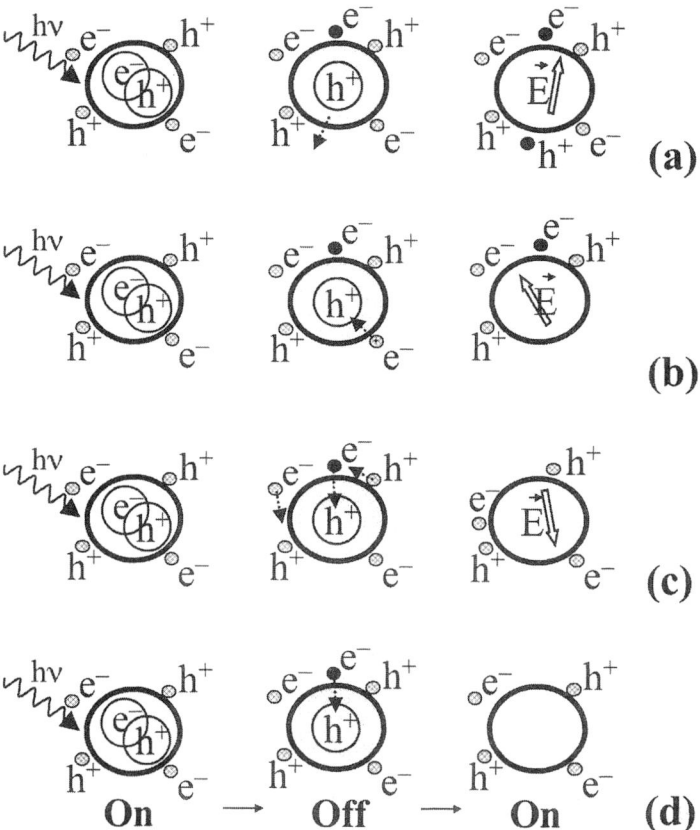

Figure 6 Four possible mechanisms for the correlation between spectral shifting and blinking. (a) An electron and hole become localized independent of the other charges surrounding the QD leading to a change in the electric field environment. (b) An electron from the core localizes to the surface, but a surrounding charge is recaptured into the core. After recombination with the delocalized hole, the net electric field has changed. (c) Although the same electron that was initially localized returns to the core to recombine with the delocalized hole, due to coulombic interaction, the charge distribution surrounding the QD has changed. (d) The same electron initially localized returns to the core to recombine with the delocalized hole and there is no change in the local electric field around the QD.

exposure to only room light. In our model, after cw laser excitation and exciton formation, an electron or hole from the exciton localizes near the surface of the QD, leaving a delocalized charge carrier inside the QD core. Following this initial charge localization or ionization, (1) the delocalized charge carrier can also be localized leading to a net neutral QD core, (2) If the QD environment is decorated by charges following process 1, then after subsequent ionization, a charge localized in the QD's environment can relax back into the QD core, recombining with the delocalized charge carrier, or (3) Coulombic interaction can lead to a permanent reorganization of the localized charge carriers present in the QD environment even after the same charge relaxes back into the core and recombines with initial delocalized charge. Mechanisms 1–3 would create, if not alter, a surface dipole and lead to a net change in the local electric field. The single-QD spectra express this change as a large Stark shift in the emission frequency. However, the model does not necessarily require that a blinking event be followed always by a shift in emission frequency. If the dark period was produced and removed by a localization and recapture of the same charge without a permanent reorganization of charges in the environment (4), the emission frequency does not change. Any changes in the emission frequency during this mechanism would be entirely thermally induced and such small spectral shifts are observed. Indeed, this pathway for recombination dominates very strongly, as most dark periods are not accompanied by large frequency shifts.

V. "POWER-LAW" BLINKING STATISTICS

The small number of QDs sampled and the short duration of each time trace limited early studies into the statistics of blinking in single CdSe QDs. Recently, Kuno et al. found that room-temperature fluorescence intermittency in single QDs exhibited power-law statistics—indicative of long-range statistical order [31]. By analyzing the power-law statistical results within a physical framework, we begin the dissection of the complex mechanism for blinking in these QDs. The statistics of both *on*- and *off*-time distributions are obtained under varying temperature, excitation intensity, size, and surface morphology conditions.

As schematically shown in Fig. 7a, we define the *on* time (or *off* time) as the interval of time when no signal falls below (or surpasses) a chosen threshold intensity value. The probability distribution is given by the histogram of *on* or *off* events:

$$P(t) = \sum \text{ events of length } t \tag{1}$$

where t refers to the duration of the blink-*off* or blink-*on* event. The *off*-time probability distribution for a single CdSe/ZnS QD at room temperature is

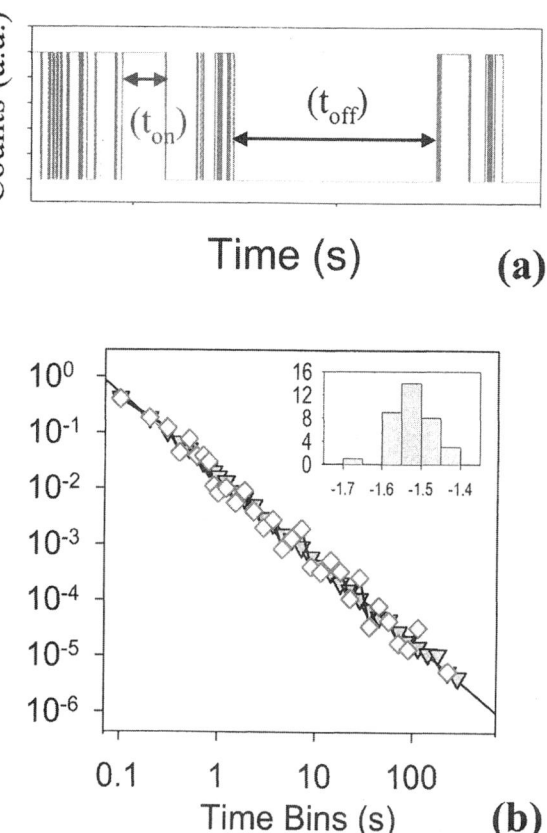

Figure 7 (a) Schematic representation of the *on*- and *off*-time per blink event for an intensity time trace of a single QD. (b) Normalized *off*-time probability distribution for 1 CdSe/ZnS QD and average of 39 CdSe/ZnS QDs. Inset shows the distribution of fitting values for the power-law exponent in the 39 QDs. The straight line is a best fit to the average distribution with exponent ~−1.5.

shown in Fig. 7b. The distribution follows a pure power law for the time regime of our experiments (~10^3 s):

$$P(t) = At^{-\alpha} \tag{2}$$

Almost all of the individual QDs also follow a power-law probability distribution with the same value in the power-law exponents ($\alpha \sim 1.5 \pm 0.1$). A histogram of α values for individual QDs is shown in the inset in Fig. 7b. The universality of this statistical behavior indicates that the blinking statistics for the *off* times are insensitive to the different characteristics (size, shape, defects, environment) of each dot. Initial experiments at RT show that the same blinking statistics are also observed in CdTe QDs, demonstrating that this power-law phenomenon is not restricted to CdSe QDs.

A. Temperature Dependence

To develop a physical model from this phenomenological power-law behavior, we probe the temperature dependence of the blinking statistics; this dependence should provide insight into the type of mechanism (tunneling versus hopping) and the energy scales of the blinking phenomenon. Qualitatively, the time traces in Figs. 2a and 2b suggest that at a low temperature, the QDs blink less and stay in the on state for a larger portion of the time observed. However, when we plot the *off*-time probability distributions at temperatures ranging from 10 to 300 K, as shown in Fig. 8b, the statistics still show power-law behavior regardless of temperature. Moreover, the average exponents in the power-law distributions are statistically identical for different temperatures (10 K: -1.51 ± 0.1; 30 K: -1.37 ± 0.1; 70 K: -1.45 ± 0.1; RT: -1.41 ± 0.1). Such a seemingly contradictory conclusion is resolved by plotting the *on*-time probability distribution at 10 K and RT, as shown in Fig. 9c. Unlike the *off*-time distribution, the *on* times have a temperature dependence that is qualitatively observed in the raw data of Fig. 2.

B. *On*-Time Truncation

The *on*-time statistics also yield a power-law distribution with the same exponent* as for the *off*-times, but with a temperature-dependent "truncation effect" that alters the long time tail of the distribution. This truncation reflects a secondary mechanism that eventually limits the maximum *on* time of the QD. The truncation effect can be seen in the *on*-time distribution of a single QD in Figs. 9a and 9b, and in the average distribution of many single QDs as a downward deviation from the pure power law. At low temperatures, the truncation effect sets in at longer times and the resulting time trace shows "long" *on* times. The extension of the power-law behavior for low temperatures on this logarithmic timescale drastically changes the time

---

* The power-law distribution for the *on* times are difficult to fit due to the deviation from power law at the tail end of the distribution. The power-law exponent with the best fit for the *on* times is observed for low excitation intensity and low temperature.

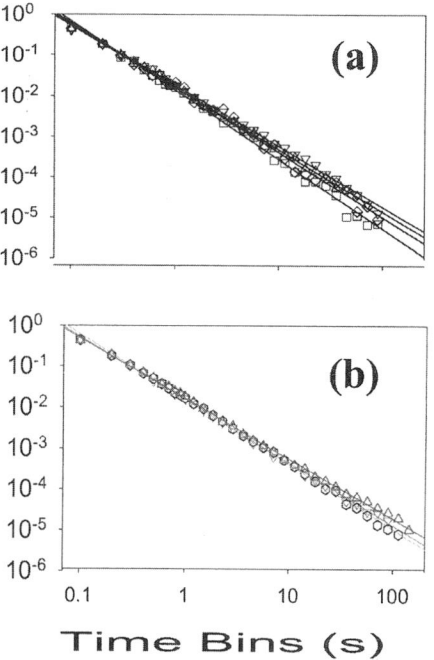

Figure 8 (a) Average *off*-time probability distribution for 25-Å-radius CdSe(ZnS) QD at 300 K ($\triangledown$), 10 K ($\triangle$), 30 K ($\diamond$), and 70 K ($\square$). The α values are 1.41, 1.51, 1.37, and 1.45, respectively. (b) Average *off*-time probability distributions for 39 CdSe(ZnS) QDs of radius 15 Å, ($\triangledown$) and 25-Å-radius CdSe(ZnS) QD ($\diamond$) and 25-Å-radius CdSe QD ($\triangle$) at RT. The α values are 1.54, 1.59, and 1.47, respectively.

trace, as seen in Fig. 2; that is, fewer *on–off* events are observed and the *on* times are longer.

As shown in Figs. 9c and 9d, varying the cw average excitation power in the range 100–700 W/cm² at 300 K and 10 K shows *on*-time probability distribution changes, consistent with earlier qualitative observations. We also compared the *on*-time statistics for QDs differing in size (15-Å versus 25-Å core radius) and QDs with and without a six-monolayer shell of ZnS overcoating shown in Fig. 9. With reduced excitation intensity, lower temperature, or greater surface overcoating, the truncation sets in at longer wait times and the power-law distribution for the *on* time becomes more evident.

Given that the exponent for the *on*-time power-law distribution is nearly the same for all of our samples, then a measure of the average truncation point (or maximum *on* time) is possible by comparing "average *on* times" for dif-

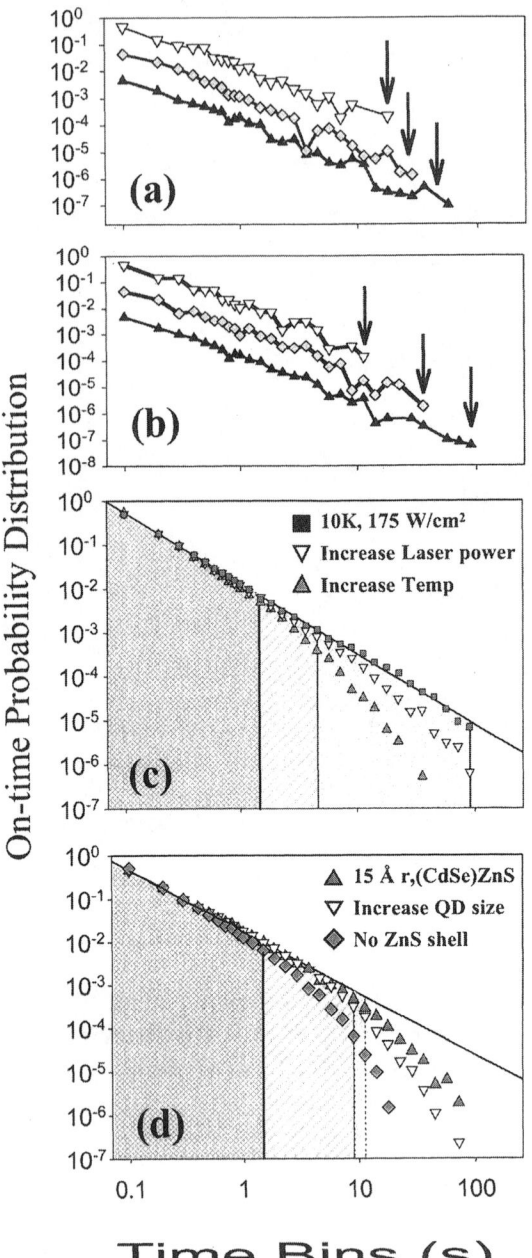

ferent samples while keeping the same overall experimental time. We calculate average *on* times of 312, 283, and 256 ms for the same CdSe/ZnS sample under 10 K and 175 W/cm^2, 10 K and 700 W/cm^2 and RT and 175 W/cm^2 excitation intensity, respectively. The effective truncation times (1.5, 4.6, and 71 s, respectively) can be extrapolated by determining the end point within the power-law distribution that corresponds to the measured average *on* time. In Figs. 9c and 9d, the vertical lines correspond to this calculated average truncation point, indicating the crossover in the time domain from one blinking mechanism to the other.

Furthermore, we can understand the consequence of this secondary mechanism in terms of single-QD quantum efficiency. For ensemble systems, quantum efficiency is defined as the rate of photons emitted versus the photons absorbed. Figure 10a shows the changes to the single-QD time trace with increasing excitation intensity: The intensity values at peak heights increase linearly with excitation power, but the frequency of the *on–off* transitions also increases. Moreover, the measured time-averaged single-QD emission photon flux at different excitation intensities, marked by empty triangles in Fig. 10b, clearly shows a saturation effect. This saturation behavior is due to the secondary blinking-*off* process shown in Fig. 9. The filled triangles in Fig. 10b plot the expected time-averaged emitted photon flux at different excitation intensities calculated from a power-law blinking distribution and truncation values similar to those in Fig. 9. The similarity of the two saturation plots (triangles) in Fig. 10b demonstrates the significance of the secondary mechanism to the overall fluorescence of the QD system. The filled circles represent the peak intensity of the single QD at each of the excitation intensities.

Modification of the surface morphology or excitation intensity showed no difference in the statistical nature of the *off* times or blinking-*on* process. The statistical data are consistent with previous work [12,22]; however, the separation of the power-law statistics from truncation effects clearly demon-

---

Figure 9 (a) Three single-QD *on*-time probability distributions at 10 K, 700 W/cm^2. The arrows indicate the truncation point for the probability distribution for each QD. (b) Four single-QD *on*-time probability distributions for CdSe(ZnS) QDs at RT, 100 W/cm^2. (c) Average *on*-time probability distribution for 25-Å-radius CdSe(ZnS) QD at 300 K and 175 W/cm^2 (▲), 10 K and 700 W/cm^2 (▽), and 10 K and 175 W/cm^2 (■). The straight line is a best-fit line with exponent ∼ −1.6. (d) Average *on*-time probability distribution for 15-Å-radius CdSe(ZnS) QD (▲), 25-Å-radius CdSe(ZnS) QD (▽) and 25-Å-radius CdSe QD (♦) at RT, 100 W/cm^2. The straight line here is a guide for the eye. The vertical lines correspond to truncation points where the power-law behavior is estimated to end.

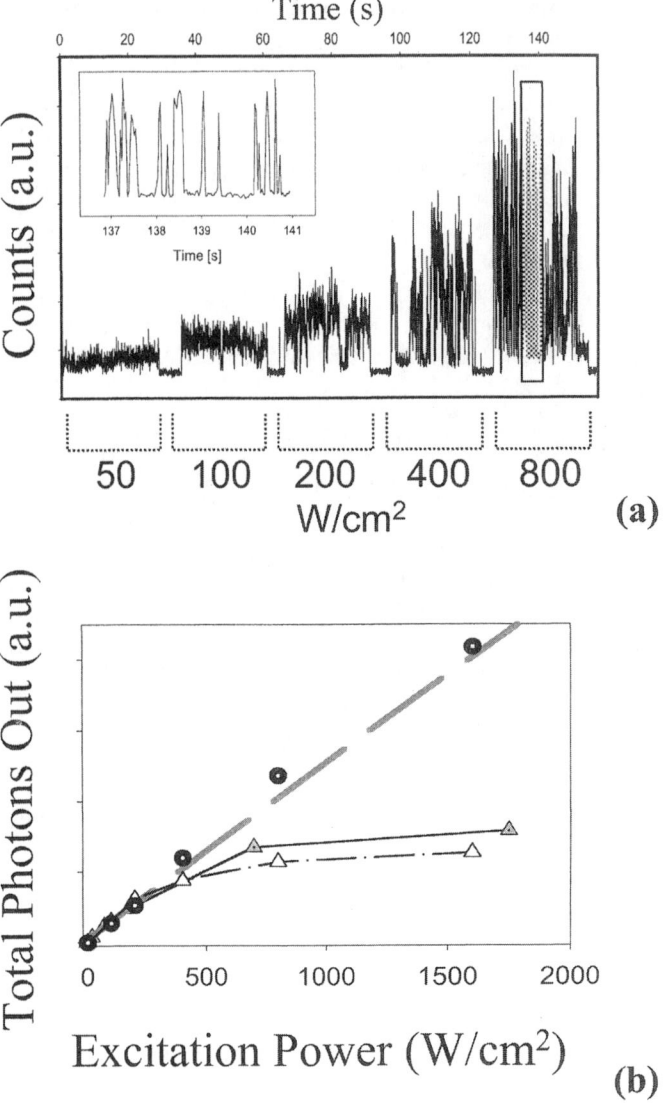

Figure 10 (a) Time trace of a CdSe(ZnS) QD with increasing excitation intensity in 30-s stages. Inset shows that the *on–off* nature still holds at high excitation intensity. (b) The emitted photon flux at the peak emission intensity (●) and average emission intensity (△) from the time trace in (a). The average emitted photon flux calculated from power-law histograms (▲). More details given in text.

strates that two separate mechanisms govern the blinking of CdSe QDs: (1) a temperature-independent tunneling process and (2) a temperature-dependent photoionization process. The truncation effect is not observed in the *off*-time statistics on the timescale of our experiments. Because the power law of the *off*-time statistics extends well beyond the truncation point of the power-law distribution of the *on*-time statistics, the *on*-time truncation/deviation is not an artifact of the experimental time.

C. Random Walk Model

The universality of the *off*-time statistics for all of the QDs indicates an intrinsic mechanism driving the mechanism of the power-law blinking behavior. Furthermore, because the power-law statistics are temperature and excitation intensity independent, the process that couples the dark to bright states is a tunneling process and not phonon assisted. As mentioned earlier, spectrally resolved emission measurements showed a correlation between blinking and spectral shifting at cryogenic temperatures. Considering the large variations in the transition energy (as large as 60 meV [13]) of the bright state, we propose a theoretical framework using a random walk–first passage time model [32] of a dark-trap state that shifts into resonance with the excited state to explain the extraordinary statistics observed here.

   In this random walk model, the *"on–off"* blinking takes place as the electrostatic environment around each individual QD, described in Fig. 6, undergoes a random walk oscillation. When the electric field changes, the total energy for a localized charge QD also fluctuates and only when the total energy of the localized-charge (*off*) state and neutral (*on*) state is in resonance, the change between the two occurs. This can be pictured as a dynamic phase space of bright and dark states. The shift from the dark to bright state (vice versa) is the critical step when the charge becomes delocalized (localized) and the QD turns *on* (*off*). The observed power-law time dependence can be understood as follows. If the system has been *off* for a long time, the system is deep within the charged state (*off* region) of the dynamic phase space and is unlikely to enter the neutral state (*on* region) of the phase space. On the other hand, close to the transition point, the system would interchange between the charged and neutral states rapidly. As the simplest random walk model, we propose an illustrative example of a one-dimensional phase space with a single trapped-charged state that is wandering in energy. At each crossing of the trap and intrinsic excited-state energies, the QD changes from dark to bright or bright to dark. Because the transition from *on* to *off* is a temperature-independent tunneling process, it can only occur when the trap state and excited core state of the QD are in resonance. In addition, a temperature-dependent hopping process, related to the movement and creation/annihilation of trapped charges surrounding the QD core, drives the trap and excited QD core states to fluctuate in a random walk. The minimum hopping time of the surrounding charge environment gives the minimum timescale for each step of the random walk to occur. This simple 1D discrete-time random walk model for blinking immediately gives the characteristic power-law probability distribution of *on* and *off* times with a power-law exponent of −1.5 [33]. The intrinsic hopping time is most likely orders of magnitude faster than our experimental binning resolution (100 ms). Although the hopping mechanism is probably temperature dependent, this temperature dependence would only be reflected in experiments that could probe timescales on the order of the hopping times, before power-law statistics set in and beyond the reach of our experimental time resolution.

Although this simple random walk model may require further development, it nevertheless explain the general properties observed. The *off*-time statistics are temperature and intensity independent because although the hopping rate of the random walker changes, the statistics of returning to resonance between the trap and excited state does not. In addition, size and surface morphology do not play a significant role in this model as long as a trap state is energetically accessible to the intrinsic excited state. Figure 11a represents a Monte Carlo simulation of the histogram of return times to the origin in a one-dimensional, discrete-time random walk. The open circles represent a histogram with a slower hopping rate than the filled circles analogous to thermally activated motion at 10 K and 300 K. The experimentally accessible region of the statisticals simulation is suggested as the area inside the dotted lines in Fig. 11a. Further experimental and theoretical work should go toward completing this model. For example, temperature- and state-dependent hopping rates as well as a higher-dimension random walk phase space and multiple transition states may be necessary.

The magnitude of truncation of the *on*-time power-law distribution depends on which QD is observed, as shown in Figs. 9a and 9b. In Fig. 9b, the

Figure 11   (a) Histogram of return times to the origin in a 1D discrete-time random walk simulation. The boxed region of the histogram represents the accessible time regime (>100 ms and <1 h) in relation to our experiments. The filled circles represent RT statistics and the unfilled circles represent the low-temperature behavior. The change from the RT to the low-temperature simulation is modeled by allowing for a probability for the random walker to remain stationary, analogous to insufficient thermal energy at the low temperatures. (b) Simulations of time traces produced from power-law distributions of *on* and *off* times. The maximum *on* times were decreased to illustrate the changes in the experimental time traces as temperature is increased, excitation intensity is increased, and/or surface passivation is decreased. Top and bottom traces are illustrative experimental traces to show the similarities between the simulation and data.

(a)

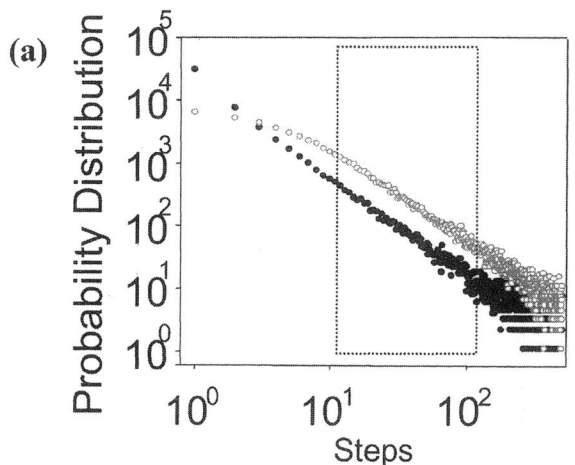

(b)

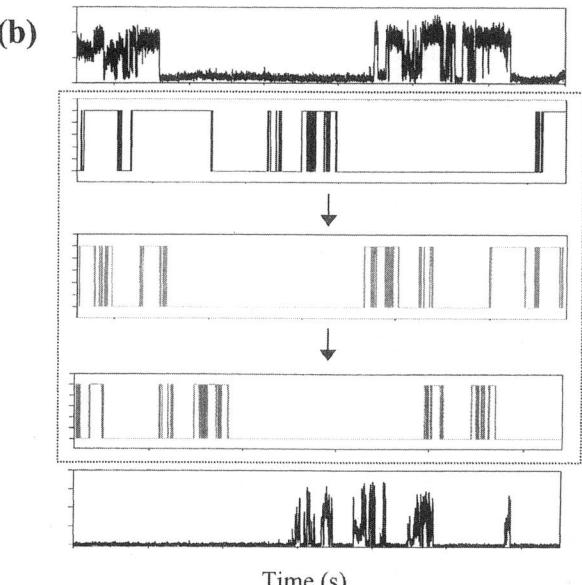

Time (s)

arrows indicate the *on*-time truncation point for four different QDs under the same excitation intensity at RT. Qualitatively, we can describe and understand the changes as a result of the interaction between the dynamic dark and bright states modeled earlier. As the excitation intensity or thermal energy is reduced, the hopping rate of the random walker slows down and the time constant for the truncation is extended within our experimental time. Surface modification in the form of ZnS overcoating also extends the power-law distribution for the *on* times. This surface modification should not change the hopping rate of the random walker but rather changes the Auger scattering rate. Hence, a mechanism such as photo-assisted ejection of a charge due to Auger ionization [12,26] may be responsible for the *on*-time truncation effect. Recently, reversible quenching of CdSe QDs was shown due to interactions with oxygen molecules in the presence of light [34]. Although the single-QC

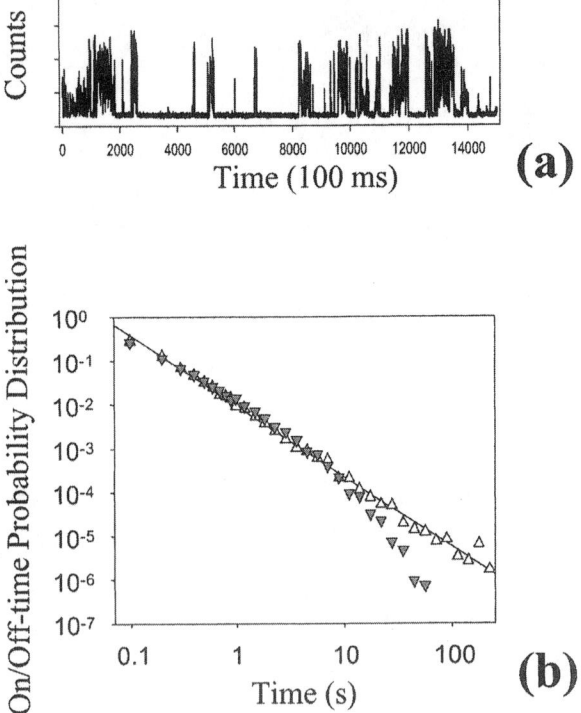

Figure 12 (a) Time trace of a single CdTe QD at room temperature, $125 \ W/cm^2$. (b) The probability distribution of the *on* time (▼) and *off* time (△) for CdTe QDs at room temperature. The best-fit line shows a power-law behavior with exponent ~ -1.6.

blinking data were not interpreted using the power-law statistics, the kinetic behavior is consistent with our above description: The *off* times are independent of the oxygen molecules introduced; however, the *on* times show a dramatic decrease in the maximum duration. The simulated time traces in Fig. 11b, taking into account shorter termination points for the maximum *on* time relative to the maximum *off* time, elucidates the difference in blinking at the higher temperatures, higher excitation intensity, and poorer surface passivation. The change in the time traces is comparable between the simulation in Fig. 11 and data in Fig. 2.

Recent results on CdTe QDs, displayed in Fig. 12, show that the power-law behavior, its exponent, and the *on*-time phenomenology is reproduced, indicating that the effects observed are not unique to the particulars of CdSe QDs, but rather reflect more universal underlying physics of nanocrystal QDs.

VI. CONCLUSIONS

In conclusion, extensive investigations into single-QD optical dynamics have uncovered uniform properties in an inherently inhomogeneous system. First, the correlation between large spectral shifting events and blinking of single QDs explains that spectral diffusion shifts, caused by electrostatic decorations present around every QD, is a direct consequence of the charging mechanism reported in the fluorescence intermittency process. Second, the probability of each QD turning *on* or *off* follows an unexpected, temperature-independent power law. These power-law statistics observed for all the CdSe QDs studied suggest a complex, yet universal, tunneling mechanism for the blinking *on* and *off* process. Third, the qualitative changes observed in the blinking behavior are due to a secondary, thermally activated, and photoinduced process that causes the probability distribution of the *on*-time statistics to be truncated at the "tail" of the power-law distribution. Although these dynamic effects are incoherent from QD to QD, these individual behaviors may encompass new technology unforeseen previously but applicable on an ensemble scale. Recently, charging devices have been fabricated showing the feasibility of controlling blinking on an ensemble film of QDs.

ACKNOWLEDGMENTS

The authors would like to thank S.A. Empedocles and R.G. Neuhauser for their contributions in designing and implementing the single-quantum-dot microscope setup. Moreover, both were instrumental in developing and interpreting key components of the previously published work in this review. The authors thank Al. L. Efros and E. Barkai for thoughtful discussions and

W. K. Woo and V. C. Sundar for assistance in materials synthesis. This work was supported in part by the NSF Materials Science and Engineering Center program under grant DMR98-08941. We also thank the M.I.T. Harrison Spectroscopy Laboratory (NSF-CHE-97-08265) for support and for use of its facilities. KTS thanks NSERC–Canada for financial support.

REFERENCES

1. Bruchez, M.; Moronne, M.; Gin, P., et al. Science 1998, *281*, 2013.
2. Wang, C.J.; Shim, M.; Guyot-Sionnest, P. Science 2001, *291*, 2390.
3. Klimov, V.I.; Mikhailovsky, A.A.; Xu, S., et al. Science 2000, *290*, 314.
4. Lee, J.; Sundar, V.C.; Heine, J.R., et al. Advanced Materials 2000, *12*, 1102.
5. Brus, L. Appl. Phys. A: Mater. Sci. Process. 1991, *53*, 465.
6. Alivisatos, A.P. Science 1996, *271*, 933.
7. Empedocles, S.A.; Neuhauser, R.; Shimizu, K., et al. Adv. Mater. 1999, *11*, 1243.
8. Moerner, W.E.; Kador, L. Phys. Rev. Lett. 1989, *62*, 2535.
9. Orrit, M.; Bernard, J. J. Lumin. 1992, *53*, 165.
10. Xie, X.S.; Dunn, R.C. Science 1994, *265*, 361.
11. Ha, T.; Enderle, T.; Chemla, D.S., et al. Phys. Rev. Lett. 1996, *77*, 3979.
12. Nirmal, M.; Dabbousi, B.O.; Bawendi, M.G., et al. Nature 1996, *383*, 802.
13. Empedocles, S.A.; Bawendi, M.G. Science 1997, *278*, 2114.
14. Mason, M.D.; Credo, G.M.; Weston, K.D., et al. Phys. Rev. Lett. 1998, *80*, 5405.
15. Zhang, B.P.; Li, Y.Q.; Yasuda, T., et al. Appl. Phys. Lett. 1998, *73*, 1266.
16. Dickson, R.M.; Cubitt, A.B.; Tsien, R.Y., et al. Nature 1997, *388*, 355.
17. Murray, C.B.; Norris, D.J.; Bawendi, M.G. J. Am. Chem. Soc. 1993, *115*, 8706.
18. Hines, M.A.; Guyot-Sionnest, P. J. Phys. Chem. 1996, *100*, 468.
19. Dabbousi, B.O.; RodriguezViejo, J.; Mikulec, F.V., et al. J. Phys. Chem. B 1997, *101*, 9463.
20. Mikulec, F.V. Ph.D. thesis, Massachusetts Institute of Technology, Cambridge, MA, 1999.
21. Empedocles, S.A.; Norris, D.J.; Bawendi, M.G. Phys. Rev. Lett 1996, *77*, 3873.
22. Banin, U.; Bruchez, M.; Alivisatos, A.P., et al. J. Chem. Phys. 1999, *110*, 1195.
23. Basche, T. J. Lumin. 1998, *76–77*, 263.
24. Ambrose, W.P.; Moerner, W.E. Nature 1991, *349*, 225.
25. Norris, D.J.; Bawendi, M.G. Phys. Rev. B 1996, *53*, 16,338.
26. Efros, A.L.; Rosen, M. Phys. Rev. Lett. 1997, *78*, 1110.
27. Ekimov, A.I.; Kudryavtsev, I.A.; Ivanov, M.G., et al. J. Lumin. 1990, *46*, 83.
28. Chepic, D.I.; Efros, A.L.; Ekimov, A.I., et al. J. Lumin. 1990, *47*, 113.
29. Roussignol, P.; Ricard, D.; Lukasik, J., et al. J. Opt. Soc. Am. B 1987, *4*, 5.
30. Krauss, T.D.; Brus, L.E. Phys. Rev. Lett 1999, *83*, 4840.
31. Kuno, M.; Fromm, D.P.; Hamann, H.F., et al. J. Chem. Phys. 2000, *112*, 3117.
32. Schrodinger, E. Phys. Z. 1915, *16*, 289.
33. Bouchaud, J.P.; Georges, A. Phys. Rep. 1990, *195*, 127.
34. Koberling, F.; Mews, A.; Basche, T. Adv. Mater. 2001, *13*, 672.

7

Electrical Properties of Semiconductor Nanocrystals

David S. Ginger and Neil C. Greenham
Cavendish Laboratory, Cambridge, England

I. INTRODUCTION

The size-dependent optical properties of semiconductor nanocrystals have been extensively studied, as described elsewhere in this volume. In this chapter, we concentrate instead on the electrical properties of nanocrystals. We show how electrons can be injected into nanocrystals, transported between nanocrystals and transferred between nanocrystals and organic molecules. We then describe the potential applications of nanocrystals in electronic and optoelectronic devices. We attempt to give an overview of the field, although we cannot mention all of the articles which have been published in this diverse and expanding area. We draw many of our examples from our own work and from the work of those other groups with which we happen to be most familiar. We restrict ourselves to chemically synthesized semiconductor nanocrystals and concentrate mostly on II–VI nanocrystals such as CdSe in the size regime of less than 10 nm, where quantum confinement effects are important.

In a nanocrystal, the electronic states are typically confined within the nanocrystal by significant potential barriers. Transport between nanocrystals can, therefore, be considered as a problem of hopping between localized states. For small nanocrystals in the strong confinement regime, charge transport is limited by interparticle electron transfer, rather than by the transport of electrons within a single particle. Therefore, the physics has more in common with that of charge transport in molecular systems, rather than with traditional semiconductor transport theory. In nanocrystals, changing the size allows the electron affinity and ionization potential of the nanocrystal to be tuned, which may, in principle, affect the rate of charge transfer. In practice, though, disorder and surface trap states are very important and may dominate the charge transport properties of nanocrystalline systems. We will discuss some of the theory relevant to charge transport and charge transfer in nanocrystalline systems in Section II.

One of the major motivations for the study of nanocrystals is the possibility that they may be useful in device applications. Some of the devices which have been demonstrated in the laboratory include light-emitting diodes (LEDs) [1–4], photovoltaic devices [5,6], and single-electron transistors [7]. For large-area applications such as LEDs and photovoltaics, nanocrystals have the advantages of being processable from solution and having size-tunable absorption and emission. Despite these advantages, nanocrystals have not yet found their way into commercial devices and we will discuss the prospects for improvements in performance which may make commercial applications more likely.

In order to illustrate some of the charge transport and charge transfer processes important in nanocrystal-based devices, we will briefly introduce the method of operation of LEDs and photodiodes. The simplest device in which light emission can occur is a thin film of nanocrystals sandwiched between metal electrodes, as illustrated in Fig. 1. One of the electrodes must be semitransparent; hence, indium–tin oxide (ITO) is typically used as the anode. When a positive voltage is applied to the anode, holes are injected from the anode and electrons from the cathode. These carriers can then move through the device under the action of the applied field by hopping from nanocrystal to nanocrystal. When an electron and hole arrive on the same nanocrystal, an exciton can be formed, which can then decay radiatively giving emission at an energy characteristic of the size of the nanocrystal. We have found that this

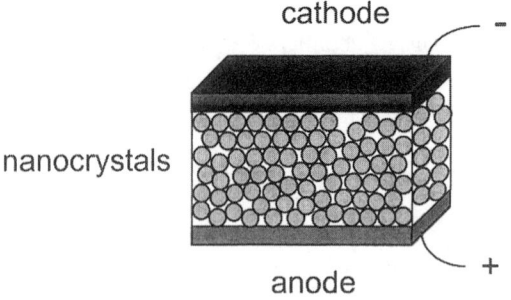

Figure 1 Schematic representation of a typical structure for electrical characterization of nanocrystalline films.

type of simple device does not work well as an LED because poor hole injection leads to low efficiencies and high turn-on voltages. Much improved performance has been achieved using a layer of a hole-transporting conjugated polymer between the anode and the nanocrystal layer [1,3], as shown schematically in Fig. 2a. Conjugated polymers have semiconducting properties due to delocalization of π-orbitals along the unsaturated polymer chain and are used in large-area LEDs and photodiodes in their own right [8,9]. In the nanocrystal LED, the polymer layer provides an intermediate energy level between the anode and the nanocrystals, which assists the overall rate of hole injection.

In a photodiode or photovoltaic device, the processes are opposite those required in an LED. Light is absorbed in the device, producing excitons which must then be split up into free electrons and holes. These charge carriers must then be transported to opposite electrodes without recombination, thus producing a current in the external circuit. Again, conjugated polymers are useful as hole transporters in conjunction with nanocrystals [5]. By mixing nanocrystals with the polymer as shown in Fig. 2b, excitons created in the polymer may be dissociated by electron transfer from polymer to nanocrystal. Both electron and hole must then find a pathway to the appropriate electrode by hopping from polymer chain to polymer chain or from nanocrystal to nanocrystal, respectively.

We will discuss the operation of LEDs and photodiodes in more detail in Section VI, but we can already identify the physical processes which are important in device operation. Photoexcited charge transfer between conjugated polymers (or other organic molecules) and semiconductor nanocrystals will be examined in Section IV, together with recombination at the nanocrystal–polymer interface. Charge injection at metal electrodes, carrier transport between nanocrystals, and charge trapping in nanocrystalline films will be discussed in Section V.

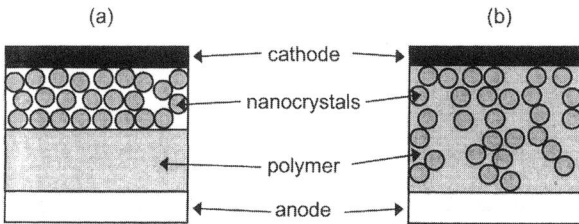

Figure 2 Schematic cross-sectional structure for (a) a light-emitting diode and (b) a photovoltaic device containing both nanocrystals and a conjugated polymer.

II. THEORY OF ELECTRON TRANSFER BETWEEN LOCALIZED STATES

Here, we introduce some of the theory relevant to determining the rate of electron transfer between two nanocrystals. In particular, we are interested in electron transport in a solid film of nanocrystals, where the particles are fixed in position and separated typically by the organic ligand present on the nanocrystal surface. We will examine how the charge transport is expected to vary with the size and spacing of the particles and will return to this question in Section V, where we present experimental results.

The theory of electron transfer between localized states has been developed to predict rates of electron transfer in molecular systems and has been extended to treat electron transfer to and from molecules at a metallic or semiconductor surface [10]. We will first give a brief review of this theory, which was initially developed by Marcus [11]. In this model the energy of the initial and final states are considered as a function of nuclear coordinates, as shown in Fig. 3. The nuclear coordinates cannot change during the electron tunneling process from initial to final state, and electron transfer must therefore occur at nuclear coordinates which correspond to an intersection of the initial and final potential energy surfaces. In Fig. 3, $\Delta G°$ represents the free-

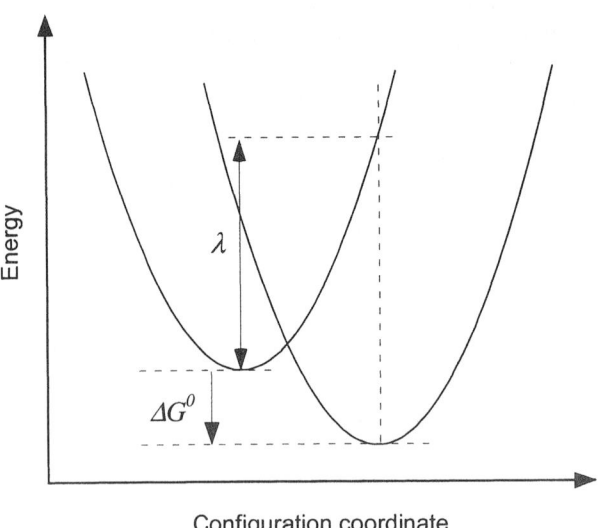

Configuration coordinate

Figure 3   Energy as a function of configuration coordinate for the initial and final states in an electron transfer reaction. The reorganization energy, λ, and the free-energy change for the reaction, $\Delta G°$, are shown.

energy change between initial and final states and λ is the reorganization energy required to change from initial to final equilibrium nuclear coordinates. For charge transport between nanocrystals, the tunneling rate is slow compared with typical vibrational frequencies and the system must, on average, sample the crossing point many times before transfer can occur. In this nonadiabatic limit, the rate of electron transfer is given by

$$k = \left(\frac{2\pi}{\hbar}\right) |V_{12}|^2 (FC) \tag{1}$$

where V_{12} is the electronic coupling matrix element between initial and final states on resonance and FC is the Frank–Condon factor, given by the sum of overlap integrals of the vibrational wave functions of the initial and final states, weighted by the Boltzmann probability of occupation of the initial vibrational state.

At high temperatures, when the vibrational frequencies are all small compared with kT/h, the vibrations may be treated classically and Eq. (1) reduces to [12]

$$k = \left(\frac{2\pi}{\hbar}\right) |V_{12}|^2 (4\pi\lambda kT)^{-1/2} e^{-\Delta E^*/kT} \tag{2}$$

where

$$\Delta E^* = \frac{(\Delta G^\circ - \lambda)^2}{4\lambda} \tag{3}$$

To calculate the total reorganization energy λ in Eq. (3), both internal vibrations in the molecules and changes in the configuration of the solvent or dielectric surrounding the molecules must be taken into account. Denoting the internal and dielectric contributions as λ_i and λ_o respectively, we have

$$\lambda = \lambda_i + \lambda_o \tag{4}$$

The contribution from the dielectric can be estimated using a dielectric continuum model, where the slow (nonelectronic) part of the response is responsible for the reorganization barrier to electron transfer. For electron transfer between two spheres, λ_o is given by

$$\lambda_o = \frac{e^2}{4\pi\varepsilon_0} \left(\frac{1}{2r_1} + \frac{1}{2r_2} - \frac{1}{d}\right) \left(\frac{1}{\varepsilon_{op}} - \frac{1}{\varepsilon_s}\right) \tag{5}$$

where r_1 and r_2 are the radii of the two spheres, d is their center-to-center separation, and ε_{op} and ε_s are respectively the optical and static dielectric constants of the surrounding dielectric.

For the specific case of charge transfer between two nanocrystals, we will now attempt to identify the main contributions to λ_i and λ_o. This problem has been studied by Brus [13], who applied the Marcus theory of electron transfer to silicon nanocrystals surrounded by solvent. To calculate the internal reorganization energy, we must consider how the lattice in the nanocrystal responds to the addition (or removal) of an electron. In general, changing the charge state of the nanocrystal changes the equilibrium nuclear configuration and we denote the energy associated with the change from the uncharged configuration to the charged configuration on a single nanocrystal as λ_s. Because reorganization is necessary in both nanocrystals for electron transfer to occur, the total value of λ_I in Eq. (4) is given by the sum of the deformation energies λ_s on both nanocrystals.

To understand the origin of λ_s in more detail, we must consider the vibrational modes which transform between the charged and uncharged configurations. These modes are the phonon modes of the nanocrystal, which may be either acoustic or optical. The strength of coupling of phonon modes to the electronic state of the nanocrystal is dependent on the chemical composition of the semiconductor and on the nanocrystal size. In a nonpolar material such as silicon, the relevant coupling is with acoustic phonons. Acoustic phonons couple to the energy of the system through the deformation potential (a change in the electronic energy due to strain in the lattice). In nanocrystals, the contribution to λ_s from coupling to acoustic phonons through the deformation potential is expected to scale approximately as r^{-3}, where r is the nanocrystal radius [14]. Brus estimated a value of $\lambda_s = 12$ meV in silicon nanocrystals of diameter 2 nm [13].

In polar nanocrystals such as CdSe, the situation is more complicated because optical phonons may also couple to the electronic state through the Fröhlich interaction. This interaction involves a polarization of the crystal lattice in response to an internal electric field. The Fröhlich interaction is responsible for the vibrational structure seen in emission spectra of CdSe nanocrystals, although, in that case, the scaling of coupling strength with nanocrystal size is complicated because both an electron and hole are present in the nanocrystal and partially compensate each other's charge [15]. The simpler case of coupling of optical phonons to a change in the overall charge state of the nanocrystal (polaron formation) has been modeled in CdSe by Oshiro et al. [16], who find that the relaxation energy associated with polaron formation increases rapidly as the nanocrystal size becomes less than the exciton Bohr radius, reaching a value of $\lambda_s \approx 30$ meV for a diameter of 2 nm. We are not able to make an accurate estimate of the additional reorganization energy due to deformation potential coupling in CdSe nanocrystals; however we expect the contribution from Fröhlich interaction to dominate. Because the longitudinal optical (LO) phonon energy in CdSe (26.5 meV) is comparable with thermal energies at room temperature, the high temperature approximation made in Eq. (2) is unlikely to be valid even at room temperature. In this case, it will be necessary to use more detailed theories which account for quantization of vibrational energy and allow nuclear tunneling between different vibrational states [17].

As explained earlier, an additional reorganization energy arises when the nanocrystals are surrounded by a dielectric which relaxes in response to the change in charge state of the nanocrystal. For the case considered by Brus, where silicon nanocrystals are surrounded by water, the highly polar nature of the solvent leads to a value of λ_o as large as 400 meV for transfer between touching nanocrystals of 2 nm diameter [13]. In a close-packed film of CdSe nanocrystals coated with tri-*n*-octylphosphine oxide (TOPO), the TOPO ligands (and the surrounding nanocrystals) play the role of the solvent. For example, we can use Eq. (5) to estimate a value of $\lambda_o = 100$ meV for 2-nm-diameter CdSe particles with a center-to-center separation of 3.3 nm (corresponding to the separation of close-packed TOPO-coated particles [18]). In this estimate, we have assumed that the local effective dielectric constant is dominated by the TOPO, which has a static dielectric constant $\varepsilon_s = 2.61$ [19], and we approximate the optical frequency dielectric constant with that of TOPO, which is $\varepsilon_{opt} = 2.07$ [20].

The $|V_{12}|^2$ tunneling term in Eq. (1) can be treated in a simple approximation as tunneling through a one-dimensional potential barrier in the presence of an applied field. This approach was used by Leatherdale et al. in the context of electron transfer from a photoexcited nanocrystal containing an electron and a hole [21]. Using the Wentzel–Kramer–Brillouin (WKB) approximation to obtain the tunneling probability through a square tunnel barrier of height ϕ and width d in an applied field E gives a tunneling probability of

$$S = \exp\left[-\frac{4}{3}\sqrt{\frac{2m^*}{\hbar^2}}\frac{1}{eE}\left[\phi^{3/2} - (\phi - eEd)^{3/2}\right]\right] \qquad (6)$$

where m^* is the effective mass of the carrier within the barrier. For thin, high barriers, the tunneling probability is found to decrease exponentially with increasing barrier width.

It is often useful to describe the charge transport properties of an extended solid in terms of the carrier mobility, μ. For electron transport, for example, the electron mobility μ_e is defined by

$$J_e = ne\mu_e E \qquad (7)$$

where J_e is the electron drift current density, n is the number density of electrons, and E is the applied field. In systems where hopping transport dominates, the mobility is often field dependent. The mobility is related to the net hopping rate for individual carriers, $R(E)$, by

$$\mu(E) = \frac{R(E)}{E} d \qquad (8)$$

where d is a typical hopping distance in the direction of the applied field and $R(E) = R_{forward}(E) - R_{back}(E)$, the difference in forward and backwards hopping rates.

In some materials, a significant proportion of the carriers may occupy trap states which are much less mobile than the "free" carriers. It is useful to define an effective mobility, μ_{eff}, such that

$$J = n_{total} e \mu_{eff} E \qquad (9)$$

where n_{total} is the sum of both trapped and free-charge carrier densities. Where the trapped carriers are completely immobile and the free carriers have a mobility μ_{free}, we may define the effective mobility as

$$\mu_{eff} = \mu_{free} \frac{n_{free}}{n_{total}} \qquad (10)$$

In nanocrystalline systems it is well known that trap states exist at the surface of the particles and that the number and depth of traps is highly sensitive to the surface passivation. Because transport of both trapped carriers and "free" carriers (those occupying core states) involves tunneling between particles, carriers in shallow traps may have mobilities which are not much less than those of "free" carriers. It is therefore not possible to draw a clear distinction between mobile (free) and immobile (trapped) carriers. In this case, where there are populations of carrier n_i with different mobilities μ_i, the appropriate definition of effective mobility is

$$\mu_{eff} = \frac{\sum\limits_{i} \mu_i n_i}{\sum\limits_{i} n_i} \qquad (11)$$

In our discussion so far we have considered only a single core electronic level as being involved in electron transport. We have also neglected the effect of disorder, which leads to a distribution of energy levels and interparticle spacings. In this simple model for electron transport between identical nanocrystals, one would expect the activation barrier to decrease with applied field until the transfer becomes activationless at $\Delta G = -\lambda$. At higher fields, the activation energy would then increase as the "inverted Marcus region" is entered. In practice, though, there is a series of quantum-confined electron levels in a nanocrystal, and at high fields, electrons are likely to be injected into higher-lying electronic states, followed by rapid relaxation to the lowest state

[13]. The presence of these higher-lying states allows electron transfer to occur with a smaller activation energy than would be necessary for direct transfer to the lowest electron state in the inverted region.

Electron transfer in a nanocrystalline film is a local process sensitive to the electronic structure of an individual nanocrystal and to its immediate environment. Therefore, disorder can play an important role in determining electron transport though the film. The first source of disorder in the electron energy levels arises from the distribution of particle sizes, leading to a distribution of quantum confinement energies. This disorder is also responsible for the inhomogeneous broadening observed in absorption and emission. For typical CdSe nanocrystals, this broadening is in the range 50–100 meV and is mostly due to variation in the electron confinement energy. At a given electric field, electron transfer will occur most readily from particles where the neighboring particle has an electron affinity which gives the lowest activation energy. Further disorder arises from the distribution in interparticle separations in a solid film. Because the tunneling rate is exponentially sensitive to the tunneling distance, a distribution of distances will lead to a distribution of tunneling rates. Also, the effective dielectric constant experienced by a charged particle will be sensitive to the exact arrangement of other particles around it. Because the electron affinity depends not only on the electron-confinement energy but also on the local dielectric environment surrounding the nanocrystal [22] spatial disorder will give rise to additional disorder in the electron affinity of individual particles.

The effects of disorder on the transport properties of nanocrystalline films have been studied for metallic particles [23], but they have not yet been modeled in detail for semiconducting particles. However, the results of work on charge transport in molecularly doped polymer films are particularly relevant because much of the physics is similar. Bässler and co-workers have used Monte Carlo simulations to model the field and temperature dependence of the mobility in systems with a Gaussian distribution of transport levels [24]. At low fields, electrons on nanocrystals with particularly low energies within the broadened distribution will find it more difficult to hop to a neighboring nanocrystal, because the neighbors are likely to be higher in energy. The presence of these low-energy sites leads to behavior which is in some ways analogous to the effect of traps. Bässler and co-workers assume microscopic hopping rates given by the simple Miller–Abrahams model [25] (activationless where the final state is lower in energy than the initial state and with an activation energy given by ΔG where the hop is "up hill" in energy). In its simplest form, their model predicts mobilities which follow

$$\mu = \mu_0 \exp\left[-\left(\frac{A}{kT}\right)^2\right] \exp\left[\gamma(T)\sqrt{E}\right] \tag{12}$$

where the constant A and the temperature dependence of γ depend on the details of the disorder that is present. This model (for systems where the disorder has a Gaussian width of about 0.1 eV) corresponds well with experiments in molecularly doped polymers [26], although it is difficult to measure over a large enough temperature range to distinguish the temperature dependence predicted by Eq. (12) from the simple Arrehnius behavior which might be expected if structural reorganization energies controlled the transport. Similar behavior to that predicted by Eq. (12) might be expected in nanocrystalline systems if disorder is the dominant effect.

In summary, we have shown that in addition to the electronic tunneling process, both internal and dielectric relaxation energies are likely to be important in determining electron transfer dynamics in nanocrystals. Both of these relaxation energies are size dependent and can be in the range 50–100 meV for small CdSe particles. We have also identified surface trapping, higher-lying states, and disorder as factors likely to complicate the transport in real samples, and we will return to these issues in the light of the experimental results discussed in Section V.

III. EXPERIMENTAL TECHNIQUES

In this section, we give a brief introduction to some of the experimental techniques used in characterizing electron transfer and electron transport in nanocrystalline materials.

Measurement of the luminescence spectrum is a standard technique for characterizing nanocrystals, because it gives information about the particle size in quantum-confined systems. Less standard, especially in solid films, is measuring the quantum efficiency of luminescence (i.e., the ratio of the number of photons emitted to the number of photons absorbed). This measurement is particularly useful in studying charge transfer to or from a nanocrystal, because charge separation typically quenches the luminescence of the photoexcited species, which may be either the nanocrystal itself or some neighboring fluorescent molecule or polymer [5,27]. In solution, luminescence efficiency can be measured by comparison with a known standard solution where the geometry, absorbance, and solution refractive indices are known. In a solid film, however, it is difficult to determine the total luminescence from a measurement in a particular direction. To overcome this problem, it is necessary to use an integrating sphere to collect the emitted light. An integrating sphere is a hollow sphere coated with a diffusely reflecting coating, which has the property that the intensity measured by a detector in the wall of the sphere is proportional to the total amount of light generated inside the sphere, irrespective of its direction. Using a laser to excite a solid film placed inside an integrating sphere allows accurate luminescence efficiencies to be calculated, provided that the efficiency is more than about 1%. The integrating sphere is also used without the sample present to measure the incident laser power. With the sample present, use of filters or spectrographic detection enables luminescence and nonabsorbed laser light to be distinguished, which can be used to calculate the fraction of laser power absorbed by the sample, as described in Refs. 28 and 29.

Transient absorption provides a powerful tool to study charge transfer processes on fast timescales. Typically, the sample is excited with a laser pulse of ~100 fs duration and is then probed with a pulse of white light after a variable delay of up to several nanoseconds. The probe beam can be used to measure the concentration of neutral and charged species because they introduce new subgap absorptions (from excitons and charged states), stimulated emission (from excitons), and bleaching (due to depletion of the ground state when excited states are present). By varying the delay between pump and probe, the populations of these various species can be studied as a function of time. Several reviews of femotsecond pump-probe techniques are available [30–32].

We find that charged states formed after charge transfer to nanocrystals often have lifetimes in the microsecond to millisecond range [27]. The population and decay of these states can be studied without the need for pulsed lasers. In the quasi-steady-state technique often known as photoinduced absorption, the sample is excited by a continuous-wave (CW) laser beam modulated with a mechanical chopper at frequencies up to a few kilohertz. Absorption is measured at energies between 0.5 and 3 eV using monochromated light from a tungsten lamp together with an appropriate detector, as shown in Fig. 4. A lock-in amplifier is used to measure the small change in absorption at the chopping frequency, allowing fractional changes in transmission as low as 10^{-6} to be measured. The lock-in amplifier measures the

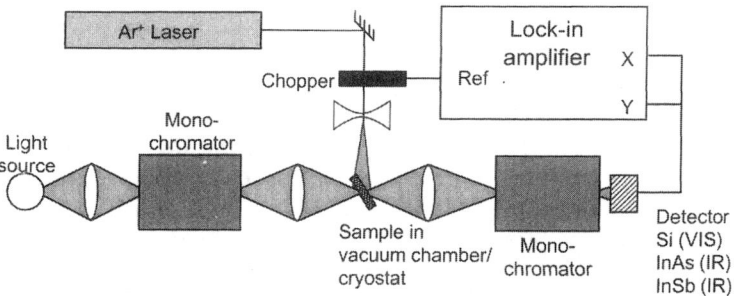

Figure 4 Experimental arrangement for measurement of photoinduced absorption.

components of modulation which are in phase and 90° out of phase with the excitation. Monitoring the signal as a function of chopping frequency and pump intensity provides information about the lifetime and recombination mechanism of the excited species.

Electrical characterization of films containing nanocrystals can use various different measurement geometries; however, we have concentrated on measurements in simple planar structures, similar to that shown in Fig. 1, where a thin film is placed between metal electrodes. Spin-coating is a convenient technique for producing uniform films of nanocrystals or of nanocrystal/polymer blends, and the top contact can be deposited on the active film by vacuum evaporation at rates of $1-2$ Å/s. Use of a nitrogen-filled glove box allows device preparation and measurement to be performed without exposure to air. Measurement of current–voltage (IV) curves is straightforward, although we have found that time-dependent effects complicate the interpretation [33]. Where the device acts as an LED, the light output may be measured simultaneously. The electroluminescence quantum efficiency is then defined as the ratio of the number of photons produced within the device to the number of charges flowing in the external circuit. Quantum efficiencies may be defined as either "internal" or "external," depending on whether all the generated photons are considered or just those which escape through the front surface of the device. In the context of applications, the brightness of a device in the forward direction is often measured in candelas per square meter, where the candela (cd) is a photometric unit where the radiant power is weighted according to the response of the eye. Efficiencies are therefore often quoted in units of candelas per ampere.

Measurement of photocurrent in a planar device requires one contact that is semitransparent, typically either a thin metal film or indium–tin oxide. Measurement of the current–voltage curve under illumination allows the short-circuit current, I_{sc} and open-circuit voltage, V_{oc}, to be defined as shown in Fig. 5. A quantum efficiency (QE) may be defined (usually under short-circuit conditions) as the ratio of electrons flowing in the external circuit to photons incident on the device. The quantum efficiency will depend on the wavelength of the incident radiation; this dependence defines the "action spectrum" of the device. For photovoltaic applications, it is the power conversion efficiency which is the appropriate figure of merit, defined as the ratio of electrical power extracted to the optical power incident. Maximum power output is given not under short-circuit or open-circuit conditions, but where the load is chosen to maximize the product of current and voltage. For illumination at a single wavelength, the power efficiency (PE) is related to the quantum efficiency by

$$PE = QE\left(\frac{eV_{oc}}{E_{photon}}\right)FF \tag{13}$$

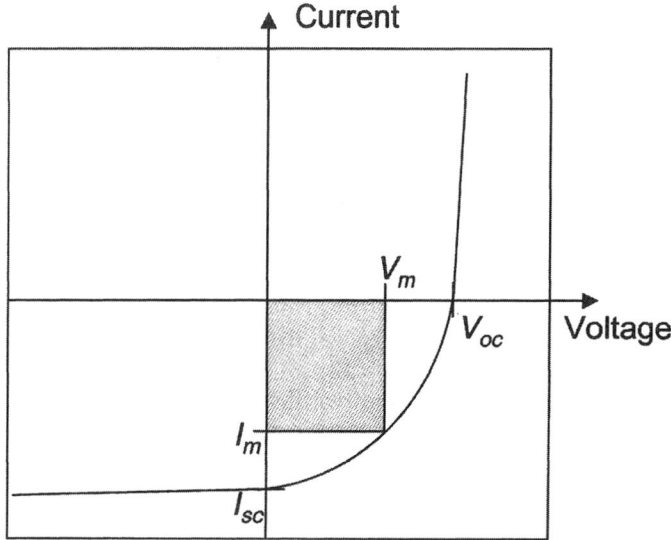

Figure 5 Current–voltage characteristic for a photovoltaic device showing the open-circuit voltage (V_{oc}) and the short-circuit current (I_{sc}) together with the maximum power rectangle.

where E_{photon} is the energy of the incident photons and FF is the fill factor, defined as

$$FF = \frac{V_m I_m}{V_{oc} I_{sc}} \tag{14}$$

where V_m and I_m are chosen so as to maximize the area of the rectangle shown in Fig. 5. Power efficiencies are often defined under conditions of solar illumination, typically with a standardized spectrum known as AM1.5, which represents the spectrum measured on the Earth's surface at a latitudes of 45°.

IV. NANOCRYSTALS AND PHOTOINDUCED ELECTRON TRANSFER

Photoinduced charge transfer at nanoscale semiconductor interfaces is fundamentally important to a variety of emerging technologies. These include applications ranging from LEDs [1–4,18], to photorefractive materials [34, 35], to photodetectors and photovoltaic cells [5,6,27–33,36–40]. In addition, with photoionization events as the most promising explanation behind both the photoluminescence blinking and spectral diffusion observed in single nanocrystals [41], understanding photoinduced charge transfer between nanocrystals and their environment is of central importance to a complete understanding of the photophysics of these colloidal quantum dots.

The possible photoinduced interactions between a quantum dot and its surroundings are determined by the relative energy levels of electrons in the dot, on the surface of the dot, and in the surrounding environment (on ligands or nearby molecules). These are depicted schematically in Fig. 6. In general, photoinduced electron transfer will be energetically favorable when the offset between the electron affinities (EAs) or ionization potentials (IPs) of a nanoparticle and a nearby molecule is larger than the Coulombic binding energy of the excited electron–hole pair. This binding energy can be as large as 0.5 eV in organic materials and is on the order of 10–100 meV for colloidal quantum dots, making it important to distinguish between the optical and single-particle energy levels when constructing diagrams like that of Fig. 6.

Charge transfer is only one of many possible relaxation pathways for a neutral photoexcited state at a quantum-dot interface. Radiative decay, För-

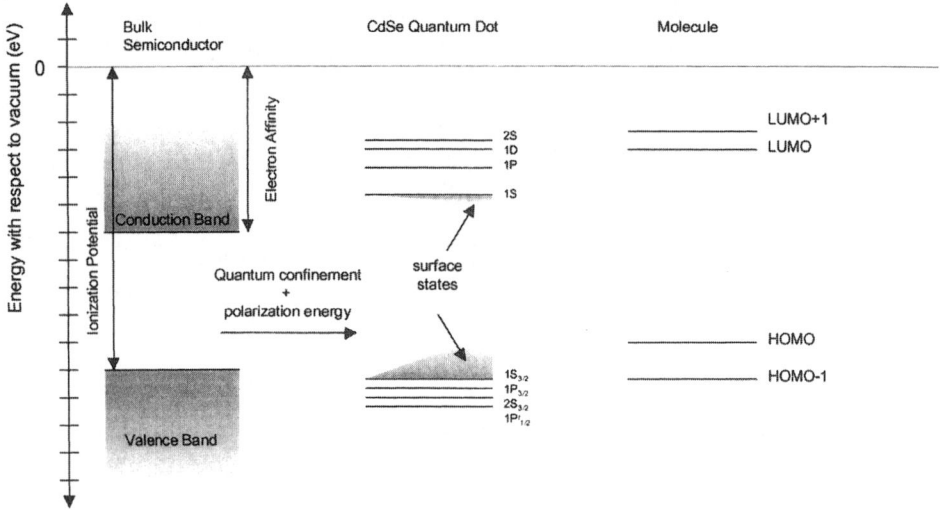

Figure 6 Energy level diagram comparing the bands of a bulk semiconductor crystal, the discrete quantum-confined states in a semiconductor nanocrystal, and the energy levels in a small molecule. In all cases, the electron affinity (EA) is the distance from the vacuum level to the lowest unoccupied level, whereas the ionization potential (IP) is the difference between vacuum and the highest occupied level. Any dangling bonds at the surface of the quantum dot may create midgap states that can act as electron- or hole-accepting traps. In molecular systems, the lowest unoccupied molecular orbital and highest unoccupied molecular orbital are denoted LUMO and HOMO respectively.

ster energy transfer, and nonradiative decay (including trapping) all compete with charge transfer. In addition, not all of these processes are rigorously distinct or mutually exclusive. Trapped charges can still transfer to more energetically favorable acceptors or even recombine radiatively (giving rise to deep-trap luminescence). Likewise, a reasonantly transferred excitation can still undergo charge separation or radiative recombination on the energy-acceptor site. For illustration, several energetically permissible paths to charge generation at a conjugated polymer–CdSe nanocrystal interface are depicted in Fig. 7.

It is the possibility of rationally controlling these processes by size tuning the energy levels of the quantum dots that is particularly exciting. Decreasing the size of a semiconductor nanoparticle alters its electron affinity

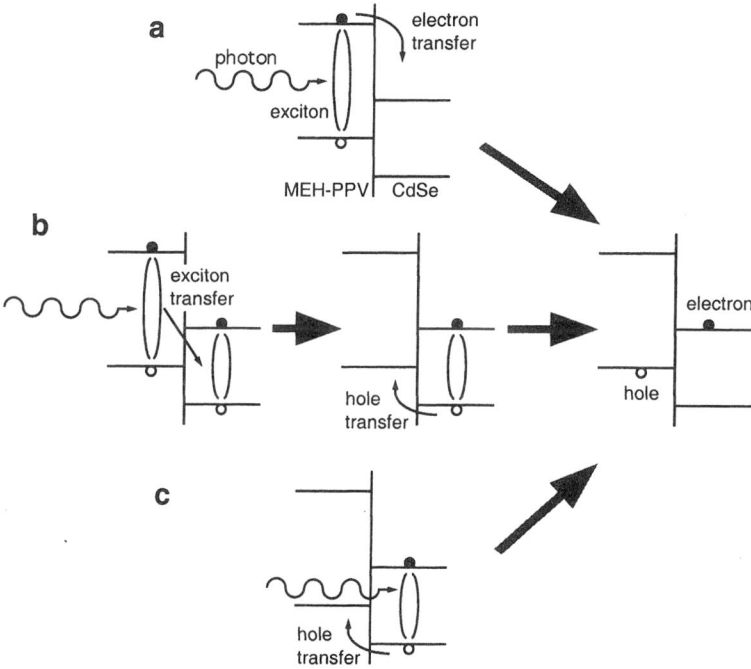

Figure 7 Possible paths to charge generation at an interface between a quantum dot and a conjugated polymer (MEH-PPV). (a) Electron transfer from photoexcited polymer to quantum dot; (b) Förster transfer of exciton from polymer to dot, followed by hole transfer to polymer; (c) hole transfer from photoexcited dot to polymer. Over each path, radiative and nonradiative recombination will compete with the charge and exciton transfer processes.

and ionization potential through two main pathways [22]. Both processes tend to destabilize excess charge on a quantum dot with respect to a bulk semiconductor (the absolute value of the electron affinity is decreased, whereas that of the ionization potential is increased). The first effect is that of quantum confinement, which requires an increase in the kinetic energy of the carriers as the particle diameter is decreased. The second contribution is a classical polarization effect. Because the organic surface ligands and matrix surrounding a chemically prepared nanocrystal have a dielectric constant that is usually considerably smaller than that of the inorganic semiconductor ($\varepsilon_r \sim 2$ versus $\varepsilon_r \sim 10$), the dielectric stabilization provided by the semiconductor decreases and hence the energy required to charge the nanocrystal increases for smaller particles. Detailed calculations of the effects of both quantum confinement and dielectric confinement on the excitonic and single-particle energy levels can be found in the literature [42–46]. For the experimentalist interested in a quick "back-of-the-envelope" calculation of a particle's redox potentials based on optical data, an estimate for the shifts of the lowest single-particle electron and hole levels can be obtained from a knowledge of the bulk semiconductor EA and IP by partitioning the experimentally observed bandgap change between the conduction and valence band states using the ratio of the carriers' effective masses:

$$\text{EA}_{\text{QD}} = \text{EA}_{\text{Bulk}} - \frac{m_h}{m_e + m_h} \Delta E \tag{15}$$

$$\text{IP}_{\text{QD}} = \text{IP}_{\text{Bulk}} + \frac{m_e}{m_e + m_h} \Delta E \tag{16}$$

where EA_{QD} and EA_{Bulk}, and IP_{QD} and IP_{Bulk} are electron affinities and ionization potentials of the quantum dot and bulk material, respectively, and ΔE is the experimentally determined increase in the optical gap with respect to the bulk. Equations (15) and (16) will likely provide a *lower* limit for the change in the EA and IP of the quantum dot material compared to the bulk semiconductor. This is because we have ignored the polarization effects discussed earlier, as well as the electron–hole Coulomb contribution to the shift of the optical gap (which will tend to reduce the optical gap with respect to the single-particle gap). Because a straightforward infinite-barrier effective-mass approximation will invariably overestimate the kinetic energy of confinement, combining polarization and particle-in-a-sphere terms [22] provides a rough *upper* limit for the difference between the quantum-dot EA or IP and the corresponding bulk conduction or valence band edge of the form

$$\frac{\hbar^2 \pi^2}{2m^* r^2} + \frac{1}{2r} \frac{e^2}{4\pi\varepsilon_0} \left(\frac{1}{\varepsilon_{\text{matrix}}} - \frac{1}{\varepsilon_{\text{sc}}} \right) \tag{17}$$

where r is the quantum dot effective radius, m^* is the carrier effective mass, and ε_{matrix} and ε_{sc} are the optical frequency dielectric constants of the surrounding matrix and the semiconductor, respectively.

A. Nanoporous TiO₂ Electrodes

Electron transfer to TiO_2 nanocrystals is of particular interest due to its potential application in photovoltaic devices. Although quantum-confinement effects are not typically observed in these systems, we mention charge transfer in nanocrystalline TiO_2 here because of its similarities with charge transfer in quantum-confined nanocrystal systems. In a photovoltaic "Grätzel cell" [36–47], a monolayer of a ruthenium-based dye is adsorbed onto the surface of a sintered thin film of colloidal TiO_2. Because the internal surface area of an ~10-μm-thick TiO_2 film can be nearly 1000 times larger than its geometrical area, a high optical density is achieved while still maintaining the intimate contact between the adsorbed dye and the semiconductor needed for ultrafast electron transfer and efficient charge separation. Following electron transfer to the TiO_2, the electrons hop across grain boundaries until they reach the back electrode, whereas the holes on the adsorbed dye are scavenged by an aqueous redox couple (typically I^-/I_3^-), which is regenerated at the opposite contact. Total solar power conversion efficiencies of over 10% can be achieved [36] and many variations on this theme are under investigation in order to optimize th spectral response, open-circuit voltage, and especially to eliminate the aqueous electrolyte in these cells. This has led to the study of charge transfer to nanocrystalline TiO_2 from a variety of ruthenium dyes [36], semiconducting conjugated polymers [48–52], and even between chemically synthesized quantum dots and TiO_2 particles [53,54].

The key advantage of using colloidally derived TiO_2 in these cells is that the exceptionally high surface-to-volume ratio provided by the nanocrystalline material yields an enormous amount of interfacial area to support photoinduced charge transfer from adsorbates to the semiconductor particles. Charge transfer in TiO_2 and other semiconductor nanocrystal systems share significant similarities because of the importance of the surface. For instance, adjusting the coupling between the particle surface and an adsorbate will affect the rate of electron transfer [55,56]. In addition, midgap surface states exist, and their exact role in the charge transfer process must be assessed [57,58].

Beyond their surface similarities, important differences between nanocrystalline TiO_2 electrodes and other semiconductor nanocrystal systems remain. Because of the large effective carrier masses and comparatively large particle sizes, typical preparations of TiO_2 nanocrystals do not exhibit the strong quantum-confinement effects characteristic of chemically synthesized quantum dots. Furthermore, because of its large bandgap, TiO_2 is almost always found as the charge acceptor, with absorption taking place in the adsorbed sensitizer. Lower-bandgap nanocrystals can be used both as tunable, sensitizing chromophores, as well as electron or hole acceptors.

B. II–VI Nanocrystal Systems

CdSe and CdS are prototypical colloidal quantum-dot materials, and much of the work on photoinduced charge transfer in chemically synthesized quantum dots has occurred in these two systems. Although many charge transfer experiments involving these particles are driven by applications, others have used charge transfer as a tool to understand the fundamental properties of the nanoparticles. In this regard, one useful probe of charge transfer from luminescent quantum dots (or from luminescent molecules to quantum dots) is photoluminescence quenching. In the absence of a nearby electron or hole acceptor, a high percentage of photoexcitations can relax through radiative recombination. However, if charge transfer from a photoexcited dot to a nearby acceptor is fast enough, the majority of electron–hole pairs will be separated before they recombine and the photoluminescence of the sample will be quenched.

An elegant example of photoluminescence quenching as a probe of charge transfer was provided by Weller and co-workers in an experiment to measure the trap distributions in CdS quantum dots (Fig. 8) [59]. Their samples exhibited both band-edge and trap luminescence and they used nitromethane and methylviologen as electron acceptors. With a relatively high electron affinity, methylviologen was found to quench both the excitonic and trap luminescence in samples of both large- and small-sized CdS dots. However, they found that although nitromethane quenched the excitonic photoluminescence of both sizes of particles, it quenched the trap luminescence only in the smaller nanocrystals. From the difference in EAs of the quantum-dot samples and the known reduction potentials of the electron acceptors, they were able to estimate the depth and width of the electron trap distribution. They found that the energy difference between the excitonic and trap photoluminescence originated from deep trapping of the hole.

Electron transfer from photoexcited CdS to methylviologen is known to occur on ultrafast timescales (200–300 fs), with the charge-separated state then persisting for microseconds; El-Sayed and co-workers used this phenomenon to isolate the hole trapping dynamics from those of the electron in the CdS system [60]. Similar experiments showed that ultrafast (200–400 fs) electron transfer occurs from photoexcited CdSe nanocrystals to adsorbed quinones, but with the quinones acting as electron shuttles that facilitate back-electron transfer in a few picoseconds, faster than the native CdSe

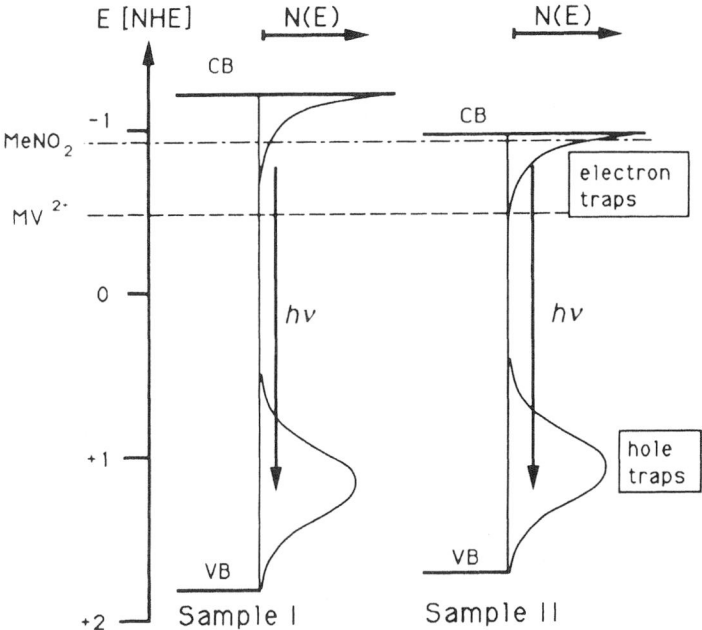

Figure 8 Energy level diagram for the photoluminescence quenching experiment of Weller and co-workers showing the reduction potentials of the methylviologen and nitromethane electron acceptors with respect to the quantum confined conduction-band (CB) and valence-band (VB) states of the nanocrystals and the inferred trap distributions. Sample I and Sample II represent small and large CdS particles, respectively. The zero of the electrochemical energy scale with respect to the normal hydrogen electrode (NHE) corresponds to 4.5 eV below the vacuum level. (From Ref. 59.)

carrier relaxation processes [61]. Although this prevented the isolation of electron and hole dynamics using quinones, both Guyot-Sionnest [62] and Klimov [63] have used pyridine as a hole-accepting surface ligand to facilitate studies of electron relaxation in CdSe dots. The formation of the pyridine cation is complete roughly 450 fs after photoexcitation [63], whereas the lifetime of the charge-separated state can approach 1 ms [64].

Although pyridine is only weakly bound to the nanocrystal surface, a novel diazaperylene that binds strongly to CdSe surfaces has recently been reported [65]. Fluorescence quenching data suggest that both hole transfer from photoexcited CdSe nanocrystals to the adsorbed perylene takes place, as well as electron transfer from the photoexcited perylene to the CdSe nanocrystals.

For many applications, a host material is needed which can not only interact electronically with the nanocrystals but can also transport charge. It is all the more desirable if this host matrix can be processed into thin films. Conducting and semiconducting polymers have the potential to fill this role, and for that reason, composites of quantum dots and electronically active polymers have been investigated by a number of researchers.

Perhaps the first report of this kind was by Wang and Herron [38], who grew CdS clusters inside a polyvinylcarbazole (Fig. 9) matrix. Polyvinylcarbazole (PVK) is a hole conductor that is transparent in the visible region, with an absorption edge near 380 nm. The CdS/PVK composites are yellow in color and have an absorption edge near 440 nm. Upon irradiation with visible light, holes are transferred from the CdS to the PVK, imparting photoconductivity to the composite with an action spectrum similar to the CdS absorption spectrum. The same authors ultimately explored clusters of CdS, PbI_2, HgS, InAs, Ga_2S_3, and In_2S_3, to produce photoconductive films with photoresponses through the near infrared and investigated the dependence of charge generation efficiency and residual quantum dot fluorescence on the electric field applied to the composites [39,40].

More recent experiments have provided additional evidence for both electron transfer from photoexcited PVK to CdS clusters, as well as hole transfer from excited CdS to PVK [66,67] and have also studied the photorefractive properties [34] of the PVK/CdS composite.

With the explosion of interest in semiconducting polymers over the last decade, a wide range of additional polymer hosts have been synthesized. Among the most prominent and intensely studied families of conjugated polymers are those derived from poly(p-phenylenevinylene) (PPV), several of which are shown in Fig. 9. Unlike PVK, these polymers strongly absorb visible light. Nevertheless, single-component polymer photodiodes generally exhibit a low efficiency in converting incident photons into electrical charges. This is because the dominant photogenerated species in most conjugated polymers is a strongly bound neutral exciton. Because these neutral excitations can be dissociated at an interface between the polymer and an electron-accepting species, charge separation is often facilitated via inclusion of a high-electron-affinity substance such as C_{60} [39,68] or another polymer with a higher electron affinity [69,70]. Common features of all such successful charge-separation-enhancing materials include both an electron affinity high enough to make charge transfer energetically favorable and the ability to

---

Figure 9 Chart showing the structures of several polymers and small molecules that exhibit interesting charge transfer behavior when adsorbed onto or blended with CdS and CdSe nanocrystals.

Compound	Literature Report
  PVK	Forms photoconductive blends with a variety of materials by accepting holes from photoexcited quantum dots. [38]
  MEH-PPV	Transfers photoexcited electrons to pyridine treated CdSe nanocrystals. Accepts photoexcited holes from CdSe. Long-lived charge-separated state. [5, 27]
  MEH-CN-PPV	Transfers photoexcited electrons to pyridine-treated CdSe nanocrystals. Accepts photoexcited holes from CdSe. [27]
  DHeO-CN-PPV	Does not undergo photoexcited charge transfer with CdSe nanocrystals, despite electronic similarities to MEH-CN-PPV. [27]
  1,12-diazaperylene	Transfers photoexcited electrons to CdSe nanocrystals and accepts photoexcited holes. [65]
  1,2-naphthoquinone	Accepts photoexcited electrons from CdSe nanocrystals and then rapidly shuttles them back to the CdSe valence band. [61]

form blends with morphologies that allow a high percentage of the polymer excitons to encounter an interface within their typical diffusion range of ~10 nm [71]. In addition, the charge-separation process must be fast enough to compete with the radiative and nonradiative decay pathways of the singlet exciton, which typically occur on timescales of 100–1000 ps [72].

The use of TiO_2 nanocrystals as electron acceptors in this situation has been briefly mentioned earlier [48–52]. The use of CdSe nanocrystals as electron acceptors in polymer blends provides several advantages to the study of photoinduced charge separation. Because the nanocrystal surfaces can be modified through the addition or removal of organic ligands without altering the intrinsic electronic properties of the nanocrystals, there exists the possibility of altering the blend morphology or of introducing a controlled spatial barrier to charge transfer while still retaining the size-tunable properties of the quantum dots. More importantly, because the energy levels of the host polymers can be tuned through chemical derivatization of the backbone chains and the energy levels of the nanocrystals can be tuned through size-dependent quantum-confinement effects, blends of the two materials offer the possibility of careful and independent positioning of both donor and acceptor levels.

Greenham et al. [5] studied blends of CdS and CdSe particles in an MEH-PPV polymer host. Unlike much of the work on PVK/CdS blends, the quantum dots and polymers were synthesized separately and the composites were formed by spin-coating films from a common solvent, allowing for precise control of both the nanocrystal and polymer chemistry. Photoinduced electron transfer from MEH–PPV to 5-nm-diameter CdSe particles was demonstrated via quenching of the polymer photoluminescence, as well as by dramatic increases in the photoconductivity of the MEH–PPV. It was found, however, that charge transfer could only take place once the protective TOPO monolayer on the particle surfaces had been exchanged with pyridine, thereby allowing the electronically active polymer backbone and nanocrystal core to interact closely. This surface exchange was also found to have a strong beneficial influence on the local film morphology. At high concentrations, pyridine-treated nanoparticles also tend to form aggregates in the blends, as seen in Fig. 10.

We have recently studied charge transfer between different sizes of CdSe nanocrystals and various PPV derivatives with different EAs and different side chains [27]. By monitoring quenching of the polymer fluorescence (Fig. 11), we were able to establish that electron transfer occurs from MEH–PPV to several smaller sizes of CdSe, despite the lower electron affinity of the smaller nanocrystals. We also measured fluorescence quenching between CdSe nanocrystals and two cyano-substituted PPV derivatives, MEH–CN–PPV and DHeO–CN–PPV. Grafting electron-withdrawing CN groups onto the poly-

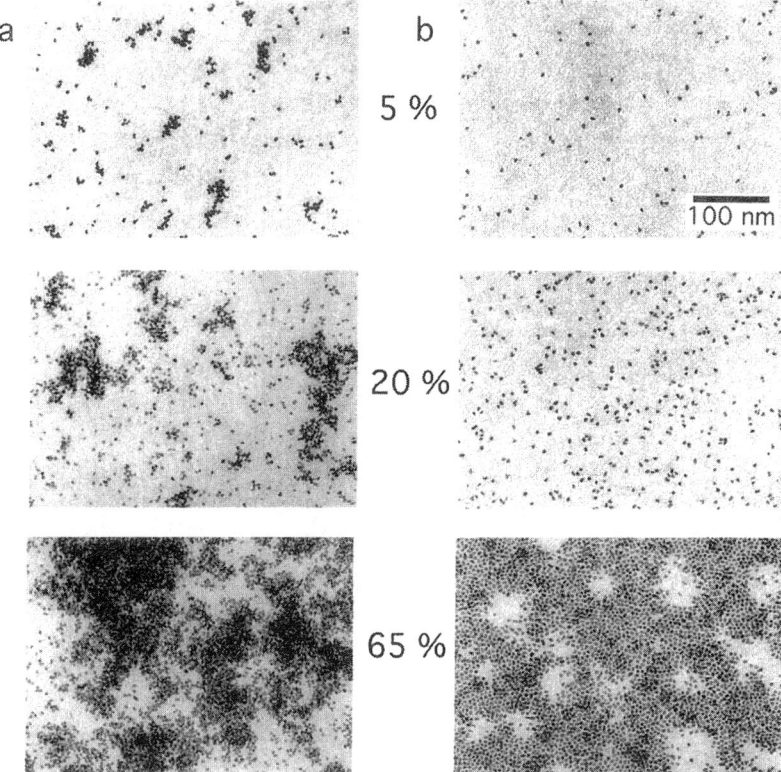

Figure 10 Transmission electron microscope images of blends of MEH–PPV with 5-nm-diameter CdSe nanocrystals at concentrations of 5%, 20%, and 65% by weight: (a) pyridine-treated nanocrystals; (b) TOPO-coated nanocrystals. Distinct aggregation can be seen in the blends containing pyridine-treated nanocrystals.

mer backbone increases the electron affinity of the polymer by roughly 0.5 eV. [73]. Nevertheless, all sizes of CdSe nanocrystals studied quenched the fluorescence of MEH–CN–PPV to nearly the same degree as they had quenched the emission of MEH–PPV. On the other hand, DHeO–CN–PPV, with HOMO and LUMO levels similar to those of MEH–CN–PPV, did not exhibit fluorescence quenching in the presence of the nanocrystals. Ultimately we attributed this difference to the fact that the symmetric alkyl side chains of DHeO–CN–PPV provide an insulating spacer between the nanocrystals and the conjugated polymer core, which prevents rapid electron transfer.

In the same article [27], we also used photoinduced absorption (PIA) spectroscopy to provide direct evidence for the presence or absence of photo-

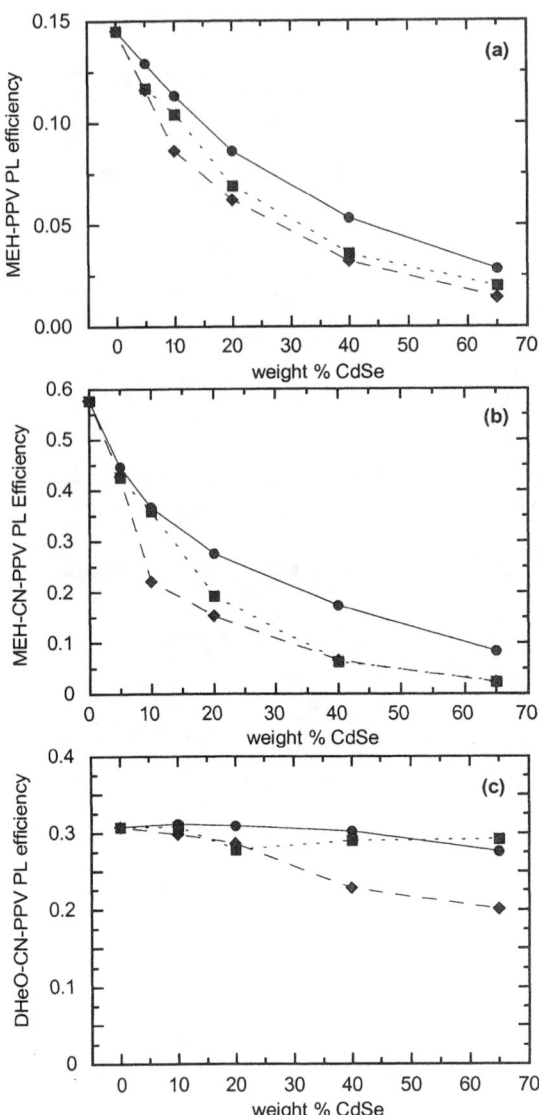

Figure 11 Photoluminescence efficiencies of blends of (a) MEH–PPV, (b) MEH–CN–PPV, and (c) DHeO–CN–PPV with pyridine-treated CdSe nanocrystals of 2.5 (squares) 3.3 (circles) and 4.0 (diamonds) nm in diameter. The structures of the three PPV derivatives are shown in Fig. 9.

induced electron transfer by monitoring the subgap absorptions characteristic of positive charges (polarons) on the polymer backbone. Consistent with the fluorescence quenching data, the only long-lived photoinduced species we observed in the DHeO–CN–PPV/CdSe composite was the neutral triplet exciton in the polymer (which has an infrared absorption to a higher-lying triplet state). However, in MEH–PPV/nanocrystal composites, we were able to observe the low- and high-energy absorption signatures characteristic of long-lived positive polarons on the polymer (Fig. 12a). Because parts of the MEH-PPV polaron absorption occur in the same region as absorptions due to the polymer triplet exciton, we characterized the absorption spectrum as a function of temperature and pump modulation frequency at various wavelengths. Whereas the triplet lifetime was very sensitive to temperature, the polaron lifetime was only weakly temperature dependent (Fig. 12b). This allowed us to resolve the contributions to the PIA spectra from the polaron and triplet species, as well as to demonstrate that triplet and polaron excitations could coexist in the composites. This observation was consistent with the strong, but not complete, fluorescence quenching in the composites, and both effects were interpreted in the context of the aggregated blend morphology (Fig. 10). Because the polymer–nanocrystal phase separation occurs on length scales comparable to the exciton diffusion range in MEH–PPV, some excitons will decay radiatively or undergo intersystem crossing to form triplet states, before they have a chance to reach an interface where they might be dissociated.

By varying the pump modulation frequency, we were able to measure the lifetime of the charge-separated state, which was found to span a distribution of recombination times from microseconds to milliseconds. The recombination process is therefore much slower than the initial charge-separation process. We will see in Section V that recombination is an important loss mechanism in polymer–nanocrystal photovoltaic devices. Measurement of the recombination time gives a rough estimate of the timescale on which charges must be removed from a photovoltaic device in order to avoid recombination.

The timescale on which the initial charge-separation event takes place in semiconducting polymer–quantum dot composites is not yet accurately known, and time-resolved studies are needed to address this issue. Clearly, electron transfer is fast enough to compete with radiative recombination in these polymers, but different kinetics would be expected depending on whether exciton diffusion to the nanocrystal–polymer interface or the intrinsic charge transfer step is the rate-limiting factor. There is also the possibility that a fraction of the charges could be generated by hole transfer from the nanocrystal to the polymer following transfer of the entire exciton from the polymer to the nanocrystals via Förster transfer. The partial concentration-

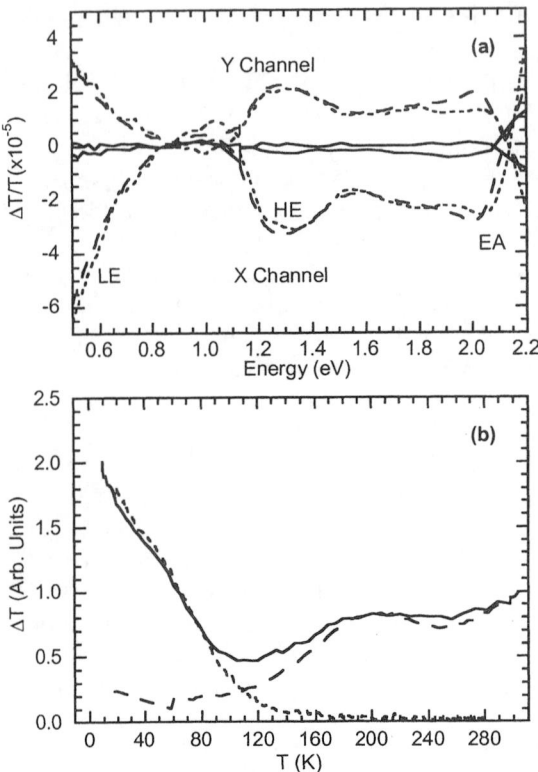

Figure 12 (a) Room-temperature PIA spectra of MEH–PPV (solid line), a blend of MEH–PPV containing 40% weight of 4.0-nm CdSe nanocrystals (long dashes), and a blend of MEH–PPV and 40% weight of 2.5-nm CdSe nanocrystals (short dashes). The nanocrystal–polymer blends exhibit characteristic low-energy (LE) and high-energy (HE) polaronic absorption features, in addition to electroabsorption (EA) due to the fields of the photogenerated charges. (b) Temperature dependence of the PIA signals for MEH–PPV at 1.34 eV (short dashes), and for the MEH–PPV/40% weight 4.0-nm CdSe nanocrystal blend at 1.34 eV (solid line) and at 0.5 eV (long dashes). At low temperature, the 1.34-eV signal in both samples is dominated by the polymer triplet–triplet absorption, but it can be seen that the polaron absorptions persist up to room temperature in the nanocrystal–polymer blend.

dependent fluorescence quenching and the coexistence of both triplet and polaron species at long times indicate that a rich variety of competing physical processes could be investigated with ultrafast transient absorption experiments.

We have shown above that chemically sythesized quantum dots can act as good electron acceptors from organic materials. We will return to photo-induced charge transfer in Section VI, where we describe photovoltaic devices based on composites of nanocrystals and conjugated polymers.

V. CHARGE TRANSPORT IN NANOCRYSTAL FILMS

Charge transport in quantum-dot systems has been a particularly vibrant area of condensed-matter physics research. Coulomb blockade effects, resonant tunneling, and single-electron transistors have all been studied in quantum dots forged with sophisticated electron beam lithography and molecular beam epitaxy (MBE) techniques [74,75]. To observe these unique effects however, the available thermal energy, $k_B T$, must be smaller than the relevant energy level scale (Coulomb charging energy and quantum level spacing) in the quantum dot. For lithographically patterned dots, this often demands working at sub-Kelvin temperatures. Nanocrystals are particularly attractive for these studies because chemical routes can prepare smaller quantum dots with larger energy spacings than are otherwise obtainable (the single-electron charging energy of a small CdSe nanocrystal can exceed 100 meV, whereas the spacings between the quantized conduction band levels can exceed 500 meV). There have been several exciting articles exploring the charge transport properties of single nanocrystals, including the demonstration of a CdSe nanocrystal single-electron transistor [7,76] and STM conductance spectroscopy on the levels in CdSe [77–79] and InAs [80,81] nanocrystals.

Arrays of chemically synthesized quantum dots are also interesting for transport studies. Not only can the properties of the constituent particles be tuned through quantum-size effects, but the collective properties of the array can be adjusted by controlling the interparticle coupling and order. A dramatic demonstration of this was provided by Health and co-workers when they drove a reversible metal–insulator transition in a monolayer of silver nanocrystals through compression and rarefaction on a Langmuir trough [82,83]. In addition, collective charge transport in nanoscale arrays has been proposed as an experimentally accessible model for emergent phenomena in complex granular systems [23]. It has even been suggested that charge transport in quantum-dot arrays might ultimately be used to perform computations [84]. In yet another application, nanocrystal–nanocrystal charge hopping is critical to the operation of the highly disorded TiO_2 electrodes discussed briefly in Section IV. Charge transport in that system exhibits a number of unusual nonexponential relaxation phenomena which have been analyzed in the context of a continuous-time random walk [85]. In the prototypical CdSe system, the efficiency of nanocrystal–nanocrystal charge transport is central to the viability of the polymer–nanocrystal-based light-emitting and photovoltaic diodes discussed in Section VI. We will eventually focus on transport in close-packed CdSe nanocrystal films, but we first turn our attention to the better studied topic of transport through arrays of chemically prepared metal nanoparticles to introduce many concepts relevant to transport through films of ligand-separated quantum dots.

A. Metallic Nanocrystals: Charging, Disorder, Coupling

Although a metallic nanoparticle may not exhibit an inheret HOMO–LUMO gap unless it is ~1 nm in diameter or smaller [86], the capacitance of a metal particle and the quantization of charge will still create a "Coulomb gap" in the single-particle density of states and endow metal nanoparticles with single-electron Coulomb blockade properties. Although more detailed reviews of these effects are available elsewhere [74,87], we briefly mention these properties here. Figure 13 depicts a single metal particle connected to two identical electrodes by tunnel junctions. An electron can only tunnel between states of identical energy. However, if the total capacitance between the particle and its surroundings is C, then the energy of a single additional charge on the particle will be $e^2/2C$ away from the Fermi level of the electrons in the metal nanoparticle. Thus, a bias must be applied sufficient to raise the energy of the electrons in the lead to the energy they would occupy on the charged dot before any can tunnel onto the particle. Furthermore, when one electron does tunnel onto the dot, its presence will prevent the addition of a second charge, either until the first charge tunnels off or until the bias voltage is raised: additional electrons are thus "blockaded" by the presence of the first. The two-terminal current–voltage curve will exhibit discontinuous increases as the bias is made large enough to allow larger numbers of electrons to exist on the dot at the same time. Finally, the addition of a third terminal to the structure in Fig. 13 allows the energy levels of the particle to be gated so as to allow or prohibit single-electron tunneling events, thus forming a single-electron transistor. As already mentioned, the preceding discussion is only relevant when $k_B T \ll e^2 2C$. At higher temperature, the thermal distribution of carriers in the leads will mask the discrete structure. Coulomb blockade effects have been the subject of extensive experimental study in both lithographically [74,87] and chemically [88–91] defined nanostructures.

Transport through ensembles of particles can exhibit more complex phenomena. Middleton and Wingreen calculated that in the Coulomb

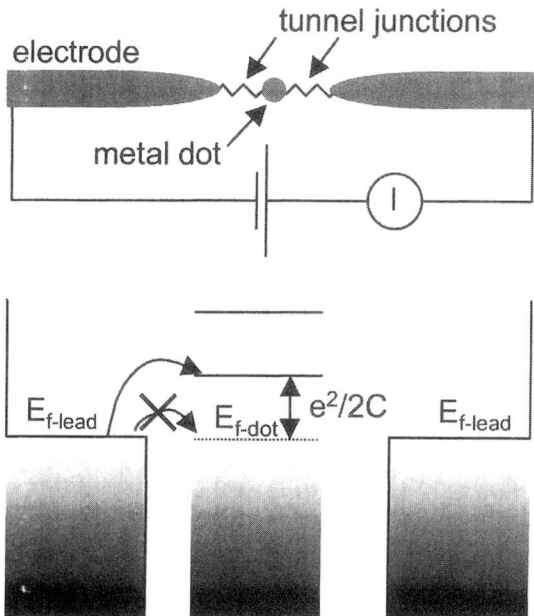

Figure 13 Illustration of a two-terminal Coulomb blockade device with a metal nanoparticle separated from electrical leads by insulating barriers. Because of the small capacitance of the system, the nearest state accessible for charge tunneling is $e^2/2C$ away from the Fermi level of the dot.

blockade regime, disorder in one- and two-dimensional arrays of metallic dots would lead to an additional voltage threshold for conductivity, V_T, with above-threshold current, I, scaling as a power law with voltage, V [23]:

$$I \sim (V/V_T - 1)^\zeta \tag{18}$$

where ζ is an exponent that depends on the dimensionality of the array. Initially observed in arrays of lithographically patterned Al dots [92], threshold behavior and power-law scaling have also been observed in arrays of chemically synthesized Co particles [93].

Following the ready availability of size-selected, thiol-capped Au nanocrystals [94], several groups have investigated transport in solids and films composed of size-selected gold clusters spaced by monolayers of organic ligands [95–100]. These monolayer-protected clusters can be dried to produce 2-D arrays, thin films, or pellets for electrical measurements in a number of configurations. For a variety of ligands, ranging from alkane [95] and ayrl [98] dithiolates to thiol-terminated DNA [101], these gold cluster solids exhibit a nonmetallic, Arrhenius-like, dependence of conductivity, σ, on temperature

$$\sigma(T) = \sigma_0 \exp[-E_A/(k_B T)] \tag{19}$$

where σ_0 is a constant, similar to the behavior seen in metal island films [102]. This simple Arrhenius form is obeyed, rather than the $\ln(\sigma) \propto T^{-0.5}$ of sputtered granular films [103] only if the cores are relatively monodisperse, and the tunnel junction lengths are not correlated with the particle sizes. The activation energies for charge transport correspond well to the capacitive charging energies of the particles embedded in the array [95,98] which can be approximated by the electrostatic energy derived by Abeles et al. [103] for carrier generation in granular metal films (cermets) in the low-field regime:

$$E_A = \frac{1}{2} E_C = \frac{1}{2} \frac{e^2}{4\pi\varepsilon\varepsilon_0} \left(\frac{1}{r} - \frac{1}{r+s} \right) \tag{20}$$

In this case, E_C is the Coulomb charging energy required to generate two charged nanoparticles from two originally neutral particles embedded in a material of relative permittivity ε. The average particle radius is r, and the average spacing between particle surfaces is s. It is important to note that this is the activation energy for carrier generation, as in an intrinsic semiconductor, and not for electron transport (tunneling) through the array.

The degree of electronic coupling between the metal cores strongly influences the tunneling rate. For alkanethiolate ligands, the conductivity decreases exponentially with increasing ligand length [97,99,100] as

$$\sigma(s) = \sigma_0 \exp(-\beta_s) \tag{21}$$

The electronic coupling, β, through the alkanethiolate chains falls in the range from 0.6 to 1.25 Å^{-1} [97,100]. Murray and co-workers illustrated the effects of both carrier density and tunneling rate on sample conductivity [100]. They found that for small particles with large charging energies, the carrier concentration in the array can be dominated by intentional "doping" (mixed valency) of the nanoparticle cores rather than through thermal activation, and they considered the possibility that Marcus theory activation energies (Sect. II) could describe the activated transport observed in their samples.

Gold is not the only metal that can be used to synthesize arrays of nanocrystals suitable for charge transport studies. Black et al. observed a number of the above effects, including power-law scaling of current with voltage and Arrhenius-like activated conductivity in arrays of Co nanocrystals [93]. Because they used a ferromagnetic metal, they were also able to observe magnetoresistance indicative of spin-dependent tunneling in their samples.

B. Close-Packed CdSe Films

Like monolayer-protected metal clusters, semiconductor quantum dots encapsulated by a monolayer of passivating surfactant can be processed from solution into a variety of "quantum-dot solids" consisting of close-packed semiconductor nanocrystals separated by the surfactant layers (Fig. 14).

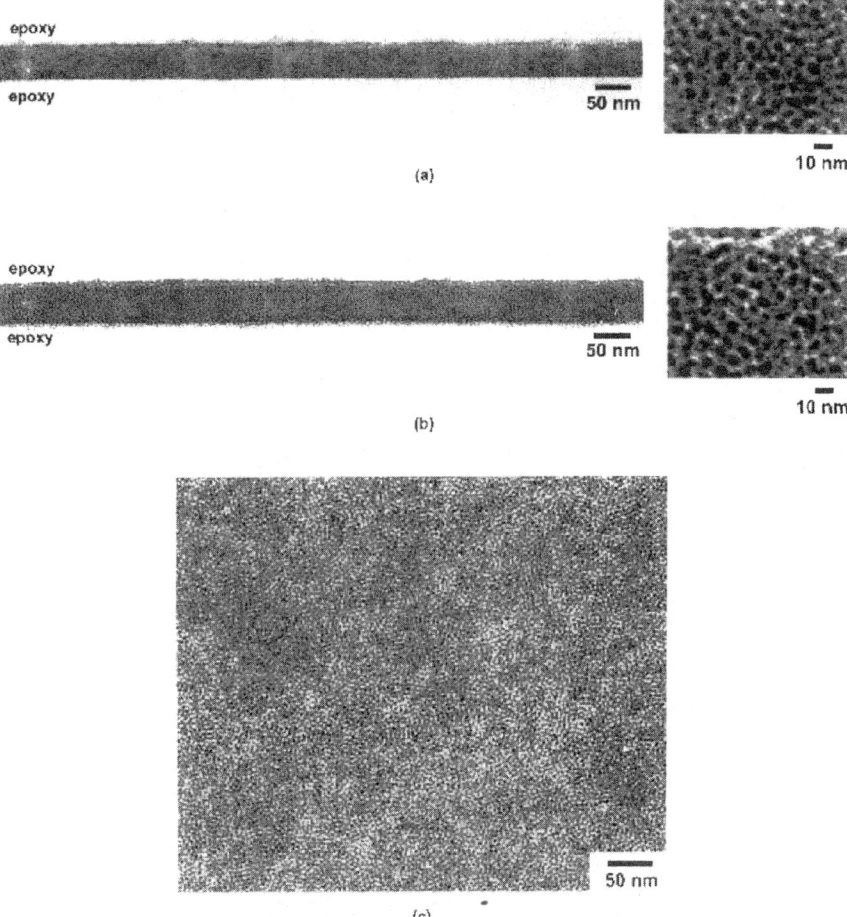

Figure 14 Cross-sectional TEM images of spin-coated CdSe (a) and CdSe/ZnS core–shell nanocrystal films (b). (c) Plan view of a thin film of CdSe nanocrystals cast on a mica plate. The images demonstrate both the close-packed structure and interparticle separation due to the surfactant monolayers. (From Ref. 18.)

Films of passivated CdSe nanocrystals are sufficiently robust to be incorporated into thin-film devices, and their transport properties are described in this section.

Many of the factors governing charge transport in arrays of metal nanocrystals are clearly important to transport in these arrays of nanocrystalline semiconductors, but there are also significant differences between transport in the two systems. The most obvious and important difference is the presence of the semiconductor bandgap. In wide-bandgap semiconductors, the thermally generated charges which dominate the transport properties of most metal particle arrays will be almost entirely suppressed, and the conductivity will be dominated by carriers injected from the electrodes [33], created by photoexcitation [21], or perhaps through chemical doping [104]. Another difference is that metal nanoparticles have a near continuum of states, modified only by the Coulomb charging energies, into which electrons can tunnel. For semiconductor nanocrystals, the spacing between the quantized conduction-band states is on the order of a few hundred milli-electron volts, and both the intrinsic density of states and Coulomb effects will determine the tunneling density of states and the charge transport properties. Furthermore, the charge carrier wave functions do not extend far beyond the surfaces of CdSe dots, and strong electronic overlap between dots is unlikely to occur, even at high pressure [105]. Finally, the charge screening length will be significantly longer in an array of semiconductor particles than it is in an array of metal particles. It can be expected that long-range Coulomb interactions and space-charge effects [33] will play a much greater role in arrays of semiconductor dots.

In the dark, charge carriers must be injected from an electrode before they can be transported through an array of CdSe nanocrystals. Because the particles are both undoped and covered with an insulating monolayer, injection can be viewed as a tunneling problem at a metal–insulator–semiconductor interface. For the case of electrons, injection must therefore occur by tunneling from states near, or below, the Fermi level of the metal contact into the conduction band states of the nanocrystal. If the Fermi level of the metal lies lower in energy than the lowest quantized conduction-band states in the nanocrystals, then no tunneling can occur. Heath and co-workers utilized this behavior to fabricate tunnel diodes from monolayers of CdSe nanocrystals [106]. From the I–V characteristics of their devices, they estimated that the lowest quantum-confined conduction-band state lies ~3.6 eV below the vacuum level for a 3.8-nm-diameter nanocrystals, in fair agreement with the estimate of ~4.0 eV obtained via Eq. (17). We have measured the efficiency of electron injection into 200-nm-thick close-packed CdSe films sandwiched between electrodes of several different metals [33]. Results from a similar set of our experiments are depicted in Fig. 15. It can be seen that the

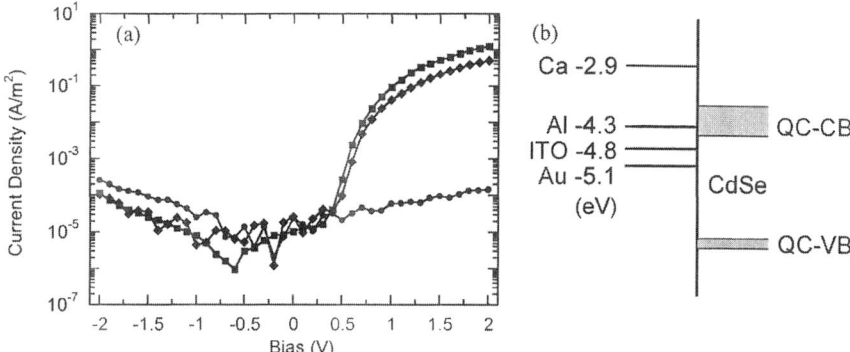

Figure 15 (a) Current–voltage (I–V) Curves for three different ITO/CdSe/metal devices. The CdSe layer for each device is composed of a 190-nm-thick film of 3.4-nm-diameter nanocrystals separated by TOPO surfactant. Diamonds indicate the I–V curve for the Al cathode device, squares for the Ca cathode device, and circles for the Au cathode device. The anode was ITO in all cases. (b) Diagram showing the relative positions of the Fermi levels of the metals used as electrodes in our nanocrystal devices and the quantum-confined conduction band (QC-CB) and valence-band (QC-VB) edge states. The range indicated is obtained from the upper and lower estimates of Eqs. (15)–(17) as described in Section IV.

low-work-function metals Ca and Al produce high currents at low bias, whereas the high-work-function metal Au and ITO exhibit currents that are smaller by several orders of magnitude at the same voltage. The currents for the Ca and Al devices turn on at the same voltage (despite work functions differing by nearly 1.5 eV), suggesting that they provide equivalent electron injection capabilities and that the observed current–voltage curves are thus limited by the bulk properties of the samples. As shown in Fig. 15, these results are entirely consistent with both the model of charge injection described above as well as the EA estimates provided by Eqs. (15)–(17).

A second means of introducing carriers into a quantum-dot film is by photoexcitation. Strictly speaking, this involves photoinduced electron transfer from one nanocrystal to another, but we include its discussion here, rather than in Section IV, because its is essentially a nanocrystal–nanocrystal charge transfer problem. Using different experimental geometries, both we [33] and Leatherdale et al. [21] have demonstrated that photocurrents in films of CdSe follow a spectral response curve (Fig. 16) that nearly matches the quantum-confined absorption spectrum of the nanocrystals. Our experiments measured photocurrents through thin (200 nm) films of CdSe sandwiched between ITO and Al electrodes, with active areas of a few square millimeters that were

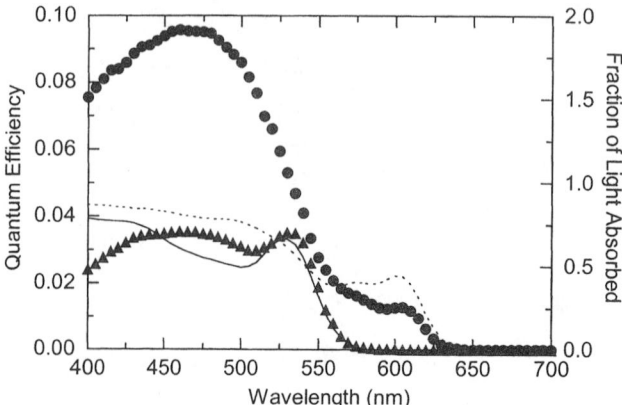

Figure 16 Photocurrent action spectra for ITO/CdSe/Al sandwich structures with 200-nm-thick layers of 2.7-nm-diameter nanocrystals (triangles) and 4.5-nm-diameter nanocrystals (circles). Also plotted is the fraction of incident light that is absorbed in each film for the small (solid line) and large (dashed line) nanocrystals, including a simple correction for interference effects.

illuminated through the ITO contact. Leatherdale et al. employed Au bar electrodes, with spacings of 1 to 20 μm. In our devices, we were able to measure external quantum efficiencies (collected charges/incident photons) of 1–10% at room temperature under short-circuit conditions. In the short-circuit mode, charges are collected by the built-in field that arises from the work-function difference between the Al and ITO contacts. This amounts to fields of $\sim 3 \times 10^6$ V/m in our devices. On the other hand, Leatherdale et al. measured quantum efficiencies on the order of 10^{-4} electrons/photon, at applied fields of $\sim 10^7$ V/m. Because the fields are similar between the two experiments, we would expect similar charge generation rates, and the different quantum efficiencies are therefore surprising. One possibility is that the mismatch results from different recombination loss rates caused by the differing distances traveled by carriers in each experiment.

Finally, in both sets of experiments, it is found that the photocurrent increases linearly with illumination intensity. This is consistent with a single-photon mechanism for carrier generation and with first-order recombination kinetics. This is the expected behavior if dissociation of the exciton is the rate-limiting step, as most electron–hole pairs will recombine while still correlated (geminate recombination). First-order kinetics are also expected if recombination is dominated by a high density of trap sites in the sample [107].

One particularly unusual electrical property of CdSe quantum-dot films is the time and history dependence of their current–voltage characteristics.

For samples measured in darkness, the current in response to a fixed-voltage step is found to decay monotonically with a nonexponential form [33,108]. We find that our data [33] consistently follow Kohlrausch's [109] stretched exponential relaxation function:

$$I(t) = I_0 \exp\left[-\left(\frac{t}{\tau}\right)^{\beta}\right]$$

although others have observed power-law decays under different conditions [108]. Stretched exponential functions are characteristic of many relaxation processes in molecular and electronic glasses [110] and have previously been observed for various carrier relaxation processes in TiO_2 [85] and CdSe [111] nanocrystals.

Our devices exhibit strong history-dependent "memory" effects, as shown in Fig. 17. Several experimental facts need to be accounted for in any description of this phenomenon. First, if allowed to "rest" in the dark at room temperature, the films will slowly regain their original current–voltage characteristics over a period of days. In addition, the original dark conductivity can be restored by even a very brief exposure to light of energy above the nanocrystal bandgap. Light below the bandgap has no effect, ruling out light-induced detrapping. Longer exposure will raise the dark conductivity of a film to levels that exceed the original dark conductivity levels of the device. Finally, under zero bias, this "persistent photoconductivity" is found to

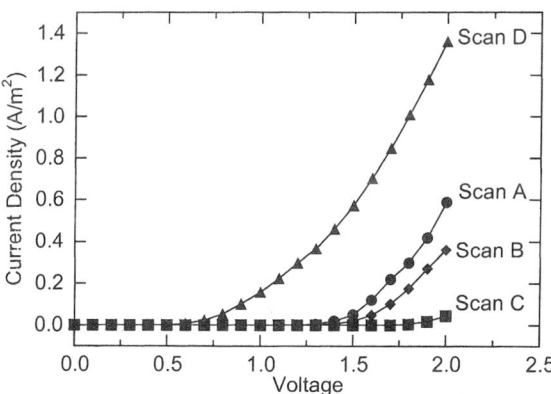

Figure 17 Current–voltage scans showing the effect of device history. Scan A (circles) is the initial scan. Scan B (diamonds) was taken after two subsequent current–voltage sweeps. Scan C (squares) was taken after holding the device at +2 V bias for 10 min. Scan D (triangles) was taken after illumination with light just above the bandgap.

decay with stretched exponential kinetics, characteristic of persistent photo-conductivity in many materials [112,113].

We have simulated this behavior with a model of space-charge-limited current in the presence of a fixed number of deep-trap sites and found that the calculations are able to describe the essential features of the experimental data [33]. In the space-charge limit, the field at any point in the device is strongly modified by the presence of the injected charge, whereas the total amount of charge is fixed by the applied bias. In our simplified model, with a fixed number of deep traps, the effective mobility [Eq. (11)] of the carriers is gradually reduced as more and more of them fall into deep traps. When trap densities are comparable to the space-charge density, the trapping rate begins to fall as the trap sites become occupied, leading to a decreasing rate of trapping and a nonexponential decay of the current with time as in the depletive trap model [110,114]. Exposure to above-bandgap light creates free carriers that can move to neutralize the trapped space charge, accounting for the restored conductivity. The persistent photoconductivity can be accounted for if the hole mobility is assumed to be much lower than the electron mobility, so that positive charge slowly builds up under irradiation, [115]. The presence of positive charge allows the maintenance of a higher density of negative space charge and hence a higher current. This conductivity then decays as the positive charge is swept out of the device, or neutralized through recombination.

Finally, the assumption of space-charge-limited currents allowed us to calculate the electron mobility in these films. We found a large sample-to-sample variation, but the mobilities fell in the range of $\sim 10^{-4}$–10^{-6} cm^2/V^{-1}/s^{-1} for the "untrapped" electrons in our samples. Because the space-charge limit represents the maximum single-carrier current that can be passed through any device, these values represent good lower limits for the electron mobility in CdSe quantum-dot solids, even in the event that the devices were not truly operating in the space-charge-limited regime.

Both additional measurements and more sophisticated modeling of the transport and relaxation phenomena are required. Introduction of a distribution of traps in which carriers slowly relax to intrinsically deeper, less mobile sites or in which they are gradually localized through disorder-induced "trapping," as in a Coulomb glass, would both follow naturally from the data. These possibilities are perhaps more realistic than assuming a fixed density of deep traps, as we have done, but the work to date provides a needed starting point.

The temperature dependence of the conductivity often provides valuable insight into the microscopic physics behind carrier transport in a material system. Although the time and exposure dependence of our current–voltage curves complicate the interpretation of the temperature data, we believe that by observing identical experimental histories for each sample, we are able to extract useful information from conductivity data at various temperatures [33].

Figure 18 shows an Arrhenius plot of $\ln(\sigma)$ versus $1/T$ for a sample of 2.7-nm-diameter CdSe nanocrystals and a sample of 4.5-nm-diameter CdSe nanocrystals. Although the temperature range is limited, a linear region of the dark conductivity, $\sigma(T)$, is evident in the Arrhenius plot at high temperatures (~300–180 K), which is typical of a simple activated hopping process [Eq. (19)]. Some form of hopping transport has been assumed from the field dependence of the current in similar structures of CdS nanocrystals [116] and seems natural given the film morphology in which the nanocrystals are separated by ~12–14 Å of insulating organic surfactant [18]. At lower temperatures (~180–60 K), a second region with a nearly linear dependence of $\ln(\sigma)$ versus $1/T$ is also observed. In this intermediate range, the temperature dependence is much smaller than at high temperatures, suggesting that a second, smaller activation energy dominates the transport in this temperature region. The current typically reaches a constant value below ~60 K at currents comparable to those observed in reverse bias at room temperature. We therefore believe that leakage currents may dominate the low-temperature (< 60 K) conductivity values in our samples.

From the slopes of the Arrhenius plots in the ~180–300-K region, we find activation energies in the range from 0.10 to 0.20 eV. These are con-

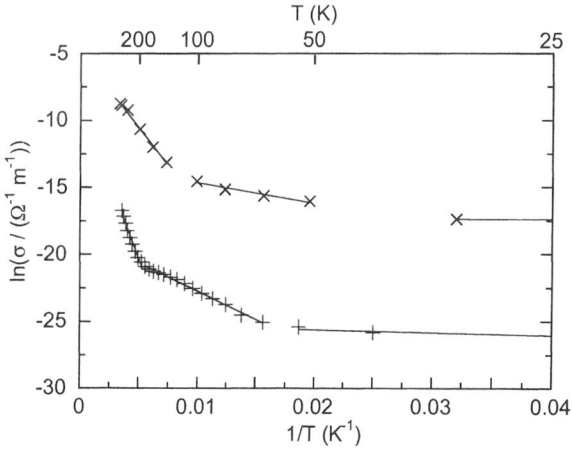

Figure 18 Natural logarithm of the dark conductivity at +2 V plotted as a function of $1/T$ for 200-nm-thick films of 4.5-nm (×'s)- and 2.7-nm (+'s)-diameter, TOPO-capped, CdSe nanocrystals.

siderably larger than might be estimated from the results of electron transfer theory as described in Section II. Furthermore, the sample-to-sample variations in activation energy are larger than any size-dependent trend that may be present or that may be predicted by Eqs. (2)–(5). For these reasons, we conclude that, at least at temperatures between 180 and 300 K, transport in the samples studied to date has been dominated by the effects of trapping and disorder. In the future, studies of more ordered films, samples with narrower size distributions, or samples with improved surface passivation might allow size-dependent trends to be identified.

VI. NANOCRYSTAL-BASED DEVICES

As has been noted in the previous sections of this chapter, an understanding of charge transfer and charge transport is fundamentally important to the rational design of nanocrystal-based optoelectronic devices. In this section, we briefly review the construction and operation of various thin-film light-emitting and photovoltaic devices based around II–VI semiconductor nanocrystals. The nearly universal geometry employed for these devices is shown in Fig. 1 and is identical to that used for the charge transport studies described in Section V. The thin-film "active layer" between the metal electrodes can be deposited by a variety of methods, including spin-coating, drop-casting, and electrochemical deposition, as well as by controlled layer-by-layer self-assembly. Many of these devices also incorporate an organic semiconductor as some part of the active layer.

A. Light-Emitting Diodes

Nanocrystal light-emitting diodes (LEDs) have been fabricated in a variety of configurations (Fig. 2). These include nanocrystal–polymer bilayer heterojunctions [1,3,18], nanocrystal–polymer intermixed composites [2,117–119], close-packed nanocrystal films [116,120], and even self-assembled stacks of nanocrystal and organic monolayers [4,121,122]. Although some of these devices exhibit broad, and even white electroluminescence spectra [116,117, 121,122], we focus on those device preparations which yield well-defined electroluminescence peaks which can be controlled by adjusting the size of the nanocrystals used to fabricate the device.

This quantum-confined electroluminescence arises from the radiative recombination of excitons formed by the injection of electrons and holes into the quantum dot or from exciton transfer to the dot from the host material. For this reason, the electroluminescence spectrum of a nanocrystal LED generally resembles the solution photoluminescence spectrum of the nanocrystals from which it was prepared (Fig. 19). However, both blue shifts [1] and red shifts [3] of the electroluminescence have been observed with respect to the photoluminescence. In almost all cases, the electroluminescence spectrum is broader than that of the photoluminescence. These effects can variously be attributed to the combined influence of local electric fields, optical interference effects from the photonic structure of the device, reabsorption of the emitted light, differences in the local dielectric environments of the emitting nanocrystals, and size-dependent charge injection and exciton transfer rates, but have not been investigated in significant detail.

A good LED design requires efficient radiative recombination of injected electrons and holes within the nanocrystals, as well as balanced electron and hole injection at low applied bias. The first condition requires that the electrons and holes can combine to form excitons on emissive centers and that the local environment allows these excitons to decay radiatively with high efficiency.

These design considerations can account for the wide range of electroluminescence efficiencies observed in the various nanocrystal LED structures that have been investigated to date. For instance, early LEDs formed by

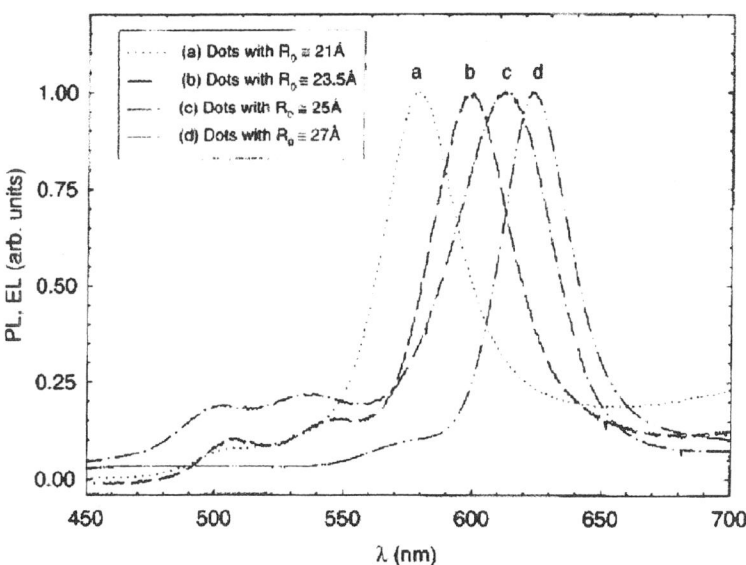

Figure 19 Spectra showing the size-tunable quantum-confined electroluminescence of several sizes of TOPO-coated CdSe nanocrystals as labeled in the inset. The weaker peaks between 500 and 550 nm are from electroluminescence of the PPV hole-transport layer. (From Ref. 18.)

dispersing CdSe nanocrystal chromophores into a matrix of the hole-conducting polymer polyvinylcarbazole (PVK) along with the small-molecule electron transporter t-Bu-PBD [2-(4-biphenylyl)-5-(4-$tert$-butylphenyl)-1,3,4-oxadiazole] were found to give electroluminescence (EL) quantum efficiencies of only 0.0005% [2]. The addition of the nanocrystals did not influence the current-voltage behavior of these devices, and it was concluded that charge was injected into the PVK and PBD components, with the nanocrystals serving as trapping and recombination centers. The low EL efficiency of these devices can thus be attributed to the energetics of the CdSe–polymer interface. Although electron transfer from PBD to CdSe should be energetically favorable (the electron affinities are 2.6 eV and ~4.4 eV, respectively), hole transfer from PVK to the CdSe is not (the ionization potentials are 5.3 eV and ~6.7 eV for PVK and CdSe). Therefore, in the randomly intermixed morphology of such a device, it is likely that most electrons and holes pass through the film without forming an exciton on the nanocrystals.

Other semiconducting polymers and device morphologies have also been utilized. For instance, PPV has been used by several groups to fabricate hybrid nanocrystal–polymer LEDs [1,3,18,118,121,122]. Section IV discussed extensive evidence for charge separation at PPV–CdSe interfaces. Just as exciton dissociation is energetically favorable, one would expect exciton formation at such an interface to be unfavorable. However, the use of bilayer structures similar to that shown in Fig. 2a can give reasonably efficient luminescence from the nanocrystals. In this type of device, holes and electrons build up on opposite sides of the interface between the polymer and the nanocrystal layers. Holes are then injected over the barrier into the nanocrystal layer, where the chance of recombining with an electron to form an exciton is high. The increased field due to the charge buildup may also enhance carrier transport across the interface. Once an exciton is formed on a nanocrystal, it may either decay radiatively, decay nonradiatively, or undergo charge separation at the interface, leaving the hole back on the polymer. Although charge separation at the interface might be energetically favorable, we have shown (see Sect. IV) that the presence of a TOPO surfactant layer on CdSe nanocrystals is sufficient to reduce the rate of charge transfer considerably [5,27], and we believe that the presence of some surface barrier to rapid charge transfer is important in allowing radiative exciton decay in nanocrystal–polymer LEDs.

Bilayer LEDs with a PPV layer at the anode and a plain TOPO-coated CdSe nanocrystal layer at the cathode have been reported with external quantum efficiencies from 0.01% [1] up to 0.1% [18]. An approximate energy level diagram for such a device is shown in Fig. 20. The PPV has a lower ionization potential than the nanocrystals, which facilitate hole

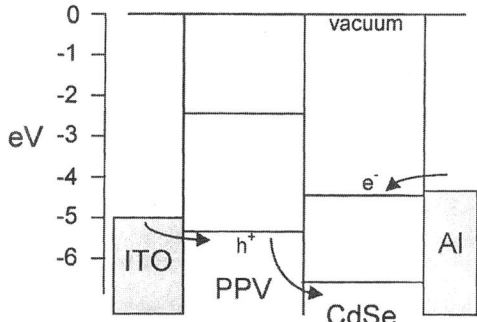

Figure 20   Energy level diagram for nanocrystal–PPV bilayer LEDs.

injection into the device. Because the band offsets generally favor hole transfer from PPV to CdSe, as opposed to electron transfer from CdSe to PPV, the electroluminescence spectrum is dominated by the nanocrystals at low drive voltages. At higher biases however, electrons can tunnel from the nanocrystal layer into the PPV layer and emission from both components can be observed.

As mentioned earlier, the surface treatment of the nanocrystals is important in determining the efficiency of LED operation. Reducing the electronic interaction between nanocrystals and their surroundings is beneficial because it both removes nonradiative decay pathways and reduces the rate of charge separation at the nanocrystal–polymer interface. This can be achieved by overcoating CdSe nanocrystals with high-gap materials such as ZnS or CdS, which is known to increase the photoluminescence efficiency. However, in the case of ZnS overlayers, coating the CdSe nanocrystals is found to decrease the nanocrystal electroluminescence efficiency. Although the ZnS layer passivates the nanocrystal surface and reduces nonradiative decay, it simultaneously introduces a larger barrier to charge injection and exciton formation [18]. On the other hand, CdSe cores capped with CdS layers exhibit both enhanced photoluminescence and enhanced electroluminescence efficiencies (external quantum efficiencies of up to 0.22%) [3]. This is partly because the CdSe core of a CdSe/CdS nanocrystal is still electronically accessible because there is a smaller conduction-band offset at the CdSe/CdS interface compared to the CdSe/ZnS interface [3,123].

The performance of nanocrystal LEDs is also sensitive to many of the injection and transport factors discussed in Section V. Thus, low-work-function metals such as Al and Mg are commonly used to facilitate electron injection into the CdSe layers. In addition, nanocrystal-based LEDs can show memory and charge storage effects [3] caused by the same trapping processes described in Section V.

The best PPV–nanocrystal LEDs generally exhibit electroluminescence efficiencies on the order of 0.1 cd/A. This compares to efficiencies in excess of 10 cd/A, which can now be achieved with entirely organic LEDs. However, these high efficiencies are the result of a huge amount of research and development work devoted to organic LEDs, whereas nanocrystal LEDs have not been so extensively studied. There are thus many unexplored routes that could improve the performance of nanocrystal-based LEDs. Among these is the optimization of heterojunction-band offsets to reduce the large barriers to carrier injection into the nanocrystals, which might be achieved through the use of new polymers and of different nanocrystal materials. In addition, a method is needed to prepare thin films of nanocrystals that are both highly luminescent and electronically accessible: a difficult task because isolation from the environment is a general precondition for high photo-luminescence efficiency. If accomplished, it may then be possible to utilize the superior photostability and continuously tunable emission of inorganic nanocrystals in LEDs. If not, nanocrystal–polymer composites may still find use as tunable phosphors in other electroluminescent devices [124].

B. Photodiodes

Photovoltaic blends of conjugated polymers and semiconductor nanocrystals can be fabricated by spin-coating films from solutions containing both the polymer and nanocrystal components. These blends offer the possibility to tune the sensitivity to incident light through both the polymer and nanocrys-tal components. Furthermore, the band offsets that cause many nanocrystal–conjugated polymer interfaces to perform poorly in LEDs have exactly the opposite effect in photovoltaic applications, where efficient separation, rather than generation, of excitons at the polymer–nanocrystal interface is required. As for the case of LEDs, the performance of polymer-composite photovoltaic devices is known to be extremely sensitive to device structure and blend morphology [70,125]. The potential to optimize the charge transfer and charge transport properties of these composites through nanoparticle surface modification (Fig. 10) [5], control of particle shape [6], or through incorpo-ration of nanocrystal-binding functionality on the polymer backbone [118,126] is thus particularly promising. Furthermore, the high atomic-number contrast between inorganic nanocrystals and conjugated polymers, and the conductive nature of the polymers facilitates electron microscopy studies of these materials [5,127].

The optimum morphology for a photovoltaic composite is determined by two principal requirements (Fig. 21). The first is that the majority (ideally

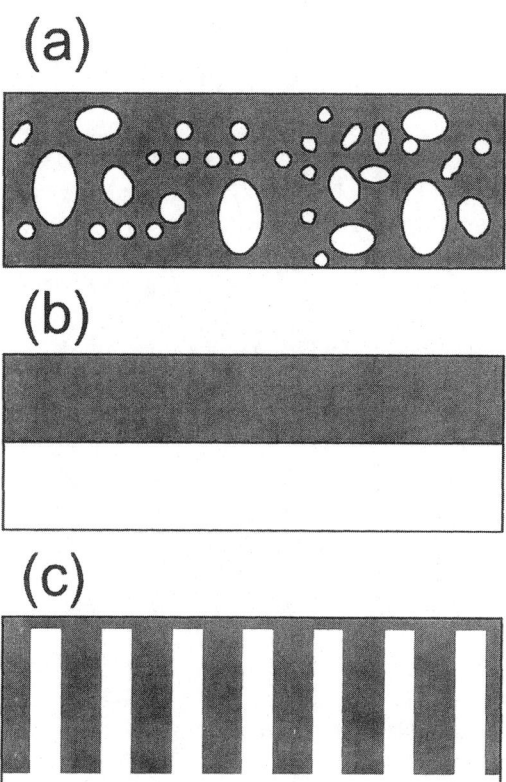

Figure 21 Morphological extremes of a composite polymer–nanocrystal photovoltaic device. Device (a) has a high charge generation efficiency because the phase separation occurs on a small scale and thus light absorbed anywhere in the film can lead to charge generation; however, transport of charges to the electrodes is difficult and the isolated domains will act as traps and recombination sites, thus reducing the overall efficiency. At the other extreme, device (b) has an efficient structure for charge collection, but will also be inefficient as only the small fraction of light absorbed near the heterojunction will contribute to the photocurrent. Device (c) shows an imaginary "ideal" morphology, in which all light is absorbed near an interface and in which all carriers can follow unimpeded paths toward the electrodes.

all) of the optically excited electron–hole pairs should diffuse to an interface and experience charge separation instead of recombining. The second requirement is that once separated, the electrons and holes should be efficiently extracted from the device with minimal losses to recombination. In most devices, a large fraction of the incident light is absorbed by the polymer component, even at high nanocrystal concentrations. Because the typical exciton diffusion length in a conjugated polymer is on the order of 10 nm, the first condition implies that a large-area distributed interface is required so that no exciton is formed farther than one diffusion length from an interface. The second condition, however, requires that a continuous path to the appropriate electrode be readily accessible from every segment of the distributed interface. These morphological considerations are summarized in Fig. 21.

The performance of early MEH–PPV/CdSe nanocrystal photovoltaic devices as a function of composition serves to illustrate the dual importance of charge separation and charge transport [5]. Although the charge generation rate (as monitored by photoluminescence quenching) plateaued at lower nanocrystal concentrations, device efficiency continued to improve at higher concentrations. This can be explained by the growth of the aggregated domains of pyridine-treated nanocrystals to provide more effective electron transport pathways at the higher nanocrystal concentrations (Fig. 10). These devices operated with short-circuit quantum efficiencies of up to 5% and power conversion efficiencies of ~0.25% at 514 nm [5]. Although charge generation in a MEH–PPV/CdSe composites is very efficient, the short-circuit quantum efficiencies of the composites are far from 100%. Indeed, we note that similar efficiencies are obtained in devices containing only nanocrystals [33] (although the low open-circuit voltages in these devices preclude photovoltaic applications). This suggests that in the nanocrystal-polymer devices, recombination losses due to inefficient transport are high, perhaps due to charge trapping at "dead ends" within the phase-separated morphology of the blends.

Consistent with this hypothesis, Huynh et al. demonstrated polymer–nanocrystal composites with improved efficiencies by using blends of anisotropic nanocrystal rods and the conjugated polymer poly(3-hexylthiophene) (P3HT) [6]. The external quantum efficiencies of these devices were 16% and power conversion efficiencies were 2% under 514 nm illumination. The improved efficiencies of these devices can be attributed to enhanced carrier transport in both the P3HT phases and the nanocrystal phases, as well as improved device morphology. First, P3HT has a significantly higher hole mobility than MEH–PPV. In addition, the long nanorods tend to align themselves end to end, thus reducing the number of interparticle hops required for an electron to cover a fixed distance. The 3-fold increase in quantum efficiency and 10-fold increase in power conversion was achieved by using rods with only moderate aspect ratios. This suggest that further gains may be possible by using even longer, wirelike rods [128,129] or by using other nonspherical nanostructures which can now be reproducibly synthesized [130]. Indeed, the authors found that devices with moderately sized (8×13-nm) nanorods could convert five times as much optical radiation into electrical power as devices made from smaller (4×7-nm) nanorods.

In some cases, polymer–nanocrystal blends have also been shown to undergo phase separation in the direction perpendicular to the plane of the film [118]. Such vertically graded structures can help optimize charge generation and charge collection efficiency, as has been shown in solution-processed molecular films [125]. Furthermore, adjusting the surface energy interactions of nanorods and a host polymer can be used to control film morphology in a rational manner [131].

Finally, although we have largely neglected quantum-confinement effects in our discussion of nanocrystal photovoltaic devices, one can envision scenarios where they could be used to tune the response of a detector to a specific wavelength or to optimize energy level offsets in order to maximize open-circuit voltages. Even though these effects have not yet been fully exploited, the study of quantum-confinement effects has led to the ability to control the size, shape, and surface of these chemically synthesized quantum dots, properties which offer compelling advantages to the development of solution-processable photovoltaic devices.

VII. CONCLUSIONS

Nanocrystals provide an interesting system in which to study the physics of charge transport at the nanoscopic level. Both electronic tunneling and structural reorganization play important roles in determining the rate of charge transport, and we have shown earlier how the charge transport is expected to change as the nanocrystal size and spacing are varied. Experimental measurements of charge transport in nanocrystalline films have allowed mobilities to be measured and have identified the importance of disorder and trapping in determining the macroscopic charge transport properties.

Nanocrystals also allow the study of photoinduced electron transfer from organic molecules and polymers to semiconductors, because a large interfacial area is present at the nanocrystal surface where charge transfer can take place. We have shown that CdSe nanocrystals act as good electron acceptors from many conjugated polymers, providing rapid electron transfer from the photoexcited polymer to the nanocrystal, followed by slow recombination of the charge-separated state. This charge-separation process can be exploited as the first step in the operation of a photovoltaic device based on composites of nanocrystals and conjugated polymers, and we have reviewed recent progress in developing both these devices and related structures which act as light-emitting diodes. Nanocrystal-based electronics remains an exciting area of research since it allows fine control of electronic energy levels through changing the nanocrystal size, together with the possibility of structural control on nanometer length scales by exploiting the ability of nanocrystals to assemble into useful structures.

REFERENCES

1. Colvin, V.L.; Schlamp, M.C.; Alivisatos, A.P. Nature 1994, *370*, 354.
2. Dabbousi, B.O.; Bawendi, M.G.; Onitsuka, O.; Rubner, M.F. Appl. Phys. Lett. 1995, *66*, 1316.
3. Schlamp, M.C.; Peng, X.; Alivisatos, A.P. J. Appl. Phys. 1997, *82*, 5837.
4. Gao, M.; Lesser, C.; Kirstein, S.; Mohwald, H.; Rogach, A.L.; Weller, H. J. Appl. Phys. 2000, *87*, 2297.
5. Greenham, N.C.; Peng, X.; Alivisatos, A.P. Phys. Rev. B 1996, *54*, 17,628.
6. Huynh, W.U.; Peng, X.; Alivisatos, A.P. Adv. Mater. 1999, *11*, 923.
7. Klein, D.L.; Roth, R.; Lim, A.K.L.; Alivisatos, A.P.; McEuen, P.L. Nature 1997, *389*, 699.
8. Greenham, N.C.; Friend, R.H. In *Solid State Physics*; Ehrenreich, H. Spaepen, F., Eds; Academic Press: San Diego, CA, 1995; Vol. 49, 1.
9. Friend, R.H.; Gymer, R.W.; Holmes, A.B.; Burroughes, J.H.; Marks, R.N.; Taliani, C.; Bradley, D.D.C.; dos Santos, D.A.; Brédas, J.L.; Lögdlund, M.; Salaneck, W.R. Nature 1999, *397*, 121.
10. Miller, R.D.; McLEndon, G.L.; Nozik, A.J.; Schmickler, W.; Willig, F. *Surface Electron Transfer Processes*; VCH Publishers: New York, 1995.
11. Marcus, R.A. Annu. Rev. Phys. Chem. 1964, *15*, 155.
12. Marcus, R.A.; Sutin, N. Biochim. Biophys. Acta 1985, *811*, 265.
13. Brus, L. Phys. Rev. B 1996, *53*, 4649.
14. Schmitt-Rink, S.; Miller, D.A.B.; Chemla, D.S. Phys. Rev. B 1987, *35*, 8113.
15. Shiang, J.J.; Risbud, S.H.; Alivisatos, A.P. J. Chem. Phys. 1993, *98*, 8432.
16. Oshiro, K.; Akai, K.; Matsuura, M. Phys. Rev. B 1998, *58*, 7986.
17. Jortner, J. J. Chem. Phys. 1976, *64*, 4860.
18. Mattoussi, H.; Radzilowski, L.H.; Dabbousi, B.O.; Thomas, E.L.; Bawendi, M.G.; Rubner, M.F. J. Appl. Phys. 1998, *83*, 7965.
19. Borovikov, Y.Y.; Ryltsev, E.V.; Bodeskul, I.E.; Feshchenko, N.G.; Makovetskii, Y.P.; Egorov, Y.P. J. Gen. Chem. USSR 1970, *40*, 1942.
20. Wolfarth, C.; Wohlfarth, B. In *Landolt-Börnstein*; Lechmer, M.D., Ed.; Spring-Verlag, Heidelberg, 1996; Vol. 38.A.
21. Leatherdale, C.A.; Kagan, C.R.; Morgan, N.Y.; Empedocles, S.A.; Kastner, M.A.; Bawendi, M.G. Phys. Rev. B 2000, *62*, 2669.

22. Brus, L.E. J. Chem. Phys. 1983, *79*, 5566.
23. Middleton, A.A.; Wingreen, N.S. Phys. Rev. Lett. 1993, *71*, 3198.
24. Schönherr, G.; Bässler, H.; Silver, M. Phil. Mag. B 1981, *44*, 47.
25. Miller, A.; Abrahams, E. Phys. Rev. 1960, *120*, 745.
26. Van der Auweraer, M.; De Schryver, F.C.; Borsenberger, P.M.; Bässler, H. Adv. Mater. 1994, *6*, 199.
27. Ginger, D.S.; Greenham, N.C. Phys. Rev. B 1999, *59*, 10,622.
28. Greenham, N.C.; Samuel, I.D.W.; Hayes, G.R.; Phillips, R.T.; Kessener, Y.A.R.R.; Moratti, S.C.; Holmes, A.B.; Friend, R.H. Chem. Phys. Lett. 1995, *241*, 89.
29. de Mello, J.C.; Wittmann, H.F.; Friend, R.H. Adv. Mater. 1997, *9*, 230.
30. Zewail, A.H. J. Phys. Chem. 1996, *100*, 12,701.
31. Diels, J.-C.; Rudolph, W. *Ultrashort Laser Pulse Phenomena*; Academic Press: San Diego, CA, 1996.
32. Demtröder, W. Laser Spectroscopy. *Basic Concepts and Instrumentation*; Springer-Verlag: Berlin, 1996; 11.
33. Ginger, D.S.; Greenham, N.C. J. Appl. Phys. 2000, *87*, 1361.
34. Winiarz, J.G.; Zhang, L.; Lal, M.; Friend, C.S.; Prasad, P.N. J. Am. Chem. Soc. 1999, *121*, 5287.
35. Kraabel, B.; Malko, A.; Hollingsworth, J.; Klimov, V.I. Appl. Phys. Lett. 2001, *78*, 1814.
36. Hagfeldt, A.; Grätzel, M. Acc. Chem. Res. 2000, *33*, 269.
37. Hagfeldt, A.; Grätzel, M. Chem. Rev. 1995, *95*, 49.
38. Wang, Y.; Herron, N. Chem. Phys. Lett. 1992, *200*, 71.
39. Wang, Y.; Herron, N.; Caspar, J. Mater. Sci. Eng. 1993, *B19*, 61.
40. Wang, Y.; Herron, N. J. Lumin. 1996, *70*, 48.
41. Empedocles, S.; Bawendi, M.G. Acc. Chem. Res 1999, *32*, 389.
42. Brus, L.E. J. Chem. Phys. 1984, *80*, 4403.
43. Rabani, E.; Hetenyi, B.; Berne, B.J.; Brus, L.E. J. Chem. Phys. 1999, *110*, 5355.
44. Franceschetti, A.; Williamson, A.; Zunger, A. J. Phys. Chem. B 2000, *104*, 3398.
45. Franceschetti, A.; Zunger, A. Appl. Phys. Lett. 2000, *76*, 1731.
46. Efros, A.L.; Rosen, M. Annu. Rev. Mater. Sci. 2000, *30*, 475.
47. O'Regan, B.; Grätzel, M. Nature 1991, *353*, 737.
48. Savenije, T.J.; Warman, J.M.; Goosens, A. Chem. Phys. Lett. 1998, *287*, 148.
49. Salafsky, J.S.; Lubberhuizen, W.H.; Schropp, R.E.I. Chem. Phys. Lett. 1998, *290*, 297.
50. van Hal, P.A.; Christiaans, M.P.T.; Wienk, M.M.; Kroon, J.M.; Janssen, R.A.J. J. Phys. Chem. B 1999, *103*, 4352.
51. Arango, A.C.; Carter, S.C.; Brock, P.J. Appl. Phys. Lett. 1999, *74*, 1698.
52. Arango, A.C.; Johnson, L.R.; Bliznyuk, V.N.; Schlesinger, Z.; Carter, S.A.; Horhold, H.H. Adv. Mater. 2000, *12*, 1689.
53. Weller, H. Ber. Bunsen-Ges. Phys. Chem. Chem. Phys. 1991, *95*, 1361.
54. Weller, H. Adv. Mat. 1993, *5*, 88.

55. Asbury, J.B.; Hao, E.C.; Wang, Y.Q.; Lian, T.Q. J. Phys. Chem. B 2000, *104*, 11957.
56. Asbury, J.B.; Hao, E.; Wang, Y.; Ghosh, H.N.; Lian, T. J. Phys. Chem. B 2001, *105*, 4545.
57. Asbury, J.B.; Ellingson, R.J.; Ghosh, H.N.; Ferrere, S.; Nozik, A.J.; Lian, T.Q. J. Phys. Chem. B 1999, *103*, 3110.
58. Huber, R.; Sporlein, S.; Moser, J.E.; Grätzel, M.; Wachtveitl, J. J. Phys. Chem. B 2000, *104*, 8995.
59. Hässelbarth, A.; Eychmüller, A.; Weller, H. Chem. Phys. Lett. 1993, *203*, 271.
60. Logunov, S.; Green, T.; Marguet, S.; El-Sayed, M.A. J. Phys. Chem. A 1998, *102*, 5652.
61. Burda, C.; Green, T.C.; Link, S.; El-Sayed, M.A. J. Phys. Chem. B 1999, *103*, 1783.
62. Guyot-Sionnest, P.; Shim, M.; Matranga, C.; Hines, M. Phys. Rev. B 1999, *60*, R2181.
63. Klimov, V.I. Phys. Rev. B 2000, *61*, R13,349.
64. Ginger, D.S.; Dhoot, A.S.; Finlayson, C.E.; Greenham, N.C. Appl. Phys. Lett. 2000, *77*, 2816.
65. Schmelz, O.; Mews, A.; Basche, T.; Herrmann, A.; Mullen, K. Langmuir 2001, *17*, 2861.
66. Cheng, J.X.; Wang, S.H.; Li, X.Y.; Yan, Y.J.; Yang, S.H.; Yang, C.L.; Wang, J.N.; Ge, W.K. Chem. Phys. Lett. 2001, *333*, 375.
67. Yang, C.L.; Wang, J.N.; Ge, W.K.; Wang, S.H.; Cheng, J.X.; Li, X.Y.; Yan, Y.J.; Yang, S.H. Appl. Phys. Lett. 2001, *78*, 760.
68. Sariciftci, N.S.; Smilowitz, L.; Heeger, A.J.; Wudl, F. Science 1992, *258*, 1474.
69. Halls, J.J.M.; Walsh, C.A.; Greenham, N.C.; Marseglia, E.A.; Friend, R.H.; Moratti, S.C.; Holmes, A.B. Nature 1995, *376*, 498.
70. Halls, J.J.M.; Arias, A.C.; Mackenzie, J.D.; Wu, W.; Inbasekaran, M.; Woo, E.P.; Friend, R.H. Adv. Mater. 2000, *12*, 498.
71. Halls, J.J.M.; Pichler, K.; Friend, R.H.; Moratti, S.C.; Holmes, A.B. Appl. Phys. Lett. 1996, *68*, 3120.
72. Samuel, I.D.W.; Crystal, B.; Rumbles, G.; Burn, P.L.; Holmes, A.B.; Friend, R.H. Chem. Phys. Lett. 1993, *213*, 472.
73. Greenham, N.C.; Cacialli, F.; Bradley, D.D.C.; Friend, R.H.; Moratti, S.C.; Holmes, A.B. Mater. Res. Soc. Symp. Proc. 1994, *328*, 351.
74. Kastner, M.A. Rev. Mod. Phys. 1992, *64*, 849.
75. Kastner, M.A. Phys. Today 1993, *46*, 24.
76. Klein, D.L.; McEuen, P.L.; Bowen Katari, J.E.; Roth, R.; Alivisatos, A.P. Appl. Phys. Lett. 1996, *68*, 2574.
77. Alperson, B.; Cohen, S.; Rubinstein, I.; Hodes, G. Phys. Rev. B 1995, *52*, 17,017.
78. Alperson, B.; Rubinstein, I.; Hodes, G.; Porath, D.; Millo, O. Appl. Phys. Lett. 1999, *75*, 1751.
79. Bakkers, E.P.A.M.; Vanmaekelbergh, D. Phys. Rev. B 2000, *62*, R7743.
80. Banin, U.; Cao, Y.-W.; Katz, D.; Millo, O. Nature 1999, *400*, 542.

81. Millo, O.; Katz, D.; Levi, Y.; Cao, Y.-W.; Banin, U. J. Low Temp. Phys 2000, *118*, 365.
82. Collier, C.P.; Saykally, R.J.; Shiang, J.J.; Henrichs, S.E.; Heath, J.R. Science 1997, *277*, 1978.
83. Markovich, G.; Collier, C.P.; Henrichs, S.E.; Remacle, F.; Levine, R.D.; Heath, J.R. Acc. Chem. Res. 1999, *32*, 415.
84. Roychowdhury, V.P.; Janes, D.B.; Bandyopadhyay, S.; Wang, X.D. IEEE Trans. Electron Devices 1996, *43*, 1688.
85. Nelson, J. Phys. Rev. B 1999, *59*, 15374.
86. Chen, S.W.; Ingram, R.S.; Hostetler, M.J.; Pietron, J.J.; Murray, R.W.; Schaaff, T.G.; Khoury, J.T.; Alvarez, M.M.; Whetten, R.L. Science 1998, *280*, 2098.
87. Averin, D.V.; Likharev, K.K. In *Mesoscopic Phenomena in Solids*; Altshuler, B.L.; Lee, P.A., Webb, R.A., Eds.; Elsevier: Amsterdam, 1991; Vol. 30, 176.
88. Feldheim, D.L.; Keating, C.D. Chem. Soc. Rev. 1998, *27*, 1.
89. McConnell, W.P.; Novak, J.P.; Brousseau, L.C.; Fuierer, R.R.; Tenent, R.C.; Feldheim, D.L. J. Phys. Chem. B 2000, *104*, 8925.
90. Sato, T.; Ahmed, H. Appl. Phys. Lett. 1997, *70*, 2759.
91. Sato, T.; Ahmed, H.; Brown, D.; Johnson, B.F.G. J. Appl. Phys. 1997, *82*, 696.
92. Rimberg, A.J.; Ho, T.R.; Clarke, J. Phys. Rev. Lett. 1995, *74*, 4714.
93. Black, C.T.; Murray, C.B.; Sandstrom, R.L.; Sun, S.H. Science 2000, *290*, 1131.
94. Brust, M.; Walker, M.; Bethell, D.; Schiffrin, D.J.; Whyman, R. J. Chem. Soc. Chem. Commun. 1994, *801*.
95. Brust, M.; Bethell, D.; Schiffrin, D.J.; Kiely, C.J. Adv. Mater. 1995, *7*, 795.
96. Janes, D.B.; Kolagunta, V.R.; Osifchin, R.G.; Bielefeld, J.D.; Andres, R.P.; Henderson, J.I.; Kubiak, C.P. Superlattices Microstruct. 1995, *18*, 275.
97. Terrill, R.H.; Postlethwaite, T.A.; Chen, C.H., et al. J. Am. Chem. Soc. 1995, *117*, 12,537.
98. Andres, R.P.; Bielefeld, J.D.; Henderson, J.I.; Janes, D.B.; Kolagunta, V.R.; Kubiak, C.P.; Mahoney, W.J.; Osifehin, R.G. Science 1996, *273*, 1690.
99. Brust, M.; Bethell, D.; Kiely, C.J.; Schiffrin, D.J. Langmuir 1998, *14*, 5425.
100. Wuelfing, W.P.; Green, S.J.; Pietron, J.J.; Cliffel, D.E.; Murray, R.W. J. Am. Chem. Soc. 2000, *122*, 11,465.
101. Park, S.J.; Lazarides, A.A.; Mirkin, C.A.; Brazis, P.W.; Kannewurf, C.R.; Letsinger, R.L. Angew. Chem. Int. Ed. 2000, *39*, 3845.
102. Neugebauer, C.A.; Webb, M.B. J. Appl. Phys. 1962, *33*, 74.
103. Abeles, B.; Sheng, P.; Coutts, M.D.; Arie, Y. Adv. Phys. 1975, *24*, 407.
104. Shim, M.; Guyot-Sionnest, P. Nature 2000, *407*, 981.
105. Kim, B.S.; Islam, M.A.; Brus, L.E.; Herman, I.P. J. Appl. Phys. 2001, *89*, 8127.
106. Kim, S.H.; Markovich, G.; Rezvani, S.; Choi, S.H.; Wang, K.L.; Heath, J.R. Appl. Phys. Lett. 1999, *74*, 317.
107. Bube, R.H. *Photoelectronic Properties of Semiconductors*; Cambridge University Press: Cambridge; 1992.

108. Morgan, N.Y.; Leatherdale, C.A.; Kastner, M.A.; Bawendi, M.G. Private communication, 2000.

109. Kohlrausch, R. Ann. Phys. Leipzig 1847, *12*, 393.

110. Phillips, J.C. Rep. Prog. Phys. 1996, *59*, 1133.

111. Beadie, G.; Sauvain, E.; Gomes, A.S.L.; Lawandy, N.M. Phys. Rev. B 1995, *51*, 2180.

112. Dulieu, B.; Wery, J.; Lefrant, S.; Bullot, J. Phys. Rev. B 1998, *57*, 9118.

113. Tsai, L.C.; Fan, J.C.; Chen, Y.F.; Lo, I. Phys. Rev. B 1999, *59*, 2174.

114. Rasaiah, J.C.; Zhu, J.; Hubbard, J.B.; Rubin, R.J. J. Chem. Phys. 1990, *93*, 5768.

115. Krauss, T.D.; O'Brien, S.; Brus, L.E. J. Phys. Chem. B 2001, *105*, 1725.

116. Artemyev, M.V.; Sperling, V.; Woggon, U. J. Appl. Phys. 1997, 81, 6975.

117. Yang, Y.; Xue, S.H.; Liu, S.Y.; Huang, J.M.; Shen, J.C. Appl. Phys. Lett. 1996, *69*, 377.

118. Mattoussi, H.; Radzilowski, L.H.; Dabbousi, B.O.; Fogg, D.E.; Schrock, R.R.; Thomas, E.I.; Rubner, M.F.; Bawendi, M.G. J. Appl. Phys. 1999, *86*, 4390.

119. Gaponik, N.P.; Talapin, D.V.; Rogach, A.L.; Eychmuller, A. J. Mater. Chem. 2000, *10*, 2163.

120. Ginger, D.S. Ph.D. Thesis, University of Cambridge, 2001.

121. Gao, M.; Richter, B.; Kirstein, S. Adv. Mater. 1997, *9*, 802.

122. Gao, M.; Richter, B.; Kirstein, S.; Mohwald, H. J. Phys. Chem. B 1998, *102*, 4096.

123. Peng, X.G.; Schlamp, M.C.; Kadavanich, A.V.; Alivisatos, A.P. J. Am. Chem. Soc. 1997, *119*, 7019.

124. Lee, J.; Sundar, V.; Heine, J.R.; Bawendi, M.G.; Jensen, K. Adv. Mater. 2000, *12*, 1102.

125. Schmidt-Mende, L.; Fechtenkotter, A.; Mullen, K.; Moons, E.; Friend, R.H.; Mackenzie, J.D. Science 2001, *293*, 1119.

126. Fogg, D.E.; Radzilowski, L.H.; Blanski, R.; Schrock, R.R.; Thomas, E.L.; Bawendi, M.G. Macromolecules 1997, *30*, 8433.

127. Kadavanich, A.V.; Kippeny, T.C.; Erwin, M.M.; Pennycook, S.J.; Rosenthal, S.J. J. Phys. Chem. B 2001, *105*, 361.

128. Peng, X.; Manna, L.; Yang, W.; Wickham, J.; Scher, E.; Kadavanich, A.; Alivisatos, A.P. Nature 2000, *404*, 59.

129. Peng, Z.A.; Peng, X. J. Am. Chem. Soc. 2001, *123*, 1389.

130. Manna, L.; Scher, E.C.; Alivisatos, A.P. J. Am. Chem. Soc. 2000, *122*, 12,700.

131. Peng, G.; Qiu, F.; Ginzburg, V.V.; Jasnow, D.; Balazs, A.C. Science 2000, *288*, 1802.

8

Tunneling and Optical Spectroscopy of Semiconductor Nanocrystal Quantum Dots: Single-Particle and Ensemble Properties

Uri Banin and Oded Millo
The Hebrew University, Jerusalem, Israel

I. INTRODUCTION

Semiconductor nanocrystals are novel materials lying between the molecular and solid-state regime with the unique feature of properties controlled by size [1–5]. Containing hundreds to thousands of atoms, 20–200 Å in diameter, nanocrystals maintain a crystalline core with the periodicity of the bulk semiconductor. However, because the wave functions of electrons and holes are confined by the physical nanometric dimensions of the nanocrystals, the electronic level structure and the resultant optical and electrical properties are greatly modified. Upon reducing the size of direct-gap semiconductors into the nanocrystal regime, a characteristic blue shift of the bandgap appears, and a discrete level structure develops as a result of the "quantum-size effect" in these quantum dots (QDs) [6]. In addition, because of their small size, the charging energy associated with the addition or removal of a single electron is very high, leading to pronounced single-electron tunneling effects [7–9]. Due to the unique optical and electrical properties, nanocrystals may play a key role in the emerging new field of nanotechnology in applications ranging from lasers [10,11] and other optoelectronic devices [12,13] to biological fluorescence marking [14–16].

The approaches to fabrication of semiconductor quantum dots can be divided into two main classifications: In the top-down approach, nano-lithography is used to reduce the dimensionality of a bulk semiconductor. These approaches are presently limited to structures with dimensions on the order of tens of nanometers [17]. In the bottom-up approach, two important fabrication routes of QDs are presently used: molecular beam epitaxy (MBE) deposition utilizing the strain-induced growth mode [18,19], and colloidal synthesis [20–24]. In this chapter, we focus on colloidal-grown nanocrystal QDs. These samples have the advantage of continuous size control, as well as chemical accessibility due to their overcoating with organic ligands. This chemical compatibility enables the use of powerful chemical or biochemical means to assemble nanocrystals in a controlled manner [25–27]. Artificial solids composed of nanocrystals have been prepared, opening a new domain of physical phenomena and technological applications [28–30]. Nanocrystal molecules and nanocrystal–DNA assemblies were also developed [31].

Colloidal synthesis has been extended to several directions, allowing further powerful control, in addition to size, on optical and electronic properties of nanocrystals. Heterostructured nanocrystals were developed, where semiconductor shells can be grown on a core [22,32]. One important class of such particles, are core–shell nanocrystals [33–38]. Here, the core is over-coated by a semiconductor shell with a gap enclosing that of the core semiconductor materials. Enhanced fluorescence and increased stability can be achieved in these particles, compared with cores overcoated by organic lig-ands. Recently, shape control was also achieved in the colloidal synthesis route [39]. By proper modification of the synthesis, rod-shaped particles can be prepared—quantum rods [40,41]. Such quantum rods manifest the transition from zero-dimensional (0D) quantum dots to 1D quantum wires [42].

From the early work on the quantum-confinement effect in colloidal semiconductor nanocrystals, electronic levels have been assigned according to the spherical symmetry of the electron and hole envelope functions [6,43]. The simplistic "artificial atom" model of a particle in a spherical box predicts discrete states with atomiclike state symmetries (e.g., s and p). To probe the electronic structure of II–VI and III–V semiconductor nanocrystals, a variety of "size-selective" optical techniques have been used, mapping the size dependence of dipole allowed transitions [44–48]. Theoretical models based on an effective mass approach with varying degree of complexity [45,49] as well as pseudopotentials [50,51] were used to assign the levels.

Tunneling transport through semiconductor nanocrystals can yield com-plementary new information on their electronic properties, which cannot be probed in the optical measurements. In the optical spectra, allowed valence band (VB) to conduction band (CB) transitions are detected, whereas in tunneling spectroscopy, the CB and VB states can be separately probed. In addition, the tunneling spectra may show effects of single-electron charging of the QD. Such interplay between single-electron charging and resonant tunneling through the QD states can provide unique information on the degeneracy and, therefore, the symmetry of the levels.

The interplay between single-electron tunneling (SET) effects and quantum-size effects in isolated nanoparticles can be experimentally observed most clearly when the charging energy of the dot by a single electron, E_c, is comparable to the electronic level separation ΔE_L, and both energy scales are larger than $k_B T$ [7,52,53]. These conditions are met by semiconductor nanocrystals in the strong quantum-confinement regime, even at room temperature, whereas for metallic nanoparticles, E_c is typically much larger than ΔE_L. SET effects are relevant to the development of nanoscale electronic devices, such as single-electron transistors [54,55].

However, for small colloidal nanocrystals, the task of wiring up the QD between electrodes for transport studies is exceptionally challenging. To this end, various mesoscopic tunnel junction configurations were employed, such as the double-barrier tunnel junction (DBTJ) geometry, where a QD is coupled via two tunnel junctions to two macroscopic electrodes [7,8,56,57]. Klein et al. achieved this by attaching CdSe quantum dots to two lithographically prepared electrodes and have observed SET effects [58]. In this device, a gate voltage can be applied to modify the transport properties. An alternative approach to achieve electrical transport through single QDs is to use scanning probe methods. Alperson et al. observed SET effects at room temperature in electrochemically deposited CdSe nanocrystals using conductive atomic force microscopy (AFM) [59].

A particularly useful approach to realize the DBTJ with nanocrystal QDs is demonstrated in Fig. 1. Here, a nanocrystal is positioned on a conducting surface providing one electrode and the STM tip provides the second electrode. Such a configuration has been widely used to study SET effects in metallic QDs and in molecules [7,60–63]. In this geometry, in addition to the QD level structure, the parameters of both junctions, in particular the capacitances (C_1 and C_2) and tunneling resistances (R_1 and R_2) strongly affect the tunneling spectra [64,65]. Therefore, a detailed understanding of the role played by the DBTJ geometry and the ability to control it are essential for the correct interpretation of tunneling characteristics of semiconductor QDs, as well as for their implementation in electronic nano-architectures, as demonstrated by Su et al. for semiconducting quantum wells [66].

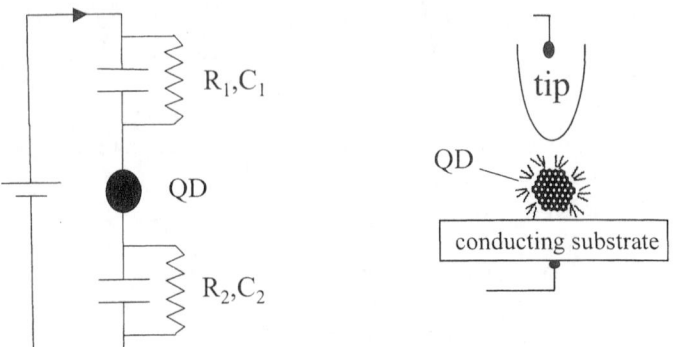

Figure 1 Experimental realization of the DBTJ using the STM (left) and an equivalent electrical circuit (right). Linker molecules (not shown) can be used to provide chemical binding of the QD to the substrate, thus enhancing the QD–substrate tunnel barrier (R_2).

The elegant artificial atom analogy, borne out from optical and tunneling spectroscopy for QDs, can be tested directly by observing the shapes of the electronic wave functions. Recently, the probability density of the ground and first excited states for epitaxially grown InAs QDs embedded in GaAs was probed using magnetotunneling spectroscopy with inversion of the frequency domain data [67]. The unique sensitivity of STM to the local density of states can also be used to directly image electronic wave functions, as demonstrated for "quantum corrals" on metal surfaces [68], carbon nanotubes [69], and d-wave superconductors [70]. This technique was also applied to image the quantum-confined electronic wave functions in MBE and in colloidal-grown QDs [71,72]. The information extracted from such imaging measurements provides a detailed test for the theoretical understanding of the QD electronic structure.

In this review, we concentrate on the application of tunneling and optical spectroscopy to colloidal-grown QDs, with particular focus on our own contributions in the study of InAs nanocrystals and InAs cores coated by a semiconducting shell (core–shells). Section II gives a general comparison between tunneling and optical spectroscopy of QDs, and in Section III, the specific application of this approach to InAs nanocrystals will be presented. In Section IV, we discuss the effects of the tunnel junction parameters on the measured tunneling spectra. Section V focuses on the synthesis of core–shell QDs with InAs cores and presents their optical and tunneling properties. The wave-function imaging of electronic states in QDs is discussed in Section VI and concluding remarks are given in Section VII.

II. GENERAL COMPARISON BETWEEN TUNNELING AND OPTICAL SPECTROSCOPY OF QDs

Tunneling and optical spectroscopy are two complementary methods for the study of the electronic properties of semiconductor QDs. In photolumines-cence excitation (PLE) spectroscopy, a method that has been widely used to probe the electronic states of QDs, one monitors allowed transitions between the VB and CB states [44,47]. Size selection is achieved by opening a narrow detection window on the blue side of the inhomogeneously broadened PL peak. Pending a suitable assignment of the transitions, the intraband level separations can be extracted from spacings between the PLE peaks. In tunneling spectroscopy, on the other hand, it is possible to separately probe the CB and VB states, and, practically, there are no selection rules [9]. Here, one measures the dI/dV versus V characteristics of single QDs that yield direct information on the tunneling density of states (DOS). For a discrete QD level structure, the spectra exhibit a sequence of peaks corresponding to resonant tunneling through the states.

Seemingly, it should be possible to directly compare the PLE and tunneling spectra. However, in tunneling spectroscopy, the QD is charged and, therefore, the level structure may be perturbed compared to the neutral dot monitored in PLE. Furthermore, even if charging does not intrinsically perturb the level structure significantly, the peak spacing and peak structure in the tunneling experiment depend extrinsically on the DBTJ parameters. We now present a simple theoretical framework for the interpretation of the dependence of the tunneling spectra on the junction parameters starting with a qualitative explanation.

As shown in Fig. 1, a DBTJ is realized by positioning the STM tip over the QD. The QD is characterized by a discrete level spectrum with degen-eracies reflecting the symmetry of the system. The DBTJ is characterized by a capacitance and a tunneling resistance for each junction. The capacitance and tunneling resistance (inversely proportional to the tunneling rate) of the tip–QD junction (C_1 and R_1) can be easily modified by changing the tip–QD distance, usually through the control over the STM bias and current settings (V_s and I_s). On the other hand, the QD–substrate junction parameters (C_2 and R_2) are practically stable for a specific QD. They can be controlled in different experiments by the choice of the QD–substrate linking chemistry.

Adding a single electron to a QD requires a finite charging energy, E_c, which in the equivalent circuit of the DBTJ is given by $e^2/[2(C_1 + C_2)]$. In a typical STM realization of a DBTJ with nanocrystals, E_c is on the order of ~100 meV, similar to that expected for an isolated sphere, $e^2/2\varepsilon r$, with a radius r of a few nanometers and a dielectric constant $\varepsilon \sim 10$. The capacitance values determine also the voltage division between the junctions, $V_1/V_2 = C_2/C_1$.

Due to this voltage division, the measured spacings between the resonant tunneling peaks do not coincide with the real level spacings. In the case where tunneling is onset in junction 1, $C_1 < C_2$ (and the Fermi levels of the outer electrodes are aligned around the QD midgap), the apparent level spacing in the tunneling spectra is larger than the real level spacing by a factor of

$$\frac{V_B}{V_1} = \left(1 + \frac{C_1}{C_2}\right) \tag{1}$$

as long as there is no simultaneous tunneling from both the valence and conduction bands. An important example, shown in Fig. 2, is the case of a DBTJ of extreme assymetry, where C_1 is much smaller than C_2 and the applied voltage V_B largely drops on the tip–QD junction. Here, tunneling through the discrete QD levels is onset in the tip–QD junction, and the regime of simultaneous tunneling through the valence- and conduction-band states is pushed to higher voltages. Meaningful level structure can thus be extracted from the tunneling spectra.

Another important parameter in the DBTJ is the ratio between the tunneling resistances, R_1/R_2. This ratio may affect the degree of QD charging during the tunneling process through the DBTJ. We will consider the important case of the asymmetric DBTJ discussed earlier, where tunneling is onset in junction 1. At a positive sample bias V_1, a peak will appear in the dI/dV versus V spectra corresponding to tunneling through the first CB state. We first discuss the case in which R_1 is much larger than R_2 (Fig. 2a). Here, an electron tunneling from the tip to the QD would escape to the substrate before the next electron could tunnel into the QD. Consequently, resonant tunneling through the QD states without charging would take place. The next peak in the dI/dV versus V spectrum will thus appear when an electron can tunnel through the excited CB level, at $V_B \sim eV_1 + \Delta_{CB}$. An equivalent process would occur at negative bias for tunneling through the VB states. On the other hand, when R_1 is on the order of R_2, charging effects start to become significant. As demonstrated in Fig. 2b, for R_2 much larger than R_1, two electrons can reside in the CB ground state at the same time and the second tunneling peak would appear at a bias, $V_B \sim eV_1 + E_c$. Equivalent resonant tunneling and charging processes would occur for the other states in the CB and VB.

A more quantitative understanding of the effects of the DBTJ parameters on the tunneling spectra is gained from the simulations solving rate equations, presented in Fig. 3. These simulations are based on the "orthodox model" for single-electron tunneling through metallic systems [56], modified to take into account the discrete level spectrum [52]. The tunneling rate onto the QD from electrode $i = 1, 2$ is given by

$$\Gamma_i^+(n) = \frac{2\pi}{\hbar} \int |T_i(E)|^2 D_i(E - E_i) f(E - E_i) D_d(E - E_d) \left[1 - f(E - E_d)\right] \, dE \tag{2}$$

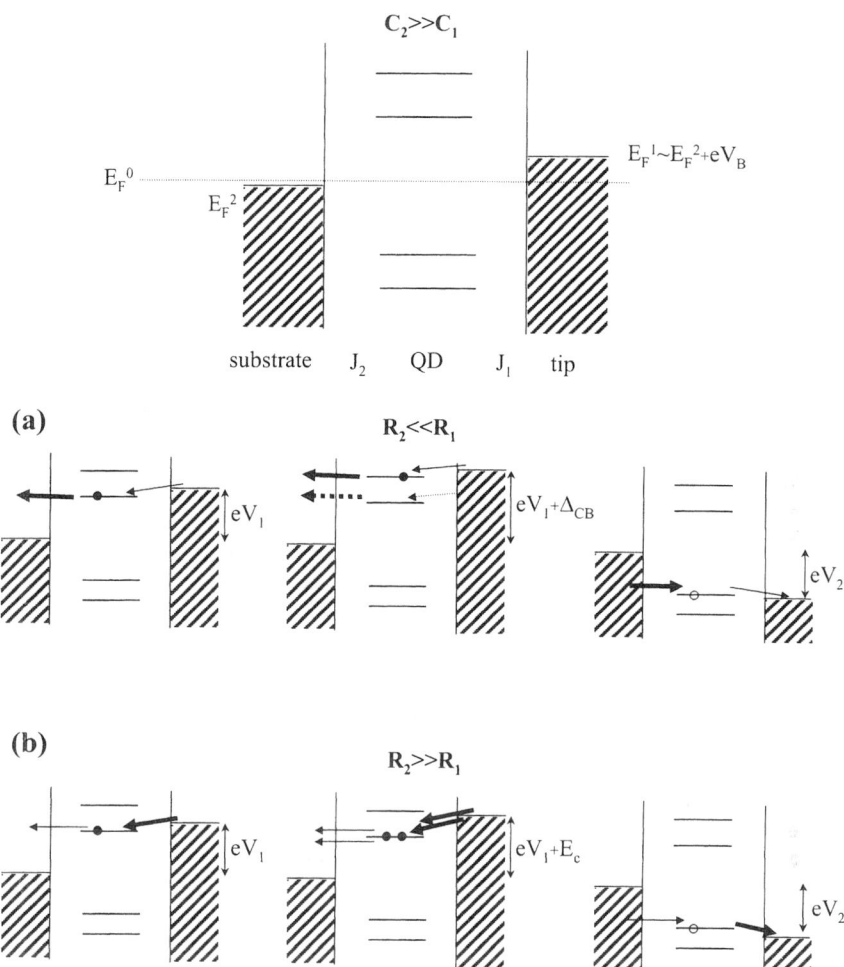

Figure 2 Schematic description of the tunneling process in the DBTJ in the case where $C_1 \ll C_2$ and therefore tunneling is onset in junction 1 (J_1). Two extreme cases are shown: (a) J_2 is the fast junction, and therefore resonant tunneling with no charging takes place for both electrons and holes; (b) J_2 is the slow junction, and charge can accumulate on the QD, resulting in resonant tunneling accompanied by single-electron charging effects. In this case, the charging multiplicity can provide information on the level degeneracy. Bold arrows correspond to high tunneling rates and the thin arrows correspond to slow rates.

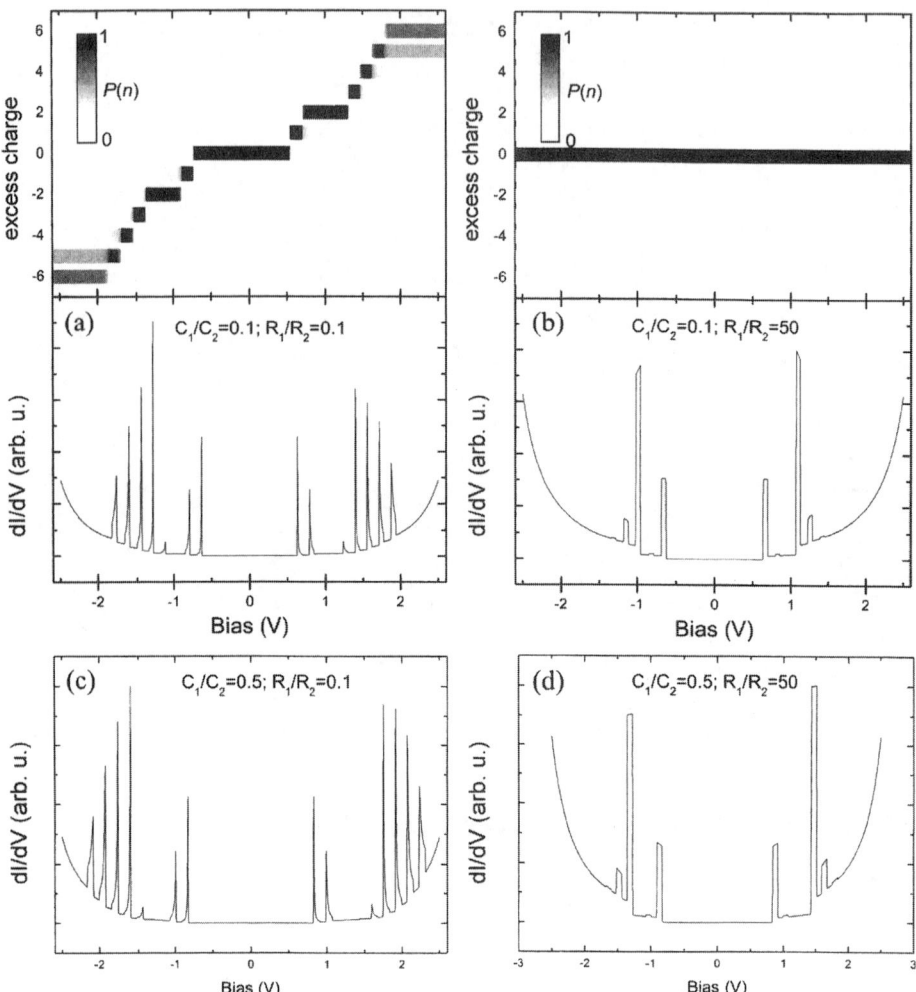

Figure 3 Simulated tunneling spectra of a QD for different values of capacitance and tunneling resistance ratios, as indicated in the figure. The calculations were performed for a QD having twofold (ground) and fourfold (excited) degenerate levels in both conduction and valence bands. The upper panels depict, in gray scale, the probability distribution of the number of excess electrons on the QD.

Similar expressions are used for tunneling off the dot, $\Gamma_i^-(n)$, where n is the number of excess electrons on the QD. $T_i(E)$ is the tunneling matrix element across junction i. We do not explicitly calculate $T_i(E)$ but, rather, assign each junction with a phenomenological "tunneling resistance" parameter [7], $R_i \sim T_i(E)^{-1}$. We take into account the dependence of $T_i(E)$ on applied bias due to the reduction of the average tunneling barrier height using the WKB approximation [73,74]. The tunneling resistances thus determine the average tunneling rate between the two junctions, $R_1/R_2 = \Gamma_2/\Gamma_1$. D_i and D_d are the density of states in the electrode and the dot, respectively, E_i and E_d are the corresponding Fermi levels whose relative positions after tunneling, $[E_d(n \pm 1) - E_i(n)]$, depend on n, C_1, C_2, and the level spectrum, and $f(E)$ is the Fermi function. The condition for resonant tunneling is the lineup of the Fermi level of the dot after tunneling (on or off the dot), with the Fermi level of the outer electrode before tunneling. D_d is taken to be proportional to a set of broadened discrete levels corresponding to the positions of the discrete energy levels.

First, one determines the probability distribution of n, $P(n)$, from the condition that at steady state, the net transition rate between two adjacent QD charging states is zero [56]:

$$P(n)[\Gamma_i^+(n) + \Gamma_2^+(n)] = P(n+1)[\Gamma_i^-(n+1) + \Gamma_2^-(n+1)] : \qquad (3)$$

The tunneling current is then calculated self-consistently from

$$I(V) = e \sum_n P(n)[\Gamma_2^+(n) - \Gamma_2^-(n)] = e \sum_n P(n)[\Gamma_1^-(n) - \Gamma_1^+(n)] \qquad (4)$$

As a working example for the purpose of our explanation, we assume that our QD has a bandgap $E_g = 1$ eV, with two discrete states in the VB and CB. The ground states are twofold spin degenerate, whereas the excited states have a fourfold degeneracy (including spin). The spacings between the states, Δ_{VB} and Δ_{CB}, are 0.3 and 0.4 eV for the VB and CB, respectively. In the present simulation, for simplicity, we did not allow for simultaneous tunneling through the VB and CB states. The I–V curves are calculated as described earlier, and the tunneling conductance curves (dI/dV versus V), are obtained by differentiation.

We present in Fig. 3 several limiting situations, where the relevant factors to describe the junction asymmetry are the ratios C_1/C_2 and R_1/R_2. In all of the curves, the current in the bandgap region around zero bias is suppressed. However, considerable differences can be seen in the peak structure after the onset of the tunneling current. In Figs. 3a and 3b, $C_1/C_2 = 0.1$ and, therefore, most of the bias drops at junction 1. In Fig. 3a, $R_1/R_2 = 0.1$, and the calculated dI/dV curve exhibits resonant tunneling accompanied by single-electron charging. In both positive and negative bias, a doublet of peaks is observed at the onset of the current. The doublet corresponds to charging of the first CB and VB states, respectively. This can be clearly discerned from the representation of $P(n)$ versus bias shown (in gray scale) above the dI/dV curves. The intradoublet spacing corresponds to the single-electron charging energy, E_c, and the spacing between the first VB and CB peaks is nearly $E_g + E_c$, up to a small correction due to the voltage division. At higher positive bias ($V_B > 1$ V), peaks arising from tunneling through the fourfold degenerate excited CB state are seen. The first small peak corresponds to a situation where the most probable n is still 2, but the second electron tunnels through the excited state, rather than the ground state. The magnitude of this peak is reduced when the ratio R_1/R_2 is decreased. The spacing between the second and third peaks is the interlevel spacing modified slightly by voltage division. The following fourfold multiplet corresponds to sequential addition of electrons to the excited state, with intramultiplet spacing corresponding to E_c. Similar behavior is observed in the negative bias side, with tunneling through the VB states.

A very different peak structure is seen for the opposite situation where $R_1/R_2 = 50$ (Fig. 3b). Here, charging effects are nearly suppressed; only a hint of the second peak in the doublets can be seen. The next large peak, at $V_B \sim$ 1 V corresponds to tunneling through the excited state, with the spacing to the first large peak equal to the interlevel separation. The most probable n is essentially zero [see the distribution $P(n)$], reflecting the situation that (at positive bias) an electron that tunnels onto a CB state through J_1 rapidly escapes through J_2. Again, the negative bias side corresponds to similar behavior for tunneling through the VB states.

The cases where $C_1/C_2 = 0.5$ are shown in Figs. 3c and 3d, with R_1/R_2 corresponding to the same ratios as in Figs. 3a and 3b, respectively. The above discussion holds also for both these cases, where charging effects can be clearly observed in Fig 3c, where $R_1/R_2 = 0.1$, and are nearly suppressed in the opposite case (Fig. 3d). However, the peak spacing becomes larger due to the enhanced effect of voltage division [Eq. (1)].

III. CORRELATION BETWEEN OPTICAL AND TUNNELING SPECTRA OF InAs NANOCRYSTAL QDs

The first detailed investigation employing a combined optical-tunneling approach was performed by us on InAs nanocrystal QDs [12]. InAs is an almost ideal system for such a study. It belongs to the family of tetrahedral semiconductors, and colloidal techniques allow for the preparation of nanocrystals that are nearly spherically shaped, over a broad range of sizes with narrow size distribution (less then 10%) [21,23]. Furthermore, InAs is a narrow-gap semiconductor (E_g = 0.418 eV) with a large Bohr radius a_0 of 340 Å (as compared to CdSe, with E_g = 1.84 eV and a_0 = 55 Å) and serves as a prototypical system for the study of quantum-confinement effects. For scanning tunneling spectroscopy (STS), the narrow gap allows one to probe excited levels in a highly charged state, as will be described in Section III.B. It is also important to note that InAs is presently perhaps the only system that can be fabricated both by epitaxial growth techniques as well as by colloidal chemistry techniques, thus providing an important point of comparison.

A. Photoluminescence Excitation Spectroscopy

The InAs QDs were prepared using a solution-phase pyrolitic reaction of organometallic precursors. These nanocrystals are nearly spherical in shape, with size controlled between 1 and 4 nm in radius and size distribution better than 10% [21,23]. As a demonstration of the size and shape homogeneity of these samples, we present in Fig. 4 the transmission electron microscopic (TEM) images of superlattices of InAs nanocrystals. The nanocrystal surface is passivated by organic ligands. For the low-temperature optical experiments, dilute samples were embedded in free-standing, optically clear films of polyvinylbutyral and cooled to 10 K.

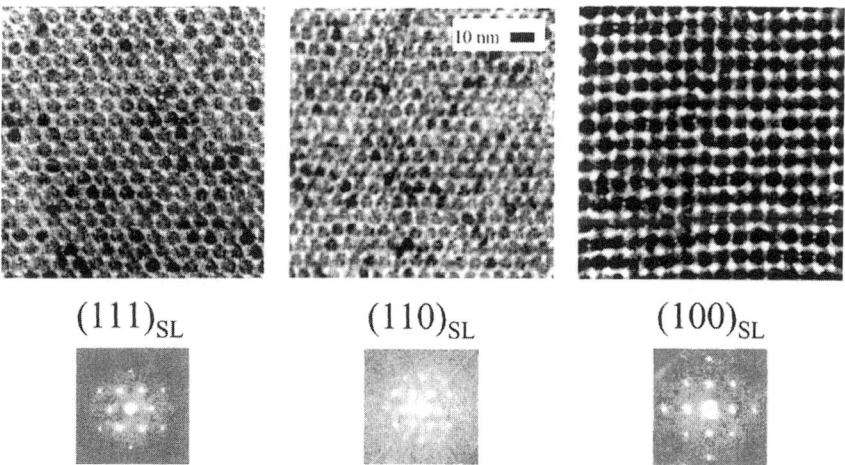

Figure 4 Transmission electron microscopic images of superlattices of InAs QDs (upper frames). Three different facets of an face-centered cubic (fcc) structure of the superlattice can be identified, as can be seen from the optical diffraction of the TEM negatives shown in the lower frames.

Figure 5 shows the typical features and spectra of the samples used in this study for InAs nanocrystals with a mean radius of 2.5 nm. The absorption onset exhibits an ~0.8-eV blue shift from the bulk bandgap. A pronounced first peak and several features at higher energies are observed in the absorption spectra. Band-edge photoluminescence (PL) is observed with no significant red-shifted (deep-trapped) emission. We utilize the size-selective PLE method to examine the level structure. In this particular example, the de-

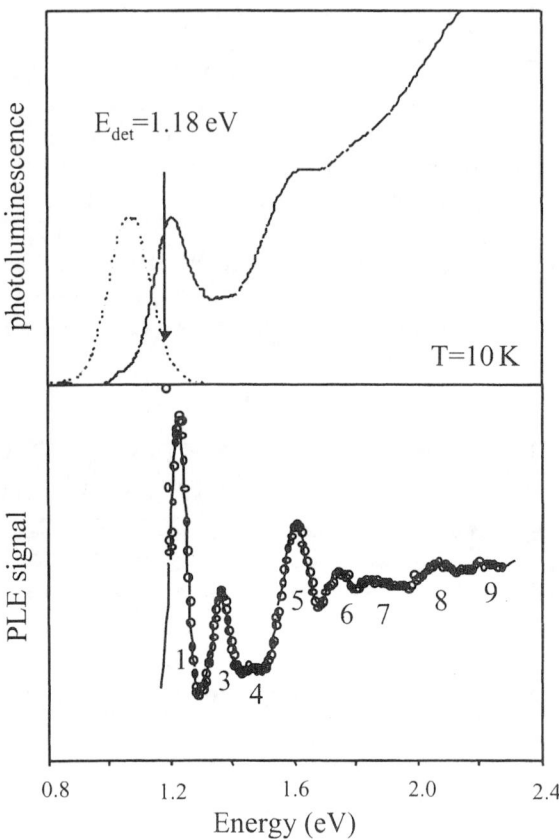

Figure 5 Optical spectroscopy of InAs nanocrystals with mean radii of 2.5 nm. The top frame shows the absorption (solid line) and the PL (dotted line) spectra of the sample. The lower frame shows a size-selected PLE spectrum measured with a narrow detection window positioned as indicated by the arrow in the top panel. Eight transitions are resolved and the positions are extracted by peak fitting (the solid line is the best fit). The weak transition E_2 is not resolved in this QD size.

tection window, E_{det}, was set to 1.18 eV corresponding to a radius of ~2.2 nm and a set of up to eight transitions are resolved (Fig. 5). The full-size dependence was measured by changing the detection window and by using different samples. A representative set of such PLE spectra is shown in Fig. 6.

Figure 7 shows the map of excited transitions for InAs nanocrystals extracted from the PLE data, plotted relative to and as a function of the bandgap transition. The left panel compares the observed transitions with those calculated within the eight-band effective mass model. Briefly, in this approach, each electron (e) and each hole (h) state is characterized by its parity and its total angular momentum $\mathbf{F} = \mathbf{J} + \mathbf{L}$, where $\mathbf{J}$ is the Bloch band-edge angular momentum (1/2 for the CB, 3/2 for the heavy and light hole bands, and 1/2 for the split-off band) and $\mathbf{L}$ is the angular momentum associated with the envelope function. We use the standard notation for the

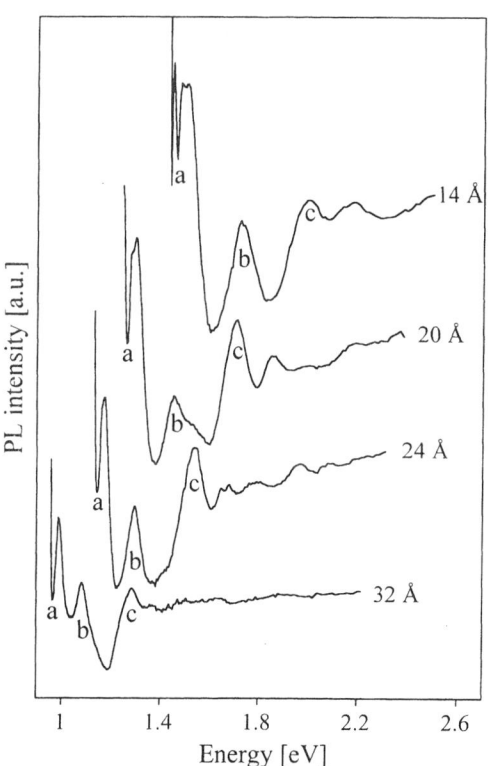

Figure 6 Size-dependent PLE spectra for four representative InAs QD radii. The bandgap transition (a) and two strong excited transitions (b, c) are indicated.

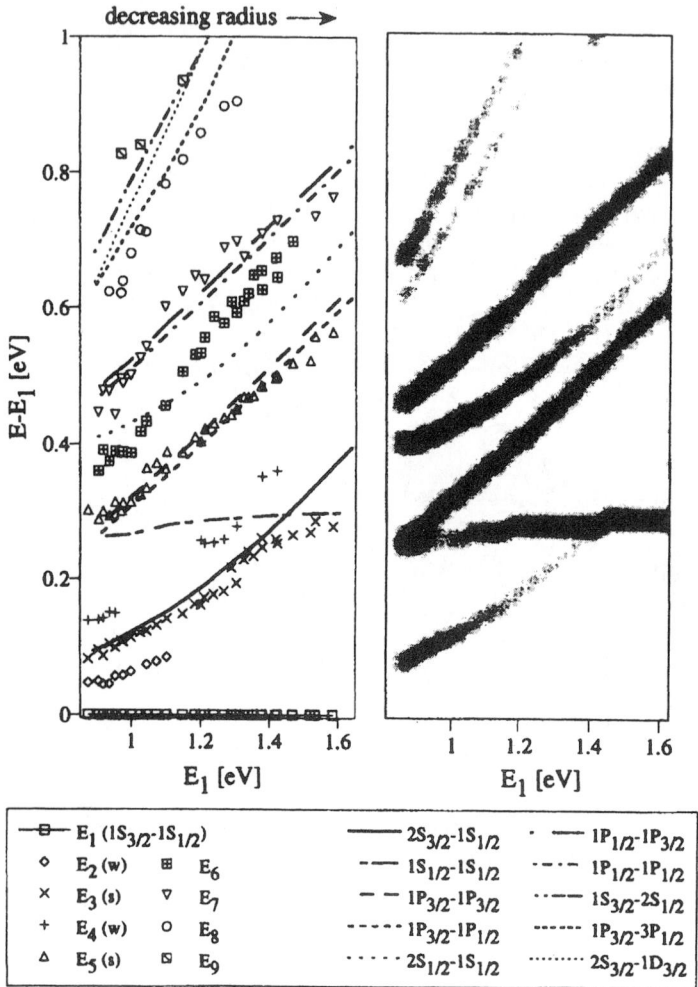

Figure 7 Map of levels for InAs nanocrystals extracted from PLE experiments. The transition energies relative to the lowest (bandgap) transition are plotted as a function of E_g. Left panel: The experimental results are compared with the levels calculated using the eight-band effective mass model. Right panel: Calculated oscillator strengths, represented in gray scale, for the transitions in the left frame.

electron and hole states: nQ_F, where n is the main quantum number, and $Q = $ S, P, D, ... denotes the lowest L in the evelope wave function. The right panel shows the calculated relative oscillator strength for the optically active transitions, usually between electron and hole states with the same Q. The calculated level separations closely reproduce the observed strong transitions. With respect to the comparison between tunneling and PLE spectra, particular focus will be given to the first three strong transitions: the bandgap transition $1S_{3/2}(h)1S_{1/2}(e)$, and excited transitions E_3 and E_5. The first was identified as the $2S_{3/2}(h)1S_{1/2}(e)$ transition, and the second follows the $1P_{3/2}(h)1P_e$ transition [in fact, $1P_e$ is split into $1P_{1/2}(e)$ and $1P_{3/2}(e)$, which are nearly degenerate].

Although the agreement between the PLE data and the theory is relatively good, it relies on the comparison of differences between transitions. A further independent probe of the level structure and symmetry may be highly beneficial, in particular to provide separate information on the CB and VB states. Such information is obtained by scanning tunneling microscopy and spectroscopy.

B. Scanning Tunneling Spectroscopy

For the tunneling measurements we link the nanocrystals to a gold film via hexane dithiol (DT) molecules [25,75], enabling the realization of a DBTJ. Figure 8a (left inset) shows a scanning tunneling microscope (STM) topographic image of an isolated InAs QD, 32 Å in radius. Also shown in Fig. 8a is a tunneling current–voltage (I–V) curve that was acquired after positioning the STM tip above the QD and disabling the scanning and feedback controls (Fig. 8a, right inset). A region of suppressed tunneling current is observed around zero bias, followed by a series of steps at both negative and positive bias. In Fig. 7b, we present the dI/dV versus V tunneling conductance spectrum, which is proportional to the tunneling DOS. A series of discrete single-electron tunneling peaks is clearly observed, where the separations are determined by both the single-electron charging energy (addition spectrum) and the discrete level spacings (excitation spectrum) of the QD. The I–V characteristics were acquired with the tip retracted from the QD to a distance where the bias predominantly drops on the tip–QD junction. Under these conditions, as discussed in Section II, CB (VB) states appear at positive (negative) sample bias, for which electrons tunnel from the tip to the QD, and the excitation peak separations are nearly equal to the real QD level spacings [60]. Similarly, at negative sample bias, the VB states can be resolved, as electrons tunnel from the dot to the tip.

On the positive bias side of Fig. 8b, two closely spaced peaks are observed right after current onset, followed by a larger spacing and a group of six

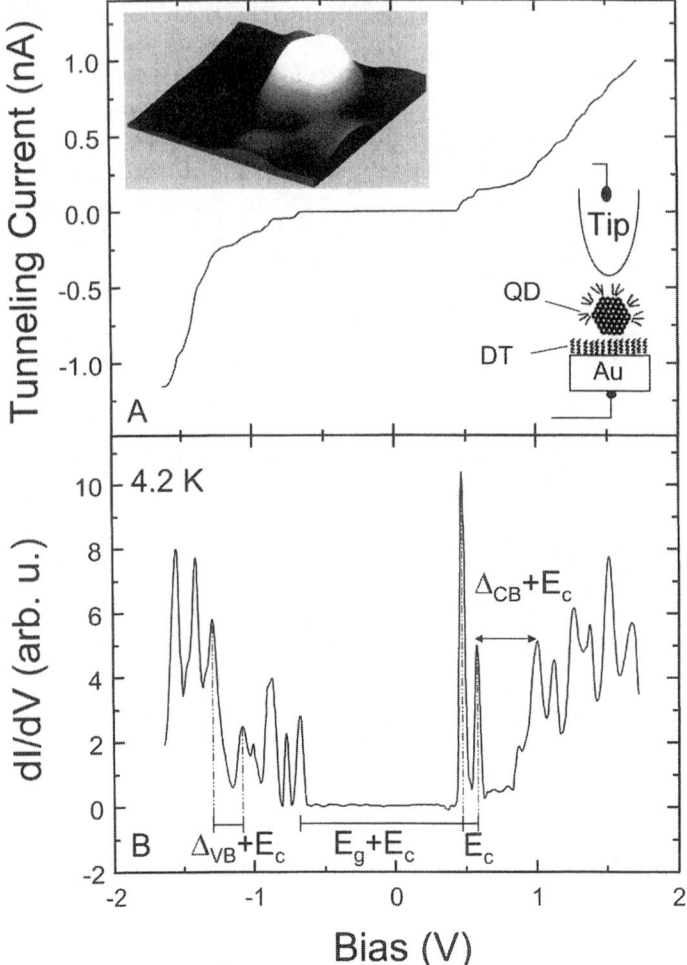

Figure 8 Scanning electron microscopy measurements on a single InAs nano-crystal of 3.2 nm radius acquired at 4.2 K. The QDs are linked to the gold substrate by hexane–dithiol (DT) molecules, as shown schematically in the right inset. The left inset presents a 10×10-nm STM topographic image of the QD. The tunneling I–V characteristic is presented in (A). The tunneling conductance spectrum, dI/dV versus V, obtained by numerical differentiation of the I–V curve, is shown in (B). The arrows depict the main energy separations: E_e is the single-electron charging energy, E_g is the nanocrystal bandgap, and Δ_{VB} and Δ_{CB} are the spacings between levels in the valence and conduction bands, respectively.

nearly equidistant peaks. We attribute the doublet to tunneling through the lowest CB QD state, where the spacing corresponds to the single-electron charging energy, $E_c = 0.11$ eV. The observed doublet is consistent with the degeneracy of the envelope function of the first CB level, $1S_e$ (here, we revert to a simpler notation for the CB states), which has s character. In this case, where charging takes place corresponding the situation described in Fig. 3a, a direct relationship between the degeneracy of a QD level and the number of addition peaks is expected. This is further substantiated by the observation that the second group consists of six peaks, corresponding to the degeneracy of the $1P_e$ state, spaced by values close to the observed E_c for the first doublet. The sixfold multiplet is seen more clearly in the spectrum presented in Fig. 9, focusing only on the CB side. This sequential level filling resembles the Aufbau principle of building up the lowest-energy electron configuration of an atom, directly demonstrating the atomiclike nature of the QD.

The separation between the two groups of peaks, 0.42 eV, is a sum of the level spacing $\Delta_{CB} = 1P_e - 1S_e$ and the charging energy E_c. A value of

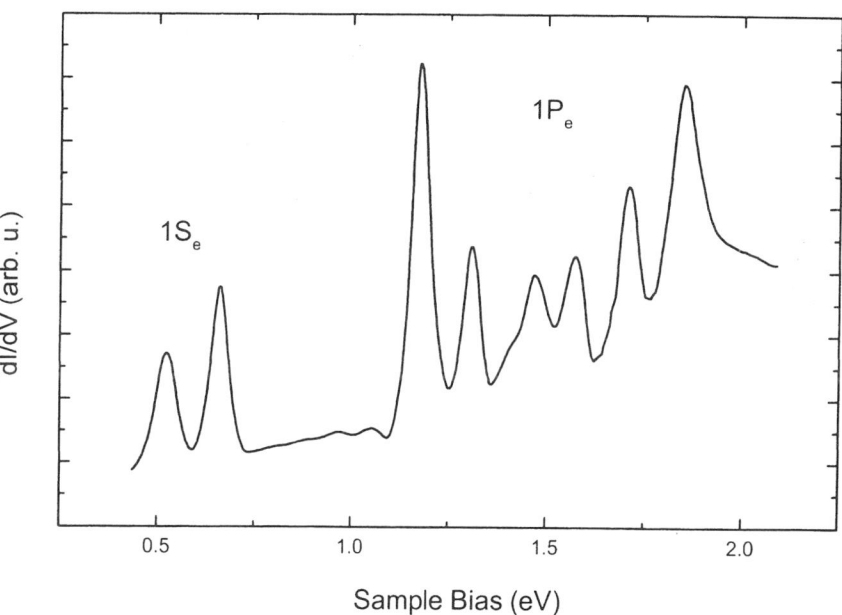

Figure 9 A tunneling conductance spectrum (positive bias side), measured for an InAs QD with radius of 2.8 nm. A doublet and a sixfold multiplet are resolved, assigned to tunneling through the $1S_e$ and $1P_e$ QD states, respectively.

$\Delta_{CB} = 0.31$ eV is thus obtained. On the negative bias side, tunneling through filled dot levels takes place, reflecting the tunneling DOS of the QD valence band. Again, two groups of peaks are observed. The multiplicity in this case, in contrast with the CB, cannot be clearly assigned to a specific angular momentum degeneracy. In a manner similar to that described for the CB states, we extract a value of $\Delta_{VB} = 0.10$ eV for the level separation between the two observed VB states.

In the region of null current around zero bias, the tip and substrate Fermi energies are located within the QD bandgap, where the tunneling DOS is zero. Tunneling is onset when the bias is large enough to overcome both the bandgap and charging energy. E_g is thus extracted by subtracting E_c from the observed spacing between the highest VB and lowest CB peaks and is equal to 1.02 eV.

The tunneling conductance spectra for single InAs nanocrystals spanning a size range of 10–35 Å in radius are presented in Fig. 10. Two groups of peaks are observed in the positive bias side (CB). The first is always a doublet, consistent with the expected s symmetry of the $1S_e$ level, whereas the second has higher multiplicity of up to six, consistent with $1P_e$. The separation between the two groups, as well as the spacing of peaks within each multiplet, increase with decreasing QD radius. This reflects quantum-size effects on both the nanocrystal energy levels and its charging energy, respectively. In some cases (e.g., Fig. 8), one can observe a small peak or shoulder just before the onset of the p multiplet, which may be related to the situation of tunneling into the p level without fully charging the s level, as discussed in Section II. On the negative bias side, generally two groups of peaks, which exhibit similar quantum-confinement effects are also observed. Here, we find variations in the group multiplicities between QDs of different size as well as variations of peak energy spacings with each group. This behavior is partly due to the fact that E_c is very close to Δ_{VB}, the level spacing in the valence band, as shown in Fig. 11b. In this case, sequential single-electron tunneling may be either addition to the same VB state or excitation with no extra charging to the next state. An atomic analogy for this situation can be found in the changing order of electron occupation when moving from the transition to the noble metals within a row of the periodic table.

C. Comparison Between Optical and Tunneling Spectra

The comparison between tunneling and PLE data can be used to decipher the complex QD level structure. This correlation is also important for examining possible effects of charging and tip-induced electric field in the tunneling measurements on the nanocrystal level structure.

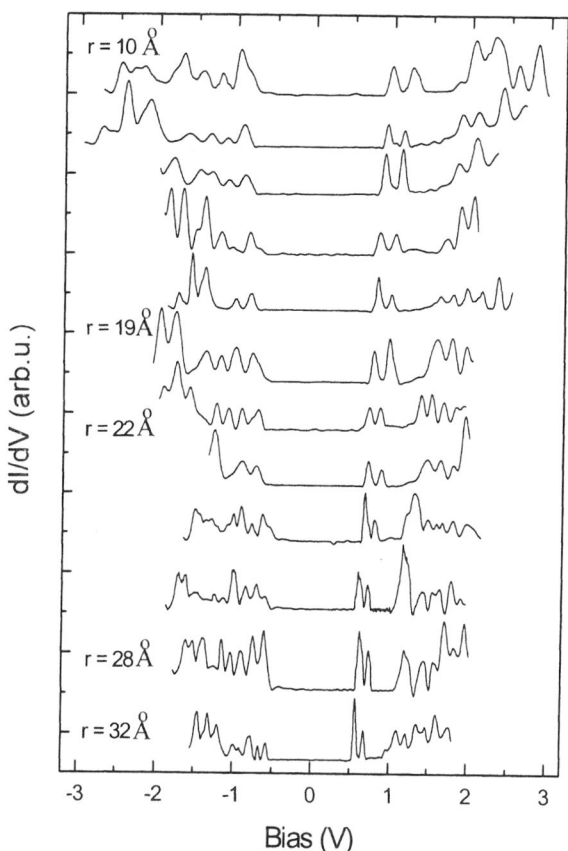

Figure 10 Size evolution of the tunneling dI/dV versus V characteristics of single InAs QDs displaced vertically for clarity. The position of the centers of the zero-current gap showed nonsystematic variations with respect to the zero bias, of the order of 0.2 eV, probably due to variations of local offset potentials. For clarity of presentation, we offset the spectra along the V direction to center them at zero bias. Representative nanocrystal radii are denoted. All spectra were acquired at $T = 4.2$ K.

We first compare the size dependence of the bandgap E_g as extracted from the tunneling data, with the nanocrystal sizing curve (Fig. 11a). The sizing curve (open diamonds) was obtained by correlating the average nanocrystal size, measured using TEM, with the excitonic bandgap of the same sample [48]. To compare these data with the tunneling results, we have added a correction term, $1.8e^2/\varepsilon r$, to compensate for the electron–hole excitonic

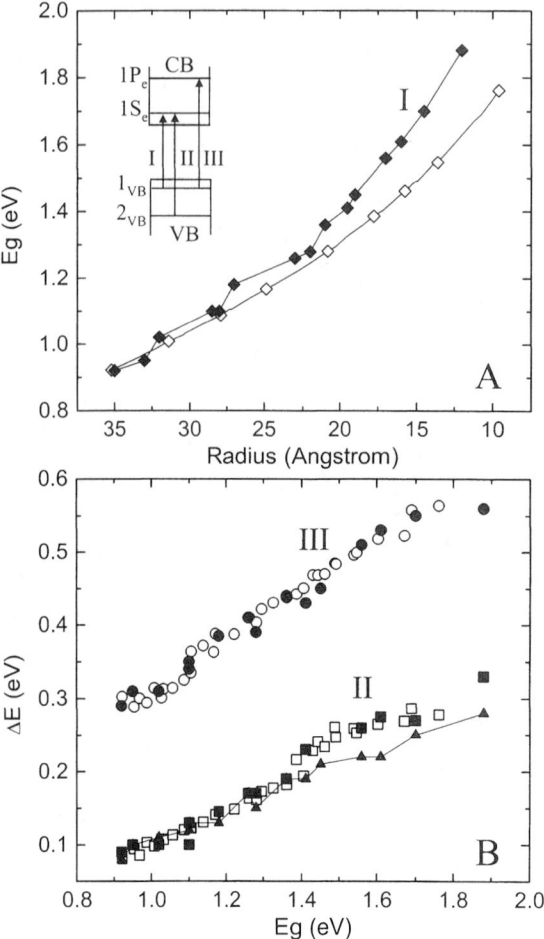

Figure 11 Correlation of optical and tunneling spectroscopy data for InAs QDs. The inset shows a schematic of the CB and VB level structure and the relevant strong optical transitions I, II, and III. (A) Comparison of the size dependence of the low-temperature optical bandgap (transition I) from which the excitonic Coulomb interaction was substracted (open diamonds), with the bandgap measured by the STM (filled diamonds). (B) Excited transitions plotted versus the bandgap for tunneling and optical spectroscopy: The two lower datasets (II) depict the correlation between $\Delta_{VB} = 1_{VB} - 2_{VB}$, detected using the tunneling spectroscopy (filled squares), with the difference between transition II and the band gap transition I (open squares). The two upper datasets (III) depict the correlation between $\Delta_{CB} = 1P_e - 1S_e$ determined using the tunneling spectroscopy (filled circles), with the difference between optical transition III and I (open circles). Also shown is the size dependence of the single-electron charging energy from the tunneling data (filled triangles).

Coulomb interaction that is absent in the tunneling data [6]. Agreement is good for the larger nanocrystal radii, with increasing deviation for smaller nanocrystals. The deviation occurs because the TEM sizing curve provides a lower limit to the nanocrystal radius due to its insensitivity to the (possibly amorphous) surface layer. On the other hand, the size extracted from the STM topographic images is overestimated because of the tip–nanocrystal convolution effect [74]. These differences should be more pronounced in the small size regime, as is indeed observed.

Next, in Fig. 11b, we compare the size dependence of the higher strongly allowed optical transitions, with the level spacings measured by tunneling spectroscopy. We plot excited level spacings versus the observed bandgap for both PLE and tunneling spectra, thus eliminating the problem of QD size estimation discussed earlier. The two lower datasets (II) in Fig. 11b compare the difference between the first strong excited optical transition and the bandgap from PLE (E_3-E_1 in Figs. 5 and 7), with the separation $\Delta_{VB} = 2_{VB}-1_{VB}$ in the tunneling data (open and full squares, respectively). The excellent correlation observed enables us to assign this first excited transition in the PLE to a $2_{VB} - 1S_e$ excitation, as shown schematically in the inset of Fig. 11a. Strong optical transitions are allowed only between electron and hole states with the same envelope function symmetry. Employing this optical selection rule, we thus infer that the envelope function for state 2_{VB} should have s character and this state can be directly identified as the $2S_{3/2}$ valence-band level.

Another important comparison is depicted by the higher pair of curves in Fig. 11b (set III). The second strong excited optical transition relative to the bandgap (E_5-E_1 in Fig. 7) is plotted along with the spacing $\Delta_{CB} = 1P_e-1S_e$ from the tunneling spectra. Again, excellent correlation is observed, which allows us to assign this peak in the PLE to the $1_{VB}-1P_e$ transition (Fig. 11a, inset). The topmost VB level, 1_{VB}, should thus have some p character for this transition to be allowed. From this and considering that the bandgap optical transition $1_{VB}-1S_e$ is also allowed, we conclude that 1_{VB} has mixed s and p character. Pseudopotential calculations of the level structure in InAs QDs show mixed s and p characteritics for the topmost VB state, whereas the effective-mass-based calculations predict that the $1S_{3/2}$ and the $1P_{3/2}$ states are nearly degenerate.

D. Theoretical Descriptions

The theoretical treatments for both optical and tunneling experiments on QDs first require the calculation of the level structure. Various approaches have been developed to treat this problem, including effective-mass-based models, with various degrees of band-mixing effects [47,49,76], and a more atomistic approach based on pseudopotentials. Both have been successfully applied to various nanocrystal systems [50,51,77–79]. To model the PLE data, one has to calculate the oscillator strength of possible transitions and take into account the electron–hole Coulomb interaction which modifies the observed (excitonic) bandgap. In the tunneling case, as discussed earlier, the device geometry should be carefully modeled and, in addition, the effect of charging on the level structure needs to be considered. The charging may affect the intrinsic level structure and also determines the single-electron addition energy.

Zunger and co-workers treated the effects of electron charging for a QD embedded in a homogeneous dielectric medium characterized by ε_{out}. The addition energies and quasi-particle gap were calculated as a function of ε_{out} [80,81]. Although this isotropic model does not represent the experimental geometry of the tunneling measurements, the authors were able to find good agreement between the energetic positions of the peaks for several QD sizes using one value of ε_{out}. These authors also noted that the charging energy contribution associated with the bandgap transition may be different from that within the charging multiplets in the excited states. This difference is, however, on the order of the peak width in our spectra.

In another approach, Niquet et al. modeled the junction parameters C_i and R_i and used a tight-binding model for the level structure [82]. The tunneling spectra were calculated using a rate equation method, extended over the more simplistic approach represented in Section II.C, by allowing for simultaneous tunneling of electrons and holes. The authors were able to reproduce the experimental tunneling spectra, attributing part of the tunneling peaks at negative bias to tunneling through the CB.

E. Detecting Surface States

The surface plays an important role in determining physical and chemical properties of nanocrystals. In particular, the PL is extremely sensitive to the surface passivation and special care is required to remove potential trap sites and to achieve high fluorescence quantum yields. Detailed investigation of surface states is needed for the understanding of such defects and optically detected magnetic resonance has been extensively applied to address these issues [83]. STM can also be used to probe surface states as demonstrated in Fig. 12. Here, the InAs nanocrystals were treated with pyridine, which partly removes the capping TOP ligands. The dI/dV curves measured on two such dots show peaks in the subgap region close to the CB edge. These peaks, absent in ligand-passivated QDs, are tentatively assigned to surface states [84]. Subgap peaks were also observed on unpassivated electrodeposited CdSe

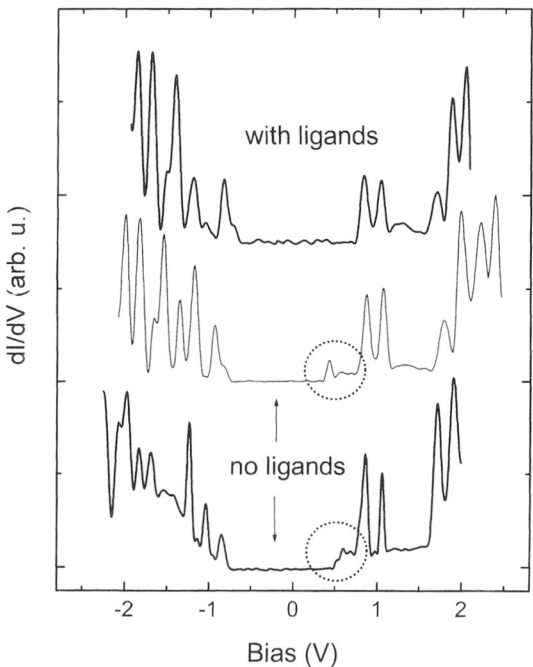

Figure 12 Tunneling spectra measured for InAs nanocrystals with ligands (upper curve) and after pyridine treatment (partly removing TOP ligands). Subgap peaks, marked by the circles, appear close to the CB edge, associated with surface states.

QDs and the relative intensity of the peaks increased with the surface-to-volume ratio, as shown by Alperson et al. [85].

IV. JUNCTION SYMMETRY EFFECTS ON THE TUNNELING SPECTRA

A detailed understanding of the role played by the DBTJ geometry and the ability to control it are essential for the correct interpretation of tunneling characteristics of semiconductor QDs, as well as for their implementation in electronic nanoarchitectures.

The tunneling data presented in Section III were acquired on InAs nanocrystals linked to gold by hexane–dithiol molecules realizing a capacitively highly asymmetric DBTJ ($C_2/C_1 \sim 10$). The observation of QD charging indicated that the tunneling rate $\Gamma_2 \propto 1/R_2$ was on the order of or smaller than $\Gamma_1 \propto 1/R_1$. Otherwise, for positive sample bias, an electron tunneling from the tip to the QD would escape to the substrate before the next electron could tunnel into the QD. Consequently, merely resonant tunneling through the QD states without charging would take place. By varying the tip–QD distance, we were able to modify the voltage division between junctions up to the distance that allowed us to obtain meaningful (well above the noise level) tunneling spectra. Bakkers and Vanmaekelbergh also reported an STM study of CdS and CdSe QDs, focusing on the role of voltage division [64]. We demonstrated that by working without linker molecules, charging-free resonant tunneling, as well as a transition back to tunneling accompanied by QD charging can be achieved for a single QD by controlling R_2 [20].

Figure 13 shows InAs nanocrystals deposited without any linker molecules directly on highly oriented pyrolitic graphite (HOPG) [86]. In Fig. 14a, we plot a tunneling spectrum measured on an InAs QD, ~2 nm in radius, along with a representative spectrum measured on a QD of similar radius, but anchored to a gold substrate via linker molecules, as described in the previous section (dashed line). There is a profound difference between these two spectra. In the spectrum measured in the QD/linker molecule/Au geometry, resonant tunneling accompanied by QD charging is clearly seen, as discussed earlier. In contrast, the charging multiplets are absent in the spectrum measured in the QD/HOPG geometry, and each multiplet is replaced by a single peak, indicating charging-free resonant tunneling through the s- and p-like CB states. Typically, the peaks observed in the QD/HOPG configuration are broadened as compared to those seen for the QD/DT/Au

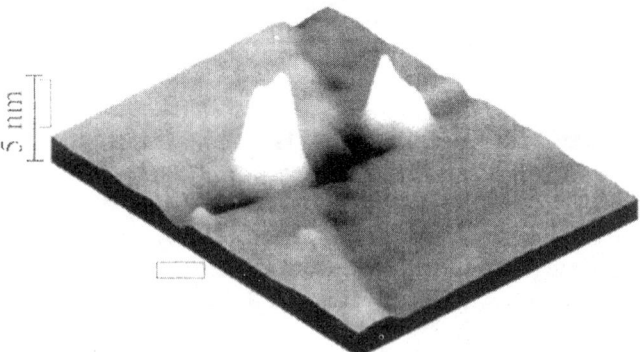

Figure 13 A 30-nm × 30-nm STM topographic showing two single InAs QDs positioned near a monolayer step on HOPG. In this case, no linker molecules separate the QD from the substrate, thus reducing the QD–substrate tunneling barrier.

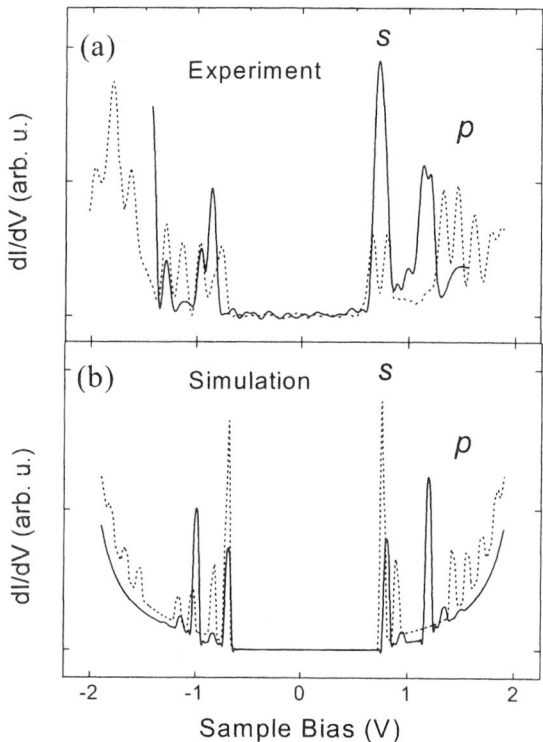

Figure 14 (a) Tunneling spectra measured on InAs QDs ~2 nm in radius. The solid curve was measured in the QD/HOPG geometry, and the dashed curve was measured in the QD/linker molecule/Au geometry. (b) Calculated spectra showing the effect of tunneling-rate ratio. The dashed and solid curves were calculated with Γ_2/Γ_1 ~1 and 10, respectively.

geometry, possibly due to small degeneracy lifting within the s and p states. We note that the charging multiplets were absent even when the peaks did not exhibit significant broadening (e.g., the s peak in the upper curve of Fig. 16). A similar behavior is seen also for the more complex VB. We attribute the difference between the two spectra to the different tunneling-rate ratios Γ_2/Γ_1 achieved in either of the DBTJs. A significantly lower tunnel barrier of the QD–substrate junction is expected in the QD/HOPG configuration.

To confirm this interpretation, we performed theoretical simulations using the method described in Section II. The solid and dashed theoretical curves presented in Fig. 14b were calculated assuming twofold and sixfold degenerate (s and p) CB levels and two fourfold degenerate VB states. The capacitance values were also kept the same for the two curves, $C_1 = 0.1$ aF and $C_2 = 1.1$ aF, resulting in an ~90% voltage drop on the tip–QD junction and $E_C \sim 100$ meV. The two curves differ in the ratio between the tunneling rates: $\Gamma_2/\Gamma_1 \sim 1$ and 10 for the dashed and solid curves, respectively. The dashed curve shows strong charging multiplets, typical for resonant tunneling taking place along with QD charging. The solid curve, on the other hand, exhibits only a signature of charging effect (e.g., one small charging peak in the p multiplet). It is evident that the curve for $\Gamma_2/\Gamma_1 \sim 1$ resembles the experimental spectrum obtained for the QD/linker molecule/Au system, whereas the $\Gamma_2/\Gamma_1 = 10$ curve better corresponds to the QD/HOPG config-uration, consistent with our above interpretation. Note also that the apparent s–p level separation (both in theory and experiment) is smaller for the QD/ HOPG configuration, due to the absence of charging contribution.

A transition from charging-free tunneling to resonant tunneling in the presence of charging is demonstrated by Fig. 15. Here, we plot two tunneling spectra acquired on the same QD of radius 2.5 nm, with different tip–QD separations. The dashed curve was measured with an STM setting $V_S = 1.5$ V and $I_S = 0.1$ nA, whereas the solid curve was taken with $I_S = 0.8$ nA, moving the tip closer to the QD. The apparent gap in the density of states around zero bias is larger for the curve measured with the tip closer to the QD. This is attributed to the effect of voltage division between the two junctions (see Section II). In these measurements C_1 is smaller than C_2; therefore, the applied voltage V_B largely drops on the tip–QD junction and tunneling

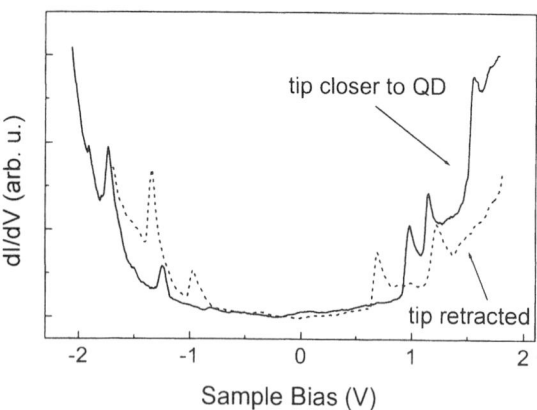

Figure 15   Tunneling spectra measured on a single QD ($r \sim 2.5$ nm) with two different tip–QD separations, exhibiting effects of both the apparent QD level spacing and single-electron charging.

through the discrete QD levels is onset in this junction. Hence, the apparent level spacing in the tunneling spectra is larger than the real level spacing by a factor of $V_B/V_1 = (1 + C_1/C_2)$. Therefore, upon reducing the tip–QD distance, C_1 increases and so does the measured gap. An additional difference is that in the dashed curve, a doublet is observed at the onset of tunneling into the CB, in contrast to a corresponding single peak seen in the solid curve. The second peak in the dashed curve cannot be associated with the p state because the apparent s–p separation must be larger here as compared to the solid curve due to the effect of voltage division, and the observed spacing is smaller. The peak spacing within this doublet is 170 meV, comparable to E_C values measured for InAs QDs of similar size. Hence, we attribute this doublet to single-electron charging of the $1S_e$ level. As the tip approaches the QD, Γ_1 increases towards the value of Γ_2 and thus the process of resonant tunneling becomes accompanied by QD charging.

Figure 16 represents the size dependence on the tunneling spectra for QDs on HOPG in the absence of charging effects. The measured VB–CB gaps (1.57, 1.37, and 1.2 eV) and s–p level separations (0.49, 0.43, and 0.32 eV) for the nanocrystals of radii 1.8, 2.5, and 3.4 nm, respectively, are in relatively good agreement with the values obtained in the QD/linker/Au case (Section III.B) and exhibit the expected quantum-size effect. The absence of charging multiplets allowed us to observe (for the larger QDs) a third peak at positive bias that may be related to the next CB state, presumably $1D_e$ [82]. This peak

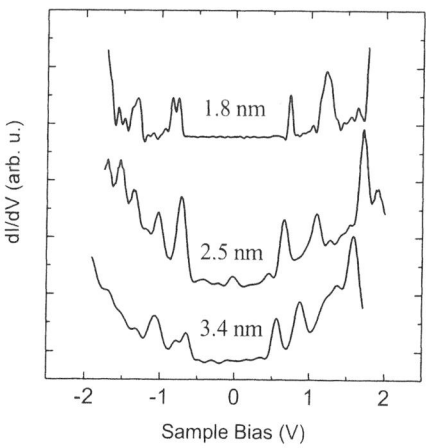

Figure 16 Size evolution of tunneling spectra of InAs QDs on HOPG. The QD radii are denoted in the figure. For clarity, the spectra are offset vertically and shifted along the bias axis to center the measured gaps at zero bias.

could not be detected in the QD/DT/Au system, where, due to the effect of charging, it was pushed out to voltages beyond the limit of current saturation or the onset of field emission.

V. TUNNELING AND OPTICAL SPECTROSCOPY OF CORE–SHELL NANOCRYSTAL QDs

A. Synthesis of Highly Luminescent Core–Shell QDs with InAs Cores

Harnessing the size-tunable emission of nanocrystals for real-world applications such as biological fluorescence marking [14–16], lasers [10,11], and other optoelectronic devices [87,88] is an important challenge, which imposes stringent requirements of a high fluorescence quantum yield (QY) and of stability against photodegradation. These characteristics are difficult to achieve in semiconductor nanocrystals coated by organic ligands due to imperfect surface passivation. In addition, the organic ligands are labile for exchange reactions because of their weak bonding to the nanocrystal surface atoms [89]. A proven strategy for increasing both the fluorescence QY and the stability is to grow a shell of a higher-bandgap semiconductor on the core nanocrystal [32–38]. In such composite core–shell structures, the shell type and shell thickness provide further control for tailoring the optical, electronic, electrical, and chemical properties of semiconductor nanocrystals.

The preparation of the InAs core–shell nanocrystals is carried out in a two-step process. In the first step, the InAs cores are prepared using the injection method with TOP as solvent, which allowed us to obtain hundreds of milligrams of nanocrystals per synthesis. We used size-selective precipitation to improve the size distribution of cores to $\alpha \sim 10\%$. In the second step, shells of various materials were grown on these cores. A complete report on the synthesis method and characterization can be found elsewhere [24,36].

An example for the flexible control on the QD optical properties afforded by shell growth is presented in Fig. 17. Here, two types of core–shell, InAs/ZnSe and InAs/CdSe, which emit strongly at 1.3 μm, were prepared. The absorption and emission spectra for a CdSe shell overgrown on a core with a radius of 2.5 nm is shown in the left frame. The core bandgap emission is at 1220 nm, and with shell growth, the emission shifts to the red. This is accompanied by substantial enhancement of the QY, up to a value of 17% achieved at 1306 nm. For ZnSe shells (right frame), the bandgap hardly shifts. Using a larger core, with radius of 2.8 nm, we could achieve a high QY of 20% at 1298 nm by growing the ZnSe shell. In both cases, the shell provides improved surface passivation, leading to an enhanced emission QY. The

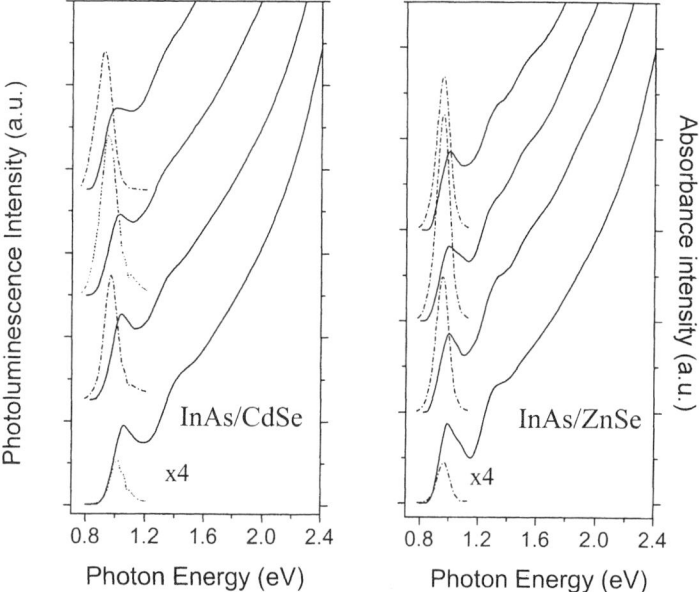

Figure 17 Absorption (solid lines) and emission spectra (dash-dotted lines) for two types of core–shell for different shell thickness. Left frame: InAs/CdSe core–shells; the nominal shell thickness and QY from bottom to top are core = 0.9%, 0.7 ML = 11%, 1.2 ML = 17%, and 1.6 ML = 14%. Right frame: InAs/ZnSe core–shells; the nominal shell thickness and QY from bottom to top are core = 0.9%, 0.7 ML = 13%, 1.3 ML = 20%, and 2.2 ML = 15%.

difference with respect to the bandgap shift can be attributed to the different band offsets of the two shell materials compared with InAs. For ZnSe, large band offsets lead to confinement of the CB and VB ground states to the core region and the bandgap therefore remains intact upon shell growth. For CdSe, due to the significantly smaller CB offset and the light electron effective mass in InAs, the $1S_e$ state is delocalized from the core into the shell region and is therefore red-shifted upon shell growth.

B. Tunneling and Optical Spectroscopy Of InAs/ZnSe Core–Shell

The combined tunneling and optical spectroscopy approach has been applied to further investigate the effect of shell growth on the electronic structure [72]. Figure 18 shows tunneling–conductance spectra measured on two InAs/ZnSe core–shell nanocrystals with 2 and 6 ML shells, along with a typical curve for

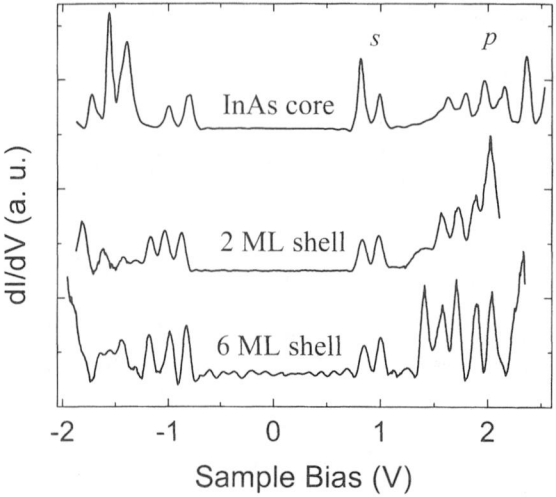

Figure 18 Tunneling conductanceF spectra of an InAs core QD and two core–shell nanocrystals with two and six ML shells with nominal core radii ~1.7 nm. The spectra were offset along the V direction to center the observed zero current gaps at zero bias.

an InAs QD of radius similar to the nominal core radius ~1.7 nm. The general appearance of the spectra of core and core–shell nanocrystals is similar. The gap in the density of states around zero bias, associated with the QD bandgap, is nearly identical, as observed in the optical absorption measurements. In contrast, the $s–p$ level separation is substantially reduced. Both effects are consistent with a model in which the s state is confined to the InAs core region, whereas the p level extends to the ZnSe shell. In this case, the p state is red-shifted upon increasing shell thickness, whereas the s level does not shift, yielding a closure of the CB $s–p$ gap.

Optical spectroscopy also provides evidence for the reduction of the $s–p$ spacing upon shell growth, as manifested in PLE spectra presented in Fig. 19. The three spectra, for cores (solid line) and core–shells with 4 ML and 6 ML shell thickness (dotted and dashed lines, respectively), were measured using the same detection window (970 nm), corresponding to the excitonic bandgap energy for InAs cores 1.7 nm in radius. The peak labeled III, which as discussed earlier, corresponds in the cores to the transition from the VB edge state to the CB $1P_e$ state, is red-shifted monotonically upon shell growth. The dependence of the difference between peak III and the bandgap transition I on shell thickness is depicted in the inset of Fig. 19 (circles), along with the $1S_e - 1P_e$ level spacing extracted from the tunneling spectra (squares).

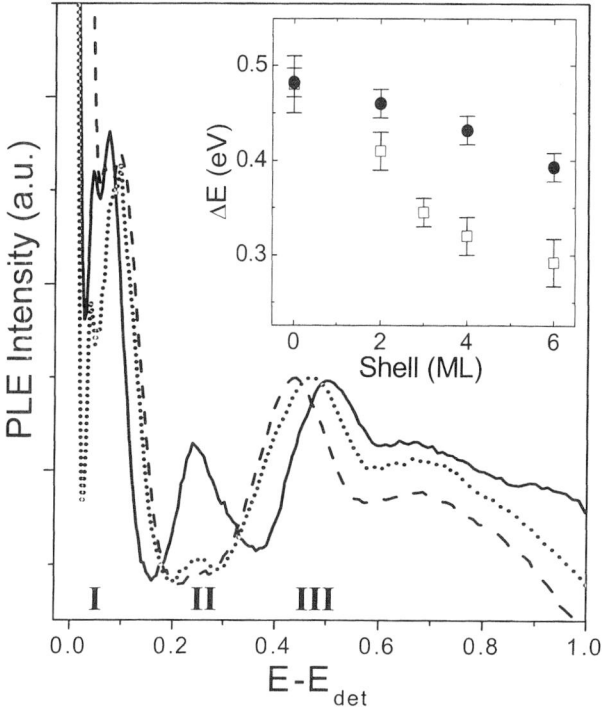

Figure 19 Photoluminescence excitation spectra, normalized to peak III, for InAs cores (solid line) and InAs/ZnSe core–shell nanocrystals of four (dotted line) and six (dashed line) ML shells, with the zero of the energy scale taken at the detection window (970 nm). The inset depicts the dependence of the s–p level spacing on shell thickness, as determined by tunneling (open squares) and PLE (solid circles). The error bars in the tunneling data represent the minimum to maximum spread in the s–p spacings measured on 5–10 QDs for each sample, most likely arising from the distribution in core radii and shell thickness. The PLE data points represent the difference between transition III and transition I (transition I, which hardly shifts upon shell growth, is taken as the central point between the detection energy and the energy of the first PLE peak), averaged over three detection windows (950, 970, and 990 nm).

Although the qualitative trend of red shift is similar for both datasets, there is a quantitative difference, with the optical shift being considerably smaller. This is in contrast to the good correlation between the optical and tunneling spectra observed for InAs cores, providing an opportunity to examine the intricate differences between these two complementary methods. Whereas the tunneling data directly depict the spacing between the two CB states, the PLE data in the inset of Fig. 19 represent the energy difference between two VB to CB optical transitions. Therefore, evolution of the complex quantum-dot VB edge states upon shell growth will inevitably affect the PLE spectra. In particular, a blue shift of the *p*-like component of the VB edge state upon shell growth will reduce the net observed PLE shift compared to the tunneling data, consistent with the experimental observations. This possibility of degeneracy lifting between the VB edge states of the core–shells gains support from further comparing their tunneling spectra with that of the core (Fig. 18). In the negative bias side, associated with the tunneling through VB states, additional peaks appear for the core–shell particles.

VI. QD WAVE–FUNCTION IMAGING

The elegant artificial atom analogy for QDs, borne out from optical and tunneling spectroscopy, can be tested directly by observing the shapes of the QD electronic wave functions. Recently, probability densities of the CB ground and first excited states for epitaxially grown InAs QDs embedded in GaAs were directly probed using cross-sectional scanning tunneling micros-copy [71]. To access the embedded QD with the tip, the sample was cleaved in a plane perpendicular to the growth direction modifying the strain field compared with that of the original embedded dots. Magnetotunneling spec-troscopy with inversion of *k*-space data was also used to probe the spatial profiles of states of such QDs [67]. This noninvasive probe revealed the elliptical symmetry of the ground state in an embedded QD.

For colloidal free-standing nanocrystal QDs, the unique sensitivity of the STM to the electronic density of states on the nanometer scale seems to provide an ideal probe of the wave functions. A demonstration of this capability is given by recent work on the InAs/ZnSe core–shells discussed earlier. Here, the different extent of the CB *s* and *p* states, implied by the spectroscopic results, can be directly probed by using the STM to image the QD atomiclike wave functions. To this end, bias-dependent current imaging measurement [74] were performed, as shown in Fig. 20 for a core–shell nanocrystal with 6 ML shell. The $dI/dV–V$ spectrum is shown in Fig. 20a, and the bias values for tunneling to the *s* and *p* states are indicated. A topographic image was measured at a bias value above the *s* and *p* states, $V_B = 2.1$ V (Fig. 20b), simultaneously with three current images. At each point along the topography scan, the STM feedback circuit was disconnected momentarily and the current was measured at three different V_B values: 0.9 V, corresponding to the CB *s* state (Fig. 20c), 1.4 V, within the *p* multiplet (Fig. 20d), and 1.9 V, above the *p* multiplet (Fig. 20e). With this measurement

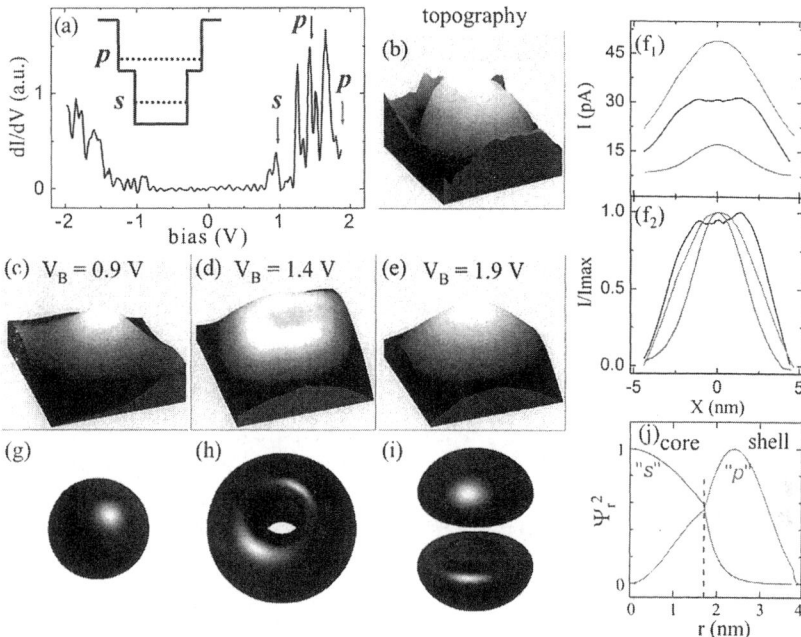

Figure 20 Wave-function imaging and calculation for an InAs/ZnSe core–shell QD having a 6 ML shell. (a) A tunneling spectrum acquired for the nanocrystal. (b) An 8×8-nm^2 topographic image taken at $V_B = 2.1$ V and $I_\delta = 0.1$ nA. (c–e) Current images obtained simultaneously with the topographic scan at three different bias values denoted by arrows in (a). (f_1) Cross sections taken along the diagonal of the current images at 0.9 V (lower curve), 1.4 V (middle curve), and 1.9 V (upper curve). (f_2) The same cross sections normalized to their maximum current values. (g–j) Envelope wave functions calculated within a "particle in a sphere" model. The radial potential and the energies of the s and p states are illustrated in the inset of frame (a). (g–i) Isoprobability surfaces, showing s^2 (g), $p_x^2 + p_y^2$, and p_z^2 (i). (j) The square of the radial parts of the s and p wave functions normalized to their maximum values. For the core–shell potential offset, we used the bulk InAs–ZnSe value, 1.26 eV. The shell–matrix potential offset was taken as 8 eV. Bulk InAs and ZnSe electron effective masses were used.

procedure, the topographic and current images are all measured with the same constant local tip–QD separation. Thus, the main factor determining each current image is the local (bias-dependent) density of states, reflecting the shape of the QD electronic wave functions.

Upon comparing the current images, pronounced differences are observed in the extent and shape of the s and p wave functions. The image corresponding to the s-like wave function (Fig. 20c) is localized to the central region of the core–shell nanocrystal, whereas the images corresponding to the p-like wave functions extend out to the shell (Figs. 20d–20e), consistent with the above-discussed model. This can also be seen in the cross sections presented in frame (f_1) taken along a common line through the center of each current image and most clearly in (f_2), which shows the current normalized to its maximum value along the same cuts. Image 20e, taken at a voltage above the p multiplet, manifests a nearly spherical geometry similar to that of image 20c for the s state but has a larger spatial extent. Image 20d, taken with V_B near the middle of the p multiplet, is also extended but has a truncated top with a small dent in its central region.

An illustrative model aids the interpretation of the current images, assuming a spherical QD shape, with a radial core–shell potential taken as shown in the inset of Fig. 20a [90]. The energy calculated for the s state is lower than the barrier height at the core–shell interface and has about the same values for core and core–shell QDs. In contrast, the energy of the p state is above the core–shell barrier and it red shifts with shell growth, in qualitative agreement with our spectroscopic result discussed in Section V. Isoprobability surfaces for the different wave functions are presented in Figs. 20g–20i, with Fig. 20g showing the s state, Fig. 20h showing the in-plane component of the p wave functions, $p_x^2 + p_y^2$, that has a toruslike shape, and Fig. 20i depicting the two lobes of the perpendicular component, p_z^2. The square of the radial parts of the s and p wave functions are presented in Fig. 20j. The calculated probability density for the s state is spherical in shape and mostly localized in the core, consistent with the experimental image taken at a bias where only this level is probed (Fig. 20c). The p components extend much further to the shell, as observed in the experimental images taken at higher bias. Moreover, the different shapes observed in the current images can be assigned to different combinations of the probability density of the p components.

A filled torus shape, similar to the current image (Fig. 20d) taken at the middle of the p multiplet, can be obtained by a combination with larger weight of the in-plane p component $p_x^2 + p_y^2$, parallel to the gold substrate, and a smaller contribution of the perpendicular p_z component. The nonequal weights reflect preferential tunneling through the in-plane components. This may result from a perturbation due to the specific geometry of the STM experiment leading to a small degeneracy lifting. A spherical shape for the isoprobability surfaces results from summing all of the p components with equal weights, consistent with the current image measured at a bias above the p manifold (Fig. 20e). This example of wave-function imaging combined with the tunneling and optical spectra allowed us to visualize the atomiclike character of nanocrystal quantum dots.

VII. CONCLUDING REMARKS

The combination of optical spectroscopy and scanning tunneling microscopy is proven to be a highly effective approach for studying the elctronic structure and tunneling transport properties of semiconductor nanocrystal QDs. The atomiclike nature of the QDs is borne out both from the observation of the Aufbau principle for sequential single-electron tunneling through the QD states as well as from direct imaging of the quantum-confined envelope wave functions. Extending the atomic analogy further to include spin correlation effects (e.g., Hund's rule) will require the incorporation of magnetic fields in the tunneling experiments. The methodology of combining optical and tunneling spectroscopy can also be extended to the study of artificial QD solids such as close-packed superlattices of nanocrystals. Here, the discrete atomiclike QD states could evolve into miniband bulklike structures. Understanding the level structure and tunneling transport properties is also essential for nanocrystal-based-device applications. Of particular relevance is the implementation of nanocrystals in room-temperature single-electron optoelectronic tunneling devices. Due to the small size, these QDs lie well within the strong confinement regime, and both the level spacings and the single electron charging energies are larger than $k_B T$ even at room temperature. The control of the relative contributions of the level structure and the charging effects will be an important ingredient in future QD devices.

ACKNOWLEDGMENTS

We would like to thank Y.-W. Cao, S.-H. Kan, and D. Katz for their important contributions to the work presented in this chapter. We also thank O. Agam, U. Landman, Y. Levi, Y.-M. Niquet, A. Sharoni, and A. Zunger for stimulating discussions and suggestions. The work was supported in part by the Israel Academy of Science and Humanities and by Intel-Israel.

REFERENCES

1. Alivisatos, A.P. Science 1996, *271*, 933.
2. Brus, L.E. Appl. Phys. A 1991, *53*, 465.
3. Weller, H. Angew. Chem. Int. Ed. Engl. 1993, *32*, 41.
4. Nirmal, M.; Brus, L. Acc. Chem. Res. 1999, *32*, 407.
5. Collier, C.P.; Vossmeyer, T.; Heath, J.R. Annu Rev. Phys. Chem. 1998, *49*, 371.
6. Brus, L.E. J. Chem. Phys. 1984, *80*, 4403.

7. Grabert, H.; Devoret, M.H., Eds. In *Single Charge Tunneling*; Plenum: New York, 1992.

8. Averin, D.V.; Likharev, K.K. In *Mesoscopic Phenomena in Solids*; Altshuler, B.L., Lee, P.A., Webb, R.A., Eds.; Elsevier: Amsterdam, 1991; 173 pp.

9. Banin, U.; Cao, Y.W.; Katz, D.; Millo, O. Nature 1999, *400*, 542.

10. Klimov, V.I.; Mikhailovsky, A.A.; Xu, S.; Malko, A.; Hollingsworth, J.A.; Leatherdale, C.A.; Eisler, H.J.; Bawendi, M.G. Science 2000, *290*, 314.

11. Kazes, M.; Lewis, D.Y.; Ebenstein, Y.; Mokari, T.; Banin, U. Adv. Mater. 2002, *14*, 317.

12. Colvin, V.L.; Schlamp, M.C.; Alivisatos, A.P. Nature 1994, *370*, 354.

13. Dabboussi, B.O.; Bawendi, M.G.; Onitsuka, O.; Rubner, M.F Appl. Phys. Lett. 1995, *66*, 1316.

14. Bruchez, M.P.; Moronne, M.; Gin, P.; Weiss, S.; Alivisatos, A.P. Science 1998, *281*, 2013.

15. Chan, W.C.W.; Nie, S. Science 1998, *281*, 2016.

16. Mitchell, G.P.; Mirkin, C.A.; Letsinger, R.L. J. Am. Chem. Soc. 1999, *121*, 8122.

17. Brodie, I.; Muray, J.I. *The Physics of Nano-Fabrication*; Plenum: New York, 1992.

18. Leon, R.; Petroff, P.M.; Leonard, D.; Fafard, S. Science 1995, *267*, 1966.

19. Grundmann, M.; Christen, J.; Ledentsov, N.N., et al. Phys. Rev. Lett. 1995, *74*, 4043.

20. Murray, C.B.; Norris, D.J.; Bawendi, M.G. J. Am. Chem. Soc. 1993, *115*, 8706.

21. Guzelian, A.A.; Banin, U.; Kadavanich, A.V.; Peng, X.; Alivisatos, A.P. Appl. Phys. Lett. 1996, *69*, 1432.

22. Mews, A.; Eychmüller, A.; Giersig, M.; Schoos, D.; Weller, H. J. Phys. Chem. 1994, *98*, 934.

23. Peng, X.; Wickham, J.; Alivisatos, A.P. J. Am. Chem. Soc. 1998, *120*, 5343.

24. Cao, Y.W.; Banin, U. Angew. Chem. Int. Ed. Engl. 1999, *38*, 3692.

25. Katari, J.E.B; Colvin, V.L.; Alivisatos, A.P. J. Phys. Chem. 1994, *98*, 4109.

26. Murray, C.B., Kagan, C.R., Bawendi, M.G. Science, *270*, 1335.

27. Collier, C.P.; Vossmeyer, T.; Heath, J.R. Annu. Rev. Phys. Chem. 1998, *49*, 371.

28. Whetten, R.L.; Khoury, J.T.; Alvarez, M.M.; Murthy, S.; Vezmar, I.; Wang, Z.L.; Stephens, P.W.; Cleveland, C.L.; Luedtke, W.D.; Landman, U. Adv. Mater. 1996, *8*, 428.

29. Black, C.T.; Murray, C.B.; Sandstrom, R.L.; Sun, S. Science 2000, *260*, 1131.

30. Pileni, M.P. J. Phys. Chem. B 2001, *105*, 3358.

31. Alivisatos, A.P.; Johnson, K.P.; Peng, X.; Wilson, T.E.; Loweth, C.J.; Bruchez, M.P.; Schultz, P.G. Nature 1996, *382*, 609.

32. Hines, M.A.; Guyot-Sionnest, P.J. J. Phys. Chem. 1996, *100*, 468.

33. Peng, X.; Schlamp, M.C.; Kadavanich, A.V.; Alivisatos, A.P. J. Am. Chem. Soc. 1997, *119*, 7019.

34. Dabbousi, B.O.; Rodriguez-Viejo, J.; Mikulec, F.V.; Heine, J.R.; Mattoussi, H.; Ober, R.; Jensen, K.F.; Bawendi, M.G. Phys. Chem. B 1997, *101*, 9463.

35. Tian, Y.; Newton, T.; Kotov, N.A.; Guldi, D.M.; Fendler, J.H. J. Phys. Chem. 1996, *100*, 8927.

36. Cao, Y.W.; Banin, U. J. Am. Chem. Soc. 2000, *122*, 9692.
37. Kershaw, S.V.; Burt, M.; Harrison, M.; Rogach, A.; Weller, H.; Eychmuller, A. Appl. Phys. Lett. 1999, *75*, 1694.
38. Harrison, M.T.; Kershaw, S.V.; Rogach, A.L.; Kornowski, A.; Eychmuller, A.; Weller, H. Adv. Mater 2000, *12*, 123.
39. Peng, X.G.; Manna, L.; Yang, W.D.; Wickham, J.; Scher, E.; Kadavanich, A.; Alivisatos, A.P. Nature 2000, *404*, 59.
40. Manna, L.; Scher, E.C.; Alivisatos, A.P. J. Am. Chem. Soc. 2000, *122*, 12,700.
41. Peng, Z.A.; Peng, X. J. Am. Chem. Soc. 2001, *123*, 1389.
42. Hu, J.; Li, L.S.; Yang, W.; Manna, L.; Wang, L.W.; Alivisatos, A.P. Science 2001, *292*, 2060.
43. Vahala, K.J.; Sercel, P.C. Phys. Rev. Lett. 1990, *65*, 239.
44. Norris, D.J.; Sacra, A.; Murray, C.B.; Bawendi, M.G. Phys. Rev. Lett. 1994, *72*, 2612.
45. Norris, D.J.; Bawendi, M.G. Phys. Rev. B 1996, *53*, 16,338.
46. Bertram, D.; Micic, O.I.; Nozik, A.J. Phys. Rev. B 1998, *57*, R4265.
47. Banin, U.; Lee, J.C.; Guzelian, A.A.; Kadavanich, A.V.; Alivisatos, A.P.; Jaskolski, W.; Bryant, G.W.; Efros, Al.L.; Rosen, M. J. Chem. Phys. 1998, *109*, 2306.
48. Banin, U.; Lee, J.C.; Guzelian, A.A.; Kadavanich, A.V.; Alivisatos, A.P. Superlattices Microstruct. 1997, *22*, 559.
49. Ekimov, A.I.; Hache, F.; Schanne-Klein, M.C.; Ricard, D.; Flytzanis, C.; Kudryavtsev, I.A.; Yazeva, T.V.; Rodina, A.V.; Efros, A.L. J. Opt. Soc. Am. B 1993, *10*, 100.
50. Fu, H.; Wang, L.W.; Zunger, A. Appl. Phys. Lett. 1997, *71*, 3433.
51. Williamson, A.J.; Zunger, A. Phys. Rev. B 2000, *61*, 1978.
52. Porath, D.; Levi, Y.; Tarabiah, M.; Millo, O. Phys. Rev. B 1997, *56*, 9829.
53. Porath, D.; Millo, O. J. Appl. Phys. 1997, *85*, 2241.
54. Kastner, M.A. Phys. Today, 1993, *46*, 24.
55. Kouwenhoven, L. Science 1997, *257*, 1896; Service, R.F. Science 1997, *275*, 303.
56. Amman, M.; Mullen, K.; Ben-Jacob, E. J. Appl. Phys. 1989, *65*, 339.
57. Hanna, A.E.; Tinkham, M. Phys. Rev. B 1991, *44*, 5919.
58. Klein, D.L.; Roth, R.; Lim, A.K.L.; Alivisatos, A.P.; McEuen, P.L. Nature 1997, *389*, 699; Klein, D.L., et al., Appl. Phys. Lett. 1996, *68*, 2574.
59. Alperson, B.; Cohen, S.; Rubinstein, I.; Hodes, G. Phys. Rev. B 1995, *52*, 17,017.
60. Bar-Sadeh, E.; Goldstein, Y.; Zhang, C.; Deng, H.; Abeles, B.; Millo, O. Phys. Rev. B 1994, *50*, 8961.
61. Bar-Sadeh, et al. J. Vac. Sci. Technol. B 1995, *13*, 1084.
62. Dubois, J.G.A.; Gerritsen, J.W.; Shafranjuk, S.E.; Boon, E.J.G.; Schmid, G.; van Kempen, H. Europhys. Lett. 1995, *33*, 279.
63. Schoenenberg, C.; van Houten, H.; Donkerlost, H.C. Europhys. Lett. 1992, *20*, 249.
64. Bakkers, E.P.A.M.; Vanmaekelbergh, D. Phys. Rev. B 2000, *62*, 7743.
65. Katz, D.; Millo, O.; Kan, S.H.; Banin, U. Appl. Phys. Lett. 2001, *79*, 117.

66. Su, B.; Goldman, V.J.; Cunningham, J.E. Phys. Rev. B 1992, 46, 7664.
67. Vdovin, E.E., et al. Science 2000, 290, 122.
68. Crommie, M.F.; Lutz, C.P.; Eigler, D.M. Science 1993, 262, 218.
69. Venema, L.C., et al. Science 1999, 283, 52.
70. Pan, S.H.; Hudson, E.W.; Lang, K.M.; Eisaki, H.; Uchida, S.; Davis, J.C. Nature 2000, 403, 746.
71. Grandidier, B., et al. Phys. Rev. Lett. 2000, 85, 1068.
72. Millo, O.; Katz, D.; Cao, Y-W.; Banin, U. Phys. Rev. Lett. 2001, 86, 5751.
73. Wolf, E.L. *Principles of Electron Tunneling Spectroscopy*; Oxford University Press: Oxford, 1989.
74. Wiesendanger, R. *Scanning Probe Microscopy and Spectroscopy*; Cambridge University Press: London, 1994.
75. Colvin, V.L.; Goldstein, A.N.; Alivisatos, A.P. J. Am. Chem. Soc. 1992, 114, 5221.
76. Efros, A.L.; Rosen, M. Annu. Rev. Phys. Chem. 2000, 30, 475.
77. Rabani, E.; Hetenyi, B.; Berne, B.J.; Brus, L.E. J. Chem. Phys. 1999, 110, 5355.
78. Franceschetti, A.; Fu, H.; Wang, L.W.; Zunger, A. Phys. Rev. B 1999, 60, 1819.
79. Zunger, A. MRS Bull. 1998, 23, 35.
80. Franceschetti, A.; Zunger, A. Phys. Rev. B 2000, 62, 2614.
81. Franceschetti, A.; Zunger, A. Appl. Phys. Lett. 2000, 76, 1731.
82. Niquet, Y.M.; Delerue, C.; Lannoo, M.; Allan, G. Phys. Rev. B 2001, 64, 3305.
83. Lifshitz, E.; Glozman, A.; Litvin, I.D.; Porteanu, H. J. Phys. Chem. B 2000, 104, 10,449.
84. Millo, O.; Katz, D.; Cao, Y.W.; Banin, U. J. Low Temp. Phys. 2000, 118, 365.
85. Alperson, B.; Hodes, G.; Rubinstein, I.; Porath, D.; Millo, O. Appl. Phys. Lett. 1999, 75, 1751.
86. Terrill, R.H.; Postlethwaite, T.A.; Chen, C-H.; Poon, C-D.; Terzis, A.; Chen, A.; Hutchison, J.E.; Clark, M.R.; Wignall, G.; Londono, J.D.; Superfine, R.; Falvo, M.; Johnson, C.S.; Samulski, E.T.; Murray, R.W. J. Am. Chem. Soc. 1995, 117, 12,537.
87. Schlamp, M.C.; Peng, X.G.; Alivisatos, A.P. J. Appl. Phys. 1997, 82, 5837.
88. Mattoussi, H.; Radzilowski, L.H.; Dabbousi, B.O.; Thomas, E.L.; Bawendi, M.G.; Rubner, M.F. J. Appl. Phys. 1998, 83, 7965.
89. Kuno, M.; Lee, J.K.; Dabbousi, B.O.; Mikulec, F.V.; Bawendi, M.G. J. Chem. Phys. 1997, 106, 9869.
90. Schooss, D., et al. Phys. Rev. B 1994, 49, 17,072.

9

III–V Quantum Dots and Quantum Dot Arrays: Synthesis, Optical Properties, Photogenerated Carrier Dynamics, and Applications to Photon Conversion

Arthur J. Nozik
The National Renewable Energy Laboratory, Golden, Colorado, and University of Colorado, Boulder, Colorado, U.S.A.

Olga I. Mićić
The National Renewable Energy Laboratory, Golden, Colorado, U.S.A

I. INTRODUCTION

As is very well known and discussed in this book, semiconductors show dramatic quantization effects when charge carriers (electrons and holes) are confined by potential barriers to small regions of space where the dimensions of the confinement are less than the deBroglie wavelength of the charge carriers or, equivalently, less than twice the Bohr radius of excitons in the bulk material. The length scale at which these effects begin to occur in III–V semiconductors is less than about 25 nm.

In general, charge carriers in semiconductors can be confined by potential barriers in one spatial dimension, two spatial dimensions, or in three spatial dimensions. These regimes are termed quantum films, also more commonly referred to as quantum wells (QWs), quantum wires, and quantum dots (QDs), respectively. These three regimes exhibiting one-, two-, and three-dimensional confinement are created in semiconductor structures that are often described as having a geometric dimensionality of 2D, 1D, and 0D, respectively. They can be formed either by epitaxial growth from the vapor phase [molecular beam epitaxy (MBE) or metallo-organic chemical vapor deposition (MOCVD) processes] or via chemical synthesis (colloidal chemistry or electrochemistry). Here, we will be discussing three-dimensional confinement (0D structures) in III–V semiconductors; the emphasis is on materials formed via colloidal chemistry, but we will also present some interesting results on QDs produced by epitaxial growth using low-pressure MOCVD. The former structures are frequently referred to as nanocrystals and the latter as quantum dots; however, for colloidal nanostructures, we will use the terms "quantum dots" (QDs) and "nanocrystals" (NCs) interchangeably.

In this chapter, we will first discuss the synthesis of various III–V colloidal QDs (InP, GaP GaInP$_2$, GaAs, GaN) (with an emphasis on InP), as well as colloidal InP–CdZnSe$_2$ core-shell QDs and GaAs QDs formed from GaAs quantum wells produced by MOCVD growth; the formation of InP QD arrays will also be discussed. Then, we will present results of interesting and unique optical properties of III–V QDs and QD arrays, including high-efficiency band-edge photoluminescence (PL), size-selective PL, efficient anti-Stokes photoluminescence (PL upconversion), PL intermittency (PL blinking), anomalies between the absorption and the photoluminescence excitation spectra, and long-range energy transfer. Next, we will discuss the photo-generated carrier dynamics in QDs, including the issues and controversies related to the cooling of hot carriers in QDs. Finally, we will discuss the potential applications of QDs and QD arrays in novel photon conversion devices, such as QD solar cells and photoelectrochemical systems for fuel production and photocatalysis.

II. SYNTHESIS OF QUANTUM DOTS

A. Colloidal Nanocrystals

The most common approach to the synthesis of colloidal QDs is the controlled nucleation and growth of particles in a solution of chemical precursors containing the metal and the anion sources (controlled arrested precipitation) [1–3]. The technique of forming monodisperse colloids is very old and can be traced back to the synthesis of gold colloids by Michael Faraday in 1857. A common method for II–VI colloidal QD formation is to rapidly inject a solution of chemical reagents containing the group II and group VI species into a hot and vigorously stirred solvent containing molecules that can coordinate with the surface of the precipitated QD particles [1,3,4]. Consequently, a large number of nucleation centers are initially formed, and the coordinating ligands in the hot solvent prevent or limit particle growth via Ostwald ripening (the growth of larger particles at the expense of smaller particles to minimize the higher surface free energy associated with smaller particles). Futher improvement of the resulting size distribution of the QD particles can be achieved through selective precipitation [3,4], whereby slow addition of a nonsolvent to the colloidal solution of particles causes precipitation of the larger-sized particles (the solubility of molecules with the same type of chemical structure decreases with increasing size). This process can be repeated several times to narrow the size distribution of II–VI colloidal QDs to several percent of the mean diameter [3,4].

The synthesis of colloidal III–V QDs is more difficult than for II–VI QDs. The reason is that III–V semiconductor compounds are more covalent and high temperatures are required for their synthesis. To use the colloidal chemical method for the synthesis of QDs, it is important that the stabilizer and solvent do not decompose during the reaction period in order to ensure good solubility of QDs after synthesis and to avoid extensive formation of trap states on the surface. The preparation method is a compromise between these two requirements. One difference compared to the synthesis described earlier for II–VI materials is that several days of heating at a reaction temperature are required to form crystalline III–V QDS, whereas II–VI QDs form immediately upon injection of the reactants into the hot coordinating solution. The synthesis must also be conducted in rigorously air-free and water-free atmospheres, and it generally requires higher reaction temperatures. The best results to date for III–V QDs have been obtained for InP QDs [5–10]. Figure 1 shows transmission electron microscopic (TEM) images of nanocrystalline InP QDs.

1. Colloidal InP Quantum Dots

In this synthesis, an indium salt [e.g., $In(C_2O_4)Cl$, InF_3, or $InCl_3$] is reacted with trimethylsilylphosphine $P[Si(CH_3)_3]_3$ in a solution of trioctylphosphine oxide (TOPO) and trioctylphosphine (TOP) to form a soluble InP organometallic precursor species that contains In and P in a 1:1 ratio [5,8]. The precursor solution is then heated at 250–290°C for 1–6 d, depending on desired QD properties. Use of TOPO/TOP as a colloidal stabilizer was first reported by Murray et al. [4], who showed the remarkable ability of TOPO/TOP to stabilize semiconductor CdSe QDs at high temperature. Different particle sizes of InP QDs can be obtained by changing the temperature at which the solution is heated. The duration of heating only slightly affects the particle size, but does improve the QD crystallinity. The precursor has a high

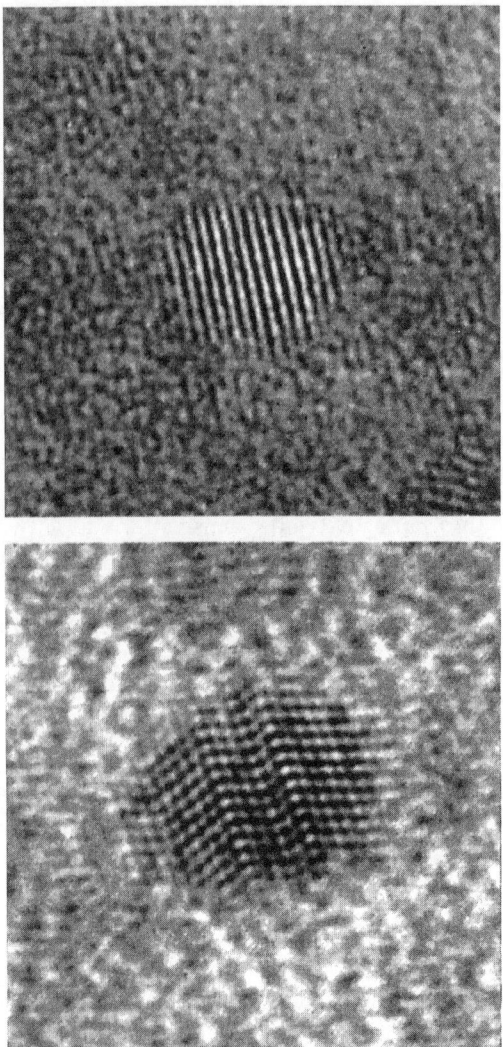

Figure 1 Transmission electron microscopic images of 60-Å InP QDs oriented with the ⟨111⟩ axis in the plane of the image. The bottom plate shows a rare dislocation defect. (From Ref. 11.)

decomposition temperature ($>200\,°C$); this is advantageous for the formation of InP quantum dots because the rate of QD formation is controlled by the rate of decomposition of the precursor. This slow process leads to InP QDs with a relatively narrow size distribution. After heating, the clear reaction mixture contains InP QDs, byproducts of the synthesis, products resulting from TOPO/TOP thermal decomposition, and untreated TOP and TOPO. Anhydrous methanol is then added to the reaction mixture to flocculate the InP nanocrystals. The flocculate is separated and completely redispersed in a mixture of 9:1 hexane and 1-butanol containing 1% TOPO to produce an optically clear colloidal solution. The process of dispersion in the mixture of hexane and 1-butanol and flocculation with anhydrous methanol is repeated several times to purify and isolate a pure powder of InP nanocrystals that are capped with TOPO. Repetitive selective flocculation by methanol gradually strips away the TOPO capping group; thus, TOPO (1%) is always included in the solvent when the QDs are redissolved in order to maintain the TOPO cap on the QDs. Fractionation of the QD particles into different sizes can be obtained by selective precipitation methods [4]; this technique can narrow the size distribution of the initial colloid preparation to about 10%.

The resulting InP QDs contain a capping layer of TOPO, which can be readily exchanged for several other types of capping agent, such as thiols, furan, pyridines, amines, fatty acids, sulfonic acids, and polymers. Finally, they can be studied in the form of colloidal solutions or powders or dispersed in transparent polymers or organic glasses (for low-temperature studies); capped InP QDs recovered as powders can also be redissolved to form transparent colloidal solutions.

X-ray diffraction patterns of InP QD particles formed into a film by drying the colloids show diffraction peaks from the $\langle 111 \rangle$, $\langle 220 \rangle$, and $\langle 311 \rangle$ planes of crystalline zinc-blende InP at 2θ of $26.2 \pm 0.2°$, $46.3 \pm 0.2°$ and $51.7 \pm 0.2°$, respectively [5]. The mean particle diameter can be estimated from the broadening of the diffraction peaks using the Debye–Scherer formula. These diameters are in agreement with the values obtained from TEM and from small-angle x-ray scattering (SAXS) data [5,8]. In the absence of the TOPO stabilizer, the particles grow large and the sharp peaks of bulk InP are obtained.

The shape and size distribution of the InP QDs can be determined by TEM [8]. TEM pictures of InP preparations with TOPO that were only heated to $220\,°C$ for 3 d do not show the formation of either amorphous or crystalline InP. Upon heating to $240\,°C$ for 3 days, the formation of zinc-blende nanocrystallites becomes evident, but the product is primarily amorphous. However, when the preparation is heated at $270\,°C$ for 2 days, electron diffraction patterns show the $\langle 111 \rangle$, $\langle 220 \rangle$, and $\langle 311 \rangle$ planes of zinc-blende InP [5,8]. The InP QDs are generally prolate.

The room-temperature absorption and uncorrected emission spectra of initially prepared InP QDs with a mean diameter of 32 Å are shown in Fig. 2. The absorption spectrum shows a broad excitonic peak at about 590 nm and a shoulder at 490 nm; the substantial inhomogeneous line broadening of these excitonic transitions arises from the QD size distribution. The transitions are excitonic because the QD radius is less than the exciton Bohr radius. The photoluminescence (PL) spectrum (excitation at 500 nm) shows two emission bands: a weaker one near the band edge with a peak at 655 nm, and a second, stronger, broader band that peaks above 850 nm. The PL band with deep red-shifted subgap emission peaking above 850 nm is attributed to radiative surface states on the QDs produced by phosphorous vacancies [9,12].

The room-temperature absorption spectra as a function of QD size ranging from 26 to 60 Å (measured by TEM) are shown in Fig. 3; the red-

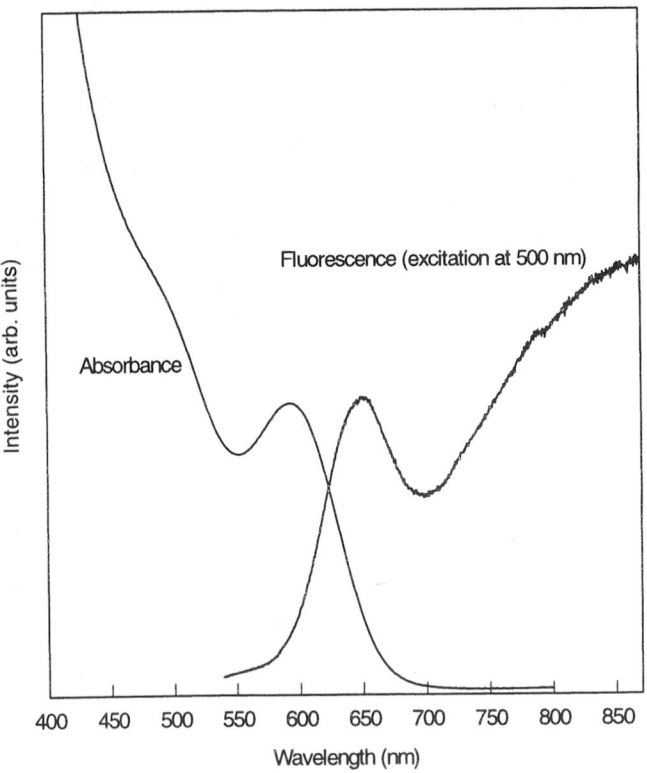

Figure 2   Absorption and emission spectra at 298 K of untreated 32-Å InP QDs. (From Ref. 9.)

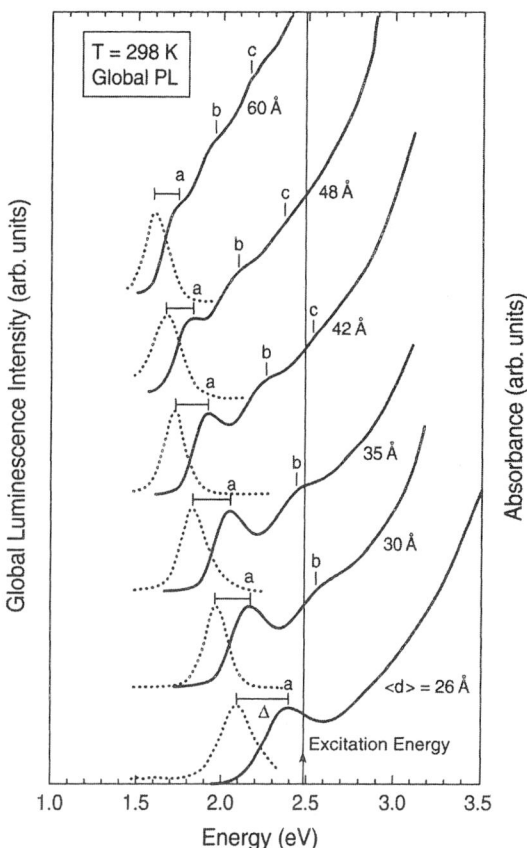

Figure 3 Absorption and emission spectra of hydrogen fluoride (HF)-treated InP QDs for different mean diameters. a, b, and c mark the excitonic transitions apparent in the absorption spectra. All samples were excited at 2.5 eV. (From Ref. 10.)

shifted deep-trap emission from the as-prepared colloidal QDs was eliminated by etching the QDs in hydrogen fluoride (HF) (see Sect. III.A). The absorption spectra show one or more broad excitonic peaks; as expected, the spectra shift to higher energy as the QD size decreases [10]. The color of the InP QD samples changes from deep red (1.7 eV) to green (2.4 eV) as the diameter decreases from 60 to 26 Å. Bulk InP is black with a room-temperature bandgap of 1.35 eV and an absorption onset at 918 nm. Higher-energy transitions above the first excitonic peak in the absorption spectra can also be easily seen in QD samples with mean diameters equal to or greater than 30 Å. The spread in QD diameters is generally about 10% and is somewhat narrower in samples with larger mean diameters; this is why higher-energy transitions can be resolved for the larger-sized QD ensembles. All of the prepared QD nanocrystallites are in the strong confinement regime because the Bohr radius of bulk InP is about 100 Å.

Figure 3 also shows typical room-temperature global emission spectra of the InP colloids as a function of QD diameters. We define global PL as that observed when the excitation energy is much higher than the energy of the absorption threshold exhibited in the absorption spectrum produced by the ensemble of QDs in the sample; that is, the excitation wavelength is well to the blue of the first absorption peak for the QD ensemble and, therefore, a large fraction of all the QDs in the sample are excited. The particle diameters that are excited range from the largest in the ensemble to the smallest, which has a diameter that produces a blue-shifted bandgap equal to the energy of the exciting photons. In Fig. 3, the excitation energy for all QD sample ensembles was 2.48 eV, well above their absorption onset in each case. The global PL emission peaks ("nonresonant") in Fig. 3 are very broad (linewidth of 175–225 meV) and are red-shifted by 100–300 meV as the QD size decreases from 60 to 26 Å [10]. The broad PL linewidth is caused by the inhomogeneous line broadening arising from the ~10% size distribution. We attribute the large global red shift and its increase with decreasing QD size to the volume dominance of the larger particles in the size distribution; the larger QDs will absorb a disproportionally larger fraction of the incident photons relative to their number fraction and will show large red shifts (because the PL excitation energy is well above their lowest transition energy) that will magnify the overall red shift of the QD ensemble.

2.   Colloidal GaP Quantum Dots

Quantum dots of GaP can be synthesized by mixing $GaCl_3$ (or the chloro-gallium oxalate complex) and $P[Si(CH_3)_3]_3$ in a molar ratio of Ga:P of 1:1 in toluene at room temperature to form a GaP precursor species, and then heating this precursor in TOPO at 400°C for 3 days [5]. Wells et al. [13,14] first synthesized and characterized the yellow GaP precursor, $[Cl_2GaP(SiMe_3)_2]_2$, formed from $GaCl_3$ and $P(SiMe_3)_3$. The mean particle diameters of GaP QD preparations can be estimated from the line broadening of their x-ray diffraction patterns and from TEM.

The absorption spectrum of a 30-Å-diameter GaP QD colloid (heated at 400°C) exhibits a shoulder at 420 nm (2.95 eV) and a shallow tail that extends out to about 650 nm (1.91 eV) [5]. For 20-Å-diameter GaP QDs (heated at 370°C), the shoulder is at 390 nm (3.17 eV) and the tail extends to about 550 nm [5].

Bulk GaP is an indirect semiconductor with an indirect bandgap of 2.22 eV (559 nm) and a direct bandgap of 2.78 eV (446 nm). Theoretical calculations [15] on GaP QDs show that the increase of the indirect bandgap with decreasing QD size is much less pronounced than that for the direct gap; for 30-Å-diameter GaP QDs, the direct and indirect bandgaps are predicted to be 3.35 eV and 2.4 eV, respectively. Below 30 Å, the direct bandgap is predicted to *decrease* with decreasing size while the indirect bandgap continues to increase. As a result, GaP is expected to undergo a transition from an indirect semiconductor to a direct semiconductor below about 20 Å.

The steep rise in absorption and the shoulder at 420 nm in the absorption spectrum of 30-Å GaP QDs [5] is attributed to a direct transition in the GaP QDs; the shallow tail region above 500 nm is attributed to the indirect transition. Also, the absorption tail extends below the indirect bandgap of bulk GaP [5]. The origin of this subgap absorption could be caused either by a high density of subgap states in the GaP QDs, by impurities created by the high decomposition temperature, or by Urbach-type band tailing produced by unintentional doping in the QDs [16]. We note that such subgap absorption below the bandgap was also observed in GaP nanocrystals that were prepared in zeolite cavities by the gas-phase reaction of trimethyl-gallium and phosphine at temperatures above 225°C [17]. This latter result implies that the subgap absorption in GaP QDs is intrinsic and is not caused by synthetic by-products or impurities.

3. Colloidal GaInP$_2$ Quantum Dots

Quantum dots of GaInP$_2$ can be synthesized by mixing chlorogallium oxalate and chloroindium oxalate complexes and P(Si(CH$_3$)$_3$)$_3$ in the molar ratio of Ga:In:P of 1:1:2.6 in toluene at room temperature, followed by heating in TOPO [5]. Heating at 400°C for 3 days is required to form 25-Å QDs. X-ray diffraction for a 65 Å sample shows that the lattice spacings of GaInP$_2$ QDs is approximately the average of that for GaP and InP [5].

The ternary Ga–In–P system forms solid solutions which can exhibit direct bandgaps ranging from 1.7 to 2.2 eV, depending on composition and growth temperature [18–24]. At the composition Ga$_{0.5}$In$_{0.5}$P, the structure can be either atomically ordered or disordered (random alloy) [18–24]; the bandgap is direct, but it can range from about 1.8 to 2.0 eV, depending on the degree of atomic ordering. An open issue of interest is how size quantization will affect the dependence of bandgap on atomic ordering.

The absorption spectrum of 25-Å GaInP$_2$ QDs does not show any excitonic structure [5]; this is caused by an exceptionally wide size distribution which masks the excitonic peaks. An estimate of the direct bandgap of the

GaInP$_2$ QDs from a plot of the square of the absorbance times photon energy versus photon energy indicates a value of about 2.7 eV; this value is blue-shifted from the bulk value of 1.8–2.0 eV.

4. Colloidal GaAs Quantum Dots

GaAs quantum dots can be formed by first reacting Ga(III) acetylacetonate and As[Si(CH$_3$)$_3$]$_3$ at reflux (216°C) in triethylene glycol dimethyl ether (triglyme) [25–28]. This produces an orange-brown turbid slurry that can be filtered through a 700-Å ultrafilter to produce the GaAs colloidal QD solution. A TEM image of GaAs QDs prepared in this way, except that quinoline was used instead of triglyme, shows perfectly spherical QDs with very well-resolved lattice planes [27]. Electron diffraction data show clear *hkl* zinc-blende GaAs patterns of (111), (220), (311), and (422). The observed lattice plane spacing of 3.2 Å in TEM images corresponds to the *d*(111) of GaAs. Optical absorption spectra of the GaAs QDs shows an onset of absorption at about 600 nm; a shallow rise with decreasing wavelength steepens at about 470 nm and peaks at about 440 nm [25–28]. The particle-size distribution of the GaAs QDs was not sufficient to observe excitonic transitions in the absorption or emission spectra [25–28].

5. Colloidal GaN Quantum Dots

Wells and co-workers [29,30] first showed that nanosize GaN can be synthesized by pyrolysis of {Ga(NH)$_{3/2}$}$_n$ at high temperature. The lack of any organic substituents in the precursor makes {Ga(NH)$_{3/2}$}$_n$ a good candidate for the generation of carbon-free gallium nitride. To produce colloidal transparent solutions of isolated GaN QDs, a method was used which is similar to that described earlier for the preparation of III–V phosphide QDs. Dimeric amidogallium Ga[N(CH$_3$)$_2$]$_6$ was synthesized by mixing anhydrous GaCl$_3$ with LiN(CH$_3$)$_2$ in hexane according to the published method [31]. This dimer was then used to prepare polymeric gallium imide {Ga(NH)$_{3/2}$}$_n$ by reacting Ga$_2$[N(CH$_3$)$_2$]$_6$ with gaseous NH$_3$ at room temperature for 24 h. Details of the preparation are given in Refs. 29 and 30. To produce GaN QDs, the resulting {Ga(NH)$_{3/2}$}$_n$ (0.2 g) was slowly heated in trioctylamine (TOA) [boiling point (b.p. 365°C, 4 mL)] at 360°C over 24 h and kept at this temperature for 1 day. Ammonia flow at ambient pressure was maintained during this heat treatment and while the solution cooled to room temperature. The solution was cooled to 220°C and a mixture of TOA (2 mL) and hexadecylamine (HDA) (b.p. 330°C, 2g) was added and stirred at 220°C for 10 h; the HDA improved hydrophobic capping of the GaN surface because HDA is less sterically hindered and creates a more dense surface cap. After that, the solution was cooled over several hours.

The synthesis was conducted in rigorously air-free and water-free atmospheres. One important aspect in the synthesis of GaN is the purity of the final product. Carbon can be left on the QD surfaces after pyrolysis, and it is difficult to remove. TOA/HDA decreases carbon adsorption and, after purification, yields a white colloidal solution. Repetitive flocculation and redispersion in a solution of hexane containing 1% HDA leads to the isolation of clean, white samples. This powder was redispersed in 2,2,4-trimethypentane, which contained 1% HDA. After that, the solution was sonicated in a high-intensity ultrasonic processor and filtered to produce an optically clear, nonscattering organic glass at 10 K. The QD surface is derivatized with TOA/HDA and ensures that the QD particles are isolated from each other in solution.

The colloidal GaN solution shows an absorption spectrum with a weak shoulder at 330 nm and a structureless emission spectrum [32]; this again indicates a broad size distribution of particles. The absorption and emission spectra are shifted to higher energies (3.65 eV) compared with bulk GaN (E_g = 3.2–3.3 eV for zinc-blende structure) [33], confirming that the GaN particles are in the quantum-confinement regime.

6. Lattice-Matched Core–Shell InP/ZnCdSe$_2$ Quantum Dots

Core–shell quantum dots with a zinc-blende structure consisting of InP cores and lattice-matched ZnCdSe$_2$ shells have been successfully prepared by colloidal chemistry [34]. The core InP QDs, with an average size of 25–45 Å, were synthesized by colloidal chemistry methods using InCl$_3$ and tris-(trimethylsilyl)phosphine ([P(SiMe$_3$)$_3$] as described earlier. Fractionation of the QD particles into different sizes was obtained by selective precipitation, collected as powders, and then redispersed in pyridine.

The CdZnSe$_2$ precursor was prepared by mixing dimethylzinc (ZnMe$_2$), dimethylcadmium (CdMe$_2$), and tributylphosphineselenium (TBPSe) in tributylphosphine (TBP) solution in a molar ratio 1:1:4, respectively. Fresh precursor solutions were always prepared before use. Excess Se was used to ensure complete formation of CdZnSe$_2$. TBPSe was prepared by dissolving Se (1 M) in TBP.

The InP QDs were dispersed in pyridine and then overcoated with CdZnSe$_2$ in pyridine by reacting the precursors at 100°C. Successful overcoating of QDs in pyridine at 100°C had been previously used for (CdSe)CdS QDs [35]. The ratio of the ZnCdSe$_2$ precursor to InP necessary to form a shell of a desired thickness was based on the ratio of the volume of the shell to that of the core, assuming that spherical cores and annular shells are formed.

High-resolution TEM (HR-TEM) images of the QDs show well-resolved lattice fringes that extend in a straight line through the whole QD crystal, indicating lattice-matched epitaxial growth of the shell onto the core. The ZnCdSe$_2$ shell passivates the surface of the InP core. Hence, whereas bare InP cores with diameters of 22 and 42 Å exhibited no photoluminescence, these cores capped with a 5-Å ZnCdSe$_2$ shell show PL quantum yields of 5–10% at room temperature (see Fig. 4). The absorption and emission spectra show a red shift of the core-shell QD compared to the core alone. The red shift was measured as a function of ZnCdSe$_2$ shell thickness (up to 50 Å) for a core diameter of 30 Å and increased with increasing shell thickness. This red shift was not as large as that between a 30-Å InP core and a larger InP QD consisting of the 30-Å InP core plus InP shells of equivalent thickness to the (InP)ZnCdSe$_2$ QDs. High-level calculations of the electronic structure of the core-shell (InP)ZnCdSe$_2$ and bare InP QDs were made using both self-consistent field and tight-binding methods [34]. The wave functions and electron radial probability density distributions were calculated, and the theoretical red shifts calculated from these functions were consistent with experiment.

B.   Quantum Dots Grown via Vapor-Phase Deposition

Semiconductor QDs can also be formed via deposition from the vapor phase onto appropriate substrates in MBE or MOCVD reactors [36,37]. There are two modes of formation. In one, termed Stranski–Krastinow (S-K) growth, nanometer-sized islands can form when several monolayers (about 3–10) of one semiconductor are deposited upon another and there is large lattice-mismatch (several percent) between the two semiconductor materials; this has been demonstrated for Ge/Si [38,39], InGaAs/GaAs [40–42], InP/GaInP [43], and InP/AlGaAs [44,45]. For these highly strained systems, epitaxial growth initiates in a layer-by-layer fashion and transforms to 3D island growth above four monolayers to minimize the strain energy contained in the film (see Fig. 5). The islands then grow coherently on the substrate without generation of misfit dislocations until a certain critical strain energy density, corresponding to a critical size, is exceeded [38,40]. Beyond the critical size, the strain of the film-substrate system is partially relieved by the formation of dislocations near the edges of the islands [40]. Coherent S-K islands can be overgrown with a passivating and carrier-confining epitaxial layer to produce QDs with good luminescence efficiency. The optical quality of such over-grown QD samples depends on the growth conditions of the capping layer.

The second approach is to first produce a near-surface quantum well (formed from 2D quantum films) and then deposit coherent S-K islands on top of the outer barrier layer of the QW that have a large lattice mismatch with the barrier that subsequently produces a compressive strain in the island [46,47]. The large resultant strain field can extend down into the QW structure

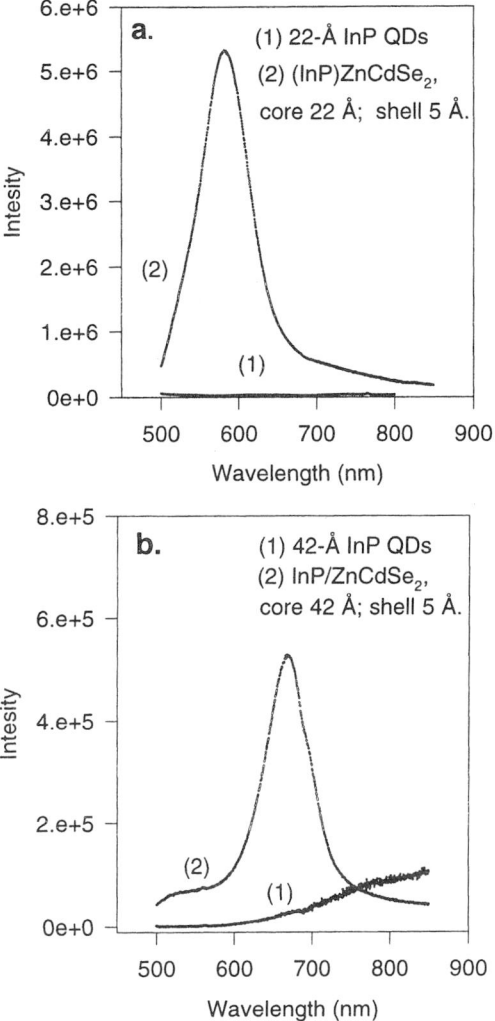

Figure 4 Photoluminescence spectra at 298 K of lattice-matched core–shell (InP) ZnCdSe$_2$ QDs compared to uncapped and untreated InP QDs with the same core diameter. In (a), the InP core is 22 Å, and in (b), the core is 42 Å; the ZnCdSe$_2$ shell is 5 Å for all core–shell QDs. (From Ref. 34.)

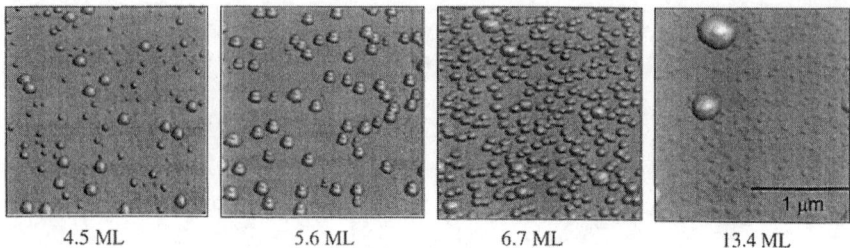

| 4.5 ML | 5.6 ML | 6.7 ML | 13.4 ML |

Figure 5 Evolution of Stranski–Krastinow InP islands grown on (100) AlGaAs at 620°C by MOCVD for increasing amounts of deposited InP [expressed as monolayers (ML)]. The scale of each scan is 2 × 2 μm.

by about one island diameter, thus penetrating through the outer barrier and well regions (see Fig. 6). This strain field will dilate the lattice of the QW and lower the bandgap beneath the S-K islands to produce a quantum dot with three-dimensional confinement. One unique aspect of this QD is that the well and barrier regions are made of the same semiconductor. The S-K islands are referred to as stressor islands; such types of stress-induced InGaAs and GaAs QDs have been reported for InP stressor islands on a GaAs/InGaAs/GaAs QW [46,47] and for InP stressor islands on an AlGaAs/GaAs/AlGaAs QW [44,45].

III. UNIQUE OPTICAL PROPERTIES

A. High-Efficiency Band-Edge PL in InP QDs

Relatively intense band-edge emission from InP QDs can be achieved by etching the particles with a dilute alcoholic solution of HF [9]. The etching is done by adding a methanolic solution containing 5% HF and 10% H_2O to a mixture of hexane and acetonitrile (1:1) that contains the InP QDs and 2–5% stabilizer. Two liquid phases are formed and the QDs are dispersed in the upper nonpolar phase of hexane, whereas HF, methanol, and H_2O are in the acetonitrile phase. The mixture is shaken and left overnight, and then the hexane phase with the colloids is separated and used.

    We believe that upon etching with HF or NH_4F, fluoride ions fill phosphorus vacancies on the surface of the InP [48]. The intensity of the band-edge emission increases by more than a factor of 10. Figure 2 shows the absorption and uncorrected emission spectra (excitation at 500 nm) at room temperature of 32-Å InP QDs before the HF treatment; Fig. 7 shows the room-temperature absorption and uncorrected emission spectra of HF-treated InP QDs with diameters of 30, 35, and 44 Å. The PL emission is at

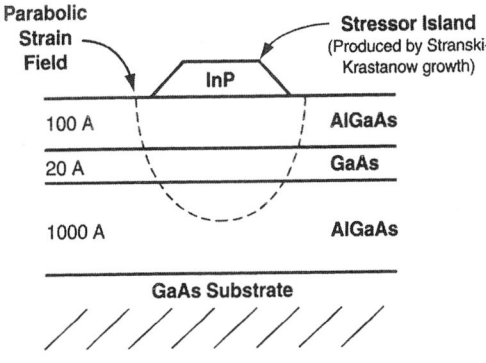

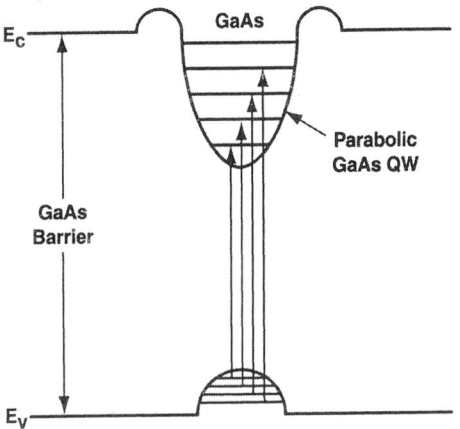

Figure 6 Diagram explaining formation of strain-induced GaAs quantum dots by depositing InP stressor islands on the thin outer barrier of an AlGaAs/GaAs/AlGaAs QW. The InP stressor island produces a compressive strain field in the lattice-mismatched QW that decreases the bandgap of the GaAs QW beneath the stressor island, producing a QD with well and barrier both made from GaAs.

the band edge of the absorption, but red-shifted from the first excitonic peak by about 60 nm. The PL quantum yield at room temperature of the HF-treated InP QDs is increased to about 30% compared to a few percent for untreated QDs. Films of the QDs made by slowly evaporating the colloidal QD solution show a quantum yield of 60% at 10 K; also the PL linewidth decreases from 161 meV at 300 K to 117 meV at 10 K. The quantum yield (QY) values are external quantum yields (photons emitted divided by incident photons).

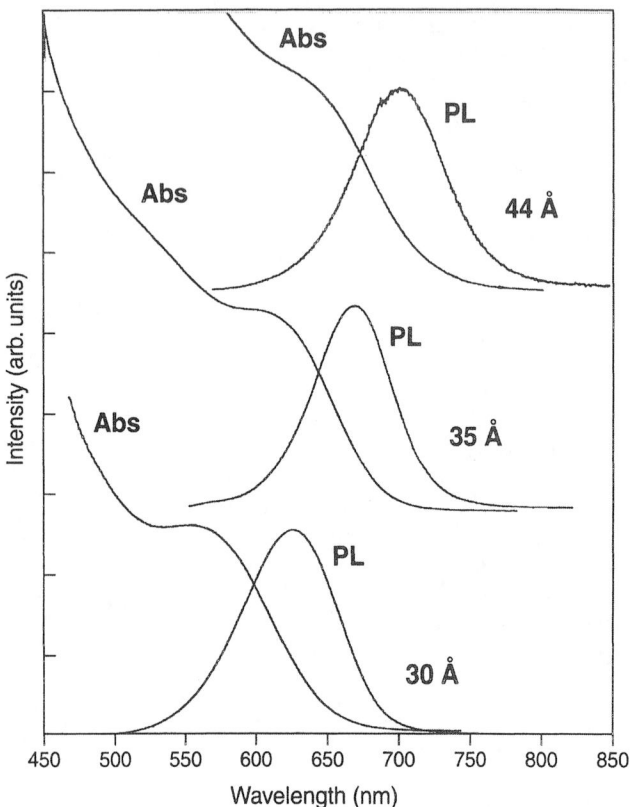

Figure 7 Photoluminescence spectra at 298 K of InP QDs of different mean diameter that have been treated with HF to enhance the PL quantum yields, enhance the band-edge emission, and inhibit the deep-trap emission that arises from radiative surface traps. (From Ref. 9.)

For untreated InP QDs, the lifetime of the deep red-shifted emission, which peaks above 850 nm, was measured to be greater than 500 ns. For the HF-treated InP QDs, the lifetime of the band-edge emission was measured to be much shorter; the decay was nonsingle exponential with lifetimes spanning a range from 5 to 50 ns.

These results show that the HF etching treatment of InP QDs removes or passivates surface states to produce band-edge luminescence with high quantum yield. The deep red-shifted emission above 850 nm for untreated InP QDs is attributed to radiative surface states produced by phosphorus vacancies; the long lifetime of this defect luminescence (500 ns) is consistent with PL from trap states. It is known that for bulk InP, a radiative transition from states deep within the band gap, which appears at 0.99 eV, is associated with phosphorus vacancies.

Recent optically detected magnetic resonance (ODMR) results [12] show that unetched InP QDs have phosphorous vacancies both at the surface and in the QD core and that these defects act as radiative traps for photo-generated electrons. Treatment of the QDs with HF eliminates the ODMR signal that results from phosphorous vacancies at the surface, but it leaves a small ODMR signal due to the small population of phosphorous vacancies in the QD core. The peaks of the ODMR and PL spectra coincide, both for the deep-trap emission and band-edge emission in untreated and HF-treated InP QDs, respectively Thus, the ODMR experiments confirm that the strong deep-trap emission of untreated InP QDs arises from phosphorous vacancies and that when the QDs are treated with HF, the phosphorous vacancies at the surface are passivated, leaving very weak or nonexistent PL emission from the low residual phosphorous vacancy population in the core. Additional electron paramagnetic resonance (EPR) experiments [48] support this conclusion and provide additional information about the nature of nonradiative hole traps near the valence band that are involved in the anti-Stokes PL (PL upconversion) that is observed with QDs (see Sect. III.C).

B. Size-Selected Photoluminescence

If the PL excitation energy is restricted to the onset region of the absorption spectrum of the QD ensemble, then a much narrower range of QD sizes is excited that have the larger particle sizes in the ensemble. Consequently, the PL spectra from this type of excitation show narrower linewidths and smaller red shifts with respect to the excitation energy. This technique is termed fluorescence line narrowing (FLN)—the resulting PL spectra being considerably narrowed.

Fluorescence line narrowing spectra at 10 K are shown in Figs. 8a–8e for InP QDs with a mean diameter of 32 Å. FLN/PL spectra are shown for a series of excitation energies (1.895–2.07 eV) spanning the absorption tail near the onset of absorption for this sample [16]. Also shown is the global PL spectrum produced when the excitation energy (2.41 eV) is deep into the high-energy region of the absorption spectrum (Fig. 8f).

Fluorescence line narrowing spectra can be combined with photoluminescence excitation (PLE) spectra to determine the resonant red shift associated with true band-edge emission [10]. The experiment is done as follows: (1) A photon energy is selected in the onset region of the absorption spectrum of the QD ensemble spectrum, and this energy is set as the detected photon energy in the PLE; (2) the PLE spectrum is then obtained, and the first peak of the PLE spectrum is taken to be the lowest-energy excitonic transition for the QDs capable of emitting photons at the selected energy; (3) an FLN

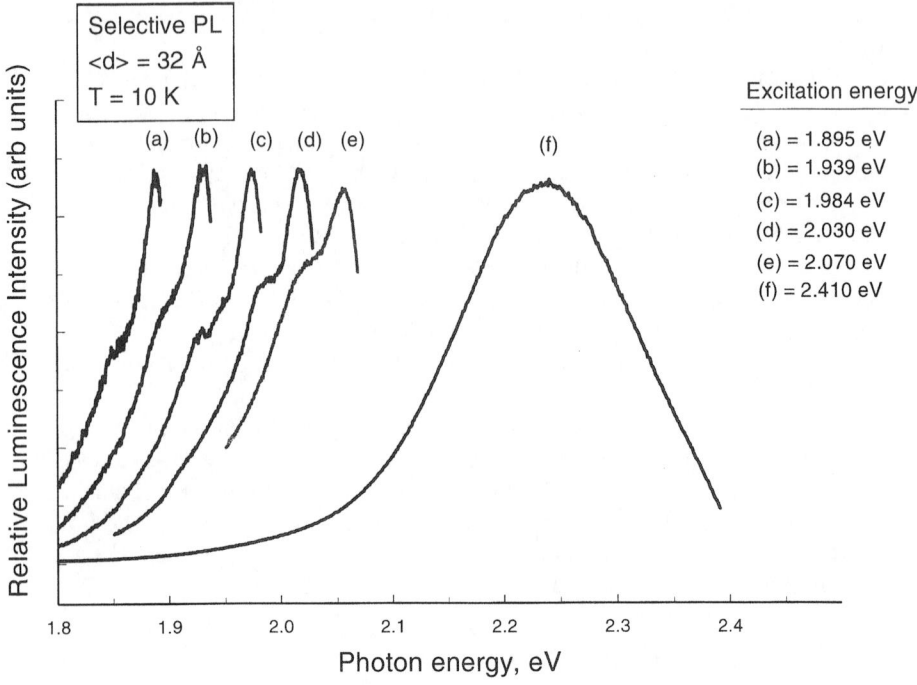

Figure 8   Photoluminescence spectra at 10 K for InP QDs with a mean diameter of 32 Å for different excitation energies. The first exciton peak in absorption is at 2.17 eV. Curve f is the global PL because all sizes in the QD ensemble are excited; curves a–e represent FLN spectra arising from size-selective excitation in the red tail of the absorption spectrum. (From Ref. 10.)

spectrum is then obtained with excitation at the first peak of the PLE spectrum. The energy difference between the first FLN peak and the first PLE peak is then defined as the resonant red shift for the ensemble of QDs represented by the selected PLE energy. This process is repeated across the red tail of the absorption onset region of the absorption spectrum to generate the resonant red shift as a function of QD size. Typical FLN and PLE spectra are shown in Fig. 9 for InP QDs with a mean diameter of 32 Å. The resultant $T = 11$ K resonant red shift as a function of PL excitation energy ranges from 6 to 16 meV [10]. Theoretical modeling of this resonant red shift in InP QDs is based on the effects of electron-hole exchange in splitting the lowest-energy excitonic transition into singlelike and triplelike states and is described in Refs. 49 and 50.

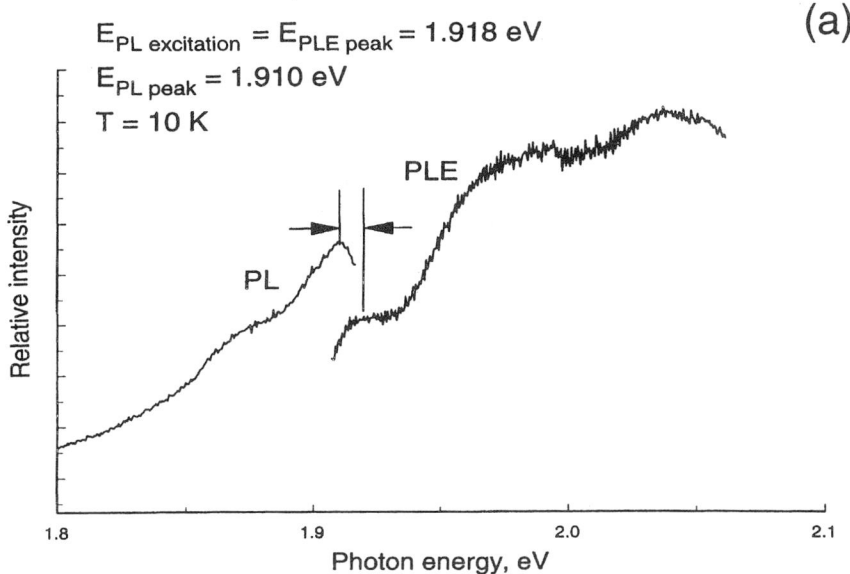

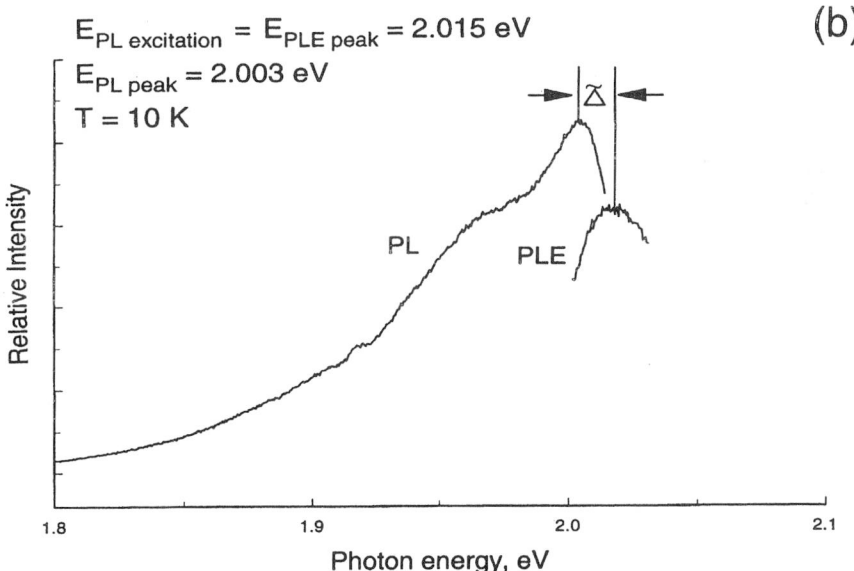

Figure 9 Representative pairs of PL (FLN) and PLE spectra for the InP sample of Fig. 7 showing how the resonant Stokes shift, Δ, is determined. (From Ref. 10.)

C. Efficient Anti-Stokes Photoluminescence (Upconversion)

Efficient anti-Stokes PL has been observed for several nanocrystals, including InP, CdSe, CdS, ZnSe, and GaN [51]. Figure 10 (curve a) shows the normal (Stokes-shifted) global PL spectrum for HF-etched colloidal InP QDs (23 Å mean diameter) excited at 400 nm, together with the upconverted PL (UCPL) emission spectra (curves b–n) for a series of excitation wavelengths that span the range of emission wavelengths of the global PL spectrum. Figure 11

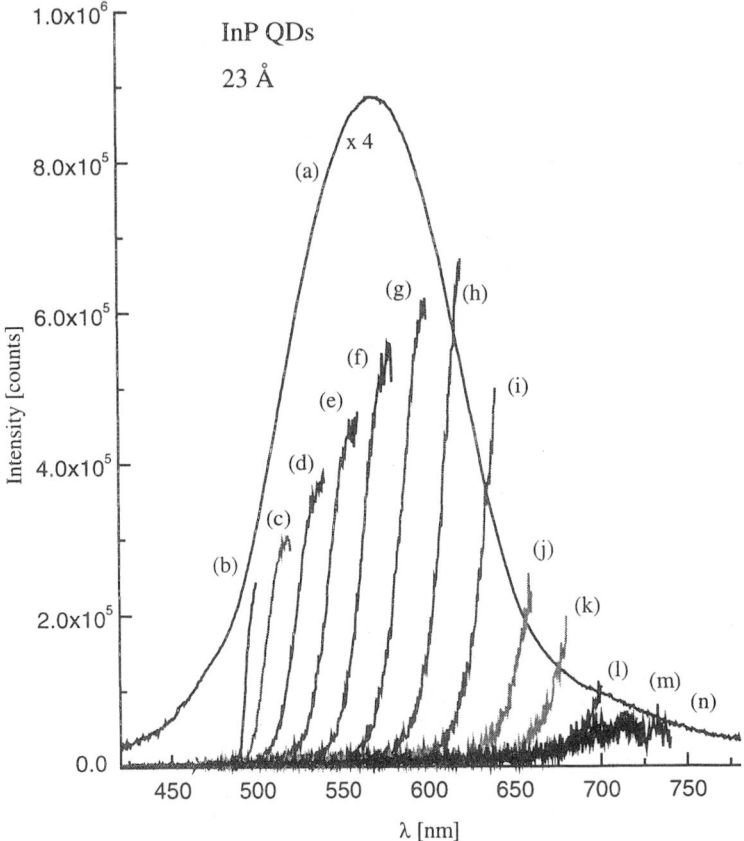

Figure 10   Global Stokes-shifted (curve a) and anti-Stokes shifted (curves b–n) PL at 298 K from InP QDs (mean diameter 23 Å). For each of the anti-Stokes PL spectra (curves b–n), the wavelength scans were terminated 20 nm to the blue of the excitation wavelength. (From Ref. 51.)

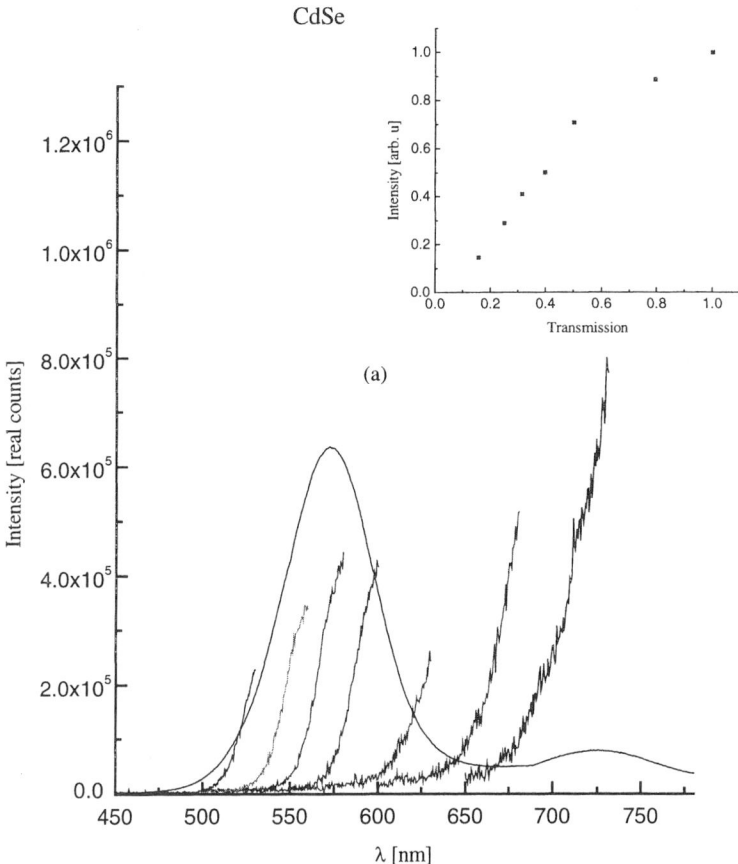

Figure 11   Global PL and anti-Stokes PL for CdSe QDs (mean diameter of 35 Å) (analogous to Fig. 9). (From Ref. 51.)

shows the same type of data for unetched CdSe QDs with a mean diameter of 35 Å.

The magnitude of the upconverted blue shift (ΔE_{UC}) is defined as the difference between the excitation energy and the energy value at which an exponential fit of the UCPL spectrum crosses the average background noise level; the error in ΔE_{UC} is defined as the spread between the crossing points of the exponential fit and the minimal and maximal noise levels. The global PL spectra for the QDs show a broad linewidth, which is expected [10] because of the QD size distribution in the QD colloid (about 10% around the mean QD diameter).

Because HF treatment has been shown to remove or passivate surface traps on InP [9], the normal PL spectra for the HF-etched InP QDs show only a small degree of deep-trap emission (manifested in Fig. 10 by a red tail in the PL spectra). For unetched and unpassivated CdSe QDs, the normal red-shifted emission from traps is more pronounced and is manifested as a peak at 725 nm in Fig. 11. In Figs. 10 and 11, it is apparent that PL upconversion is also occurring for trap emission in the red tail of the PL spectra. With increased trap density, the intensity of the UCPL from traps and from the QD band edge is greatly increased relative to the band-edge emission. It is noted that the maximum degree of UCPL occurs with QDs that were aged (sitting in ambient conditions for several weeks to months).

There are several important features in the UCPL spectra of Figs. 10 and 11. First, the UCPL cannot be detected at energies above the maximum energy exhibited in the normal PL emission. Second, the intensity of the UCPL generally follows the intensity distribution of the global PL emission across the whole PL spectrum, except that its peak intensity is red-shifted from the global PL peak intensity. Third, the UCPL spectra were obtained at low light intensity (a Xe lamp in the fluorimeter was used as the excitation source). A fourth feature of the data is that the intensity of the UCPL follows a linear dependence on excitation intensity at low excitation intensity and then begins to saturate; this is shown, for example, for the CdSe QDs as an inset in Fig. 11. The fifth feature is the critical role of surface states in the PL upconversion.

The first two features of the UCPL spectra indicate that only band-edge emission can be detected from the QDs. This means the upconversion mechanism cannot involve carrier ejection to a barrier surrounding the QDs, followed by radiative recombination in the barrier, as was proposed for upconversion in semiconductor heterojunctions [52–55]; PL from a barrier would exhibit higher energies than the QD bandgap and the UCPL would not be confined to the range of normal PL emission from the QDs. Thus, Auger processes cannot be involved here. The fact that PL upconversion is restricted only to the QD bandgap energies means that subgap states are involved as an intermediate state.

The third and fourth features indicate that nonlinear two-photon absorption (TPA) cannot be the cause of upconversion because the lamp excitation source is too weak for generating a TPA process; TPA requires very high light intensities usually generated by laser excitation, is generally inefficient, and is nonlinear with intensity.

The fourth and fifth features indicate that surface states or traps are playing a critical role in the UCPL process. The relatively intensity of UCPL is directly correlated with the surface state density. The linear dependence of UCPL intensity on excitation intensity is also consistent with photoexcitation to traps.

To explain these results, the following model for the UCPL that invokes surface states was proposed [51]. The model is based on Fu and Zunger's [50] calculated energy level structure of InP QDs as a function of QD size, including surface states produced by In and P dangling bonds; their results are reproduced in Fig. 12a. At QDs sizes above 57 Å, the In dangling bond (In-DB) energy level is coincident with the QD conduction-band minimum (CBM); however, below 57 Å, the energy separation between the CBM and the In-DB increases with decreasing size as the conduction-band energy moves up and the In-DP energy remains nearly constant. On the other hand, as seen in Fig. 12a, the P-DB energy level is always above the valence-band maximum (VBM) at all QD sizes (0.3 eV for bulk InP), and this separation increases as the VBM moves down with decreasing QD size.

All of the UCPL results can be explained within the context of the energy levels calculated by Fu and Zunger [50] and the model is shown in Fig. 12b. For upconversion of photon energies above the red-shifted trap emission energies, the first step in the process is photoexcitation from the P-DB state to the conduction band; if the QD size is significantly less than 57 Å, the electron in the conduction band then relaxes to the In-DB state, which lies below the conduction band. The second step is excitation of the photogenerated hole in the P-DB state to the valence band, followed by radiative recombination of the electrons and holes across the bandgap. For upconversion of the trap emission, the first step is photoexcitation from the P-DB state to the In-DB state, followed by excitation of the P-DB hole to the valence band and radiative recombination from the In-DB state to the VBM.

This model also predicts that as the QD size gets smaller, the UCPL decreases and goes to zero at the QD size (15 Å) where the quantity (P-DB − VBM) equals the quantity (CBM − In-DB) (i.e., the relaxation energy of electrons from the conduction band to the In-DB cancels out the upconversion energy of the holes from the P-DB to the VBM). The experimental results in Ref. 51 are consistent with this prediction. As seen in Fig. 12a, the difference between the theoretically calculated P-DB energy and the CBM − In-DB energy as a function of QD size follows the equivalent experimental value of the UCPL shift (ΔE_{UC}) added to the VBM energy; also, ΔE_{UC} goes to zero at 15 Å, as predicted. However, although the experimental results fit the In-DB and P-DB model of Fu and Zunger [50] very well, there is presently no independent experimental identification of the actual chemical nature of the surface states in the InP QD samples.

In the UCPL model, the energy of upconversion is produced by excitation of the In-DP hole to the valence band; this process could be driven either by phonon absorption or by absorption of a second photon. The former process is favored for several reasons. (1) The upconversion shows a strong decrease in intensity with decreasing temperature (while the normal Stokes PL intensity increased with decreasing temperature); a sequential two-photon

(a)

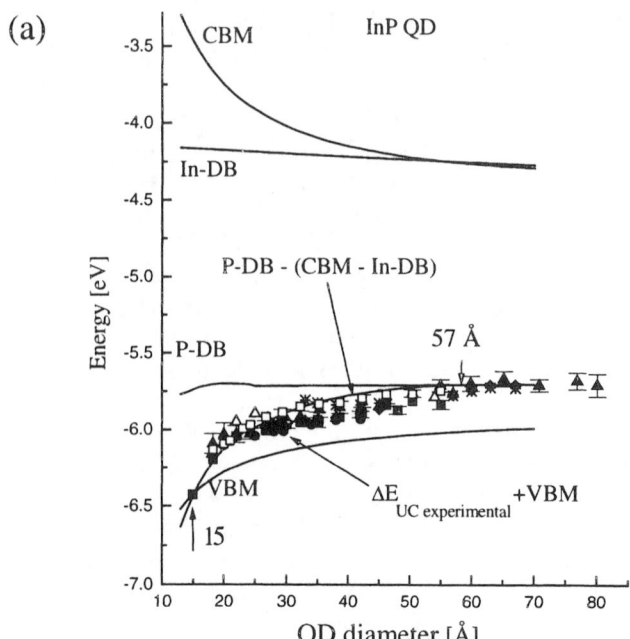

(b)

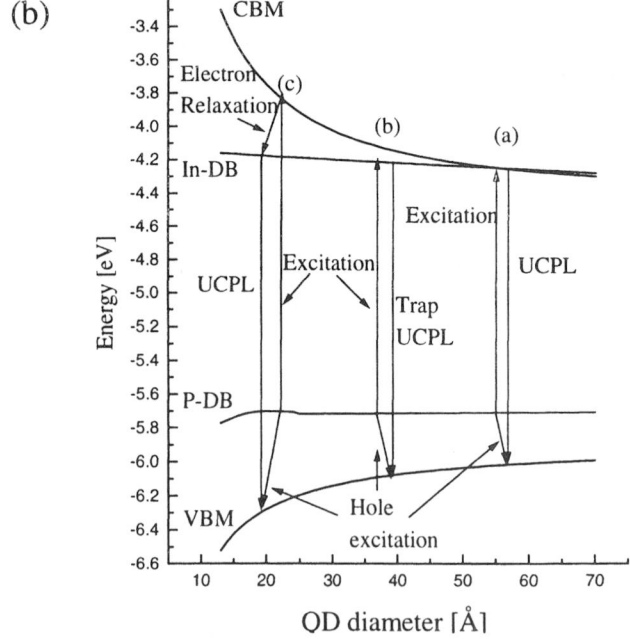

UCPL process would not show such a temperature dependence. (2) The UCPL occurs at very low light intensity, and the lifetime of the surface state would have to be in the millisecond region to permit a stepwise two-photon process; the lifetime of the trap emission was measured to be in the ns ranger. (3) Although the energy required to upconvert the hole is in the range of 300 meV, which is large compared to the bulk phonon energies of InP and might imply the need for many phonons, surface phonons associated with hydrogen-bonding to P surface atoms can be as large as 300 meV [56]; thus, only one or two phonons may be required for the upconversion if hydrogen or equivalent type of bonding is associated with the P-DB (such bonding could arise from the chemical treatment processes for the QD colloids). (4) Because the UCPL line shape does not show a peak but rather a continuous rise, this means the In-DP state has an energy distribution and the Boltzmann factor for the upconverted hole population in the UVB is not determined by the 300-meV gap. (5) Phonon localization into surface defects is a known process [57] and may contribute to the high relative efficiency of the UCPL process.

Recent ODMR [12] and EPR experiments [48] support the model for UCPL discussed earlier. The EPR results show that a nonradiative, permanent hole trap near the valence band develops at the surface of InP QDs after they have been exposed to light and aged. The EPR signal from this hole trap is removed upon electron injection into the QDs from sodium biphenyl and the UCPL is also quenched. The EPR signal from the hole trap is also absent in freshly prepared InP QDs, as is UCPL. The EPR results also show that an electron trap at the surface is present in untreated InP QDs and that this trap is removed by HF treatment; this result is consistent with the ODMR results showing surface electron traps attributed to phosphorus vacancies.

D. Photoluminescence Blinking

Fluorescence intermittency (PL blinking) in single QDs has been observed in both colloidal nanocrystals [58–64] and in epitaxially grown quantum dots [65–67]. The effect is manifested as intermittent photoluminescence with the time between light emission being on and off, varying from 10 ms to 100 s; the

---

Figure 12 (a) Plot of conduction-band minimum (CBM), valence-band maximum (VBM), In dangling bond (In-DB), and P dangling bond (P-DB) energy levels as a function of QD size (solid lines). Data points are the sum of the experimental anti-Stokes shift plus the absolute calculated VBM energy; the data points agree with the predicted plot, which is equivalent to P-DB energy minus (CBM − In-DB). At 15 Å, the anti-Stokes shift is predicted to be zero, and this agrees with the experimental result. (From Ref. 51.)

intensity of the PL when it is on also varies. Typical results are shown in Fig. 13 for 30 Å InP QDs. The PL blinking kinetics was shown [64] to follow an inverse power law:

$$P(\tau) \propto \left(\frac{1}{\tau}\right)^m \tag{1}$$

where P is the probability density of on or off times, τ is the PL-on or PL-off time period, and $m \approx 1.5$ for the off periods and ≈ 2.0 for the on periods [64].

All experimental studies [58–61,63–67] and models [58,68–70] of PL blinking in QDs invoke photoionization, whereby an electron is ejected from the QD, leaving it charged and nonemissive; the return of the electron back to the QD to neutralize it turns the PL back on. Although photoionization is

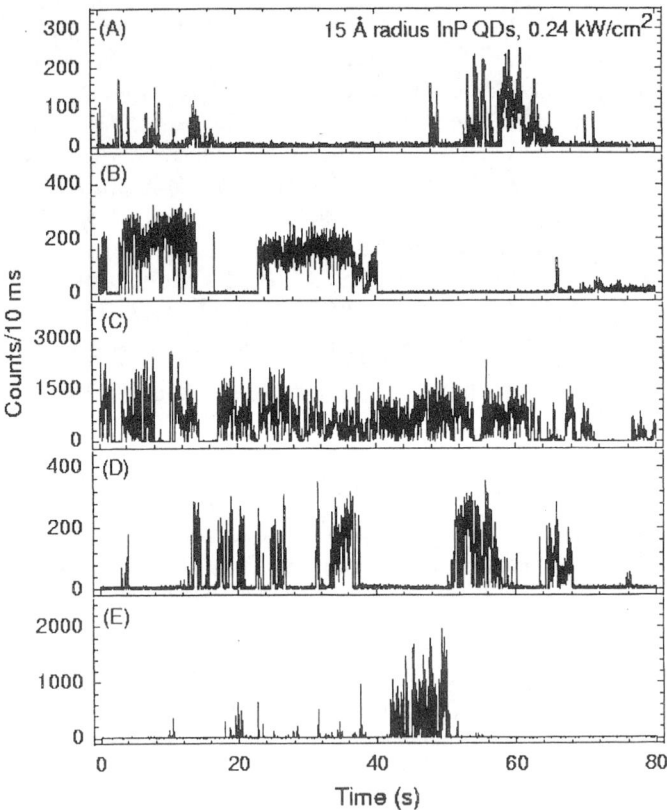

Figure 13 Photoluminescence intermittency (blinking) at 298 K in five single InP QDs (mean diameter 30 Å). (From Ref. 64.)

accepted as the underlying mechanism for blinking, there is still uncertainty about how the electrons leave the QD and where they go and reside before returning. The results of Ref. 64 and the power law of Eq. (1) support a model wherein the electrons leave by quantum mechanical tunneling (possibly Auger assisted) through the potential barrier at the surface, that the potential barrier fluctuates in height or width to affect the tunneling rate by five orders of magnitude, and that the external trap states to which the electrons transfer have an energy distribution and are relatively far from the surface of the QD core. A critical feature of this model that is generally accepted is that the local environment around the QD fluctuates and is itself affected by the photoionization. Further work is required to understand the details of PL blinking with greater certainty.

Unusual PL blinking has been reported in strain-induced S-K GaAs QDs created from GaAs/AlGaAs QWs with InP stressor islands [71]. For a sample with 140-nm-diameter InP stressor islands sitting on top of an outer 100-Å barrier of $Al_{0.3}Ga_{0.7}As$ and a 30-Å GaAs QW beneath the barrier, the survey PL spectrum of the strain-induced QDs (SIQDs) following excitation at 488 nm (shown in Fig. 14) exhibits a number of well-defined peaks. The small peak at 1.92 eV arises from the $Al_{0.3}Ga_{0.7}As$ barriers. The peak at 1.75

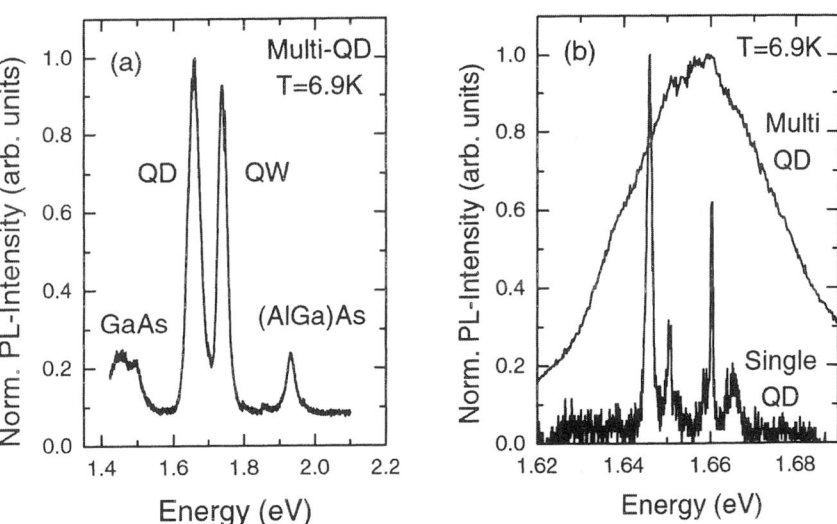

Figure 14 Photoluminescence spectra at 6.9 K of Stranski–Krastinow strain-induced GaAs QDs. The left panel shows the PL spectra from a large number of strain-induced QDs and includes peaks arising from the AlGaAs substrate, the GaAs QW, and the GaAs barrier; the right panel compares the broad PL spectrum from a large number of QDs with that from a single strain-induced GaAs QD. (From Ref. 71.)

eV stems from recombining carriers confined in the unmodulated region of the near-surface QW, whereas the peak at 1.65 is due to recombination of carriers in the SIQDs. Due to the strain imposed on the QW region under the InP S-K islands, the QW is expanded, leading to a decrease of the bandgap energy and the formation of SIQDs in the QW region. Luminescence from the GaAs substrate and/or the buffer layer appears at and below 1.5 eV. The size and shape distribution of the stressors leads to a corresponding distribution of quantized states within the SIQDs; this results in a rather broad luminescence line. The full width at half-maximum (FWHM) of the SIQD peak is 43.5 meV. The QW peak has a FWHM of 32.8 meV, which we believe is due to thickness fluctuations within the QW.

The spectrum of a single SIQD at low excitation intensity is compared to the multidot spectrum in Fig. 14b. The single-dot spectrum consists of several lines, two of which are most prominent and very narrow. The FWHM of the lowest-energy line at 1.645 eV is 1.6 meV, and the single most intense higher-energy peak, at 1.66 eV, has a FWHM of 0.6 meV. The emission energies of the QD fall well within the range shown by the averaged spectrum for emission from multiple QDs. It is unlikely that the single SIQD under investigation was actually a closely spaced double dot, because the area from which the PL for this spectrum is collected is much smaller than the imaged size of a single SIQD. Thus, the higher-energy peaks are due to recombination from carriers occupying excited states of the same SIQD.

The PL of single SIQDs shows blinking of an unusual type. Not only does the ground-state PL peak blink but also the excited state shows a temporal intensity modulation, which is phase shifted to the ground state by 180°. This effect is called two-color blinking (TCB).

To illustrate the TCB, a series of spectra from a single SIQD under constant experimental conditions were taken consecutively about every second without changing any parameter. The result is shown in Fig. 15, where the right-hand panel shows spectra at four different times as indicated and the left side shows a gray-scale plot of the intensity as a function of energy and time.

In Fig. 15b, all spectra have been vertically offset to enhance clarity. The first spectrum shows the initial luminescence detected from the SIQD with the main peak at 1.656 eV (FWHM = 1.27 meV) and a small peak at lower energy, around 1.649 eV (FWHM = 1.1 meV). The main peak at this time clearly dominates the spectrum. At a later time (10 s), the intensity ratio of the two peaks is reversed and the low-energy peak is now much stronger than the high-energy line. At around 15 s, the higher-energy peak is again much stronger than the low-energy luminescence, and then the intensities are reversed again at around 20 s.

Figure 15a gives a gray-scale representation of the whole set of data as measured for this SIQD; a spectral shift in the luminescence is not observed.

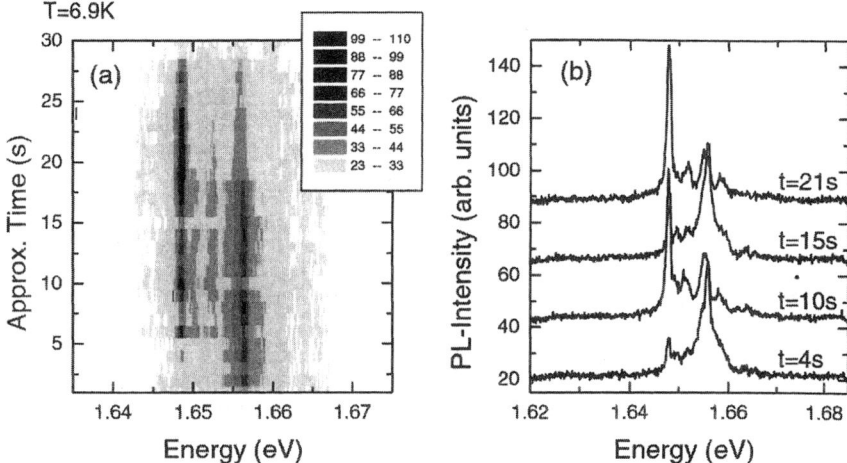

Figure 15 Two-color PL blinking in Stranski–Krastinow strain-induced GaAs QDs. (From Ref. 71.)

Figure 15 clearly shows the TCB as a beating of the luminescence intensity between the ground state and excited state, respectively. A detailed analysis of the spectra reveals that the overall intensity of the PL emission of that particular QD is almost constant over the time period the TCB is observed.

A potential model to explain TCB is shown in Fig. 16. The 488-nm excitation is absorbed preferentially in the barrier regions surrounding the SIQD, and electrons and holes from the barrier are readily trapped into the GaAs SIQD to produce band filling. When both of the first two radiative states are filled, PL can occur from both levels. Photoionization via an Auger process involving the lowest state will quench the PL from that state. The lowest state can be refilled with electrons from the higher state, thus turning on the lower-energy emission and turning off the higher-energy emission. Repopulation of the upper state by electron trapping in the SIQD from the barrier turns the higher-energy PL back on.

E. Anomaly in the Photoluminescence Excitation Spectra

Photoluminescence excitation spectra measure the emitted PL intensity detected at a fixed wavelength as a function of the excitation wavelength. If the photogenerated carriers relax efficiently to their lowest-energy levels and do not follow other parallel radiative or nonradiative channels from their higher excited states (hot carriers), then the PLE spectrum should have the same spectral shape as the absorption spectrum. This is generally true for

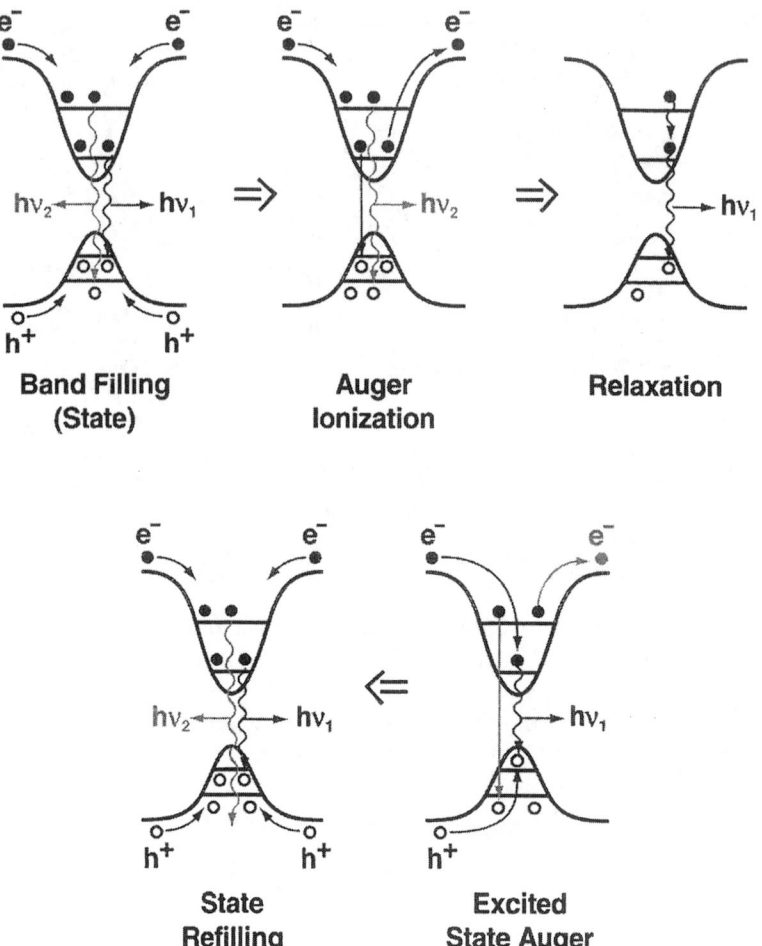

Figure 16 Model to explain two-color PL blinking in Stranski–Krastinow strain-induced QDs where the barrier is excited.

molecules and follows from Kasha's rule [72] that states that the photoexcited system always relaxes to its lowest excited state before emission and, hence, the emission energy is independent of the excitation energy. However, QDs show a deviation of the PLE spectra from the absorption spectra at excitation energies above the first excitonic transition; this has been reported for CdSe QDs [73] and InP QDs [74]. The anomaly is shown in Fig. 17 for InP QDs

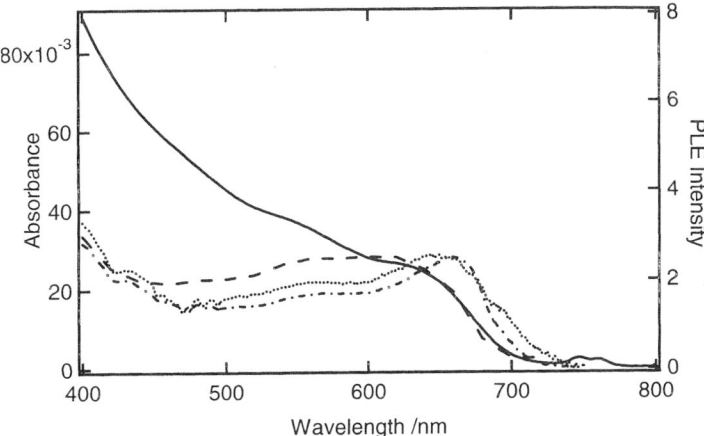

Figure 17 Comparison of absorption (curve i) and PLE spectra (curves ii, iii, and iv) of 38 Å InP QDs at 298 K. The PLE for curve ii is detecting emission at 675 nm; PLE curve iii is the total integrated emission over all wavelengths; and PLE curve iv is emission integrated under PL spectra taken at 20-nm intervals between 400 and 660 nm. (From Ref. 74.)

where the absorption spectrum of 38-Å QDs is shown along with PLE spectra obtained by three different procedures (described in the caption). Whereas the absorption spectrum continues to rise at higher photon energies above the first excitonic peak at about 630 nm, the PLE spectra saturate at energies above the first excitonic transition and do not follow the shape of the absorption spectrum. Pump-probe transient absorption measurements (exciting well above the bandgap and probing at the band edge) also show that for InP QDs, the electrons (or holes) photogenerated at energies above the first excitonic transition do not appear at the band edge in proportion to their photogenerated density [75]. Thus, it appears another nonradiative channel opens for the higher-energy carriers besides relaxation to the band edge followed by radiative emission. The nature of this new channel is presently not understood. However, recent studies indicate that the PLE anomaly is not intrinsic in QDs and can be eliminated under certain conditions [76].

IV. RELAXATION DYNAMICS OF PHOTOGENERATED CARRIERS IN QDs

When the photon energy absorbed in semiconductor QDs is greater than the lowest-energy excitonic transition (frequently termed "the QD bandgap," although this term is not rigorously correct because of electron–hole interactions), photogenerated electrons and holes (usually in the form of excitons) are created with excess energy above the lowest exciton energy; these energetic carriers are termed "hot carriers" The fate of this excess energy can follow several paths: (1) It can be dissipated as heat through electron–phonon interactions or Auger processes as the carriers relax to their lowest state; (2) a second electron–hole pair can be created by the process of impact ionization if the excess energy is at least twice the QD bandgap; (3) the electrons and holes can separate and the excess energy can be converted to increased electrical free energy via a photovoltaic effect or stored as additional chemical free energy by driving more endoergic electrochemical reactions at the surface [11]. The efficiency of photon conversion devices, such as photovoltaic and photoelectrochemical cells, can be greatly increased if path 2 or 3 can dominate over path 1. Path 1 is generally a fast process in bulk semiconductors that occurs in a few picoseconds or less if the photogenerated carrier density is less than about 5×10^{17} cm^{-3} [77–79]. The hot electron relaxation, or cooling time can be increased by two orders of magnitude in semiconductors quantum wells when the photogenerated carrier density is increased above about 5×10^{18} cm^{-3} by a process termed "hot phonon bottleneck" [77,79,80]. QDs are intriguing because it is believed that slow cooling of energetic electrons can occur in QDs at low photogenerated carrier densities [81–87], specifically at light intensities corresponding to the solar insolation on Earth.

The first prediction of slowed cooling at low light intensities in quantized structures was made by Boudreaux et al. [81]. They anticipated that cooling of carriers would require multiphonon processes when the quantized levels are separated in energy by more than phonon energies. They analyzed the expected slowed cooling time for hot holes at the surface of highly doped n-type TiO$_2$ semiconductors, where quantized energy levels arise because of the narrow space charge layer (i.e., depletion layer) produced by the high doping level. The carrier confinement is this case is produced by the band bending at the surface; for a doping level of 1×10^{19} cm^{-3} the potential well can be approximated as a triangular well extending 200 Å from the semiconductor bulk to the surface and with a depth of 1 eV at the surface barrier. The multiphonon relaxation time was estimated from

$$\tau_c - \omega^{-1}\exp\left(\frac{\Delta E}{kT}\right) \tag{2}$$

where τ_c is the hot carrier cooling time, ω is the phonon frequency, and ΔE is the energy separation between quantized levels. For strongly quantized electron levels, with $\Delta E > 0.2$ eV, τ_c could be > 100 ps according to Eq. (2).

However, carriers in the space-charge layer at the surface of a heavily doped semiconductor are only confined in one dimension, as in a quantum film. This quantization regime leads to discrete energy states which have dispersion in k-space [88]. This means the hot carriers can cool by undergoing interstate transitions that require only one emitted phonon followed by a cascade of single-phonon intrastate transitions; the bottom of each quantum state is reached by intrastate relaxation before an interstate transition occurs. Thus, the simultaneous and slow multiphonon relaxation pathway can be bypassed by single-phonon events, and the cooling rate increases correspondingly.

More complete theoretical models for slowed cooling in QDs have been proposed recently by Bockelmann and co-workers [86,89] and Benisty and co-workers [85,87]. The proposed Benisty mechanism [85,87] for slowed hot carrier cooling and a phonon bottleneck in QDs requires that cooling only occurs via longitudinal optical (LO) phonon emission. However, there are several other mechanism by which hot electrons can cool in QDs. Most prominent among these is the Auger mechanism [90]. Here, the excess energy of the electron is transferred via an Auger process to the hole, which then cools rapidly because of its larger effective mass and smaller energy level spacing. Thus, an Auger mechanism for hot electron cooling can break the phonon bottleneck [90]. Other possible mechanisms for breaking the phonon bottleneck include electron–hole scattering [91], deep level trapping [92], and acoustical–optical phonon interactions [93,94].

A. Experimental Determination of Relaxation/Cooling Dynamics and a Phonon Bottleneck in Quantum Dots

Over the past several years, many investigations have been published that explore hot electron cooling/relaxation dynamics in QDs and the issue of a phonon bottleneck in QDs [11]. The results are controversial, and it is quite remarkable that there are so many reports that both support [11,95–109] and contradict [11,92,110–122] the prediction of slowed hot electron cooling in QDs and the existence of a phonon bottleneck. One element of confusion that is specific to the focus of this chapter is that although some of these publications report relatively long hot electron relaxation times (tens of picoseconds) compared to what is observed in bulk semiconductors, the results are reported as being not indicative of a phonon bottleneck because the relaxation times are not excessively long and PL is observed [123–125] (theory predicts infinite relaxation lifetime of excited carriers for the extreme, limiting condition of a phonon bottleneck; thus, the carrier lifetime would be determined by nonradiative processes and PL would be absent). However, because the interest here is on the relative rate of relaxation/cooling compared to the rate of electron transfer, slowed relaxation/cooling of carriers can be considered to occur in QDs if the relaxation/cooling times are greater than

10 ps (about an order of magnitude greater than that for bulk semiconductors). This is because previous work that measured the time of electron transfer from bulk III–V semiconductors to redox molecules (metallocenium cations) absorbed on the surface found that ET times can be subpicoseconds to several picoseconds [36,126–128]; hence, photoinduced hot ET can be competitive with electron cooling and relaxation if the latter is greater than tens of picoseconds.

In a series of articles, Sugawara et al. [97–100] have reported slow hot electron cooling in self-assembled InGaAs QDs produced by Stranski–Krastinow growth on lattice-mismatched GaAs substrates. Using time-resolved PL measurements, the excitation power dependence of PL, and the current dependence of electroluminescence spectra, these researchers report cooling times ranging from 10 ps to 1 ns. The relaxation time increased with electron energy up to the fifth electronic state. Also, an extensive review of phonon bottleneck effects in QDs was recently published, which concludes that the phonon bottleneck effect is indeed present in QDs [42].

Gfroerer et al. report slowed cooling of up to 1 ns in strain-induced GaAs QDs formed by depositing tungsten stressor islands on a GaAs QW with AlGaAs barriers [109]. A magnetic field was applied in these experiments to sharpen and further separate the PL peaks from the excited-state transitions, and thereby determine the dependence of the relaxation time on level separation. The authors observed hot PL from excited states in the QD which could only be attributed to slow relaxation of excited (i.e., hot) electrons. Because the radiative recombination time is about 2 ns, the hot electron relaxation time was found to be of the same order of magnitude (about 1 ns). With higher excitation intensity sufficient to produce more than one electron–hole pair per dot, the relaxation rate increased.

A lifetime of 500 ps for excited electronic states in self-assembled InAs/GaAs QDs under conditions of high injection was reported by Yu et al. [104]. PL from a single GaAs/AlGaAs QD [107] showed intense high-energy PL transitions which were attributed to slowed electron relaxation in this QD system. Kamath et al. [108] also reported slow electron cooling in InAs/GaAs QDs.

Quantum dots produced by applying a magnetic field along the growth direction of a doped InAs/AlSb QW showed a reduction in the electron relaxation rate from 10^{12} to 10^{10} s^{-1} [99].

In addition to slow electron cooling, slow hole cooling was reported by Adler et al. [105,106] in S-K InAs/GaAs QDs. The hole relaxation time was determined to be 400 ps based on PL rise times, whereas the electron relaxation time was estimated to be less than 50 ps. These QDs only contained one electron state, but several hole states; this explained the faster electron cooling time because a quantized transition from a higher quantized electron state to the ground electron state was not present. Heitz et al. [101] also report relaxation times for holes of about 40 ps for stacked layers of S-K InAs QDs deposited on GaAs; the InAs QDs are overgrown with GaAs and the QDs in each layer self-assemble into an ordered column. Carrier cooling in this system is about two orders of magnitude slower than in higher-dimensional structures.

All of the above studies on slowed carrier cooling were conducted on self-assembled S-K type of QDs. Studies of carrier cooling and relaxation have also been performed on II–VI CdSe colloidal QDs by Klimov et al. [116,129] and Guyot-Sionnest et al. [95] and on InP QDs by Ellingson et. al. [130], and Blackburn et al. [131]. The Klimov group first studied electron relaxation dynamics from the first-excited $1P$ to the ground $1S$ state using interband pump-probe spectroscopy [116]. The CdSe QDs were pumped with 100-fs pulses at 3.1 eV to create high-energy electrons and holes in their respective band states and then probed with femtosecond white-light continuum pulses. The dynamics of the interband bleaching and induced absorption caused by state filling was monitored to determine the electron relaxation time from the $1P$ to the $1S$ state. The results showed very fast $1P$ to $1S$ relaxation, on the order of 300 fs, and was attributed to an Auger process for electron relaxation, which bypassed the phonon bottleneck. However, this experiment cannot separate the electron and hole dynamics from each other. Guyot-Sionnest et al. [95] followed up these experiments using femtosecond infrared pump-probe spectroscopy. A visible pump beam creates electrons and holes in the respective band states and a subsequent infrared (IR) beam is split into an IR pump and an IR probe beam; the IR beams can be tuned to monitor only the intraband transitions of the electrons in the electron states and, thus, can separate electron dynamics from hole dynamics. The experiments were conducted with CdSe QDs that were coated with different capping molecules (TOPO, thiocresol, and pyridine), which exhibit different hole trapping kinetics. The rate of hole trapping increased in the order: TOPO, thiocresol, and pyridine. The results generally show a fast relaxation component (1–2 ps) and a slow relaxation component (~200 ps). The relaxation times follow the hole trapping ability of the different capping molecules and are longest for the QD systems having the fastest hole trapping caps; the slow component dominates the data for the pyridine cap, which is attributed to its faster hole trapping kinetics.

These results [95] support the Auger mechanism for electron relaxation, whereby the excess electron energy is rapidly transferred to the hole, which then relaxes rapidly through its dense spectrum of states. When the hole is rapidly removed and trapped at the QD surface, the Auger mechanism for hot electron relaxation is inhibited and the relaxation time increases. Thus, in the above experiments, the slow 200-ps component is attributed to the phonon bottleneck, most prominent in pyridine-capped CdSe QDs, whereas the fast 1–2-ps component reflects the Auger relaxation process. The relative weight of these two processes in a given QD system depends on the hole trapping dynamics of the molecules surrounding the QD.

Klimov et al. further studied carrier relaxation dynamics in CdSe QDs and published a series of articles on the results [129,132]; a review of this work was also recently published [133]. These studies also strongly support the presence of the Auger mechanism for carrier relaxation in QDs. The experiments were done using ultrafast pump-probe spectroscopy with either two or three beams. In the former, the QDs were pumped with visible light across its bandgap (hole states to electron states) to produce excited-state (i.e., hot) electrons; the electron relaxation was monitored by probing the bleaching dynamics of the resonant HOMO to LUMO transition with visible light or by probing the transient IR absorption of the $1S$ to $1P$ intraband transition, which reflects the dynamics of electron occupancy in the LUMO state of the QD. The three-beam experiment was similar to that of Guyot-Sionnest et al. [95], except that the probe in the experiments of Klimov et al. is a white-light continuum. The first pump beam is at 3 eV and creates electrons and holes across the QD bandgap. The second beam is in the IR and is delayed with respect to the optical pump; this beam repumps electrons that have relaxed to the LUMO back up in energy. Finally, the third beam is a broadband white-light continuum probe that monitors photoinduced interband absorption changes over the range 1.2–3 eV. The experiments were done with two different caps on the QDs: a ZnS cap and a pyridine cap. The results showed that with the ZnS-capped CdSe, the relaxation time from the $1P$ to the $1S$ state was about 250 fs, whereas for the pyridine-capped CdSe, the relaxation time increased to 3 ps. The increase in the latter experiment was attributed to a phonon bottleneck produced by rapid hole trapping by the pyridine, as also proposed by Guyot-Sionnest et al. [95]. However, the timescale of the phonon bottleneck induced by hole trapping by pyridine caps on CdSe that were reported by Klimov et al. was not as great as that reported by Guyot-Sionnest et al. [95].

Similar studies of carrier dynamics have been made on InP QDs [75,130, 131]. It was found that whereas the electron relaxation time from the $1P$ to the $1S$ state was 350–450 fs for the cases where the photogenerated electrons and holes were confined to the core of a 42-Å InP QD, this relaxation time increases by about an order of magnitude to 3–4 ps when the hole was trapped at the surface by an effective hole trap such as sodium biphenyl. These results are consistent with the conclusion derived from studies of CdSe QDs that the phonon bottleneck is bypassed by an Auger cooling process, but if the Auger process is inhibited by rapidly removing the photogenerated holes from the QD core by trapping them on or near the QD surface, the electron cooling time can be slowed down significantly.

In contradiction to the above-discussed results, many other investigations exist in the literature in which a phonon bottleneck was apparently not observed. These results were reported for both self-organized S-K QDs [11,92,110–122] and II–VI colloidal QDs [116,118,120]. However, in several case [101,123,125], hot electron relaxation was found to be slowed, but not sufficiently to enable the authors to conclude that this was evidence of a phonon bottleneck. For the issue of hot electron transfer, this conclusion may not be relevant because in this case, one is not interested in the question of whether the electron relaxation is slowed so drastically that nonradiative recombination occurs and quenches photoluminescence, but rather whether the cooling is slowed sufficiently so that excited-state electron transport and transfer can occur across the semiconductor–molecule interface before cooling. For this purpose, the cooling time need only be increased above about 10 ps, because electron transfer can occur within this timescale [36,126–128].

The experimental techniques used to determine the relaxation dynamics in the above-discussed experiments showing no bottleneck were all based on time-resolved PL or transient absorption spectroscopy. The S-K QD system that were studied and exhibited no apparent phonon bottleneck include $In_xGa_{1-x}As/GaAs$ and $GaAs/AlGaAs$. The colloidal QD systems were CdSSe QDs in glass (750 fs relaxation time) [120] and CdSe [112]. Thus, the same QD systems studied by different researchers showed both slowed cooling and nonslowed cooling in different experiments. This suggest a strong sample-history dependence for the results; perhaps, the samples differ in their defect concentration and type, surface chemistry, and other physical parameters that affect carrier cooling dynamics. Much additional research is required to sort out these contradictory results.

V. QUANTUM-DOT ARRAYS

A major goal in semiconductor nanoscience is to form QD arrays and understand the transport and optical properties of these arrays. One approach to forming arrays of close-packed QDs is to slowly evaporate colloidal solutions of QDs; upon evaporation, the QD volume fraction increases and interaction between the QDs develops and leads to the formation of a self-organized QD film. Figure 18 shows a TEM micrograph of a monolayer made with InP QDs with a mean diameter of 49 Å and in which each QD is separated from its neighbors by TOPO/TOP capping groups; local hexagonal order is evient. Figure 19a shows the formation of a monolayer organized in a

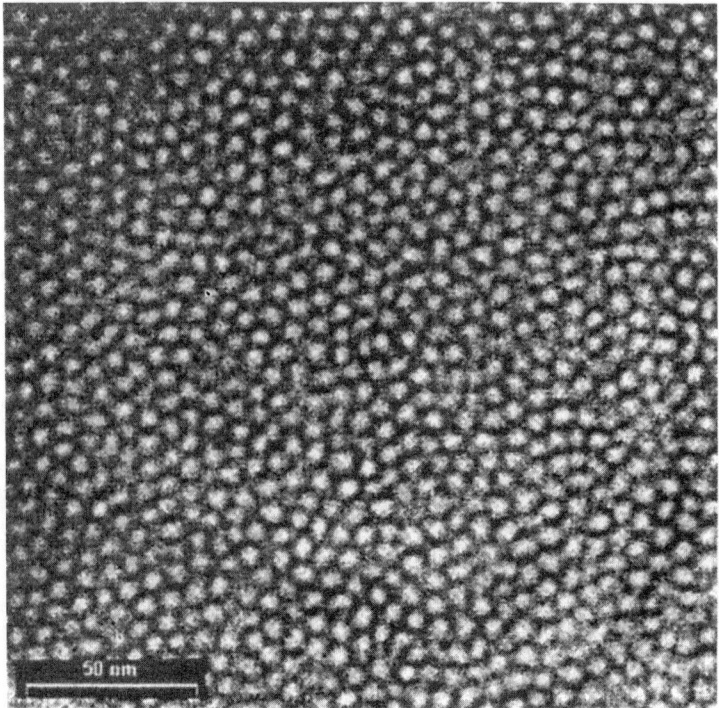

Figure 18   Transmission electron microscopic image of close-packed array of 49-Å InP QDs.

hexagonal network made with QDs 57 Å in diameter and which are capped with dodecanethiol; InP QDs capped with oleyamine can form monolayers with shorter-range hexagonal order. The QDs in these arrays have size distribution of about 10%, and with such a size distribution, the arrays can only exhibit local order. To form colloidal crystals with a high degree of order in the QD packing, the size distribution of the QD particles must have a mean deviation less than about 5% and uniform shape. Murray and co-workers [135] fabricated highly ordered 3D superlattices of CdSe QDs that have a size distribution of 3–4%. Figure 19b shows an ordered array of 60-Å-diameter InP QDs with multiple layers. A step in the TEM indicates a change in height of one monolayer.

     The critical parameters that control inter-QD electronic coupling, and hence carrier transport, include QD size, interdot distance, QD surface chemistry, the work function and dielectric properties of the matrix containing the

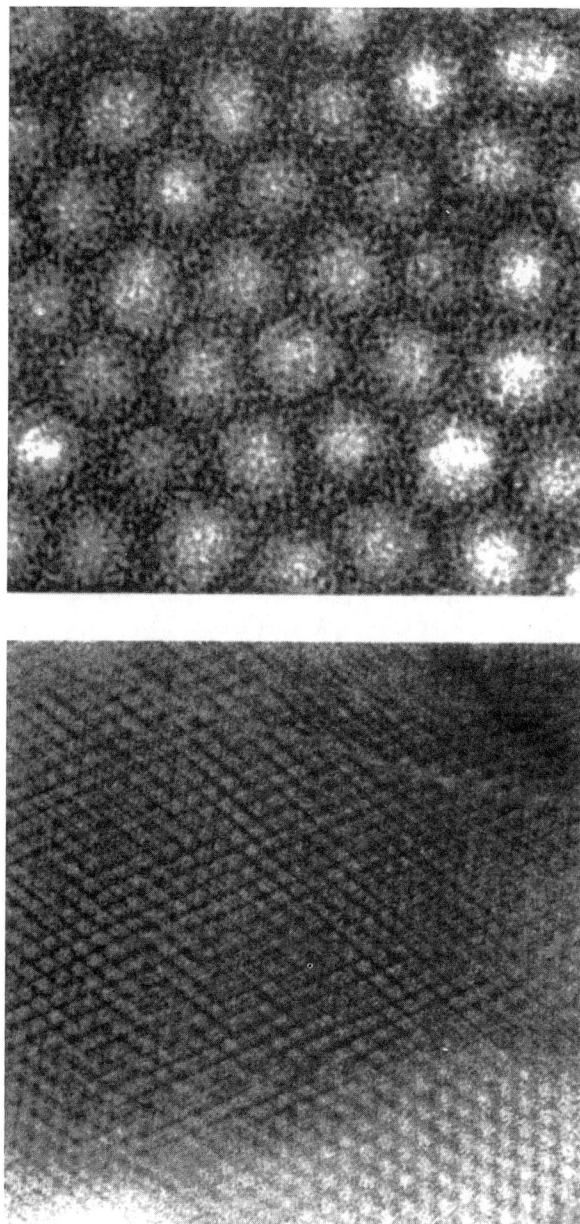

Figure 19 Transmission electron microscopic images of close-packed 3D array of 57-Å InP QDs showing hexagonal order. The bottom panel is at a lower magnification and shows a monolayer step between the darker and lighter regions. (TEM by S.P. Ahrenkiel.)

QDs, the nature of the QD capping species, QD orientation and packing order, uniformity of QD size distribution, and the crystallinity and perfection of the individual QDs in the array. Several studies of electronic coupling in colloidal QD arrays have been reported [37,134–138]. In these arrays, the semiconductor QD cores are surrounded with insulating organic ligands and create a large potential barrier between the QDs. Thus, the electrons and holes are confined to the QD, and very weak electronic communication exists between dots in such arrays. Measurements of the photoconductivity of close-packed films of colloidally prepared CdSe QDs [137] with diameters > 20 Å have shown that excitons formed by illumination are confined to individual QDs and electron transport through the array does not occur. This lack of electronic coupling between QDs is also seen from the fact that the absorption spectra are the same for both colloidal solutions and close-packed arrays.

However, arrays with very small CdSe QDs with a mean diameter of 16 Å show that significant electronic coupling between dots in close-packed solids can occur [138]. Recently, InP QDs with diameters of 15–23 Å were formed into arrays that also show evidence of electronic coupling [139]. This conclusion is based on the differences in the optical spectra of isolated colloidal QDs compared to solid films of QD arrays (see Fig. 20). For close-packed QD solids, a large red shift of the excitonic peaks in the absorption spectrum is expected if the electron or hole wave function extends outside the boundary of the individual QDs as a result of inter-QD electronic coupling.

The first indication of a difference in the absorption spectra between colloidal QD solutions and QD films appears for 32-Å InP QDs with an inter-QD spacing of 11 Å (QDs were capped with TOPO); a very small shift of about 15 meV was apparent (see Fig. 20a). Above 32 Å, no difference was observed between the spectra of QD colloidal solutions and solid films. Figures 20b and 20c show results for 18-Å-diameter InP QDs, where the interdot distance was varied between 9 and 18 Å by varying the length of a linear alkylamine molecule that was used as the colloidal stabilizer. These colloids were initially prepared in trioctylamine, which was then replaced with hexylamine (interdot distance ∼9 Å) or oleyamine (interdot distance ∼18) by a ligand-exchange method. The interdot distance for the different organic ligands was estimated from previous published data for close-packed arrays of nanocrystals capped with different organic ligands [140,141]. However, because hexylamine is a weak stabilizer with a relatively low boiling point, QDs may partially lose the organic ligand so that the interdot distance may be shorter.

Quantum-dot arrays with an interdot spacing of 9 Å have an absorption spectrum that is relatively smooth with only weak structure (Fig. 20b) and which shows a red shift for the first exciton of about 140 meV relative to the

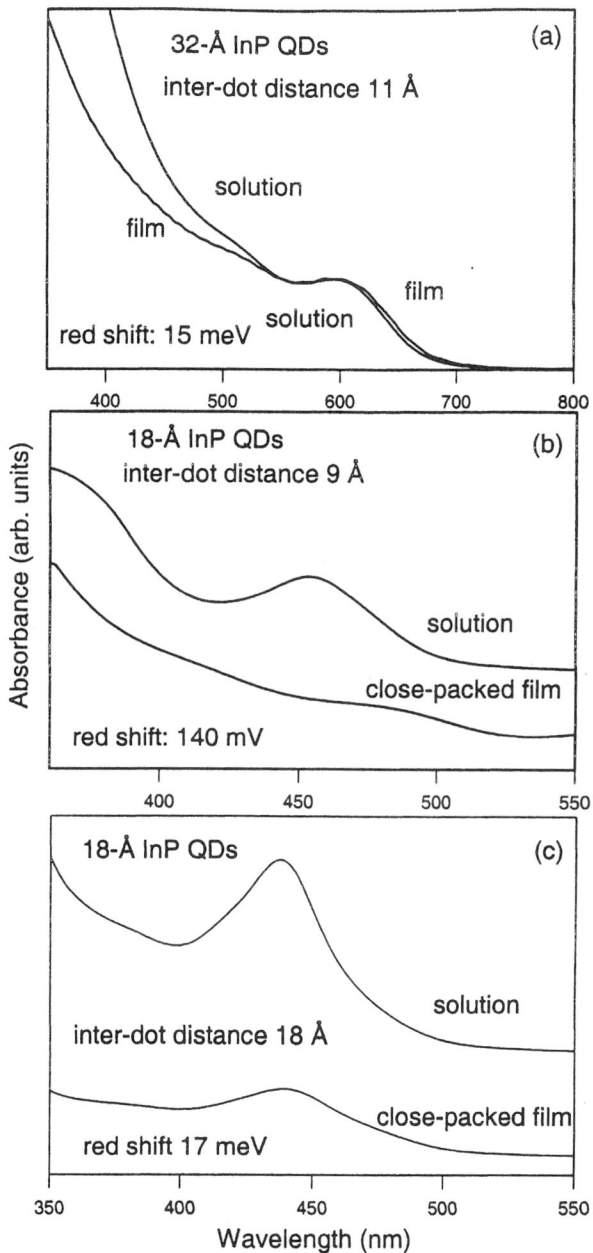

Figure 20 Evidence for inter-QD electronic coupling in small-sized InP QD arrays when the interdot distance is less than 20 Å. (From Ref. 139.)

QD solution. For an interdot distance of 18 Å, the red shift is much smaller at about 18 meV (Fig. 20c). When the films were redissolved into colloidal solution, the excitonic peak positions of the colloid solution were re-established. QDs could also contact each other through their oxide layers. However, the possibility of QDs contacting through strong chemical bonds or interdot fusion of QDs can be excluded because in that case, the solid film could not be redissolved to yield the same spectrum as that of the initial colloidal solution.

These results suggest that electronic coupling between QDs does take place and are similar to the results for close-packed solid arrays of CdSe QDs that were prepared with 16-Å-diameter dots [138]. As expected, it was found that the strength of the electronic coupling increases with decreasing QD diameter and decreasing interdot distance. When the interdot distance in solid QD arrays is large, the QDs maintain their individual identity and their isolated electronic structure, and the array behaves as an insulator. Quantum mechanical coupling becomes important when the potential barrier and distance between the dots is small [142–145]. A recent theoretical study on Si QDs showed that for small interdot distances in a perfect superlattice and also in disordered arrays, one can expect the formation of delocalized, extended states (minibands) from the discrete set of electron and/or hole levels present in the individual QDs [145]. This effect is similar to the formation of minibands in a one-dimensional superlattice of quantum wells [37]. Randomly arranged QDs in a disordered array show the coexistence of both discrete (localized) and bandlike (delocalized) states [145], and transitions are possible from completely localized electron states to a mixture of localized and delocalized states.

Long-range energy transfer between QDs in an array has also been observed in close-packed CdSe QD arrays [146–148] and in close-packed InP QD arrays [134]; multilayer films were optically transparent and the QDs were randomly ordered. Energy transfer from small QDs to large QDs was observed. The absorption spectra of uncoupled QDs in colloidal solution are virtually identical to those from a QD solid film formed from the solution. The peaks of the emission spectra of the close-packed films are red-shifted with respect to the QD solution spectra. These observations suggest that energy transfer occurs within the inhomogeneous distribution of the QDs in the solid film. The observed red shift, together with a narrowing of the emission spectra, becomes more prominent for samples with a broadened size distribution.

Energy transfer between close-packed QDs in a mixed system consisting of 20% larger dots (37 Å) and 80% smaller dots (28 Å) was also investigated. The absorption and emission spectra of the two QD samples with different sizes are presented in the bottom two plots in Fig. 21 and are labeled A

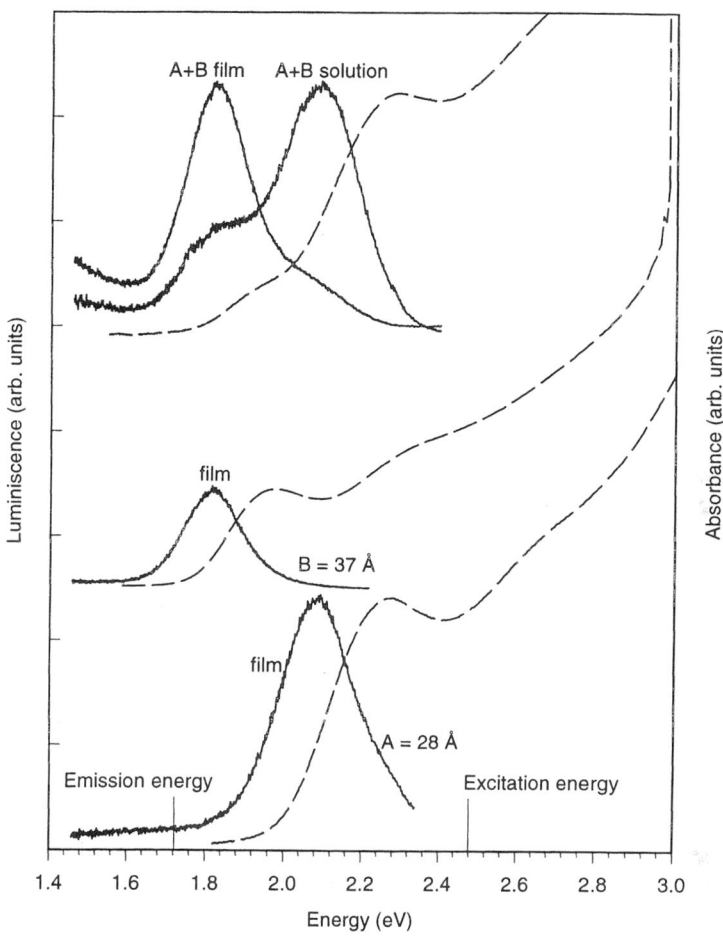

Figure 21 Absorption (dashed lines) and emission (solid lines) spectra of InP QDs with two sizes (28 and 37 Å). The bottom two curves show these spectra for QDs of a single size in close-packed films and the top curves show these spectra when the two sizes are mixed both in a close-packed film and as colloidal solutions. (From Ref. 134.)

(donor) (28 Å) and B (acceptor) (37 Å). The absorption spectra (dashed lines) for the single-sized QD solution or the QD film are identical. The blue shifts in the absorption and emission spectra for the smaller QD size are as expected. The curves at the top of Fig. 21 shows the absorption and emission spectra of the two QD sizes when they are mixed in solution and when mixed in the solid QD film. In the mixed QD solution, the emission spectrum is the sum of the emission spectra of the two individual QD sizes; the absorption spectrum of the solution or the QD film is also the sum of the individual spectra. However, for the mixed-QD film, the emission is primarily from the larger (37 Å) QDs; that is, emission from the small (28 Å) QDs decreases and that from the 37-Å QDs increases when the two sizes are mixed in the close-packed film. Photo-luminescence excitation spectra also show that the emission from the large QDs originates from excitation of both small and large QDs. These results strongly indicate that energy transfer occurs from the higher-energy states in the small QDs to the lower-energy states in the large QDs.

A series of additional PL experiments were conducted to directly measure the distance, R_0, at which the rate of energy transfer becomes equal to the rate of other mechanisms for nonradiative quenching of the excitation energy in the QD. In these experiments, an inert molecular diluent is added to the QD films to produce a known average separation between large (40 Å) and small (28 Å) QDs; the diluent was heptamethylnonane (HMN). A known amount of HMN was added to a hexane solution of the QDs, and the hexane was then evaporated to produce thin QD films containing the HMN. In all experiments, the absorbance of the small and large QDs was 0.01 at the first excitation maxima. The distance, R, between the small (donor) and large (acceptor) QDs was varied by varying the HMN concentration; it could be calculated from the total QD volume and the HMN volume. The TOPO capping layer (11 Å) was also taken into account.

Figure 22 shows that when the separation between the QDs decreases from large values associated with high concentrations of the HMN diluent to small values in the close-packed film, the PL emission from the large QDs (acceptor) increases while the emission for the small QDs (donor) decreases. The emission intensities for the small and large QDs do not change for $R >$ 108 Å in diluted films, and in close-packed films for $R < 49$ Å with TOPO caps and $R < 40$ Å with pyridine caps.

In Fig. 23, we plot the R dependence of (1) the emission intensity of the 40 Å acceptor QDs (after subtracting the emission in the most diluted sample) divided by that of the acceptor QDs in the close-packed film and (2) the emission intensity of the 28-Å donor QDs (after subtracting the emission of the close-packed film) divided by the intensity at high HMN dilution. The acceptor QD emission increases with decreasing R, and the donor emission decreases with decreasing R. Because R_0 is the separation distance at which the rate of energy transfer becomes equal to the rate of excitation energy quenching by other mechanisms, it can be estimated from Fig. 23 by equating it to the R value at which 50% of the emission intensity of the donor is transferred to the acceptor (i.e., where the two curves cross). This yields $R_0 =$ 83 Å, indicating efficient long-range energy transfer.

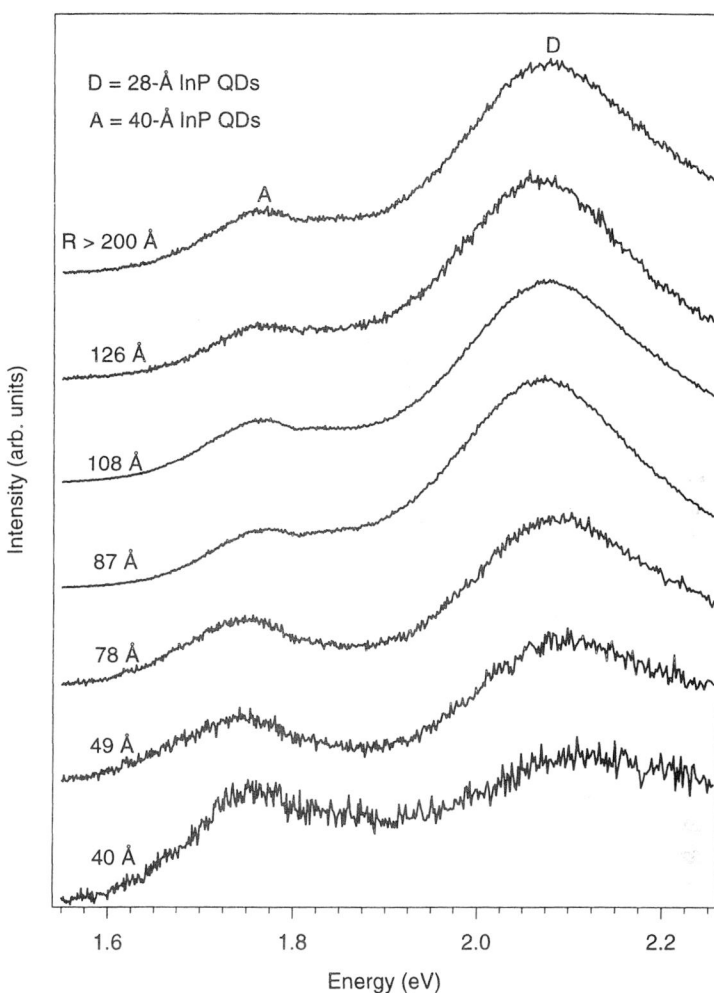

Figure 22 Photoluminescence spectra of films of InP QDs with two sizes (28 and 40 Å) for different average distances between QDs adjusted by the addition of hepta-methylnonane to the films. (From Ref. 134.)

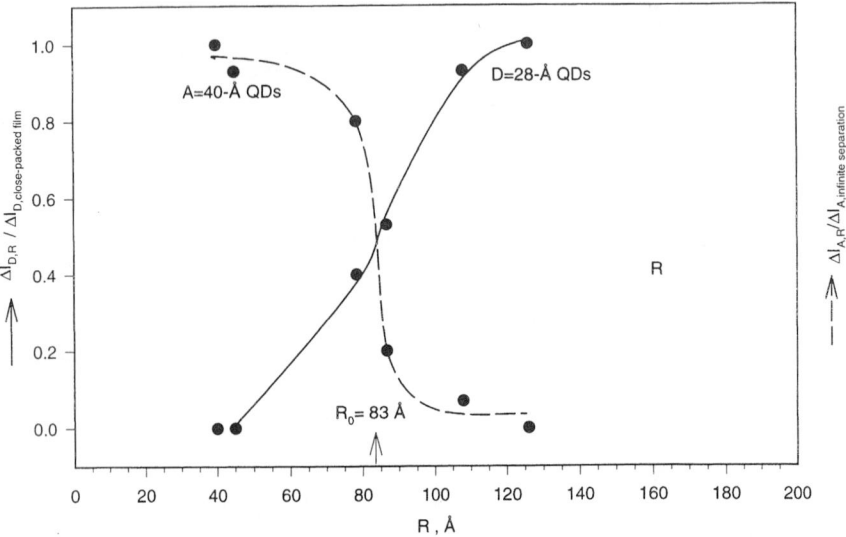

Figure 23   Change of the PL intensity in a film of 28- and 40-Å InP QDs as a function of separation between the QDs. (From Ref. 134.)

VI.   APPLICATIONS: QUANTUM-DOT SOLAR CELLS

A.   Background

The maximum thermodynamic efficiency for the photovoltaic conversion of unconcentrated solar irradiance into electrical free energy in the radiative limit assuming detailed balance and a single threshold absorber was calculated by Shockley and Queissar in 1961 [149] to be about 31%; this analysis is also valid for the conversion to chemical free energy [150,151]. Because conversion efficiency is one of the most important parameters to optimize for implementing photovoltaic and photoelectrochemical cells on a truly large scale [152], several schemes for exceeding the Shockley–Queissar (S-Q) limit have been proposed and are under active investigation. These approaches include tandem cells [153], hot carrier solar cells [37,81,154], solar cells producing multiple electron-hole pairs per photon through impact ionization [155–157], multiband and impurity solar cells [152,158], and thermophotovoltaic/thermophotonic cells [152]. Here, we will only discuss hot carrier and impact ionization solar cells and the effects of size quantization on the carrier dynamics that control the probability of these processes.

The solar spectrum contains photons with energies ranging from about 0.5 to 3.5 eV. Photons with energies below the semiconductor bandgap are not absorbed, whereas those with energies above the bandgap create hot electrons and holes with a total excess kinetic energy equal to the difference between the photon energy and the bandgap. The initial temperature can be as high as 3000 K, with the lattice temperature at 300 K.

A major factor limiting the conversion efficiency in single-bandgap cells to 31% is that the absorbed photon energy above the semiconductor bandgap is lost as heat through electron–phonon scattering and subsequent phonon emission, as the carriers relax to their respective band edges (bottom of conduction band for electrons and top of valence for holes). The main approach to reducing this loss in efficiency has been to use a stack of cascaded multiple *p–n* junctions with bandgaps better matched to the solar spectrum; in this way, higher-energy photons are absorbed in the higher-bandgap semiconductors and lower-energy photons in the lower-bandgap semiconductors, thus reducing the overall heat loss due to carrier relaxation via phonon emission. In the limit of an infinite stack of band gaps perfectly matched to the solar spectrum, the ultimate conversion efficiency at one sun intensity can increase to about 66%.

Another approach to increasing the conversion efficiency of photovoltaic cells by reducing the loss caused by the thermal relaxation of photogenerated hot electrons and holes is to utilize the hot carriers before they relax to the band edge via phonon emission [37]. There are two fundamental ways to utilize the hot carriers for enhancing the efficiency of photon conversion. One way produces an enhanced photovoltage, and the other way produces an enhanced photocurrent. The former requires that the carriers be extracted from the photoconverter before they cool [81,154], whereas the latter requires the energetic hot carriers to produce a second (or more) electron-hole pair through impact ionization [155,156]—a process that is the inverse of an Auger process whereby two electron–hole pairs recombine to produce a single highly energetic electron–hole pair. In order to achieve the former, the rates of photogenerated carrier separation, transport, and interfacial transfer across the contacts to the semiconductor must all be fast compared to the rate of carrier cooling [81–83,159]. The latter requires that the rate of impact ionization (i.e., inverse Auger effect) be greater than the rate of carrier cooling and other relaxation processes for hot carriers.

Hot electrons and hot holes generally cool at different rates because they generally have different effective masses; for most inorganic semiconductors, electrons have effective masses that are significantly lighter than holes and, consequently, cool more slowly. Another important factor is that hot carrier cooling rates are dependent on the density of the photogenerated hot carriers (viz. the absorbed light intensity) [77,79,160]. Here, most of the dynamical effects we will discuss are dominated by electrons rather than holes; therefore, we will restrict our discussion primarily to the relaxation dynamics of photo-generated electrons.

For QDs, one mechanism for breaking the phonon bottleneck that is predicted to slow carrier cooling in QDs and hence allow fast cooling is an Auger process. Here, a hot electron can give its excess kinetic energy to a thermalized hole via an Auger process, and then the hole can then cool quickly because of its higher effective mass and more closely spaced quantized states. However, if the hole is removed from the QD core by a fast hole trap at the surface, then the Auger process is blocked and the phonon bottleneck effect can occur, thus leading to slow electron cooling. This effect was first shown for CdSe QDs [95,131]; it has now also been shown for InP QDs, where a fast hole trapping species (Na biphenyl) was found to slow the electron cooling to about 3–4 ps [130,131]. This is to be compared to the electron cooling time of 0.3 ps for passivated InP QDs without a hole trap present and, thus, where the holes are in the QD core and able to undergo an Auger process with the electrons [130,131].

B. Quantum-Dot Solar Cell Configurations

The two fundamental pathways for enhancing the conversion efficiency (increased photovoltage [81,154] or increased photocurrent [155,156]) can be accessed, in principle, in three different QD solar cell configurations; these configurations are shown in Fig. 24 and they are described in the following subsections. However, it is emphasized that these potential high-efficiency configurations are speculative and there is no experimental evidence yet that demonstrates actual enhanced conversion efficiencies in any of these systems.

1. Photoelectrodes Composed of Quantum-Dot Arrays

In this configuration, the QDs are formed into an ordered 3D array with inter-QD spacing sufficiently small such that strong electronic coupling occurs and minibands are formed to allow long-range electron transport (see Fig. 24a). The system is a 3D analog to a 1D superlattice and the miniband structures formed therein [37]. The delocalized quantized 3D miniband states could be expected to slow the carrier cooling and permit the transport and collection of hot carriers to produce a higher photopotential in a photovoltaic (PV) cell or in a photoelectrochemical cell where the 3D QD array is the photoelectrode [161]. Also, impact ionization might be expected to occur in the QD arrays, enhancing the photocurrent (see Fig. 25). However, hot electron transport/collection and impact ionization cannot occur simultaneously; they are

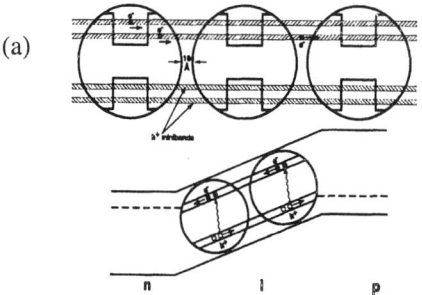

(a)

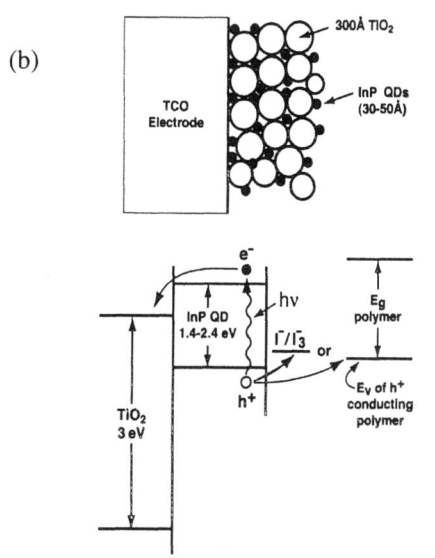

(b)

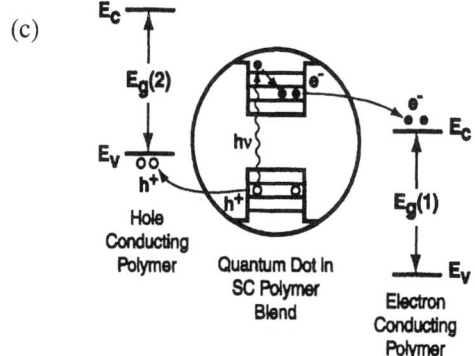

(c)

Figure 24 Possible configurations of QD solar cells. (From Ref. 157.)

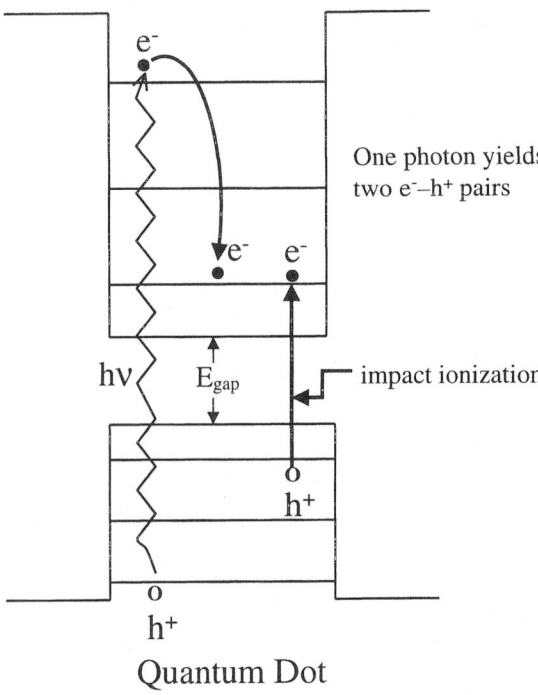

One photon yields
two e⁻–h⁺ pairs

impact ionization

Quantum Dot

Figure 25 Impact ionization in QDs.

mutually exclusive and only one of these processes can be present in a given system.

As discussed in Section V, significant progress has been made in forming 3D arrays of both colloidal [162] and epitaxial [42] II–VI and III–V QDs. The former have been formed via evaporation and crystallization of colloidal QD solutions containing a uniform QD size distribution; crystallization of QD solids from broader size distributions lead to close-packed QD solids, but with a high degree of disorder. Concerning the latter, arrays of epitaxial QDs have been formed by successive epitaxial deposition of epitaxial QD layers; after the first layer of epitaxial QDs is formed, successive layers tend to form with the QDs in each layer aligned on top of each other [42,163]. Theoretical and experimental studies of the properties of QD arrays are currently under way. Major issues are the nature of the electronic states as a function of interdot distance, array order versus disorder, QD orientation and shape, surface states, surface structure/passivation, and surface chemistry. Transport properties of QD arrays are also of critical importance, and they are under investigation.

2. Quantum-Dot-Sensitized Nanocrystalline TiO$_2$ Solar Cells

This configuration is a variation of a recent promising new type of PV cell that is based on dye-sensitization of nanocrystalline TiO$_2$ layers [164–166]. In this latter PV cell, dye molecules are chemisorbed onto the surface of 10–30 nm-size TiO$_2$ particles that have been sintered into a highly porous nanocrystalline 10–20-μm TiO$_2$ film. Upon photoexcitation of the dye molecules, electrons are very efficiently injected from the excited state of the dye into the conduction band of the TiO$_2$, affecting charge separation and producing a photovoltaic effect.

For the QD-sensitized cell, QDs are substituted for the dye molecules; they can be adsorbed from a colloidal QD solution [167] or produced in situ [168–171]. Successful PV effects in such cells have been reported for several semiconductor QDs, including InP, CdSe, CdS, and PbS [167–171]. Possible advantages of QDs over dye molecules are the tunability of optical properties with size and better heterojunction formation with solid hole conductors. Also, a unique potential capability of the QD-sensitized solar cell is the production of quantum yields greater than one by impact ionization (inverse Auger effect) [172].

3. Quantum Dots Dispersed in Organic Semiconductor Polymer Matrices

Recently, photovoltaic effects have been reported in structures consisting of QDs forming junctions with organic semiconductor polymers. In one configuration, a disordered array of CdSe QDs is formed in a hole-conducting polymer—MEH–PPV {poly[2-methoxy, 5-(2'-ethyl)-hexyloxy-p-phenylene-vinylene]} [173]. Upon photoexcitation of the QDs, the photogenerated holes are injected into the MEH–PPV polymer phase and are collected via an electrical contact to the polymer phase. The electrons remain in the CdSe QDs and are collected through diffusion and percolation in the nanocrystalline phase to an electrical contact to the QD network. Initial results show relatively low conversion efficiencies. [173,174], but improvements have been reported with rodlike CdSe QD shapes [175] embedded in poly(3-hexylthiophene) (the rodlike shape enhances electron transport through the nanocrystalline QD phase). In another configuration [176], a polycrystalline TiO$_2$ layer is used as the electron-conducting phase, and MEH-PPV is used to conduct the holes; the electron and holes are injected into their respective transport mediums upon photoexcitation of the QDs.

A variation of these configurations is to disperse the QDs into a blend of electron- and hole-conducting polymers [177]. This scheme is the inverse of light-emitting-diode structures based on QDs [178–182]. In the PV cell, each type of carrier-transporting polymer would have a selective electrical contact to remove the respective charge carriers. A critical factor for success is to prevent electron–hole recombination at the interfaces of the two polymer blends; prevention of electron–hole recombination is also critical for the other QD configurations mentioned earlier.

All of the possible QD–organic polymer photovoltaic cell configurations would benefit greatly if the QDs can be coaxed into producing multiple electron–hole pairs by the inverse Auger/impact ionization process [172]. This is also true for all the QD solar cell systems described earlier. The most important process in all the QD solar cells for reaching very high conversion efficiency is the multiple electron–hole pair production in the photoexcited QDs; the various cell configurations simply represent different modes of collecting and transporting the photogenerated carriers produced in the QDs.

ACKNOWLEDGMENTS

We acknowledge and thank our present and former colleagues at the National Renewable Energy Laboratory (NREL) for their contributions to the work reviewed here: R. J. Ellingson, M. Hanna, G. Rumbles, B. B. Smith, A. Zunger, H. Fu, D. C. Selmarten, P. Ahrenkiel, K. Jones, J. Blackburn, P. Yu, D. Bertram, H. M. Cheong, Z. Lu, E. Poles, and J. Sprague. We also acknowledge our collaborators outside of NREL for their contributions: D. Nesbitt, K. Kuno, E. Lifshitz, T. Rajh, and M. Thurnauer. The work reviewed here was supported by the U.S. Department of Energy, Office of Science, Office of Basic Energy Sciences, Division of Chemical Sciences, Geosciences, and Biosciences.

REFERENCES

1. Weller, H.; Eychmüller, A. In: *Semiconductor Nanoclusters*; Kamat, P.V., Meisel, D., Eds.; Elsevier Science: Amsterdam, 1997; Vol. 103, 5 pp.
2. Overbeek, J.T.G. Adv. Colloid Interf. Sci. 1982, *15*,251.
3. Murray, C.B. Synthesis and characterization of II–VI quantum dots and their assembly into 3D quantum dot superlattices, Ph.D. thesis, Massachusetts Institute of Technology, 1995.
4. Murray, C.B.; Norris, D.J.; Bawendi, M.G. J. Am. Chem. Soc. 1993, *115*, 8706.
5. Mićić, O.L.; Sprague, J.R.; Curtis, C.J.; Jones, K.M.; Machol, J.L.; Nozik, A.J.; Giessen, H.; Fluegel, B.; Mohs, G.; Peyghambarian, N. J. Phys. Chem. 1995, *99*, 7754.
6. Banin, U.; Cerullo, G.; Guzelian, A.A.; Bardeen, C.J.; Alivisatos, A.P.; Shank, C.V. Phys. Rev. B 1997, *55*, 7059.

7. Guzelian, A.A.; Katari, J.E.B.; Kadavanich, A.V.; Banin, U.; Hamad, K.; Juban, E.; Alivisatos, A.P.; Wolters, R.H.; Arnold, C.C.; Heath, J.R. J. Phys. Chem. 1996, *100*, 7212.

8. Mićić, O.I.; Curtis, C.J.; Jones, K.M.; Sprague, J.R.; Nozik, A.J. J. Phys. Chem. 1994, *98*, 4966.

9. Mićić, O.I.; Sprague, J.R.; Lu, Z.; Nozik, A.J. Appl. Phys. Lett. 1996, *68*, 3150.

10. Mićić, O.I.; Cheong, H.M.; Fu, H.; Zunger, A.; Sprague, J.R.; Mascarenhas, A.; Nozik, A.J. J. Phys. Chem. B 1997, *101*, 4904.

11. Nozik, A.J. Annu. Rev. Phys. Chem. 2001, *52*, 193.

12. Langof, L.; Ehrenfreund, E.; Lifshitz, E.; Mićić, O.I.; Nozik, A.J. J. Phys. Chem. B 2002, *106*, 1606.

13. Aubuchon, S.R.; McPhail, A.T.; Wells, R.L.; Giambra, J.A.; Browser, J.R. Chem. Mater. 1994, *6*, 82.

14. Wells, R.L.; Self, M.F.; McPhail, A.T.; Auuchon, S.R.; Wandenberg, R.C.; Jasinski, J.P. Organometallics 1993, *12*, 2832.

15. Rama Krishna, M.V.; Friesner, R.A. J. Chem. Phys. 1991, *95*, 525.

16. Pankove, J.I. *Optical Processes in Semiconductors*; Dover: New York, 1971.

17. MacDougall, J.E.; Eckert, H.; Stucky, G.D.; Herron, N.; Wang, Y.; Moller, K.; Bein, T.; Cox, D. J. Am. Chem. Soc. 1989, *111*, 8006.

18. DeLong, M.C.; Ohlsen, W.D.; Viohl, I.; Taylor, P.C.; Olson, J.M. J. Appl. Phys. 1991, *70*, 2780.

19. Wei, S.-H.; Zunger, A. Phys. Rev. B 1989, *39*, 3279.

20. Wei, S.H.; Ferreira, L.G.; Zunger, A. Phys. Rev. B 1990, *41*, 8240.

21. Froyen, S.; Zunger, A. Phys. Rev. Lett. 1991, *66*, 3132.

22. Mascarenhas, A.; Olson, J.M. Phys. Rev. B 1990, *41*, 9947.

23. Horner, G.S.; Mascarenhas, A.; Froyen, S.; Alonso, R.G.; Bertness, K.A.; Olson, J.M. Phys. Rev. B 1993, *47*, 4041.

24. Olson, J.M.; Kurtz, S.R.; Kibbler, A.E.; Faine, P. Appl. Phys. Lett. 1990, *56*, 623.

25. Mićić, O.I.; Nozik, A.J. J. Lumin. 1996, *70*, 95.

26. Olshavsky, M.A.; Goldstein, A.N.; Alivisatos, A.P. J. Am. Chem. Soc. 1990, *112*, 9438.

27. Uchida, H.; Curtis, C.J.; Kamat, P.V.; Jones, K.M.; Nozik, A.J. J. Phys. Chem. 1992, *96*, 1156.

28. Nozik, A.J.; Uchida, H.; Kamat, P.V.; Curtis, C. Isr. J. Chem. 1993, *33*, 15.

29. Janik, J.F.; Wells, R.L. Chem. Mater. 1996, *8*, 2708.

30. Coffer, J.L.; Johnson, M.A.; Zhang, L.; Wells, R.L. Chem. Mater. 1997, *9*, 2671.

31. Nöth, H.; Konord, P.Z. Z. Naturforsch. 1997, *30b*, 681.

32. Mićić, O.I.; Ahrenkiel, S.P.; Bertram, D.; Nozik, A.J. Appl. Phys. Lett. 1999, *75*, 478.

33. Morkoc, H.; Strite, S.; Gao, G.B.; Lin, M.E.; Sverdlov, B.; Burns, M. J. Appl. Phys. 1994, *76*, 1363.

34. Mićić, O.I.; Smith, B.B.; Nozik, A.J. J. Phys. Chem. 2000, *104*, 12,149.

35. Peng, X.; Schlamp, M.C.; Kadavanich, A.V.; Alivisatos, A.P. J. Am. Chem. Soc. 1997, *119*, 7019.

36. Miller, R.D.J.; McLendon, G.; Nozik, A.J.; Schmickler, W.; Willig, F. *Surface Electron Transfer Processes*; VCH: New York, 1995.

37. Gaponenko, S.V. Optical Properties of Semiconductor Nanocrystals, Cambridge Univ. Press Cambridge, UK, 1998.

37a. Sandmann, J.H.H.; Grosse, S.; von Plessen, G.; Feldmann, J.; Hayes, G.; Phillipps, R.; Lipsanen, H.; Sopanen, M.; Ahopelto, J. Phys. Status Solidi B 1997, *204*, 251.

38. Eaglesham, D.J.; Cerullo, M. Phys. Rev. Lett. 1990, *64*, 1943.

39. Bimberg, D.; Grundmann, M.; Ledentsov, N.N. *Quantum Dot Heterostructures*; Wiley: Chichester, 1999.

40. Guha, S.; Madhukar, A.; Rajkumar, K.C. Appl. Phys. Lett. 1990, *57*, 2110.

41. Snyder, C.W.; Orr, B.G.; Kessler, D.; Sander, L.M. Phys. Rev. Lett. 1991, *66*, 3032.

42. Sugawara, M. In *Semiconductors and Semimetals*; Willardson R.K., Weber, E.R., Eds., Academic Press: San Diego, CA, 1999; Vol. 60.

43. Yamaguchi, A.A.; Ahopelto, J.; Nishi, K.; Usui, A.; Akiyama, H.; Sakaki, H. Inst. Phys. Conf. Ser. 1992, *129*, 341.

44. Hanna, M.C.; Lu, Z.H.; Cahill, A.F.; Heben, M.J.; Nozik, A.J. J. Cryst. Growth 1997, *174*, 605.

45. Hanna, M.C.; Lu, Z.H.; Cahill, A.F.; Heben, M.J.; Nozik, A.J. Mater. Res. Soc. Symp.. Proc. 1996, *417*, 129.

46. Sopanen, M.; Lipsanen, H.; Ahopelto, J. Appl. Phys. Lett. 1995, *66*, 2364.

47. Lipsanen, H.; Sopanen, M.; Ahopelto, J. Phys. Rev. B 1995, *51*, 13,868.

48. Mićić, O.I.; Nozik, A.J.; Lifshitz, E.; Rajh, T.; Poluektov, O.G.; Thurnauer, M.C. J. Phys. Chem. 2002, *106*, 4390.

49. Fu, H.; Zunger, A. Phys. Rev. B 1997, *55*, 1642.

50. Fu, H.; Zunger, A. Phys. Rev. B 1997, *56*, 1496.

51. Poles, E.; Selmarten, D.C.; Mićić, O.I.; Nozik, A.J. Appl. Phys. Lett. 1999, *75*, 971.

52. Cheong, H.M.; Fluegel, B.; Hanna, M.C.; Mascarenhas, A. Phys. Rev. B 1998, *58*, R4254.

53. Driessen, F.A.J.M.; Cheong, H.M.; Mascarenhas, A.; Deb, S.K. Phys. Rev. B 1996, *54*, R5263.

54. Seidel, W.; Titkov, A.; André, J.P.; Voisin, P.; Voos, M. Phys. Rev. Lett. 1994, *73*, 2356.

55. Zegrya, G.G.; Kharchenko, V.A. Sov. Phys. JETP 1992, *74*, 173.

56. Dubois, L.H.; Schwartz, G.P. Phys. Rev. B 1982, *26*, 794.

57. Fu, H.; Ozolins, V.; Zunger, A. Phys. Rev. B 1999, *59*, 2881.

58. Nirmal, M.; Dabbousi, B.O.; Bawendi, M.G.; Macklin, J.J.; Trautman, J.K.; Harris, T.D.; Brus, L.E. Nature 1996, *383*, 802.

59. Tittel, J.; Gohde, W.; Koberling, F.; Mews, A.; Kornowski, A.; Weller, H.; Eychmuller, A.; Basche, T. Ber. Bunsenges. Phys. Chem. 1997, *101*, 1626.

60. Banin, U.; Bruchez, M.; Alivisatos, A.P.; Ha, T.; Weiss, S.; Chemla, D.S. J. Chem. Phys. 1999, *110*, 1195.

61. Kuno, M.; Fromm, D.P.; Hamann, H.F.; Gallagher, A.; Nesbitt, D.J. J. Chem. Phys. 2000, *112*, 3117.

62. Kuno, M.; Fromm, D.P.; Hamann, H.F.; Gallagher, A.; Nesbitt, D.J. J. Chem. Phys. 2001, *115*, 1028.
63. Neuhauser, R.G.; Shimizu, K.; Woo, W.K.; Empedocles, S.A.; Bawendi, M.G. Phys. Rev. Lett. 2000, *85*, 3301.
64. Kuno, M.; Fromm, D.P.; Gallagher, A.; Nesbitt, D.J.; Mićić, O.I.; Nozik, A.J. Nano Lett. 2001, *1*, 557.
65. Sugisaki, M.; Ren, H.W.; Nair, S.V.; Lee, J.S.; Sugou, S.; Okuno, T.; Masumoto, Y. J. Lumin. 2000, *15*, 40.
66. Sugisaki, M.; Ren, H.W.; Nishi, K.; Masumoto, Y. Phys. Rev. Lett. 2001, *86*, 4883.
67. Pistol, M.E.; Castrillo, P.; Hessman, D.; Prieto, J.A.; Samuelson, L. Phys. Rev. B 1999, *59*, 10,725.
68. Krauss, T.D.; Brus, L.E. Phys. Rev. Lett. 1999, *83*, 4840.
69. Krauss, T.D.; O'Brien, S.; Brus, L.E. J. Phys. Chem. B 2001, *105*, 1725.
70. Efros, A.L.; Rosen, M. Phys. Rev. Lett. 1997, *78*, 1110.
71. Bertram, D.; Hanna, M.C.; Nozik, A.J. Appl. Phys. Lett. 1999, *74*, 2666.
72. Kasha, M. Rad. Res. 1960, *2*(Suppl), 243.
73. Hoheisel, W.; Colvin, Y.L.; Johnson, C.S.; Alivisatos, A.P. J. Chem. Phys. 1994, *101*, 845.
74. Rumbles, G.; Selmarten, D.C.; Ellingson, R.J.; Blackburn, J.L.; Yu, P.; Smith, B.B.; Mićić, O.I.; Nozik, A.J. J. Photochem. Photobiol. A 2001, *142*, 187.
75. Ellingson, R.; Blackburn, J.L.; Yu, P.; Rumbles, G.; Mićić, O.I.; Nozik, A.J. J. Phys. Chem. B 2002, *106*, 7758.
76. Jones, M.; Nedeljkovic, J.; Ellingson, R.; Nozik, A.J.; Rumbles, G. J. Phys. Chem. 2003, Submitted.
77. Pelouch, W.S.; Ellingson, R.J.; Powers, P.E.; Tang, C.L.; Szmyd, D.M.; Nozik, A.J. Phys. Rev. B 1992, *45*, 1450.
78. Ulstrup, J.; Jortner, J. J. Chem. Phys. 1975, *63*, 4358.
79. Rosenwaks, Y.; Hanna, M.C.; Levi, D.H.; Szmyd, D.M.; Ahrenkiel, R.K.; Nozik, A.J. Phys. Rev. B 1993, *48*, 14,675.
80. Pelouch, W.S.; Ellingson, R.J.; Powers, P.E.; Tang, C.L.; Szmyd, D.M.; Nozik, A.J. Proc. SPIE 1993, *1677*, 602.
81. Boudreaux, D.S.; Williams, F.; Nozik, A.J. J. Appl. Phys. 1980, *51*, 2158.
82. Nozik, A.J.; Boudreaux, D.S.; Chance R.R.; Williams, F. In *Advances in Chemistry*; Wrighton, M., Ed.; American Chemical Society: New York, 1980; Vol. 184, 162 pp.
83. Williams, F.E.; Nozik, A.J. Nature 1984, *311*, 21.
84. Williams, F.; Nozik, A.J. Nature 1978, *271*, 137.
85. Benisty, H.; Sotomayor-Torres, C.M.; Weisbuch, C. Phys. Rev. B 1991, *44*, 10,945.
86. Bockelmann, U.; Bastard, G. Phys. Rev. B 1990, *42*, 8947.
87. Benisty, H. Phys. Rev. B 1995, *51*, 13281.
88. Jaros, M. *Physics and Applications of Semiconductor Microstructures*; Oxford University Press: New York, 1989; 245 pp.
89. Bockelmann, U.; Egeler, T. Phys. Rev. B 1992, *46*, 15,574.
90. Efros, A.L.; Kharchenko, V.A.; Rosen, M. Solid State Commun. 1995, *93*, 281.

91. Vurgaftman, I.; Singh, J. Appl. Phys. Lett. 1994, *64*, 232.
92. Sercel, P.C. Phys. Rev. B 1995, *51*, 14,532.
93. Inoshita, T.; Sakaki, H. Phys. Rev. B 1992, *46*, 7260.
94. Inoshita, T.; Sakaki, H. Phys. Rev. B 1997, *56*, R4355.
95. Guyot-Sionnest, P.; Shim, M.; Matranga, C.; Hines, M. Phys. Rev. B 1999, *60*, R2181.
96. Wang, P.D.; Sotomayor-Torres, C.M.; McLelland, H.; Thomas, S.; Holland, M.; Stanley, C.R. Surface Sci. 1994, *305*, 585.
97. Mukai, K.; Sugawara, M. Jpn. J. Appl. Phys. 1998, *37*, 5451.
98. Mukai, K.; Ohtsuka, N.; Shoji, H.; Sugawara, M. Appl. Phys. Lett. 1996, *68*, 3013.
99. Murdin, B.N.; Hollingworth, A.R.; Kamal-Saadi, M.; Kotitschke, R.T.; Ciesla, C.M.; Pidgeon, C.R.; Findlay, P.C.; Pellemans, H.P.M.; Langerak, C.J.G.M.; Rowe, A.C.; Stradling, R.A.; Gornik, E. Phys. Rev. B 1999, *59*, R7817.
100. Sugawara, M.; Mukai, K.; Shoji, H. Appl. Phys. Lett. 1997, *71*, 2791.
101. Heitz, R.; Veit, M.; Ledentsov, N.N.; Hoffmann, A.; Bimberg, D.; Ustinov, V.M.; Kop'ev, P.S.; Alferov, Z.I. Phys. Rev. B 1997, *56*, 10,435.
102. Heitz, R.; Kalburge, A.; Xie, Q.; Grundmann, M.; Chen, P.; Hoffmann, A.; Madhukar, A.; Bimberg, D. Phys. Rev. B 1998, *57*, 9050.
103. Mukai, K.; Ohtsuka, N.; Shoji, H.; Sugawara, M. Phys. Rev. B 1996, *54*, R5243.
104. Yu, H.; Lycett, S.; Roberts, C.; Murray, R. Appl. Phys. Lett. 1996, *69*, 4087.
105. Adler, F.; Geiger, M.; Bauknecht, A.; Scholz, F.; Schweizer, H.; Pilkuhn, M.H.; Ohnesorge, B.; Forchel, A. Appl. Phys. 1996, *80*, 4019.
106. Adler, F.; Geiger, M.; Bauknecht, A.; Haase, D.; Ernst, P.; Dörnen, A.; Scholz, F.; Schweizer, H. J. Appl. Phys. 1998, *83*, 1631.
107. Brunner, K.; Bockelmann, U.; Abstreiter, G.; Walther, M.; Böhm, G.; Tränkle, G.; Weimann, G. Phys. Rev. Lett. 1992, *69*, 3216.
108. Kamath, K.; Jiang, H.; Klotzkin, D.; Phillips, J.; Sosnowki, T.; Norris, T.; Singh, J.; Bhattacharya, P. Inst. Phys. Conf. Ser. 1998, *156*, 525.
109. Gfroerer, T.H.; Sturge, M.D.; Kash, K.; Yater, J.A.; Plaut, A.S.; Lin, P.S.D.; Florez, L.T.; Harbison, J.P.; Das, S.R.; Lebrun, L. Phys. Rev. B 1996, *53*, 16,474.
110. Li, X.-Q.; Nakayama, H.; Arakawa, Y. In: *Proceeding of the International Conference on Physics and Semiconductors*; Gershoni, D., Ed.; World Scientific: Singapore, 1998; 845 pp.
111. Bellessa, J.; Voliotis, V.; Grousson, R.; Roditchev, D.; Gourdon, C.; Wang, X.L.; Ogura, M.; Matsuhata, H. In *Proceedings of the International Conference on Physics and Semiconductors*; Gershoni, D., Ed.; World Scientific: Singapore, 1998; 763 pp.
112. Lowisch, M.; Rabe, M.; Kreller, F.; Henneberger, F. Appl. Phys. Lett. 1999, *74*, 2489.
113. Gontijo, I.; Buller, G.S.; Massa, J.S.; Walker, A.C.; Zaitsev, S.V.; Gordeev, N.Y.; Ustinov, V.M.; Kop'ev, P.S. Jpn. J. Appl. Phys. 1999, *38*, 674.
114. Li, X.-Q.; Nakayama, H.; Arakawa, Y. Jpn. J. Appl. Phys. 1999, *38*, 473.

115. Kral, K.; Khas, Z. Phys. Status Solidi B 1998, *208*, R5.

116. Klimov, V.I.; McBranch, D.W. Phys. Rev. Lett. 1998, *80*, 4028.

117. Bimberg, D.; Ledentsov, N.N.; Grundmann, M.; Heitz, R.; Boehrer, J.; Ustinov, V.M.; Kop'ev, P.S.; Alferov, Z.I. J. Lumin. 1997, *72–74*, 34.

118. Woggon, U.; Giessen, H.; Gindele, F.; Wind, O.; Fluegel, B.; Peyghambarian, N. Phys. Rev. B 1996, *54*, 17,681.

119. Grundmann, M.; Heitz, R.; Ledentsov, N.; Stier, O.; Bimberg, D.; Ustinov, V.M.; Kop'ev, P.S.; Alferov, Z.I.; Ruvimov, S.S.; Werner, P.; Gösele, U.; Heydenreich, J. Superlattices Microstruct. 1996, *19*, 81.

120. Williams, V.S.; Olbright, G.R.; Fluegel, B.D.; Koch, S.W.; Peyghambarian, N. J. Mod. Opti. 1988, *35*, 1979.

121. Ohnesorge, B.; Albrecht, M.; Oshinowo, J.; Forchel, A.; Arakawa, Y. Phys. Rev. B 1996, *54*, 11,532.

122. Wang, G.; Fafard, S.; Leonard, D.; Bowers, J.E.; Merz, J.L.; Petroff, P.M. Appl. Phys. Lett. 1994, *64*, 2815.

123. Heitz, R.; Veit, M.; Kalburge, A.; Xie, Q.; Grundmann, M.; Chen, P.; Ledentsov, N.N.; Hoffman, A.; Madhukar, A.; Bimberg, D.; Ustinov, V.M.; Kop'ev, P.S.; Alferov, Z.I. Physica E (Amsterdam) 1998, *2*, 578.

124. Li, X.-Q.; Arakawa, Y. Phys. Rev. B 1998, *57*, 12,285.

125. Sosnowski, T.S.; Norris, T.B.; Jiang, H.; Singh, J.; Kamath, K.; Bhattacharya, P. Phys. Rev. B 1998, *57*, R9423.

126. Meier, A.; Selmarten, D.C.; Siemoneit, K.; Smith, B.B.; Nozik, A.J. J. Phys. Chem. B 1999, *103*, 2122.

127. Meier, A.; Kocha, S.S.; Hanna, M.C.; Nozik, A.J.; Siemoneit, K.; Reineke-Koch, R.; Memming, R. J. Phys. Chem. B 1997, *101*, 7038.

128. Diol, S.J.; Poles, E.; Rosenwaks, Y.; Miller, R.J.D. J. Phys. Chem. B 1998, *102*, 6193.

129. Klimov, V.I.; Mikhailovsky, A.A.; McBranch, D.W.; Leatherdale, C.A.; Bawendi, M.G. Phys. Rev. B 2000, *61*, R13,349.

130. Ellingson, R.J.; Blackburn, J.L.; Nedeljkovic, J.; Rumbles, G.; Jones, M.; Fu, X.; Nozik, A.J. Phys. Rev. B 2003, *67*, 75308.

131. Blackburn, J.; Ellingson, R.J.; Mićić, O.I.; Nozik, A.J. J. Phys. Chem. B 2003, *107*, 102.

132. Klimov, V.I.; McBranch, D.W.; Leatherdale, C.A.; Bawendi, M.G. Phys. Rev. B 1999, *60*, 13,740.

133. Klimov, V.I. J. Phys. Chem. B 2000, *104*, 6112.

134. Mićić, O.I.; Jones, K.M.; Cahill, A.; Nozik, A.J. J. Phys. Chem. B 1998, *102*, 9791.

135. Murray, C.B.; Kagan, C.R.; Bawendi, M.G. Science 1995, *270*, 1335.

136. Collier, C.P.; Vossmeyer, T.; Heath, J.R. Annu. Rev. Phys. Chem. 1998, *49*, 371.

137. Leatherdale, C.A.; Kagan, C.R.; Morgan, N.Y.; Empedocles, S.A.; Kastner, M.A.; Bawendi, M.G. Phys. Rev. B 2000, *62*, 2669.

138. Artemyev, M.V.; Bibik, A.I.; Gurinovich, L.I.; Gaponenko, S.V.; Woggon, U. Phys. Rev. B 1999, *60*, 1504.

139. Mićić, O.I.; Ahrenkiel, S.P.; Nozik, A.J. Appl. Phys. Lett. 2001, 78, 4022.
140. Leff, D.V.; Brandt, L.; Heath, J.R. Langmuir 1996, 12, 4723.
141. Motte, L.; Pileni, M.P. J. Phys. Chem. B 1998, 102, 4104.
142. Schedelbeck, G.; Wegscheider, W.; Bichler, M.; Abstreiter, G. Science 1997, 278, 1792.
143. Vdovin, E.E.; Levin, A.; Patanè, A.; Eaves, L.; Main, P.C.; Khanin, Y.N.; Dubrovskii, Y.V.; Henini, M.; Hill, G. Science 2000, 290, 120.
144. Bayer, M.; Hawrylak, P.; Hinzer, K.; Fafard, S.; Korkusinski, M.; Wasilewski, Z.R.; Stern, O.; Forchel, A. Science 2001, 291, 451.
145. Smith, B.B.; Nozik, A.J. Nano Lett. 2001, 1, 36.
146. Kagan, C.R.; Nirmal, MurrayM.; Bawendi, M.G. Phys. Rev. Lett. 1996, 76, 1517.
147. Kagan, C.R.; Murray, C.B.; Nirmal, M.; Bawendi, M.G. Phys. Rev. Lett. 1996, 76, 3043 (erratum).
148. Kagan, C.R.; Murray, C.B.; Bawendi, M.G. Phys. Rev. B 1996, 54, 8633.
149. Shockley, W.; Queisser, H.J. J. Appl. Phys. 1961, 32, 510.
150. Ross, R.T. J. Chem. Phys. 1966, 45, 1.
151. Ross, R.T. J. Chem. Phys. 1967, 46, 4590.
152. Green, M.A. Third Generation Photovoltaics; Bridge Printery: Sydney, 2001.
153. Green, M.A. Solar Cells; Prentice-Hall: Englewood Cliffs, NJ, 1982.
154. Ross, R.T.; Nozik, A.J. J. Appl. Phys. 1982, 53, 3813.
155. Landsberg, P.T.; Nussbaumer, H.; Willeke, G. J. Appl. Phys. 1993, 74, 1451.
156. Kolodinski, S.; Werner, J.H.; Wittchen, T.; Queisser, H.J. Appl. Phys. Lett. 1993, 63, 2405.
157. Nozik, A.J. Physica E, 2002, 14, 115.
158. Luque, A.; Marti, A. Phys. Rev. Lett. 1997, 78, 5014.
159. Nozik, A.J. Phil. Trans. R. Soc. (Lond) 1980, A295, 453.
160. Pelouch, W.S.; Ellingson, R.J.; Powers, P.E.; Tang, C.L.; Szmyd, D.M.; Nozik, A.J. Semicond. Sci. Technol. 1992, 7, B337.
161. Nozik, A.J. unpublished manuscript, 1996.
162. Murray, C.B.; Kagan, C.R.; Bawendi, M.G. Annu. Rev. Mater. Sci. 2000, 30, 545.
163. Nakata, Y.; Sugiyama, Y.; Sugawara, M. In Semiconductors and Semimetals; Sugawara, M., Ed.; Academic Press: San Diego, CA, 1999; Vol. 60, 117 pp.
164. Hagfeldt, A.; Grätzel, M. Acc. Chem. Res. 2000, 33, 269.
165. Moser, J.; Bonnote, P.; Grätzel, M. Coord. Chem. Rev. 1998, 171, 245.
166. Grätzel, M. Prog. Photovolt. 2000, 8, 171.
167. Zaban, A.; Mićić, O.I.; Gregg, B.A.; Nozik, A.J. Langmuir 1998, 14, 3153.
168. Vogel, R.; Weller, H. J. Phys. Chem. 1994, 98, 3183.
169. Weller, H. Ber. Bunsen-ges. Phys. Chem 1991, 95, 1361.
170. Liu, D.; Kamat, P.V. J. Phys. Chem. 1993, 97, 10,769.
171. Hoyer, P.; Könenkamp, R. Appl. Phys. Lett. 1995, 66, 349.
172. Nozik, A.J. unpublished manuscript, 1997.
173. Greenham, N.C.; Poeng, X.; Alivisatos, A.P. Phys. Rev. B 1996, 54, 17,628.
174. Greenham, N.C.; Peng, X.; Alivisatos, A.P. In Future Generation Photovoltaic

Technologies: First NREL Conference; McConnell, R., Ed.; American Institue of Physics: Washington DC, 1997; 295 pp.

175. Huynh, W.U.; Peng, X.; Alivisatos, P. Adv. Mater. 1999, *11*, 923.
176. Arango, A.C.; Carter, S.A.; Brock, P.J. Appl. Phys. Lett. 1999, *74*, 1698.
177. Nozik, A.J.; Rumbles, G.; Selmarten, D.C. unpublished manuscript, 2000.
178. Dabbousi, B.O.; Bawendi, M.G.; Onitsuka, O.; Rubner, M.F. Appl. Phys. Lett. 1995, *66*, 1316.
179. Colvin, V.; Schlamp, M.; Alivisatos, A.P. Nature 1994, *370*, 354.
180. Schlamp, M.C.; Peng, X.; Alivisatos, A.P. J. Appl. Phys. 1997, *82*, 5837.
181. Mattoussi, H.; Radzilowski, I.H.; Dabbousi, B.O.; Fogg, D.E.; Schrock, R.R.; Thomas, E.L.; Rubner, M.F.; Bawendi, M.G. J. Appl. Phys. 1999, *86*, 4390.
182. Mattoussi, H.; Radzilowski, L.H.; Dabbousi, B.O.; Thomas, E.L.; Bawendi, M.G.; Rubner, M.F. J. Appl. Phys. 1998, *83*, 7965.

10

Synthesis and Fabrication of Metal Nanocrystal Superlattices

**R. Christopher Doty, Michael B. Sigman, Jr.,
Cynthia A. Stowell, Parag S. Shah, Aaron E. Saunders,
and Brian A. Korgel**
The University of Texas, Austin, Texas, U.S.A.

I. INTRODUCTION

Organic monolayer-coated metal nanocrystals can be synthesized by arrested precipitation with sizes ranging from 1.5 to 20 nm in diameter [1–9]. Arrested precipitation involves the reduction of a metal salt [3] or metallorganic precursor [1] in the presence of bifunctional organic molecules with a chemical binding group on one end and a bulky inert group on the other. The ligands coat the nanocrystals and provide a steric barrier to aggregation as shown in Fig. 1. The ligands also provide chemical passivation of the surface, thus decreasing the susceptibility to oxidation and other forms of chemical degradation while controlling the dispersibility in various solvents. Nanocrystal size distributions with standard deviations less than 10% can be achieved by employing size-selective precipitation [10]. These size-monodisperse sterically stabilized nanocrystals can be condensed into superlattices: the particles organize into a crystal lattice with long-range translational order simply by evaporating the solvent from a concentrated dispersion [10,11]. Figure 2 shows high-resolution scanning electron microscopy (HR-SEM) and transmission electron microscopy (HR-TEM) images of two- (2D) and three-dimensional (3D) colloidal crystals of organic monolayer coated metal nanocrystals. These images demonstrate the potential for precise, controlled self-assembly of nanocrystal superlattices: The ligands control the interparticle separation and the nanocrystal size is tunable.

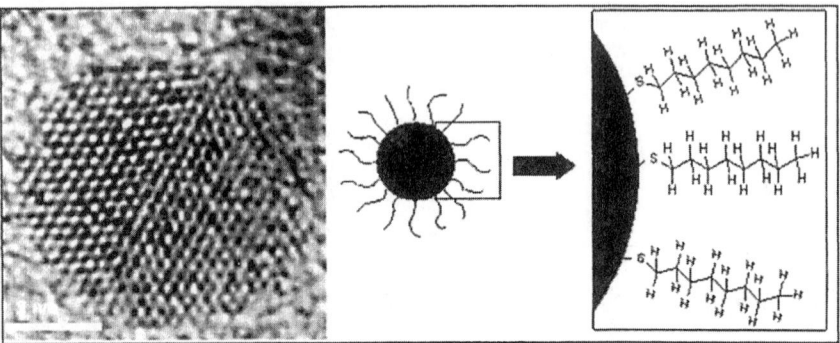

Figure 1 Left: High-resolution transmission electron microscopic image showing the crystalline structure of a gold nanoparticle. Right: Dodecanethiol molecules chemisorb to the gold surface, providing a passivation layer that prevents aggregation of the gold cores. The capping ligand length can be varied to control the interparticle spacing in a superlattice.

Metal nanocrystal superlattices provide the opportunity to develop new technologies. These granular materials have controllable domain size, separation, and interparticle spacing. Magnetic and superconductor nanocrystals offer particularly unique opportunities [1,12]. Sterically stabilized nanocrystals also provide model systems to study molecular-phase behavior [13]. The interaction energies can be tuned by particle size and capping ligand chemistry and the materials are readily characterized by electron microscopy and x-ray scattering techniques. Furthermore, due to their small size, the mass transfer limitations common to larger colloidal systems do not pose technical obstacles. Therefore, these particles provide a test bed for simple fluid models of phase behavior.

A. Early Developments

Research on submicrometer-scale colloidal superlattices (or colloidal crystals) began as early as the 1930s with x-ray studies by Levin and Ott on the crystallinity of natural deposits of opal [14]. These researchers were primarily interested in understanding the phenomena responsible for the color, or "fire," seen in opal specimens obtained from different areas of the world [15–17]. In 1946, opals were described by Copisarow and Copisarow [18] as being composed of colloidal particles, as directly confirmed in 1964 by Sanders, who

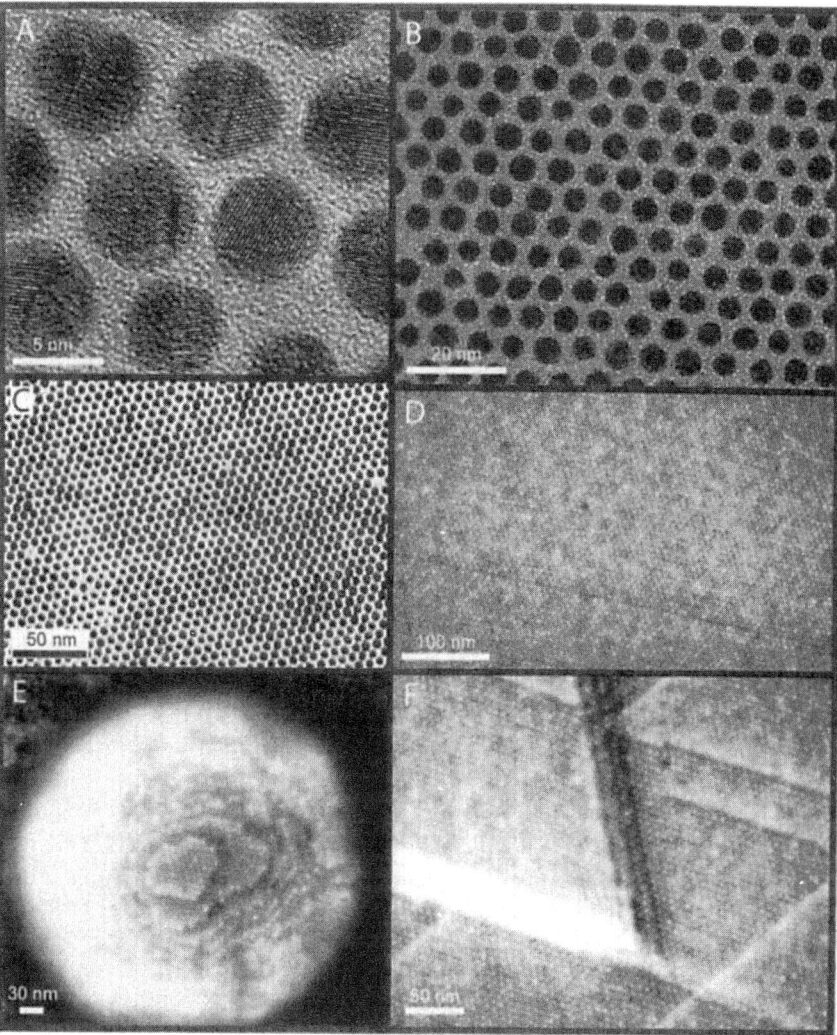

Figure 2 (A–C) Progression of HR-TEM images of decreasing magnification showing ordered superlattices formed from C_{12}-capped gold nanocrystals drop cast onto a carbon-film TEM grid. (D) HR-SEM image of 3D ordered superlattice formed from silver nanocrystals drop cast onto a glassy carbon substrate from chloroform. HR-SEM image of gold nanocrystals drop cast from (E) hexane and (F) chloroform illustrating the effect of solvent conditions on the corresponding 3D growth of the superlattice film.

imaged the internal colloidal opal structure by electron microscopy [16]. The opals are *superlattices* of uniform-sized spherical silica colloids ranging from 170 to 350 nm in diameter. These structural studies confirmed a century-old hypothesis that the brilliant color produced by opals upon light exposure results from the diffraction of light by the 3D colloidal array [16]. In 1966, Darragh et al. determined that the colloidal "building blocks" of the opal— previously considered to be amorphous silica—in fact consisted of aggregated smaller silica particles 30–40 nm in diameter [15].

In 1989, Schmid and Lehnert [19] demonstrated Au nanocrystal superlattice formation and Bentzon and co-workers [20] demonstrated hexagonal close packing of iron oxide nanocrystals, some of the first examples of self-organized metal nanocrystal superlattices. Schmid and co-workers found that 20-nm-diameter Au nanocrystals synthesized by sodium citrate reduction of an aqueous gold salt solution and stabilization with phosphane [19] with a polydispersity of 10% [20] ordered into a lattice when evaporated onto a substrate. Bentzon et al. used 6.9-nm-diameter iron oxide nanocrystals with a narrow size distribution [21]. These studies reiterated that size monodisperse colloidal particles with sufficient stabilization to aggregation at high-volume fractions organize into superlattice structures. Nanocrystals coated with organic monolayers provide an ideal material for forming superlattice structures, as the steric organic barrier prevents the metal cores from touching even after the solvent is removed. This is not the case with charge-stabilized colloids coated by ions. In a dispersion, these colloids are well stabilized by repulsive electrostatic double-layer forces; however, upon removal of the solvent, the particles aggregate irreversibly.

Nanocrystal surface passivation by organic ligands was examined for CdSe nanocrystals by Steigerwald and co-workers in 1988 [22]. The use of organic ligands to control particle size has proven very powerful for semiconductor nanocrystals, and dramatic progress in the synthesis of semiconductor nanocrystals ensued upon the development of high-temperature synthesis approaches in coordinating solvents. In 1994, Brust and co-workers demonstrated the capping ligand approach to metal nanocrystal synthesis [3]. As outlined in Fig. 3, their method relies on the room-temperature reduction of a gold salt in the presence of alkanethiol capping ligands under ambient conditions to yield crystalline particles stabilized with long-chain alkanes. Due to its simplicity, their approach has been reproduced and utilized by many different research groups to study gold and silver nanocrystals, with various improvements in size control and narrowing of the size distribution [6,13]. Most recently, the high-temperature capping solvent approaches developed for semiconductor nanocrystals have been applied to metal nanocrystals by reducing metal precursors at high temperatures in various coordinating solvents (see Fig. 4) [1].

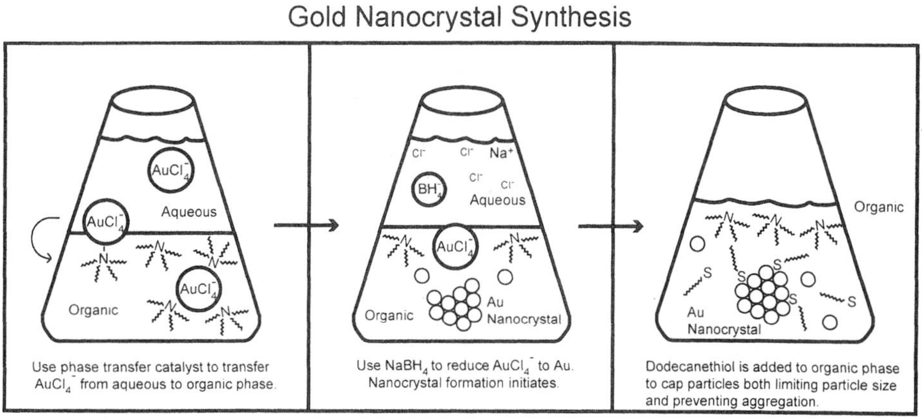

Gold Nanocrystal Synthesis
AuCl4-
Aqueous
AuCl4-
AuCl4-
Organic
Use phase transfer catalyst to transfer AuCl4- from aqueous to organic phase.
Cl- Cl- Na+
BH4-
Cl- Cl-
Aqueous
AuCl4-
Au Nanocrystal
Organic
Use NaBH4 to reduce AuCl4- to Au. Nanocrystal formation initiates.
Organic
Au Nanocrystal
Dodecanethiol is added to organic phase to cap particles both limiting particle size and preventing aggregation.

Figure 3 Schematic for gold nanocrystal synthesis using the two-phase reduction technique developed by Brust et al. [3].

N2 Gas (through Schlenk Line)
Condenser
Thermocouple
Co(CO)3NO, oleic acid, oleylamine, injected through syringe
Pt(acac)2, 1,2-hexadecanediol and dioctylether
Stir Bar
Heat Source
3 Neck Round Bottom Flask, capped with septums and a condenser with N2 gas flow

Figure 4 Synthesis of CoPt nanocrystals through the use of a coordinating solvent at high temperatures. Platinum acetylacetonate and 1,2-hexadecanediol are dissolved in dioctylether and heated to 100°C in an airless environment. $Co(CO)_3NO$, oleic acid, and oleyamine are injected into the mixture and the flask is refluxed at 298°C for 30 min. Purification of particles is achieved by flooding the reaction with ethanol to precipitate the nanocrystals, which are then dispersed in hexane in the presence of oleic acid and oleylamine. Alternating dispersions and precipitations in hexane and ethanol can purify the product further.

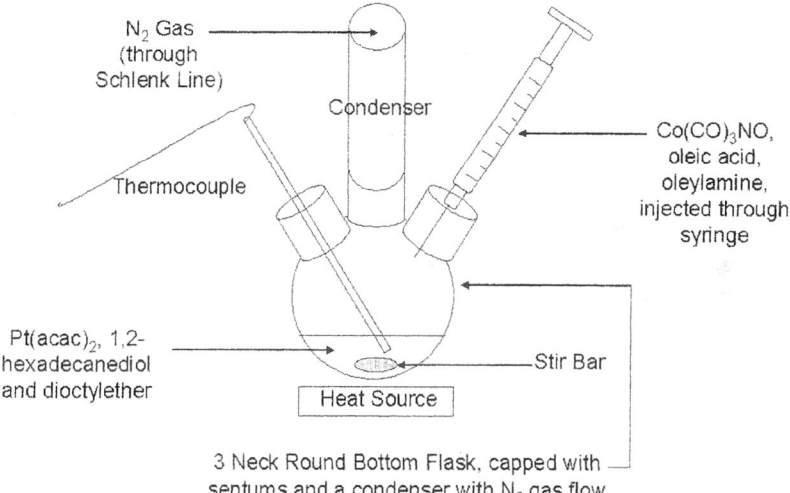

B. Metal Nanocrystal Synthesis

Organic monolayer-coated metal nanocrystals can be synthesized at room
temperature and high temperatures, depending on the material of interest.
For example, gold and silver nanocrystals with highly crystalline cores are
readily produced at room temperature using alkanethiols as capping ligands
[3]. Other materials, such as Co [1] and FePt [23], require higher temperatures
to drive the necessary metal reduction chemistry and ligand adsorption/
desorption equilibria necessary to produce reasonably monodisperse sterically stabilized particles.

Metal nanocrystal-arrested precipitation is conceptually simple. Metals
are precipitated from dissolved metal ions, or metallorganic precursors, in the
presence of capping ligands that bind to the nanocrystal surface to control
particle growth. Without the capping ligands, the metal would precipitate into
bulk material. In this synthetic process, a fundamental competition exists
between particle nucleation and growth kinetics, and ligand adsorption/de-
sorption kinetics. Nucleation kinetics in the liquid phase and ligand adsorption chemistry on small particles are both not well understood. Therefore, the
successful development of arrested precipitation procedures generally
requires trial and error. The binding strengths of different functional groups,
like thiols and amines, can vary dramatically for different metallic species. If
the ligands bind the metal too strongly, particles will not grow. In some cases,
as with thiols and cobalt, the functional group stabilizes a molecular complex
that is extremely difficult to reduce to cobalt metal. Therefore, different
chemical routes and temperatures can be required to form nanocrystals with
differing core composition and capping ligand chemistry. For example, gold
and silver are easily reduced by $NaBH_4$ in the presence of thiolates [3],
whereas copper will not reduce at room temperature upon the addition of
$NaBH_4$ due to the strong bonding between copper cations and thiolates. In
fact, in the case of copper, the thermodynamic driving force toward metal
oxidation by the thiol favors the thermolysis of copper thiolates into Cu_2S at
elevated temperatures ($\sim$150–200°C) [24]. Therefore, the chemical interaction between the chosen ligand and the metal must be considered in any
synthetic scheme explored. Amines and carboylated molecules have proven
relatively versatile for metal nanocrystal synthesis, as in the case of Co [1] and
FePt [23] nanocrystals. Once the reduction chemistry proceeds, the ligands
must provide sufficient binding and steric stabilization to prevent uncontrol-
lable growth to very large sizes and high degrees of polydispersity. A
combination of ligands in a synthesis often provides the appropriate binding
strength: One ligand binds weakly to provide particle growth under controlled
conditions, and the other ligand provides strong binding to quench growth at
a particular size.

Nanocrystal growth can occur by two different general mechanisms. The first involves the diffusion-limited growth of the particles by monomer addition to the particles. The second is growth by aggregation of nanocrystals. Aggregative growth leads to broad log-normal size distributions, whereas diffusion-limited growth leads to a narrowing of the size distribution. The moments of the size distribution, $\mu_1 = r_3/r_h$ and $\mu_3 = r_1/r_3$ provide a quantitative determination of the growth mechanism, where $r_1 = \Sigma r_1/N_\infty$ is the arithmetic mean radius, $r_3 = \sqrt[3]{\Sigma r_1^3/N_\infty}$ is the cube mean radius, and $r_h = N_\infty/\Sigma(1/r_1)$ is the harmonic mean radius, where N_∞ is the total number of particles. Nanocrystals formed through condensation of free atoms or small oligomers onto growing metal cores are relatively monodisperse with $\mu_1 = \mu_3 = 1$, whereas coagulative growth results in broad size distributions with $\mu_1 > 1.25$ and $\mu_3 < 0.905$ [25,26]. Figure 5 shows representative examples of size distributions for silver nanocrystals that formed primarily by diffusion-limited growth and aggregative growth. Note the broad size distribution that results from aggregative growth compared to diffusion-limited growth [26].

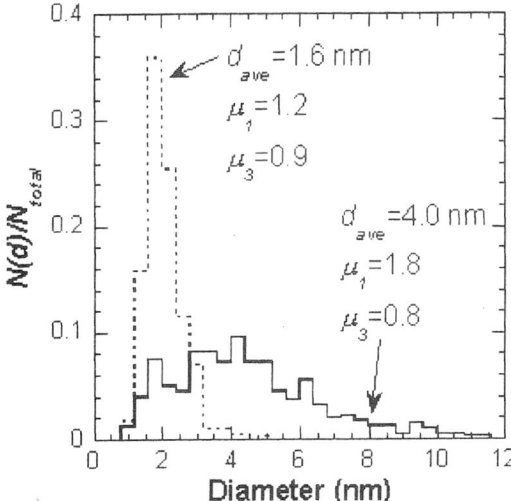

Figure 5 Histograms of perfluorodecanethiol-capped silver nanocrystals synthesized in supercritical carbon dioxide under good (80°C and 345 bar, dotted line) and poor (80°C and 207 bar, solid line) solvent conditions. Under good solvent conditions, the particle size is smaller and the size distribution moments (μ_1 and μ_3) are closer to unity.

The organic monolayer-coated, sterically stabilized nanocrystals can be dispersed in various solvents and precipitated upon the addition of antisolvent (e.g., using solvent/antisolvent pairs such as hexane/ethanol or toluene/methanol). A miscible antisolvent increases the polarity of the solution, which aggregates particles. Nanocrystal aggregation depends sensitively on particle size; larger particles require lower antisolvent concentrations to aggregate than the smaller particles [10]. The larger hydrophobic nanocrystals expose higher surface areas to the solvent and exhibit greater interparticle interaction than the smaller particles. Therefore, particles can be separated by size by titrating organic dispersions with miscible polar solvents, using a process called size-selective precipitation. Nanocrystals differing in size by one lattice shell can be isolated using this technique [10,13]. The technique is generally practiced by visual observation of the dispersion. If the dispersion becomes cloudy, the solution has been overtitrated and particles of all sizes have aggregated; the solution must be titrated to the point where opalescence just appears.

Postsynthesis thermal annealing after size-selective precipitation has proven to be very useful for further narrowing the size distribution of gold and silver nanocrystals [6,27]. By refluxing in toluene for several hours, the size and *shape* distributions narrow to produce the highest-quality superlattices. Refluxing in solution prior to size-selective precipitation simply broadens the size distribution; the distribution prior to annealing appears to be critical to the success of this postsynthesis process [27]. It is not entirely clear which physical and chemical processes the heat treatment promotes; however, it appears that the nanocrystal shape becomes more consistent within the sample and that the surface chemistry becomes more robust. In the case of gold and silver nanocrystals coated with alkanethiol monolayers, the thiols can serve as etchants. Because the surface-bound ligands are in dynamic equilibrium with free ligands, the particles may approach a preferred shape during solution-phase annealing that aids superlattice formation.

II. NANOCRYSTAL CHARACTERIZATION

Electron microscopy and small-angle x-ray scattering provide complementary materials characterization techniques for determining the average particle size, size distribution, and superlattice structure. High-resolution scanning and transmission electron microscopy provide direct imaging of the nanocrystals; the crystal structure can be closely examined and the shape probed. Microscopy, however, is practically limited to the examination of only a few hundred nanocrystals in a particular sample and is often highly dependent on the materials preparation procedures. Scattering techniques, on the other hand, provide an ensemble measurement of the entire sample. Because the scattering profiles are very sensitive to particle size, even small amounts of impurities of large particles or breadth in the size distribution shows up strongly (generally in a deteriorative way) in the scattering profiles of the materials. Therefore, these two characterization techniques, when used in concert, provide a very effective means of characterizing the sample.

A. Small-Angle X-ray Scattering Background and Theory

Small-angle x-ray scattering (SAXS) is a valuable experimental technique, which along with complementary characterization methods such as HR-TEM can provide a detailed description of a collection of nanocrystals. SAXS measurements can be conducted on dispersions and thin films to obtain information about the shape of the core, size distribution, interparticle interactions, and the symmety and range of order in a superlattice [10,13,28].

Small-angle x-ray scattering experiments may be carried out using laboratory-scale rotating copper-anode x-ray sources or high-power synchrotron beamlines. Laboratory-scale systems typically operate by bombarding a rotating copper anode with high-energy electrons; the valence electrons in the copper are excited to higher-energy states and decay to give off radiation with wavelengths in the x-ray regime. The radiation is typically filtered and collimated to produce a coherent, monochromatic beam of radiation. Synchrotron sources, in contrast, operate by accelerating electrons around a ring near the speed of light, a process that produces radiation over a wide range of wavelengths with intensities orders of magnitude greater than that of laboratory-scale equipment. For most applications, lab-scale equipment suffices, except in cases where high x-ray flux is required, as in cases where the scattering density of the sample is low or in the case of time-dependent measurements of dynamic phenomena [11,29].

Scattering techniques, in general, measure the angular dependence of (time averaged) scattered radiation from a sample. The scattering of an incident beam of electromagnetic radiation or neutrons from a sample results from density and/or concentration fluctuations; x-ray scattering, for example, arises due to fluctuations between the electron density in the scattering object (e.g., nanoparticles) and the background media (e.g., solvent). Typically, the electron density of the capping ligand does not differ significantly from the background solvent and the scattering may be assumed due only to the nanocrystal core.

The angle between an incident x-ray beam and the scattered beam is denoted as 2θ. In many cases, it is more illustrative to relate the scattering angle to the scattering wave vector, $\mathbf{q}$; the relationship between the two is given by $q = |\mathbf{q}| = (4\pi/\lambda) \sin(\theta)$. The scattering wave vector is inversely proportional to a characteristic distance, d, in the system: $q = 2\pi/d$.

Scattered radiation is collected using solid-state or gas-filled detectors and corrected for background scattering due to solvent, window materials, and so forth present in the experimental system. For systems with no preferred orientation (1D) or with 3D orientation, the scattering patterns are radially symmetric and the scattered intensity may be radially averaged and expressed as a function of the scattering vector.

For the general case of scattering from a homogeneous system, the scattered x-ray intensity $I(q)$ depends proportionally on the shape factor for individual nanocrystals $P(q)$, and the static structure factor $S(q)$, which results from multiple scattering in the sample: $I(q) \propto P(q)S(q)$. The proportionality constant depends on the number of scatterers, the difference in electron density between the nanocrystals and the background media, and the incident x-ray flux. $P(q)$ and $S(q)$ result from intraparticle and interparticle scattering, respectively [30].

$P(q)$ is determined from the distribution of scattering centers within a particle (i.e., the electron distribution within the particle for SAXS), and analytic expressions for many geometries have been developed. For most metal nanocrystals, it is a generally a good assumption to treat the metal core as a sphere of homogeneous electron density, although Murray and co-workers have found that some systems are better described as ellipsoidal particles [31,32]. Overall, very different particle shapes give rise to qualitatively similar shape factors, although scattering from spherical particles is somewhat uniquely characterized by oscillations at medium values of q. $S(q)$ contains information regarding interparticle interactions and reveals structural order in the system [30]. For a nanocrystal sample, $P(q)$ can be measured directly from a dilute dispersion of noninteracting particles, as $S(q) = 1$. For the case of ordered nanocrystal films, $S(q)$ exhibits sharp peaks due to Bragg diffraction between crystallographic superlattice planes in the superlattice and $I(q)$ deviates strongly from $P(q)$ as seen in Figure 6.

B. Nanocrystal Dispersions

Dilute nanocrystal dispersions allow direct measurement of $P(q)$ to provide information about the particle size and size distribution. Figures 6 and 7 show different representations of SAXS data that can aid the analysis, each with its advantages and disadvantages: standard form $\{\log[I(q)]$ versus $q\}$ Guinier plots $\{\ln[I(q)]$ versus $q^2\}$, and Porod plots $[I(q)q^4$ versus $q]$.

Guinier plots provide useful information at relatively low scattering angles and must be used with care (Figure 7). At low angles, the radius of gyration R_g, can be determined from a plot of $\ln[I(q)]$ versus q^2 because

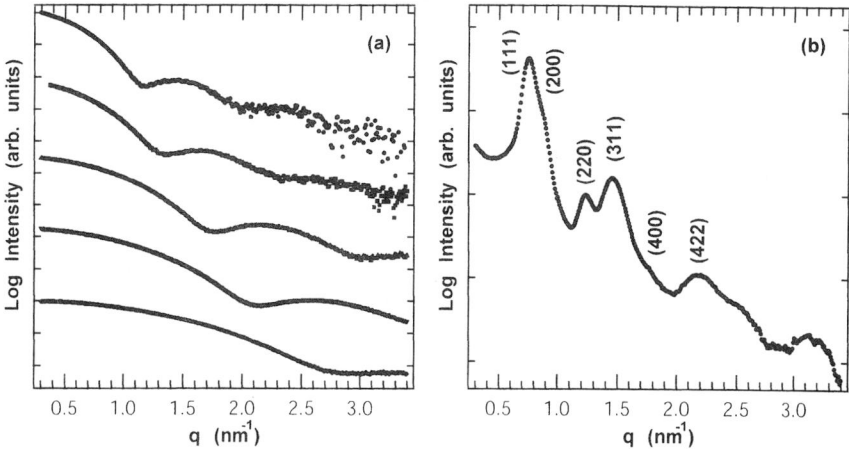

Figure 6 (a) SAXS patterns for dispersions of gold nanoparticles with average diameters of (top to bottom) 7.50, 6.60, 5.04, 4.18, and 3.04 nm. The characteristic features of the shape function shift toward higher q as the particle diameter decreases. (b) Small-angle diffraction from a superlattice of 7.5-nm-diameter gold nanoparticles capped with dodecanethiol give rise to peaks indicative of an face-centered cubic (fcc) lattice structure.

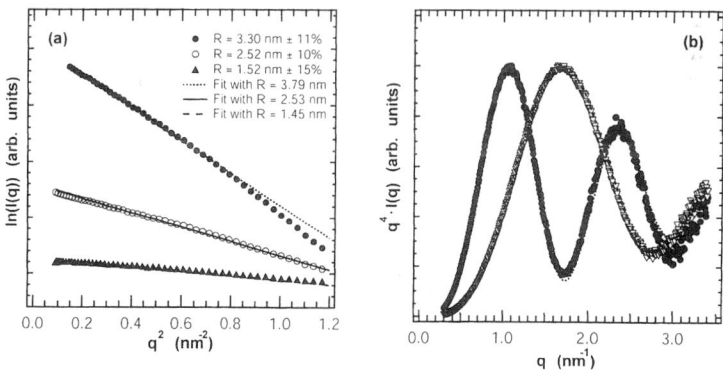

Figure 7 (a) Guinier plots for dispersions of monodisperse gold nanoparticles. The radius of gyration R_g, of the nanoparticles is obtained from the slope of the fit to the data; for spherical particles, the radius is related to the radius of gyration by $R^2 = (5/3) R_g^2$. (b) Typical Porod plots for dispersions of nanoparticles with average radii of 2.52 and 1.45 nm. The oscillations shift toward higher q for decreasing particle size.

$I(q) = I(0)\exp(-q^2 R_g^2/3)$, where $I(0)$ is the scattering intensity at zero angle [30]. For spherical paticles, the particle radius R_g relates to the particle radius as $R = R_g\sqrt{5/3}$. R_g is the volume-weighted average in a polydisperse dispersion. As a general guide, the Guinier approximation applies when $Rq \le 1$, although this may vary depending on particle geometry and polydispersity.

For values of $Rq > 1$, $I(q)$ oscillates from spherical scatterers due to constructive and destructive interference of x-rays scattered from individual particles. A representation of the scattering data as a Porod plot [$I(q)q^4$ versus q] visually enhances the oscillations for more accurate curve fitting of the experimental and modeled intensities. Fitted curves are usually generated from analytic expressions based on the shape and polydispersity of a collection of independent scatterers, although atomistic models have also been reported in the literature [31]. For example, $P(q)$ for a monodisperse sphere of constant core density is [13,30]

$$P(q) = \left[3\frac{\sin(qR) - qR\cos(qR)}{(qR)^3}\right]^2$$

The sample polydispersity strongly affects $P(q)$ and must be accounted for to fit scattering curves of real samples. The scattering intensity can be calculated when the shape of the size distribution is known using

$$I(q) = \int_0^\infty N(R)P(q)R^6 dR \tag{1}$$

where $N(R)$ is the normalized number distribution as a function of particle radius, typically taken as a Gaussian or log-normal distribution. The shape of the size distribution must be assumed to determine the nanocrystal size and size distribution from the scattering data; however, this can be checked by comparison with histograms generated from TEM images of nanocrystals. A fit of Eq. (1) to the experimental data provides the average radius and standard deviation of the particles. The polydispersity damps the scattering oscillations in the Porod region and standard deviations in particle size above ~20% nearly eliminate the oscillations, decreasing the usefulness of SAXS for sizing nanocrystal samples with broad size distributions.

For nanocrystal dispersions in which interparticle interactions are important or for concentrated solutions where the molecular motion is inhibited, $S(q) \ne 1$ and $I(q)$ deviates from $P(q)$ at low q values [10,30]. Positive deviations from the shape factor indicate interparticle attraction and may be characterized by the presence of dimers or the flocculated particles (or a negative second virial coefficient) [10]. Conversely, a negative departure of the scattering curve from the shape factor indicates strong interparticle repulsion.

Figure 8 The gold nanocrystal film shown on the left was prepared by drop-casting a concentrated dispersion onto a polymer window. A penny is shown on the right to illustrate the typical size of the film studied by SAXS.

C. Nanocrystal Thin Films

Small-angle x-ray scattering is an ideal tool to characterize condensed nano-crystal films, as it probes a relatively large region of sample and provides an averaged value of the structural order and symmetry of the superlattice, or glassy structure. Because $P(q)$ can be measured directly from dispersions,

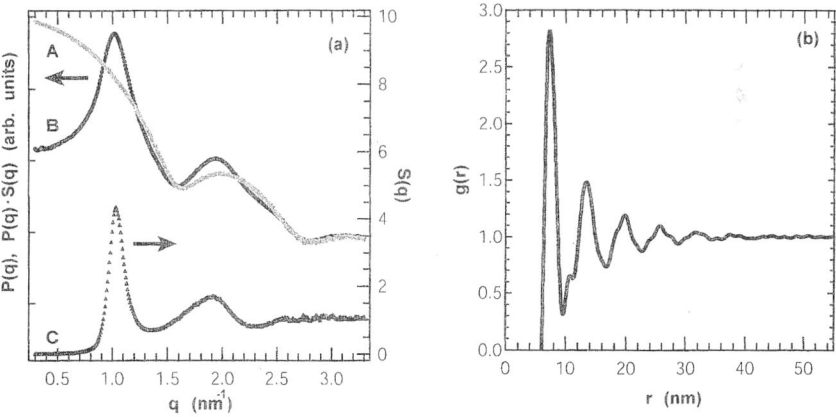

Figure 9 (a) Small-angle scattering data from a dispersion of gold nanocrystals ($R = 2.75$ nm $\pm$ 10%) in hexane (curve A) and diffraction pattern from a thin film of the dispersion drop-cast onto a polymer substrate (curve B). The structure factor $[S(q)]$ (curve C) is determined from the diffraction pattern. (b) The corresponding radial distribution function $[g(r)]$ for the thin film is calculated from a Fourier transformation of $S(q)$.

$S(q)$ can then be extracted directly from SAXS patterns of condensed nano-crystal solids. Figure 6 shows SAXS data for the metal nanocrystal super-lattice shown in Figure 8. A very useful quantity, the pair distribution function $g(r)$, can be calculated directly from $S(q)$ [33]:

$$g(r) = 1 + \frac{1}{2\pi^2 \rho r} \int q[S(q) - 1] \sin(qr) \, dq$$

$\rho g(r) 4\pi r^2 \, dr$ is the number of molecules between r and $r + dr$ about a central particle. Figure 9 shows measured values of $S(q)$, $P(q)$, and calculated values of $g(r)$. Figure 10 shows the real-space images of an ordered and disordered lattice and the corresponding values of $g(r)$.

As nanocrystals condense from an evaporating dispersion onto a substrate, their self-organization depends on the size distribution, the nature

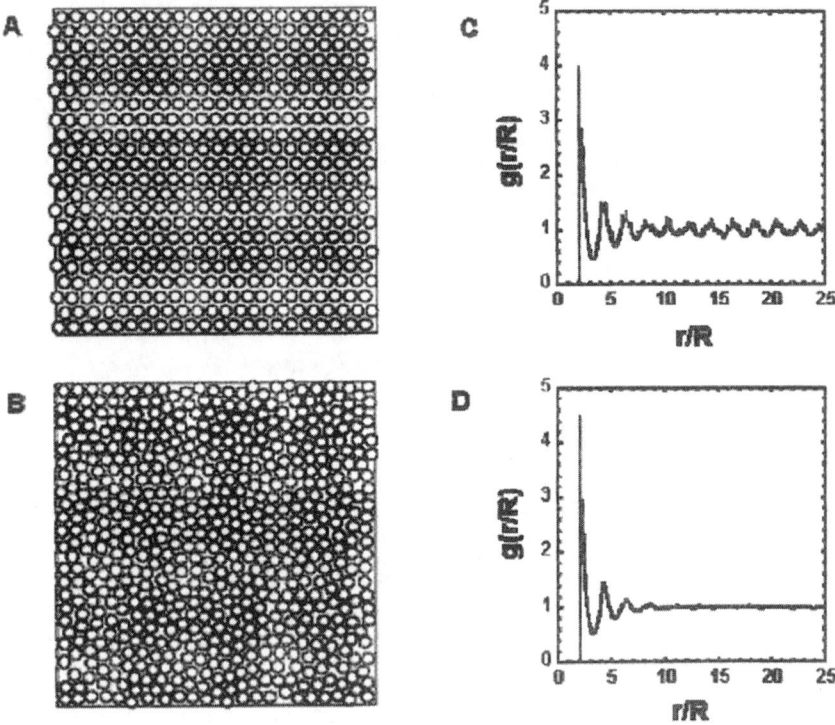

Figure 10 (A) Ordered and (B) disordered superlattices with a surface coverage of 0.608. (C) Pair distribution function of the ordered and (D) disordered superlattices as a function of disk radius, R.

of the stabilizing ligand, and the steric interactions between particles. Polydisperse nanocrystals (σ typically >10%) do not form superlattices [13,34] and the coordination number is lower than in close-packed ordered solids because long-range order does not exist. The radial distribution functions, however, for these films exhibit regular oscillations characteristic of the average particle diameter and interparticle separation, as in the case for molecular liquids [13]. Scattering from these dense fluids has enabled the calculation of pair interparticle potentials in the absence of solvent with reasonable accuracy [13]. Monodisperse nanocrystals form lattices that give rise to sharp diffraction peaks. These peaks are readily indexed to determine the superlattice crystal structure and lattice parameters [9,13].

III. SUPERLATTICE FORMATION

Nanocrystal superlattices require size-monodisperse nanocrystals that are well stabilized by organic capping ligands. Although this is a necessary condition for superlattice formation, it is not sufficient. The formation *process* is also extremely important in determining the dimensionality and the extent of order. Superlattices have been formed by essentially two different approaches: (1) Langmuir–Blodgett processing in which the density of the nanocrystal dispersion is carefully controlled through the disorder–order phase transition in two dimensions [35] and (2) evaporative crystallization, which yields superlattices in either two or three dimensions [10,36].

A. Two-Dimensional Order in Langmuir–Blodgett Techniques

Nanocrystal dispersions can be drop cast onto *liquid* surfaces to form two-dimensional superlattices, or monolayers, at an air–liquid interface. Deposition on a Langmuir–Blodgett (LB) trough provides the ability to control the surface density of the nanocrystal layer by compressing the trough. Two detailed studies have been reported for CdSe nanocrystals [37] and Ag and Au nanocrystals [36]. Bawendi and co-workers [37] used monodisperse CdSe nanocrystals capped with triotylphosphine oxide with nanocrystals that varied in size from 25 to 53 Å in diameter. Surface pressure versus area isotherms showed strong hysteresis, indicating attractive interactions between the nanocrystals in the rigid monolayer. TEM analysis of films transferred onto glass substrates and carbon-coated TEM grids showed short-range translational order and long-range nearest-neighbor orientational order—characteristics of a hexatic phase. It was also determined that the lack of long-range translational order was caused by dislocation defects, not the nanocrystal size distribution. As a result, the authors concluded that long-range translational order at the air–water interface should be possible.

Knobler and co-workers have studied the effects of nanocrystal composition, size, capping ligand length, and temperature on monolayer formation and phase behavior, focusing on monodisperse and polydisperse Ag and Au nanocrystals with diameters in the range of 20–75 Å capped with no-nanethiol, dodecanethiol, or oleylamine [35]. The horizontal lift-off technique to transfer films onto glass and carbon-coated substrates was found to give better representations of the monolayer of nanocrystals at the air–water interface, transferring cracks, fissures, holes, and other defects. Results were analyzed with respect to the geometry of a particular nanocrystal–capping ligand combination and the "excess volume" available for ligand inter-penetration was found to be the most important factor determining the phase behavior. Three different types of phase behavior were observed in this study. In Case I, involving small nanocrystals with large capping ligands, ligand interpenetration determined whether one-dimensional structures or a compressed foamlike phase formed. The surface pressure versus area isotherms were reversible, and dense, two-dimensional (2D) monolayers could be formed upon the addition of sufficient material. In Case II, for small or medium-sized nanocrystals with medium-sized capping ligands, the nanocrystal films went through a gas phase to 2D close-packed phase to 2D collapsed phase (3D phase) upon increasing surface pressure. Surface coverages were approximately 100%. This phase is the most important for forming ordered 2D monolayers of nanocrystals. In Case III, for large nanocrystals with medium-sized or small capping ligands, the surface pressure versus area isotherms were not reversible. Diffusion-limited-aggregation structures formed instead of ordered 2D monolayers. Both Ag and Au nanocrystals produced the same phase behavior. Increased temperature decreased the surface pressures necessary for transitions between the different phases.

In addition to the detailed study on the formation of ordered 2D Ag and Au nanocrystal monolayers at the air–water interface, Health and co-workers closely examined the formation of islands and stripes at the air–water interface [38,39]. Monodisperse octanethiol-capped Ag nanocrystals with diameters between 30 and 60 Å were deposited at the air–water interface at surface coverages much below those used to form ordered 2D monolayers. At the lowest surface coverages, the nanocrystals organize into 2D islands. As the surface coverage was increased, either by compression or increased nanocrystal loading on the surface, a transition from islands to stripes was observed. Stripe thickness and stripe-to-stripe spacing was observed to be a function of nanocrystal size and the dispersing solvent used. Increased surface pressure decreased the stripe-to-stripe spacing and increased the collective alignment of the stripes. Upon further compression, the stripes coalesced in an irreversible process to form a close-packed 2D monolayer. Health and co-workers successfully modeled the 2D island and stripe formation using a pair potential with an attractive interaction and a longer-ranged repulsion [38]. About the islands and stripes, they concluded that their formation results from attractive forces that bring the nanocrystals together and the longer-ranged repulsive forces that limit the aggregated domain size. About the island-to-stripe transition upon increased surface coverage, they concluded that a "spontaneous reorganization" occurs "as the repulsions between the aggregates become more important than those between the individual particles within them."

The LB approach to monolayer formation provides two technologically attractive features. The first is the ability to compress the nanocrystal monolayer to interparticle distances much closer than those allowed under normal solution drop-cast conditions, thus providing greater control over the geometric properties of nanocrystal monolayers that will be integral for future scientific inquiries and device applications. The second is the ability to transfer a nearly exact copy of the nanocrystal film at the air–water interface to virtually any substrate. This is also an immensely important feature for the design and implementation of nanocrystal-based electrical and optical devices.

B. Evaporative Self-Assembly

1. Hard-Sphere Behavior

In situ SAXS measurements of an evaporating concentrated dispersion of C_{12}-coated silver nanocrystals reveal that the particles spontaneously self-assemble during solvent evaporation provided that they are sufficiently size monodisperse and well stabilized from aggregation [11]. These nanocrystals are sufficiently small that their diffusive characteristic time is comparable to the characteristic time for solvent evaporation and they can order into their thermodynamically lowest-energy phase [10]. This is qualitatively different than large 0.1–10- µm colloids that require days to settle slowly from solution to form colloidal crystals due to their slow diffusion times. In a concentrated nanocrystal dispersion, a 5-nm-diameter particle can diffuse up to 10 nm during the time it takes for the solvent to evaporate 10 nm in thickness; the nanocrystals can sample their available phase space and it can be argued that nanocrystal ordering can be treated as an equilibrium (or quasiequilibrium) problem. As a first approximation, the statistical mechanics of particles that experience hard-sphere interactions with interparticle potential $\Phi(r)$, defined as a function of center-to-center interparticle separation r, and the particle radius R, where

$$\Phi = \begin{cases} \infty & \text{if } r \leq 2R \\ 0 & \text{if } r > 2R \end{cases}$$

provides a conceptual framework for understanding the disorder–order phase transition observed for sterically stabilized metal nanocrystals. One could hypothetically calculate the free energy of the solid and fluid phases,

$$F = -kT \ln \left(\sum_{\text{configuration}} \exp \left(- \sum_{(i,j,\text{bonds})} \frac{\Phi_{i,j}}{kT} \right) \right)$$

where k is Boltzmann's constant. Because $\Phi(r)$ is either 0 or ∞,

$$\exp \left(- \sum_{(i,j,\text{bonds})} \frac{\Phi_{i,j}}{kT} \right) = \begin{cases} 0 & \text{if } r \leq 2R \\ 1 & \text{if } r > 2R \end{cases}$$

and therefore, the free energy depends only on the possible configurations or packing geometries in the fluid; it depends only on the entropy of the system. At low nanocrystal densities, the disordered fluid can achieve many more configurations than the ordered phase and it is thermodynamically favored. However, at relatively high densities, the solid phase allows the particles greater free volume relative to the dense fluid. For example, the fcc lattice exhibits a maximum packing density of 0.74, whereas the disordered phase freezes into a closest packed glass at a volume fraction of 0.68. The disorder–order phase transition for hard spheres has therefore been termed a disorder-avoiding phase transition where the entropic driving force results from an increase in microscopic disorder that exceeds the macroscopic configurational entropy loss due to crystallization [40,41]. These considerations also hold true in two dimensions, even though true long-range order cannot exist. In fact, Gray and co-workers [34,42] found that nanocrystal monolayers can proceed through a fluid–hexatic–hexagonal close-packed (hcp) phase transition under the appropriate conditions.

Conceptually, an understanding of hard-sphere crystallization is important for understanding nanocrystal superlattice formation. However, the nanocrystals are not truly hard spheres and the interparticle interaction energies can exceed $1/2\ kT$ for particles less than 10 nm in diameter, despite their small size [10,33]. In 1995, Ohara et al. revealed that size-dependent van der Waals attractions between particles resulted in the formation of polydisperse aggregates of Au nanocrystals, with the largest particles located on the interior of the aggregate and progressively smaller particles located on the exterior [43]. The aggregation was reversible and the particles were readily redispersed with the addition of organic solvent. Molecular dynamics simulations also confirmed the importance of the van der Waals attraction on the formation of the aggregates. The interparticle attraction has been found to be very important to superlattice formation [36,44].

2. Steric Stabilization and a Soft Repulsion

Attractive van der Waals forces due to dipole-induced dipole interactions are relatively strong for metal nanocrystals due to the high polarizability of the cores, despite their small size and relatively thick organic coating. [For example, dodecanethiol (C_{12} thiol) extends approximately 1.5 nm into solution; therefore, the C_{12}-coating surrounding a 5-nm-diameter nanocrystal, for example, occupies 75% of the total volume occupied by the core and ligand!] The van der Waals attraction Φ_{vdW}, between two spheres of equal radius R, separated with a center-to-center distance of d and a Hamaker constant A [45] is

$$\Phi_{vdW} = -\frac{A}{6}\left[\frac{2R^2}{d^2 - 4R^2} + \frac{2R^2}{d^2} + \ln\left(\frac{d^2 - 4R^2}{d^2}\right)\right]$$

The Hamaker constant reflects the strength of the core–core attraction and depends on the polarizability of the sphere and the dispersing medium. The Hamaker constants for metals interacting across a vacuum have been determined. For example, the tabulated value for bulk silver is $A_{11} = 2.185$ eV [46]. To calculate the Hamaker constant of the nanocrystals dispersed in different solvents (denoted as A_{131}, where the subscript 1 represents the metal and the subscript 3 represents the solvent), the following relationship can be used:

$$A_{131} \approx \left(\sqrt{A_{11}} - \sqrt{A_{33}}\right)^2$$

where A_{11} is for the metal interacting across vacuum and A_{33} is for the solvent interacting across vacuum. The value for A_{33} can be estimated using a simplification of Lifshitz theory [46]:

$$A_{33} = \frac{3}{4}kT\left(\frac{\varepsilon_3 - \varepsilon_{vacuum}}{\varepsilon_3 + \varepsilon_{vacuum}}\right)^2 + \frac{3h\upsilon_e(n_3^2 - n_{vacuum}^2)^2}{16\sqrt{2}(n_3^2 + n_{vacuum}^2)^{3/2}}$$

where ε is the dielectric constant, n is the refractive index, h is Planck's constant, k is Boltzmann's constant, T is temperature, υ_e is the maximum electronic ultraviolet adsorption frequency typically taken to be 3×10^{15} s^{-1}, and $\varepsilon_{vacuum} = 1$ and $n_{vacuum} = 1$.

The typical Hamaker constant for silver in conventional solvents under ambient conditions such as hexane is $A_{131} = 0.91$ eV. This is very high, considering, for example, that A for polystyrene in water is 0.087 eV and is 0.125 eV for mica in water. Therefore, despite their small nanometer size, values for Φ_{vdW} can be significant compared to kT.

The adsorbed capping ligands provide a short-range steric barrier to particle aggregation. As two nanocrystal cores capped with stabilizing ligands approach each other, the ligand tails repel each other in a good solvent, allowing for nanocrystal dispersibility. According to a theory developed by Vincent et al. [47], the total steric repulsive energy consists of osmotic Φ_{osm} and elastic Φ_{elas} contributions, and the total interaction energy Φ_{total} is the sum of the forces:

$$\Phi_{total} = \Phi_{vdW} + \Phi_{osm} + \Phi_{elas}$$

Φ_{total} depends on nanocrystal size, ligand composition, length and graft density, and the solvent condition. The osmotic term Φ_{osm} results from the energetic balance between solvent–ligand tail and tail–tail interactions. The solvent conditions and ligand length l largely control Φ_{osm} [47]:

$$\Phi_{osm} = \frac{4\pi R k_b T}{v_{solv}} \phi^2 \left(\frac{1}{2} - \chi\right)\left(1 - \frac{d - 2R}{2}\right)^2, \quad l < d - 2R < 2l$$

$$\Phi_{osm} = \frac{4\pi R k_b T}{v_{solv}} \phi^2 \left(\frac{1}{2} - \chi\right)\left\{l^2\left[\frac{d - 2R}{2l} - \frac{1}{4} - \ln\left(\frac{d - 2R}{l}\right)\right]\right\}, \quad d - 2R < l$$

Here, v_{solv} is the molecular volume of the solvent, ϕ is the volume fraction profile of the stabilizer extending from the particle surface, and χ is the Flory–Huggins interaction parameter. In Flory–Huggins theory, $\chi = \frac{1}{2}$ typically represents the boundary between a good solvent ($\chi < 1/2$) and a poor solvent ($\chi > \frac{1}{2}$).

The elastic repulsive energy Φ_{elas} originates from the entropy loss that occurs upon compression of the stabilizing ligands, which is only important at interparticle separations in the range $d - 2R < l$ [47]:

$$\Phi_{elas} = \frac{2\pi R k_b T l^2 \phi \rho}{MW_2} \left\{x \ln\left[x\left(\frac{3 - x}{2}\right)^2\right] - 6\ln\left(\frac{3 - x}{2}\right) + 3(1 - x)\right\}$$

$$x = \frac{d - 2R}{l}$$

Here, ρ and MW_2 represent the ligand density and molecular weight. Because the elastic term represents the physical compression of the stabilizer, this term is repulsive at all of the solvent conditions. The dispersion stability is essentially controlled by the osmotic term because it becomes effective as soon as the ligands start to overlap, $l < d - 2R < 2l$, and the elastic term does not contribute significantly to Φ_{total} until the ligands are forced to compress (i.e., when $d - 2R < l$).

The ligand volume fraction profile extending from the nanocrystal surface can be determined for good solvent conditions in which the ligands are fully extended. Moving away from the surface, ϕ decreases significantly due to the curvature of the nanocrystal surface. This decrease is further exaggerated in smaller nanocrystals with greater surface curvature. The ϕ curves can be calculated using the geometric equation for cylinders with ligand cross-section area (SA_{thiol}) of 14.5 Å, extending radially from a curved surface with radius $R_p + z$ [26]:

$$\phi(z) = \frac{SA_{thiol} R_p}{\theta_{thiol}(R_p + z)}$$

where θ_{thiol} is the surface area per thiol head group, which represents the binding density, and z is the radial distance from the metal surface.

The most important parameters determining the strength of the steric repulsion are the ligand length l and the Flory–Huggins interaction parameter χ between the ligand and the solvent. When $\chi > \frac{1}{2}$, Φ_{osm} becomes attractive due to the poor solubility of the ligands. For conventional solvents, the Flory–Huggins parameter can be estimated using solubility parameters or comparing cohesive energy densities [48]. Nanocrystal dispersibility depends on the balance between the attractive van der Waals forces and the repulsive steric forces. At good solvent conditions, the ligands provide strong repulsive forces, whereas a weak repulsion or even an attraction can exist under poor solvent conditions, leading to flocculation of the nanocrystals.

3. Superlattice Crystallization

Energetic attraction between particles lowers the nanocrystal solubility and provides a greater thermodynamic driving force for superlattice formation. Superlattice nucleation and growth, however, requires nanocrystals to desolvate, and the activation barrier to this process is much lower on a surface than in homogeneous solution. In the case of fast evaporative assembly from a volatile solvent, nucleation and growth of nanocrystal superlattices proceeds by a heterogeneous process. For example, nanocrystals drop-cast from a concentrated dispersion in a good solvent, such as chloroform or hexane, under conditions of rapid evaporation crystallize primarily on the $(111)_{SL}$ plane (where the subscript SL denotes superlattice). Time-resolved SAXS has been used to observe the formation of nanocrystal superlattice forms during solvent evaporation from concentrated nanocrystal dispersions and the scattering intensity profile shows the transition from the characteristics shape factor for a dispersion of noninteracting particles to the organized superlattice, giving rise to the $(111)_{SL}$ diffraction peak from an fcc superlattice [11]. The significant prominence of this peak relative to the others also reveals the heterogeneous crystallization and growth of the superlattice on the $(111)_{SL}$ lattice plane.

Many features observed for superlattice growth are consistent with the model developed by Burton et al. for the growth of crystals from atoms in the vapor phase that describes the growth of crystal layers based on the steps and kinks along the crystal surface that serve as adsorption sites for additional atoms [49,50]. Slight differences in interparticle attraction can lead to rather dramatic changes in superlattice film morphology. For example, the Hamaker constant for gold in hexane is approximately 20% higher than for gold in chloroform due to the lower dielectric screening in hexane. Even though both chloroform and hexane are considered good solvents for alkanethiol capping ligands with $\chi \approx 0$, the films deposited from hexane are very rough with round superlattice domains, whereas the films from chloroform are relatively smooth, as shown in Figs. 2E and 2F. HR-SEM images of films evaporated from these two solvents, using the same nanocrystal size and capping ligands, show that the superlattice crystallizes preferably in the lowest energy $[110]_{SL}$ direction, whereas superlattice growth from hexane is isotropic: The increased interparticle attraction in the solvent leads to similar crystal growth rates in all crystallographic directions.

Homogeneous superlattice nucleation can be induced by adding a relatively small amount of antisolvent to the dispersion. The antisolvent increases the supersaturation of the dispersion and promotes interparticle attractions. If too much antisolvent is added, the nanocrystals will simply flocculate. The activation barrier to homogeneous nucleation, however, is significant. The external addition of energy, for example, by sonication can induce homogeneous nucleation as shown in Fig. 11, but the rapid evaporation of the solvent under relatively poor solvent conditions results in a disordered nanocrystal film. Apparently, the balance between solvation and desolvation of the nanocrystals during the crystallization process does not provide sufficient entropic freedom for crystallization to occur in this timescale.

In addition to isotropic van der Waals forces and steric repulsion between particles, researchers have identified the potential importance of chain–chain interactions and shape effects [9,13,51]. For example, the superlattice structure depends on the ratio of the capping ligand chain length $\langle L \rangle$, to the radius of the nanocrystal core R: $\chi = \langle L \rangle / R$. [9,13]. Whetten and co-workers [9] found that fcc packing is preferred below $\chi \approx 0.73$, whereas body-centered cubic (bcc) order occurs above this value. As χ increases, the energetic interparticle attraction decreases, which induces the structural transition to the more entropically favored bcc structure. Ligand–ligand interactions can also lead to other deviations from fcc packing. For example, Luedtke and Landman performed molecular dynamics simulations, finding that dodecane chains adsorbed to gold nanocrystals could form interlocking bundles in contrast to shorter butyl groups which did not exhibit bundling [51]. Based on TEM investigations, nanocrystals have been found to form a variety of

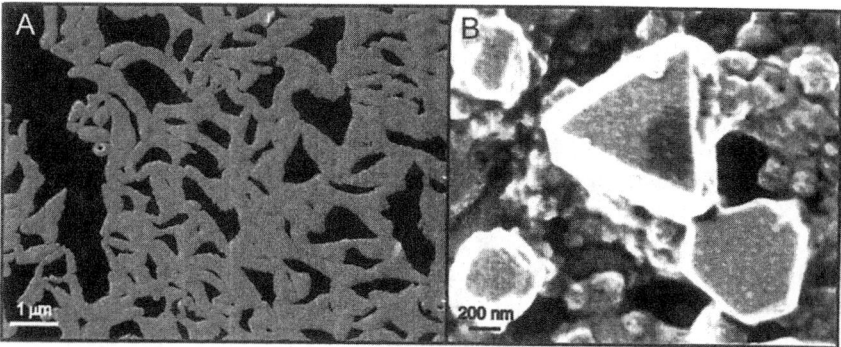

Figure 11 High-resolution SEM image of gold nanocrystals deposited from a dispersion of 10% ethanol and 90% hexane by volume (A) without and (B) with sonication. Without sonication, the superlattices exhibit rough films with significant areas of dendritic growth and some areas of homogeneous aggregation. Sonication provides the energy necessary to induce homogeneous nucleation resulting in nanocrystal superlattice crystallites that settle onto the substrate.

structures including spheres, ellipsoids, tetrahedron, truncated octahedron, decahedron, and icosahedron that affect packing geometry [32,52,53].

C. Solvent Effects During Nanocrystal Deposition and Thin-Film Formation

On a macroscopic scale, solvent evaporation can lead to convective instabilities and substrate dewetting phenomena that affect nanocrystal film formation [54–56]. On a local microscopic scale, the nanocrystals organize into close-packed superlattices according to the guidelines discussed earlier. However, very interesting macroscopic patterning can occur, particularly when depositing nanocrystals by evaporation of a volatile solvent.

An evaporating drop of a concentrated nanocrystal dispersion exhibits various flow patterns due to the temperature gradient that occurs between the center of the droplet and the outer droplet edge [57,58]. Solvent evaporates faster at the edge of the droplet due to the boundary conditions above the droplet and the evaporative flux, for example, can be expressed as $J(r) = J_0(1 - r/R)^{-\lambda}$, where $\lambda = (\pi - 2\theta_c)/(2\pi - 2\theta_c)$. In fact, the flux diverges at the edge of the droplet, leading to a significant mass, thermal, and surface tension gradients between the center of the drop and the outer rim. These gradients, in turn, pull the nanocrystals to the edge of the droplet. The evaporating droplets generally exhibit a stick–slip motion during evaporation, where they become pinned for a short time as nanocrystals accumulate at the droplet edge and deposit as the droplet snaps away suddenly as evaporation proceeds [57–60]. This evaporation process deposits a series of concentric rings, as shown in Fig. 12.

Solvent evaporation can occur under conditions of diffusive mass and heat transfer, or convective transfer. It is generally difficult to avoid convection in the system; however, on local length scales, solvent evaporation can occur in a controlled fashion, particularly when the solvent viscosity is high and volatility is low. In this case, solvent dewetting from the substrate can lead to the formation of ring patterns of nanocrystals [54,55]. An evaporating wetting film will reach a critical thickness when holes open up in the film in an attempt to maintain the equilibrium film thickness of the solvent layer, $t_e = (3A_H/S)^{1/2}$, which depends on the Hamaker constant A_H and the spreading coeffcient S [61]. The opening hole drags the nanocrystals with the solvent front, collecting more nanocrystals as it continues to widen. Once the hole collects a critical number of nanocrystals, it becomes pinned due to the frictional force resulting from the interactions between the particles and the substrate, and the nanocrystals deposit as a ring. Figure 13A shows a TEM image of evaporated rings of gold nanocrystals. Note the relatively narrow size distribution indicative of diffusion-limited ring growth. If several holes are opening within close proximity to each other, it is possible for these holes to merge together and make a loose network of nanoparticles.

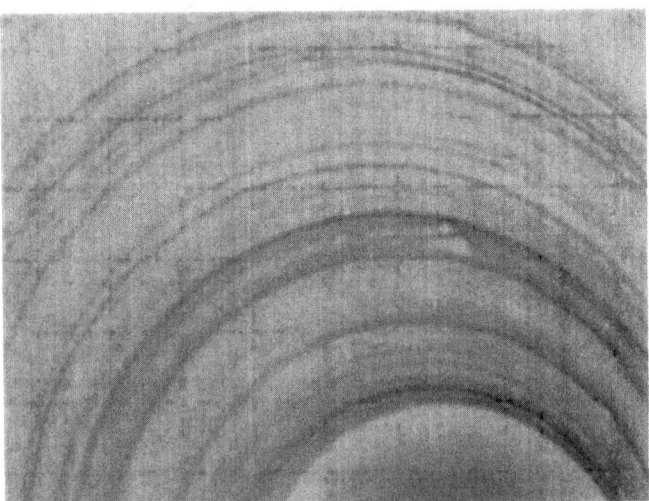

Figure 12 Nanocrystals drop-cast onto a mica substrate exhibit concentric rings characteristic of stick–slip evaporative deposition of colloidal particles resulting from mass, thermal, and surface tension gradients in the droplet.

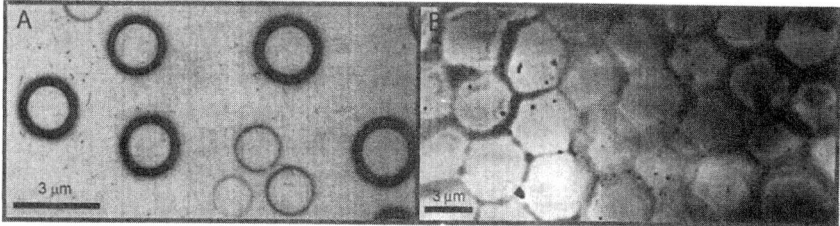

Figure 13 (A) Transmission electron microscopic image of a 1.6-μm dewetting ring formed from 35-Å gold nanocrystals in chloroform with an amorphous carbon film as the substrate; (B) TEM image of hexagonal networks formed by Marangoni convection of 35-Å diameter gold nanoparticles deposited using chloroform.

Convective instability in the evaporating fluid eliminates the possibility of hole formation due to dewetting, yet it can lead to the organization of convective cells and Marangoni flow [54,56]. Solvent evaporation at the droplet surface leads to a *vertical* temperature gradient that can under certain conditions lead to convective fluid transport. The nanocrystals become swept up in the convective currents. Marangoni flow occurs when the dimensionless Marangoni number, $Ma = \sigma_T \Delta Td/\rho v \kappa$, exceeds a critical value of 80, where σ is the solvent surface tension, $\sigma_T = |d\sigma/dT|$, ΔT is the temperature difference between the upper and lower fluid boundaries, κ is the thermal diffusivity, v is the kinematic viscosity, ρ is density, and d is the film thickness [62]. The critical lattice parameter of the convective cells is $a_c = 2\pi d/L$, where $a_c = 2$ at the onset of Marangoni convection and L is the length of the hexagonal repeating unit. Figure 13B shows an example of the hexagonal networks of ribbons of nanocrystals with a 4.3-μm lattice spacing that have settled at the boundaries separating the spatially organized convective cells [54].

IV. PHYSICAL CONSEQUENCES OF SUPERLATTICE ORDER

The tunability of superlattice structure with changes in nanocrystal composition, size, interparticle spacing, and symmetry provides unique opportunities to manipulate optical and electronic material properties. A number of recent studies have focused on electron transport through monolayers of metal nanocrystal superlattices [2,63–66]. Although these studies have been conducted on different types of nanocrystals (Co [63], Au [2,64], and Ag [65,66]), a few general trends have been observed. Most importantly, single-nanocrystal charging energies were found to dominate electron transport. Second, the currents are very low with conductivity values more characteristic of a semiconductor. This is a result of the organic capping molecules. The electrons must either hop from metal core to metal core via an activated process or tunnel through the organic medium surrounding the nanocrystals. The current was also found to decrease as the interparticle separation increased. At temperatures lower than approximately 200 K, the conductivity decreased with decreasing temperature, characteristic of a semiconductor. With one exception [64], the researchers attribute this behavior to a thermally activated hopping process. Two of the studies [63,64] found nonlinear $I-V$ curves near zero bias at low temperatures ($T < 70$ K), whereas Health and co-workers [65,66] did not observe non-Ohmic behavior until the temperature dipped below ~50 K, which they attribute to the much smaller interparticle separation in their films resulting from transfer from an LB trough.

Several researchers have now examined the effect of topological disorder on electron transport through condensed metal nanocrystal films. Parthasarathy et al. [64], for example, controlled the degree of superlattice order between the device electrodes by slowing the rate of evaporation of the nanocrystal dispersion. Fast evaporation left voids occupying approximately 15–20% of the area spanning the electrodes. Slowing the evaporation rate by adding excess capping ligand—in this case, dodecanethiol—decreased the void space to ~5% and long-range translational order in the array was observed. The topological disorder was seen in the I–V curves as a deviation from the simple power-law behavior predicted by Middleton and Wingreen [67].

Black et al. [63] and Beverly et al. [66] examined the topological disorder of a nanocrystal superlattice when the size distribution was broad, above ~10%. Black et al. [63] found that the decreasing conductivity with temperature of their Co nanocrystal superlattices could be modeled by thermally activated hopping between identical metal islands. Deviations from this model occurred when the nanocrystal size distribution exceeded 15%. Beverly et al. [66] systematically increased the size distribution of their Ag nanocrystals while keeping the average nanocrystal diameter and interparticle spacing constant. In all of their nanocrystal monolayers, Beverly et al. [66] observed a transition from an initial, room-temperature metal-like electron transport process to a thermally activated hopping mechanism (~200 K). They also observed another transition at lower temperatures (<100 K) from the thermally activated hopping mechanism to an Efros–Shklovskii variable range hopping (ES-VRH) mechanism. Whereas the temperature for the "metal–insulator transition" was not a function of nanocrystal size distribution, the temperature at which the thermally activated hopping/ES-VRH transition took place increased as the nanocrystal size distribution increased. The authors suggest that this second transition is a function of the density of localized states at the Fermi level and thus, the disorder of the nanocrystal superlattice. Beverly et al. [66] also found the conductivity increased as the nanocrystal size distribution decreased.

Sampaio et al. [65] detailed the effect of interparticle spacing on the electron transport through a nanocrystal superlattice. By lifting films of Ag nanocrystals off of a LB trough at different surface pressures, they were able to vary the interparticle spacing while keeping the nanocrystal size constant. The conductivity of the nanocrystal superlattices increased as interparticle separation decreased. Also, the temperature of the transition from a metallike electron transport to a thermally activated hopping mechanism decreased as the interparticle spacing decreased, and, along the same lines, the activation energy in the thermally activated hopping region decreased as the interparticle separation decreased. All of these results are consistent with an electron transport process that is dominated by the single-nanocrystal charging energy.

Electron transport through three-dimensional nanocrystal superlattices has also been studied. [68]. The size and size distribution both have an effect on the conductivity. Figure 14 shows a typical interdigitated array electrode fabricated by optical lithography used to measure electron transport through the nanocrystal superlattice. The electrode spacing in this image is 10 μm. Compare this to the electrodes with a 100-nm separation defined by electron beam lithography shown in Fig. 15. The silver nanocrystals were drop cast on the substrate to form an incomplete monolayer. Doty et al. [68] observed a metal–insulator transition in the temperature-dependent dc conductivity, defined as a change in sign of the slope of the resistance versus temperature, occurred for size-monodisperse nanocrystals ordered into 3D superlattices (Fig. 16). This behavior was later confirmed for monolayers of Ag nanocrystals as well [65]. Doty et al. [68] found that the metal-insulator transition does not occur for thin films with large size distributions and topological disorder (>20%). Instead, the conductivity showed semiconducting conductivity versus temperature behavior throughout the entire temperature range studied, as shown in Fig. 16. The TEM images in Fig. 17 show the significant difference in nanocrystal order in the monolayer for monodisperse and polydisperse nanocrystals. The qualitative difference in electron transport through spatially ordered and disordered superlattices occurs because the quantum mechanical exchange interactions between nanocrystals cannot overcome the disorder induced by the nanocrystal size distribution, and the *disordered* superlattice is an Anderson insulator. The authors also found that the "metal–insulator transition" temperature decreased with increasing nanocrystal diameter, as did the activation energy for transport in the thermally activated hopping regime. Deposition of 3D nanocrystal superlattices with specified film thicknesses and height, however, is difficult and quantitative device-to-device comparisons are hard to make.

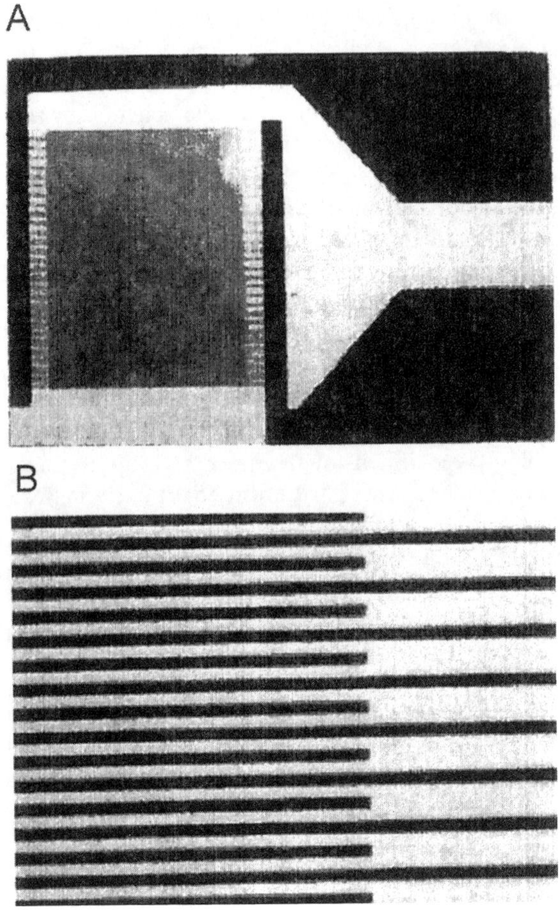

A

B

Figure 14   (A) Optical micrograph of an interdigitated array electrode; (B) the active region of the electrode consists of fingers with a 10- μm spacing.

Another less utilized, tunable property of nanocrystal superlattices is the type of molecule used to cap the individual nanocrystals. The most widely touted alternative to simple alkanethiols is an aromatic molecule so that the effect of π-orbitals on electron transport can be investigated [2]. Incorporating different kinds of capping ligands into the nanocrystal synthesis is not always easy, however. The different binding strengths of the ligands can have a noticeable effect on the final nanocrystal size, size distribution, and solubility characteristics. The most common way of circumventing this problem is through a ligand-exchange reaction with previously synthesized alkane-

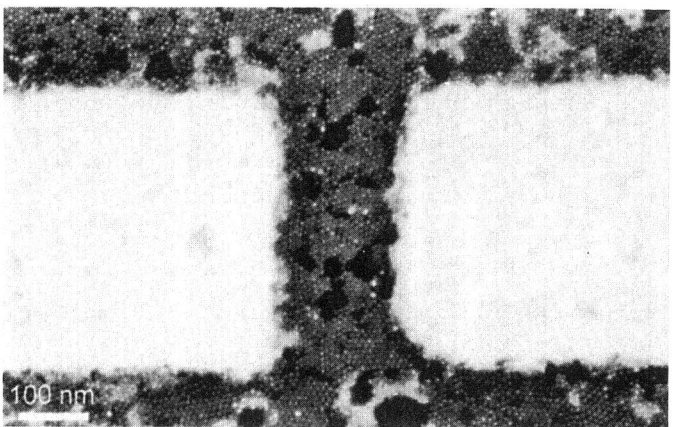

Figure 15 High-resolution SEM image of C_{12}-capped Au nanocrystals (4.92 nm ± 10%) deposited between gold electrodes with an ~150-nm gap.

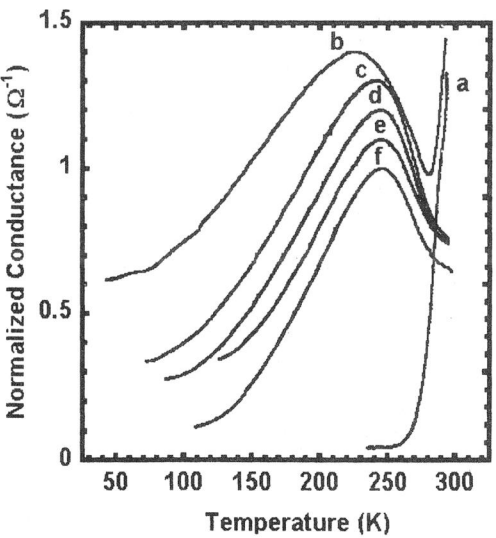

Figure 16 Normalized conductance as a function of temperature for Ag nanocrystal films: (a) original, size-polydisperse sample; size-selected, monodisperse samples with diameters of (b) 7.7 nm, (c) 5.5 nm, (d) 4.8 nm, (e) 4.2 nm, and (f) 3.5 nm.

A

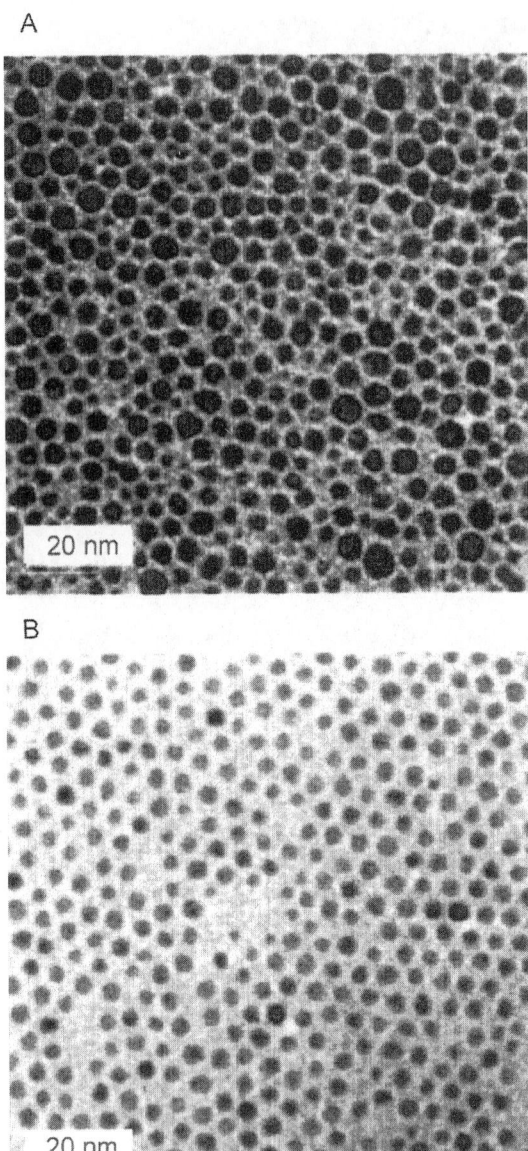

20 nm

B

20 nm

Figure 17 Transmission electron microscopic images of (A) size-polydisperse (3.8 ± 0.8 nm) dodecanethiol-capped silver nanocrystals and (B) size-monodisperse (3.7 ± 0.3 nm) dodecanethiol-capped silver nanocrystals.

thiol-capped nanocrystals [2]. The effect of different ligands is an area of nanocrystal research that has not been extensively studied and may have important implications for chemical sensor applications, particularly vapor-sensing devices.

V. CONCLUSIONS AND FUTURE WORK

A general strategy and understanding of metal nanocrystal synthesis and superlattice assembly now exists. Clearly, the structural order in the super-lattice greatly affects the properties of the materials, as in the case of electron transport, where polydispersity and structural disorder can lead to qualitatively different behavior. Unfortunately, synthesis of new materials still requires significant trial and error to determine compatible capping ligands with the material of interest and feasible reaction chemistry. One important future focus of metal nanocrystal synthesis will be in alloy and layered materials. These materials could have applications in catalysis and nanoscale electronics. Many magnetic alloys, for example, exhibit complicated phase behavior and it will be interesting to explore the effects of the nanosize of the material on the structures and metastability of different phases under these conditions. In magnetic nanocrystal films in particular, packing symmetry, interparticle spacing, and particle size will be very important in determining the properties of these materials. The potential tunability of the material properties and their future applications rests on the ability to synthesize a wide variety of materials with very narrow size and shape distribution. Also, of course, the ability to direct the assembly of superlattices to specific regions of a substrate will be required in order to employ these materials into real applications. Many imprint and soft lithographic patterning techniques should be compatible with these materials, providing a potential route to overcome these obstacles.

REFERENCES

1. Sun, S.; Murray, C.B. J. Appl. Phys. 1999, *85*, 4325.
2. Andres, R.P.; Bielefeld, J.D.; Henderson, J.I., et al. Science 1996, *273*, 1690.
3. Brust, M.; Walker, M.; Bethell, D., et al. J. Chem. Soc. Chem. Commun. 1994, 801.
4. Courty, A.; Fermon, C.; Pileni, M.P. Adv. Mater. 2001, *13*, 254.
5. Fink, J.; Kiely, C.J.; Bethell, D., et al. Chem. Mater. 1998, *10*, 922.
6. Lin, X.M.; Jaeger, H.M.; Sorensen, C.M., et al. J. Phys. Chem. B 2001, *105*, 3353.
7. Park, J.I.; Kang, N.J.; Jun, Y.W., et al. ChemPhysChem. 2002, *6*, 543.

8. Shevchenko, E.; Talapin, D.; Kornowski, A., et al. Adv. Mater. 2002, *14*, 287.
9. Whetten, R.L.; Shafigullin, M.N.; Khoury, J.T., et al. Acc. Chem. Res. 1999, *32*, 397.
10. Korgel, B.A.; Fullam, S.; Connolly, S., et al. J. Phys. Chem. B 1998, *102*, 8379.
11. Connolly, S.; Fullam, S.; Korgel, B., et al. J. Am. Chem. Soc. 1998, *120*, 2969.
12. Weitz, I.S.; Sample, J.L.; Ries, R.; Spain, E.M.; Heath, J.R. J. Phys. Chem. B 2000, *104*, 4288.
13. Korgel, B.A.; Fitzmaurice, D. Phys. Rev. B 1999, *59*, 14–191.
14. Levin, I.; Ott, E. J. Am. Chem. Soc. 1932, *54*, 828.
15. Darragh, P.J.; Gaskin, A.J.; Terrell, B.C.; Sanders, J.V. Nature 1966, *209*, 13.
16. Sanders, J.V. Phil. Mag. A 1980, *42*, 704.
17. Raman, C.V.; Jayaraman, A. Proc. Indian Acad. Sci. 1953, *38A*, 343.
18. Copisarow, A.C.; Copisarow, M.J. Nature 1946, *157*, 768.
19. Schmid, G.; Lehnert, A. Angew. Chem. Int. Ed. Engl. 1989, *28*, 780.
20. Bentzon, M.D.; van Wonterghem, J.; Morup, S.; Tholen, A.; Koch, C.J.W. Phil. Mag. B 1989, *60*, 169.
21. Turkevich, J.; Stevenson, P.C.; Hillier, J. Discuss. Faraday Soc. 1951, *55*, 55.
22. Steigerwald, M.L.; Alivisatos, A.P.; Gibson, J.M., et al. J. Am. Chem. Soc. 1988, *110*, 3046.
23. Sun, S.H.; Murray, C.B.; Weller, D.; Folks, L.; Moser, A. Science 2000, *287*, 1989.
24. Larsen, T.; Sigman, M.; Ghezelbash, A.; Doty, R.C.; Korgel, B.A., J. Am. Chem. Soc. 2003, *125*, 5638–5639.
25. Pich, J.; Friedlander, S.K.; Lai, F.S. Aerosol Sci. 1970, *1*, 115.
26. Shah, P.S.; Husain, S.; Johnston, K.P., et al. J. Phys. Chem. B 2002, *106*, 12–178.
27. Doty, R.C.; Korgel, B.A. unpublished data.
28. Korgel, B.A. Phys Rev. Lett. 2001, *86*, 127.
29. Korgel, B.A.; Zaccheroni, N.; Fitzmaurice, D. J. Am. Chem. Soc. 1999, *121*, 3533.
30. Glatter, O.; Kratky, O., Eds. *Small Angle X-ray Scattering*; Academic Press: New York, 1982.
31. Murray, C.B.; Kagan, C.R.; Bawendi, M.G. Annu. Rev. Mater. Sci. 2000, *30*, 545.
32. Korgel, B.A.; Fitzmaurice, D. Adv. Mater. 1998, *10*, 661.
33. Hansen, J.P.; McDonald, I.R. In: *Theory of Simple Liquids*; Academic Press: New York, 1976.
34. Gray, J.J.; Klein, D.H.; Korgel, B.A.; Bonnecaze, R.T. Langmuir 2001, *17*, 2317.
35. Heath, J.R.; Knobler, C.M.; Leff, D.V. J. Phys. Chem. B 1997, *101*, 189.
36. Korgel, B.A.; Fitzmaurice, D. Phys. Rev. Lett. 1998, *80*, 3531.
37. Dabbousi, B.O.; Murray, C.B.; Rubner, M.F., et al. Chem. Mater. 1994, *6*, 216.
38. Sear, R.P.; Chung, S.W.; Markovich, G., et al. Phys. Rev. E 1999, *59*, R6255.
39. Chung, S.W.; Markovich, G.; Heath, J.R. J. Phys Chem. B 1998, *102*, 6685.
40. Frenkel, D. Phys. World 1993, *6*, 24.

41. Ackerson, B.J. Nature 1993, *365*, 11.
42. Gray, J.J.; Klein, D.H.; Bonnecaze, R.T.; Korgel, B.A. Phys. Rev. Lett. 2000, *85*, 4430.
43. Ohara, P.C.; Leff, D.V.; Heath, J.R., et al. Phys. Rev. Lett. 1995, *75*, 3466.
44. Ge, G.; Brus, L. J. Phys. Chem. B 2000, *104*, 9573.
45. Hamaker, H.C. Physica IV 1937, *10*, 1058.
46. Israelachvili, J. *Intermolecular and Surface Forces*; New York: Academic Press, 1992.
47. Vincent, B.; Edwards, J.; Emmett, S., et al. Colloids Surfaces 1986, *18*, 261.
48. Shah, P.S.; Holmes, J.D.; Johnston, K.P., et al. J. Phys. Chem. B 2002, *106*, 2545.
49. Burton, W.K.; Cabrera, N.; Frank, F.C. Phil. Trans R. Soc. Lond. A 1951, *243*, 299.
50. Levi, A.C.; Kotrla, M. J. Physics: Condens. Matter 1997, *9*, 299.
51. Luedtke, W.D.; Landman, U. J. Phys. Chem. 1996, *100*, 13–323.
52. Wang, Z.L. Adv. Mater. 1998, *10*, 13.
53. Wang, Z.L.; Harfenist, S.A.; Vezmar, I., et al. Adv. Mater. 1998, *10*, 808.
54. Stowell, C.; Korgel, B.A. Nano Lett. 2001, *11*, 595.
55. Ohara, P.C.; Heath, J.R.; Gelbart, W.M. Angew. Chem., Int. Ed. Engl. 1997, *36*, 1077.
56. Maillard, M.; Motte, L.; Ngo, A.T.; Pileni, M.P. J. Phys. Chem. B 2000, *104*, 11–871.
57. Deegan, R.D. Phys. Rev. E 2000, *61*, 475.
58. Deegan, R.D.; Bakajin, O.; Dupont, T.F.; Huber, G.; Nagel, S.R.; Witten, T.A. Phys. Rev. E 2000, *62*, 756.
59. Adachi, E.; Dimitrov, A.S.; Nagayama, K. Langmuir 1995, *11*, 1057.
60. Maenosono, S.; Dushkin, C.D.; Saita, S.; Yamaguchi, Y. Langmuir 1999, *15*, 957.
61. Elbaum, M.; Lipson, S.G. Phys. Rev. Lett. 1994, *72*, 3562.
62. Ha, V.M.; Lai, C.L. Proc. R. Soc. Lond. A 2001, *457*, 885.
63. Black, C.T.; Murray, C.B.; Sandstrom, R.L.; Sun, S. Science 2000, *290*, 1131.
64. Parthasarathy, R.; Lin, X.-M.; Jaeger, H.M. Phys. Rev. Lett. 2001, *87*, 186807.
65. Sampaio, J.F.; Beverly, K.C.; Heath, J.R. J. Phys. Chem. B 2001, *105*, 8797.
66. Beverly, K.C.; Sampaio, J.F.; Heath, J.R. J. Phys. Chem. B 2002, *106*, 2131.
67. Middleton, A.A.; Wingreen, N.S. Phys. Rev. Lett. 1993, *71*, 3198.
68. Doty, R.C.; Yu, H.; Shih, C.K.; Korgel, B.A. J. Phys. Chem. B 2001, *105*, 8291.

11

Optical Spectroscopy of Surface Plasmons in Metal Nanoparticles

Stephan Link and Mostafa A. El-Sayed
Georgia Institute of Technology, Atlanta, Georgia, U.S.A.

I. INTRODUCTION

Metal nanoparticles have been used in applications long before the science of nanometer-size materials have attracted much attention. Novel properties that are different from those of individual atoms or bulk materials are interesting not only from a fundamental research standpoint but also because of a wide variety of different potential applications [1–6]. Gold and silver nanoparticles were already used as coloring pigments in stained glass back in the Middle Ages. Faraday [7] first recognized that the red color is due to a "different form" of gold and postulated correctly that the color is caused by small gold particles without having any of the modern tools such as transmission electron microscopy (TEM). The intense color of metal colloids is due to the coherent excitation of the conduction-band electrons and is known as the surface plasmon absorption [8–10]. Mie [11] was able to theoretically model the plasmon absorption about a century ago by applying the Maxwell equations to describe the interaction between a metal nanoparticle with a known dielectric function and an electromagnetic field.

Mie's theory can explain the size dependence of the surface plasmon absorption for metallic nanoparticles larger than ∼20 nm. For smaller particles, an enhanced electron-surface scattering [12,13] leads to a broadening of the plasmon absorption. The enhanced surface scattering is due to the fact that the electron mean free path becomes longer than the dimensions of the particle. The size and shape dependence of the plasmon absorption together with the effect of the particle composition are presented in the first part of this chapter. The size, shape, and composition are the parameters that allow one to tune the optical properties of metallic nanoparticles over the entire visible and near-infrared spectral ranges.

The decay of the plasmon absorption is due to ultrafast dephasing of the coherent electron motion, which occurs on a sub-10-fs timescale [14–16]. The processes leading to this dephasing are also briefly discussed in this chapter.

The interaction of metal nanoparticles with an ultrashort laser pulse has been of great interest recently, as it allows the selective excitation of the electrons in a metal nanoparticle [17–23]. The electron relaxation involves electron–electron scattering, coupling of the hot electrons to lattice vibrations (phonons), and the energy exchange between the particles and the surrounding medium. All of these processes can be studied by femtosecond transient absorption spectroscopy. The emphasis in these studies has mainly been on the investigation of the electron–phonon coupling as a function of a particle size and shape. The excitation of coherent breathing modes has also been of significant interest [24,25]. Although the latter topic is reviewed in great detail in Chapter 12, the energy relaxation after femtosecond excitation will be briefly discussed here.

We will also review the work on the shape transformation of gold nanorods into spherical nanoparticles induced by intense laser irradiation [26–30]. We have studied the rod-to-sphere shape transformation in aqueous solutions of nanoparticles as a function of a pulse energy and a pulse width and determined the minimal energy required to melt a gold nanorod as well as the melting time. We also discuss possible mechanisms for this transformation.

II. SIZE, SHAPE, AND COMPOSITION DEPENDENCE OF SURFACE PLASMON RESONANCES

As already mentioned in Section I, the intense colors of metal colloids are due to their surface plasmon absorption [8–10]. The surface plasmon absorption can be qualitatively explained in terms of collective oscillations of electrons in the conduction band. The electric field of the incident light induces a polarization of free electrons with respect to a much heavier ionic core of a spherical nanoparticle. A net charge difference is only created at the nanoparticle surface, and the surface charges generate a restoring force. In this way, dipolar oscillations of the electrons are excited. The surface plays a significant role in generating plasmons. Therefore, the surface plasmon absorption is a unique feature of small nanoparticles and it is not observed for either individual atoms (or even small clusters) or bulk metals.

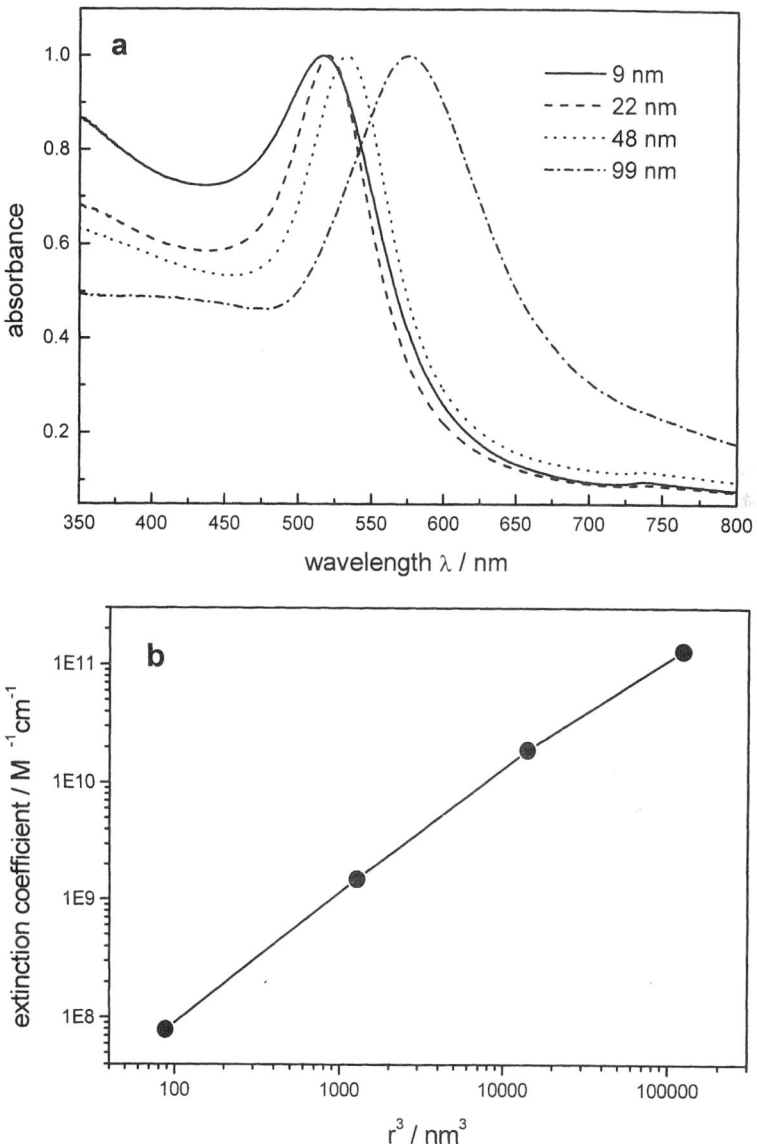

Figure 1 (a) Ultraviolet–visible absorption spectra of four colloidal gold nano-particle solutions having mean diameters of 9, 22, 48, and 99 nm. (b) Molar extinction coefficients of the gold nanoparticles shown in (a) as a function of the particle radius r.

Figure 1a shows the surface plasmon absorption of 9-,22-,48-, and 99-nm gold nanoparticles [15]. The particles were prepared in aqueous solutions by reduction with sodium citrate. From Fig. 1a, one can see that the surface plasmon resonance red shifts with increasing particle size, which is accompanied by an increase in the bandwidth. Both effects will be discussed in more detail later. Figure 1b shows the molar extinction coefficients for the same four samples. These coefficients are on the order of 1×10^8 to 1×10^{11} M^{-1} cm^{-1}; they increase linearly with the increasing volume of the particles. Note that these extinction coefficients are several orders of magnitude greater than those for strongly absorbing organic dye molecules. The large extinction coefficients are due to a large number of free electrons contributing to the polarizability and, hence, to the extinction of the particles.

Theoretically, the surface plasmon absorption can be described using the Maxwell equations. This was first done by Gustav Mie [11] in 1908. He solved the Maxwell equations for the electromagnetic wave interacting with a small sphere, assuming the same macroscopic, frequency-dependent material's dielectric constant as for a bulk metal. The solution of these equations for a spherical object with appropriate boundary conditions leads to a series of multipole oscillations for the extinction cross section. Mie obtained the following expressions for the extinction and scattering cross sections (σ_{ext} and σ_{sca}, respectively) [8–11] for a nanoparticle with radius r:

$$\sigma_{\text{ext}} = \frac{2\pi}{\eta|k|^2} \sum_{L=1}^{\infty} (2L+1) \ \text{Re}\{a_L + b_L\} \tag{1}$$

$$\sigma_{\text{sca}} = \frac{2\pi}{|k|^2} \sum_{L=1}^{\infty} (2L+1) \left(|a_L|^2 + |b_L|^2 \right) \tag{2}$$

where

$$a_L = \frac{m\psi_L(mx)\psi_L'(x) - \psi_L'(mx)\psi_L(x)}{m\psi_L(mx)\eta_L'(x) - \psi_L'(mx)\eta_L(x)}$$

$$b_L = \frac{\psi_L(mx)\psi_L'(x) - m\psi_L'(mx)\psi_L(x)}{\psi_L(mx)\eta_L'(x) - m\psi_L'(mx)\eta_L(x)}$$

n is the complex index of refraction of the metal particle, n_m is the real index of refraction of the surrounding medium, $m = n/n_m$, k is the wave vector, $x = |k|r$, ψ_L and η_L are the Ricatti–Bessel cylindrical functions, L is the summation index of the partial waves ($L = 1$ corresponds to the dipole oscillation, $L = 2$ is associated with the quadrupole oscillation, and so on), and the prime indicates differentiation with respect to the argument in the parentheses. The absorption cross section can be calculated as $\sigma_{\text{abs}} = \sigma_{\text{ext}} - \sigma_{\text{sca}}$.

For nanoparticles that are much smaller than the wavelength of light ($r < 10$ nm), only the dipole oscillations ($L = 1$) contribute significantly to the extinction cross section. In the dipole approximation, this cross section can be presented as [8–10,12]

$$\sigma_{ext}(\omega) = 9\frac{\omega}{c}\varepsilon_m^{3/2}V\frac{\varepsilon_2(\omega)}{[\varepsilon_1(\omega) + 2\varepsilon_m]^2 + \varepsilon_2(\omega)^2} \tag{3}$$

where V is the particle volume, ω is the angular frequency, c is the speed of light, ε_m is the frequency-independent dielectric constant of the surrounding medium, and $\varepsilon(\omega) = \varepsilon_1(\omega) + i\varepsilon_2(\omega)$ is the dielectric constant of the metal (it is a complex quantity that is a function of the frequency). If ε_2 is small or weakly dependent on ω, the approximate resonance condition is $\varepsilon_1(\omega) = -2\,\varepsilon_m$.

The strength of Mie's theory is due to the fact that it only requires the knowledge of the metal dielectric constant in order to calculate the nanoparticle absorption spectrum. This is true for any metal, although a strong absorption band in the visible is only observed for nanoparticles of noble metals (copper, silver, and gold) [31]. Furthermore, the absorption of mixed-alloy nanoparticles can also be well described by Mie's theory if the dielectric constant of the alloy with the same composition is known. This was verified for gold–silver-alloy nanoparticles [32]. These metals from binary mixtures over the whole composition range. Figure 2a shows the surface plasmon absorption of four gold–silver-alloy nanoparticle samples having different compositions, as indicated by the gold mole fraction. The alloy nanoparticles were prepared in a way similar to that for the pure gold nanoparticles. The reducing agent was added into a solution of a mixture of gold and silver ions with the desired mole fractions. A gradual shift of the surface plasmon absorption with changing composition (Fig. 2b) indicates that this procedure indeed produces alloy particles, but not just a mixture of pure gold and silver nanoparticles. The measured spectra agree well with the absorption spectra calculated within the dipole approximation using the alloy dielectric constant [32].

By using gold–silver-alloy nanoparticles, it is possible to tune the plasmon absorption in the range between 400 and 500 nm. This range can be further extended to a longer wavelength using elongated particles instead of spherical particles. In elongated particles, the plasmon resonance splits into two bands [33–35]. Cylindrical or rod-shaped nanoparticles are usually described by their aspect ratio (i.e., the ratio between the long- and short-axis lengths). As the aspect ratio increases, the energy separation between the resonance frequencies of the two plasmon bands increases. This effect is illustrated in Fig. 3a, which displays absorption spectra for two nanorod samples with aspect ratios of 2.7 and 3.3. The samples were prepared electrochemically using organic micelles as the rod-shaping and stabilizing

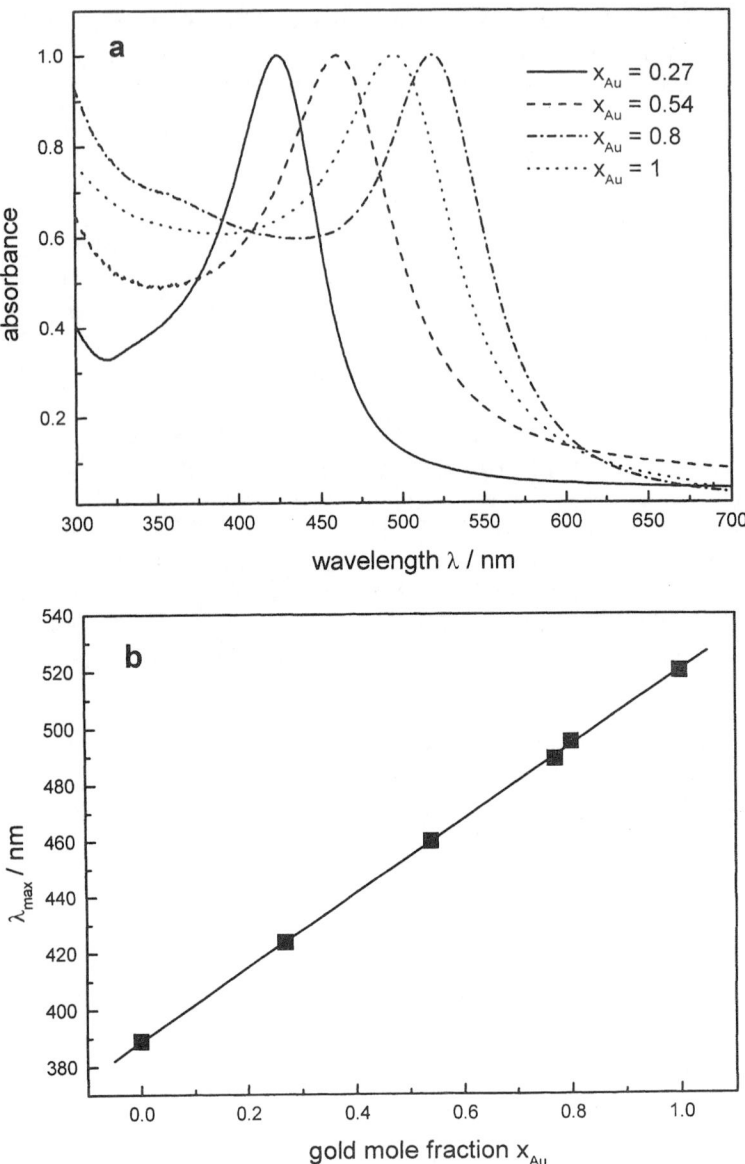

Figure 2 (a) Ultraviolet–visible absorption spectra of four colloidal gold–silver alloy nanoparticle solutions with various compositions given by the gold mole fraction x_{Au}. (b) Plot of the absorption maximum as a function of the gold mole fraction.

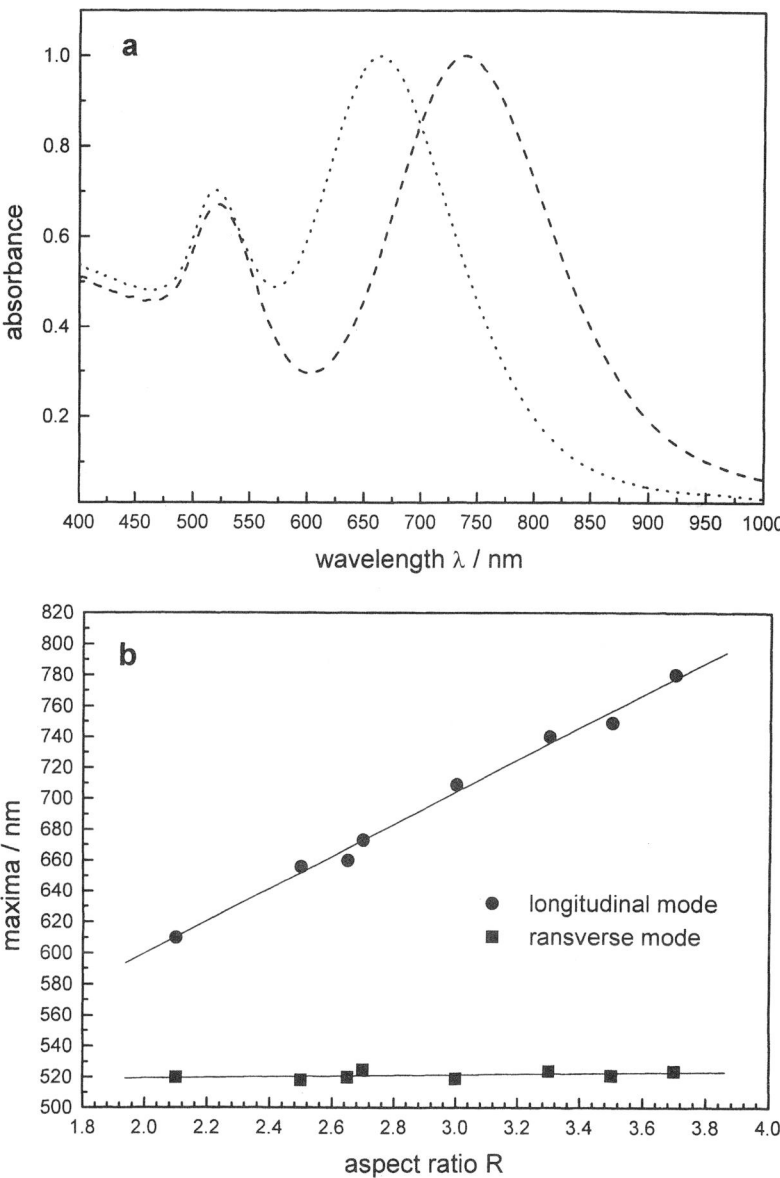

Figure 3 (a) Ultraviolet–visible absorption spectra of two colloidal gold nanorod solutions with aspect ratios of 2.7 (dotted line) and 3.3 (dashed line). (b) Plot of the absorption maxima of the transverse (squares) and longitudinal (circles) plasmon band as a function of the nanorod aspect ratio.

agents [34]. The short-wavelength band at around 520 nm corresponds to the oscillation of the electrons perpendicular to the major rod axis. It is referred to as the transverse plasmon absorption [36]. This feature is relatively insensitive to the nanorod aspect ratio and coincides spectrally with the surface plasmon band in spherical nanoparticles. The other absorption band is shifted to a longer wavelength and is due to oscillations of free electrons along the major rod axis. This band is referred to as the longitudinal surface plasmon absorption [36]. Figure 3b shows a plot of the maxima of the transverse (squares) and longitudinal (circles) surface plasmon modes as a function of the rod aspect ratio. These data nicely illustrate the strong dependence of the longitudinal plasmon maximum on the aspect ratio, which provides a means to spectrally tune the optical absorption of metal nanoparticles.

As an extension to Mie's theory, Gans [37] applied the theoretical treatment developed for spherical particles to cylinders (rods). Within the dipole approximation, the optical absorption spectrum of a collection of randomly orientated gold nanorods with aspect ratio R is given by

$$\sigma_{ext} = \frac{\omega}{3c} \varepsilon_m^{3/2} V \sum_j \frac{\left(1/P_j^2\right)\varepsilon_2}{\left\{\varepsilon_1 + [(1 - P_j)/P_j]\varepsilon_m\right\}^2 + \varepsilon_2^2}, \tag{4}$$

where P_j are the depolarization factors for the three axes,

$$P_A = \frac{1 - e^2}{e^2}\left[\frac{1}{2e}\ln\left(\frac{1 + e}{1 - e}\right) - 1\right] \tag{5}$$

$$P_B = P_C = \frac{1 - P_A}{2}, \tag{6}$$

A, B, and C are the axes lengths ($A > B = C$, $R = A/B$), and

$$e = \sqrt{1 - \left(\frac{B}{A}\right)^2} = \sqrt{1 - \frac{1}{R^2}} \tag{7}$$

Gans' theory accurately reproduces the red shift of the longitudinal surface plasmon absorption with increasing nanorod aspect ratio. The red shift of the surface plasmon absorption in spherical nanoparticles is also well described by Mie's theory for larger nanoparticles (>20 nm). When the dipole approximation is no longer valid, the plasmon resonance depends explicitly on the particle size, as x is a function of the particle radius r [see Eqs. (1) and (2)]. As the particle size increases, higher-order modes become more important, as the light cannot polarize the nanoparticles homogeneously. Retardation effects also become important, and the plasmon band red shifts with increasing particle size. At the same time, the plasmon bandwidth increases with particle size because the plasmon absorption is the convolution of several modes peaking at different wavelengths. This is illustrated by the experimental spectra in Figs 1a and 4a, which show the plasmon bandwidth as a function of a particle size.

Whereas the plasmon bandwidth increases for the nanoparticles larger than 20 nm, it also increases for smaller particles (i.e., in the range of sizes for which the dipole approximation is applicable). As can be seen from Eq. (3), the extinction coefficient within the dipole approximation does not depend on the particle dimensions, which implies that the surface plasmon absorption is size independent for particles smaller than ~20 nm. However, this conclusion contradicts experimental observations. Spectroscopic data indicate that the plasmon band is strongly damped for particles smaller than 5 nm in diameter and finally disappears completely for sizes below ~2 nm [37–39]. Furthermore, it has been shown experimentally that the plasmon bandwidth is inversely proportional to the particle size for particles smaller than ~20 nm [8,12]. It has therefore been argued that for small nanoparticles, the assumption of the same electronic and optical properties as for bulk materials is no longer valid. Especially, the use of the bulk dielectric constant, which enters the Mie equations as the material's only parameter, is not justified as the particle size is decreased. Because Mie's theory has been very successful in describing optical absorption spectra of metal nanoparticles, most approaches to describing the surface plasmon absorption in the regime of very small sizes have focused on modifying the dielectric constant (to introduce a size dependence in it) rather than on changing the Mie model [8,12].

In one of the earliest approaches, Kreibig and co-workers [12,13] argued that electron-surface scattering is enhanced in small particles as the mean free path of the conduction-band electrons becomes limited by the physical dimensions of the nanoparticle. This model will be discussed later in more detail because it illustrates the concept of introducing size dependence in the dielectric constant; it also allows some physical insight into size-dependent plasmon properties. The mean free path of the electrons in silver and gold is on the order of 40–50 nm [40]. The smaller the particle, the more frequent the electron-surface collisions are. Each elastic or inelastic electron-surface scattering event leads to a loss of coherence in the electron motion. This occurs on a faster timescale for smaller particles because of the enhanced electron-surface scattering. For a simple two-level system, a faster dephasing leads to an increased line width [41].

The size dependence of the dielectric function is introduced in Kreibig's model [12,13] by presenting the dielectric constant as a combination of an

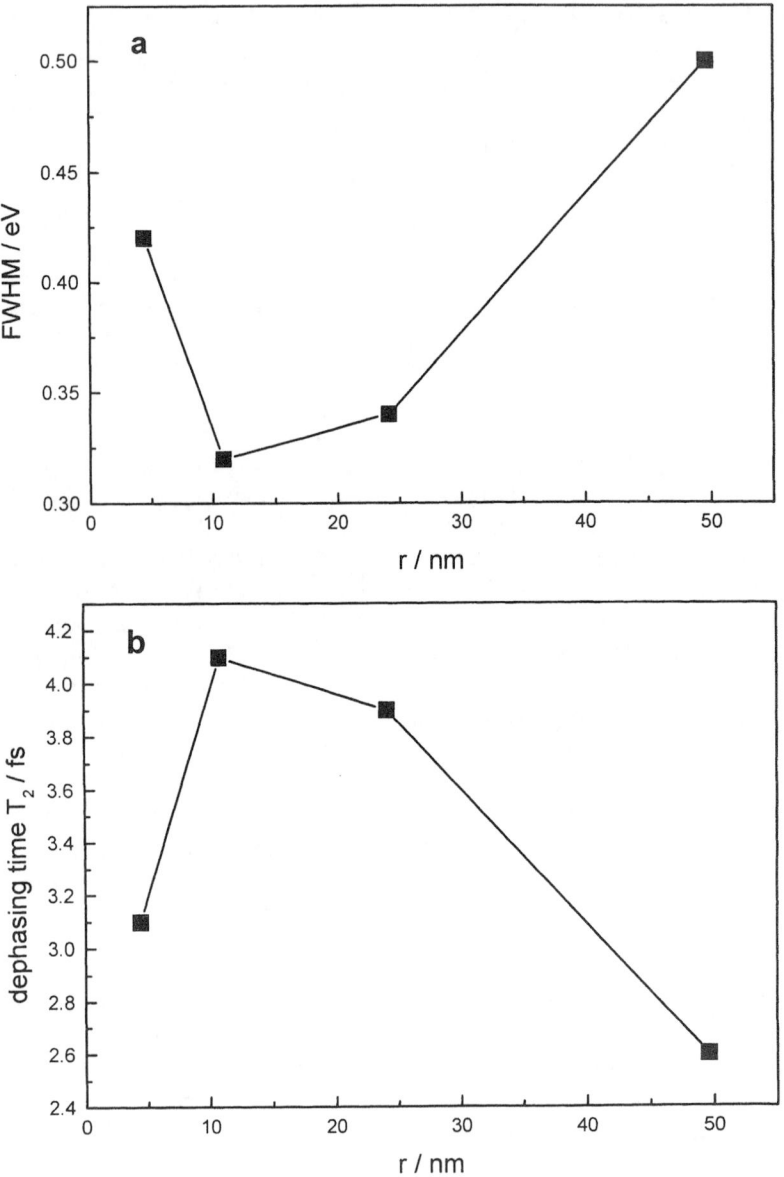

Figure 4 (a) Plot of the plasmon absorption linewidth as a function of the particle radius r. (b) Plot of the plasmon oscillation dephasing time T_2 as a function of the particle radius r. The data points were derived from the absorption spectra shown in Fig. 1.

interband term $\varepsilon_B(\omega)$ and a Drude free-electron term $\varepsilon_D(\omega)$ describing the intraband conduction-band contribution: $\varepsilon(\omega) = \varepsilon_B(\omega) + \varepsilon_D(\omega)$. The latter is given within the free-electron model [40] by

$$\varepsilon_D(\omega) = 1 - \frac{\omega_p^2}{\omega^2 + i\gamma\omega} \qquad (8)$$

where $\omega_p^2 = (ne^2/\varepsilon_0 m_{\text{eff}})$ is the bulk plasmon frequency expressed in terms of the free-electron density n, the electron charge e, the vacuum permittivity ε_0, and the electron effective mass m_{eff}. In Eq. (8), γ is a phenomenological damping constant which is equal to the plasmon band width Γ for the case of a free-electron gas in the limit of $\gamma \ll \omega$. The damping constant γ is related to characteristic times of all electron scattering processes. In bulk materials, these processes are mainly due to electron–electron (time $\tau_{e\text{-}e}$), electron–phonon (time $\tau_{e\text{-}ph}$), and electron–defect interactions (time $\tau_{e\text{-}d}$). As a result, the bulk metal damping constant γ_0 can be expressed as follows:

$$\gamma_0 = \sum_i \tau_i^{-1} = \tau_{e-e}^{-1} + \tau_{e-ph}^{-1} + \tau_{e-d}^{-1} \qquad (9)$$

For nanoparticles that are smaller than the mean free path of the electrons in the conduction band, electron-surface scattering will also contribute to the damping constant γ. Therefore, an additional term accounting for this effect is added to Eq. (9) [8,12,13]:

$$\gamma(r) = \gamma_0 + \frac{Av_F}{r} \qquad (10)$$

where A is a parameter that depends of the details of the scattering process and v_F is the velocity of the electrons at the Fermi energy. As the particle radius r decreases, the contribution from the electron-surface scattering to the damping constant γ increases. When $\gamma(r)$ is substituted back into the dielectric constant, one can then use Mie's theory in the dipole approximation to calculate the absorption spectrum. In this case, the plasmon band becomes size dependent, in agreement with experimental observations. This model has been successfully used [12,37] to explain the $1/r$ dependence of the plasmon bandwidth for nanoparticles smaller than $\sim$20 nm. The constant A is typically used as a fitting parameter when the results of calculations are compared with the experimental data [8].

   In addition to the above-described model, there are other theories for the explanation of the size dependence of the plasmon bandwidth [42–48]. These theories consider such effects as changes in the band structure, environmental changes (chemical interface damping), and the "spill out" of the conduction-band electrons. Most of the theories arrive at the size-dependent dielectric constant, and this modified constant is then used within the framework of traditional Mie's theory. For very small clusters, quantum chemical methods or a jellim model have also been used to calculate the dielectric function from first principles [45–47]. A comprehensive overview of different theories with references to the original work can be found in Ref. 8. Most of the theories predict the $1/r$ dependence of the plasmon bandwidth. However, there are discrepancies regarding the size dependence of the plasmon absorption maximum. Some theories predict a red shift, whereas others predict a blue shift of the plasmon resonance with decreasing particle size [8]. Certainly, the medium dielectric constant ε_m has a large effect on the resonance condition, as can be seen from Eq. (3). An increase in ε_m leads to a red shift of the surface plasmon absorption maximum. Comparison between theories and experiments are further complicated by the fact that both red and blue shifts are observed experimentally. Furthermore, it is difficult to prepare nearly monodisperse metal nanoparticles over a large range of sizes using the same capping agent. Because different capping agents have different values of the medium dielectric constant ε_m, it is difficult to distinguish between contributions to the plasmon shift due to the particle size and the surrounding medium.

III. DEPHASING OF THE SURFACE PLASMON RESONANCE

Although the plasmon absorption is well understood in terms of the electro-magnetic interaction between the metal particles and light, less is known about the decay mechanism of the coherent electron motion. Both the dephasing of the coherent electron motion and the energy relaxation have to be considered as decay mechanism. The dephasing of the electron motion leads to the loss of coherence, which is usually described by the time constant T_2. T_2 and the time constant for energy relaxation, T_1, are related by the pure dephasing time T_2^* [41]:

$$\frac{1}{T_2} = \frac{\Gamma}{2\hbar} = \frac{1}{2T_1} + \frac{1}{T_2^*} \qquad (11)$$

where Γ is the homogeneous spectral linewidth. The time T_2^* can originate from collisions that change the plasmon wave vector but not its energy. Relaxation pathways for the plasmon energy decay (T_1 time constant) include the radiative decay of the plasmon and the nonradiative decay into single electron–hole excitation [8,14].

For a homogeneously broadened line, the dephasing time T_2 can be directly obtained from the absorption spectrum using Eq. (11). Figure 4b shows the dephasing times T_2 derived from the plasmon bandwidth as a function of the particle radius [15]. The dephasing times are extremely short (sub-10-fs scale). Because the dephasing time is inversely proportional to the linewidth, the broader the line is, the faster the dephasing is. For small particles, the faster dephasing time can be attributed to the enhanced electron-surface scattering leading to the loss of coherence. On the other hand, for large nanoparticles, retardation effects due to inhomogeneous polarization of the nanoparticle determine the dephasing time.

The above analysis assumes a monodispersity of metal nanoparticles, whereas real samples of metal colloids are always polydisperse (for gold colloids, the size dispersion is typically ~10%). However, from the obvious shift of the plasmon band maximum for spherical gold nanoparticles with average sizes that only differ by a few nanometers, one can conclude that the linewidth is mainly determined by dephasing processes in the individual particles despite the sample-size polydispersity. Single-particle studies [14,16,49,50] are well suited to measure the homogeneous linewidth and, hence, T_2, without the problems associated with the inhomogeneous size distribution. In Ref. 14, a conventional microscope with illumination from a halogen lamp was used to measure homogeneous linewidths for individual gold nanodots and nanorods [14] (these experiments took advantage of large scattering cross sections of metal nanoparticles). Dephasing times of 1–5 fs for spherical gold nanoparticles were found, in agreement with results in Fig. 4b. On the other hand, much longer dephasing times (up to 18 fs) for gold nanorods were derived from the homogeneous width of the longitudinal surface plasmon resonance. These results were explained by the decreased spectral overlap between the longitudinal plasmon resonance at lower energies and the interband transition in gold. The results of single-particle studies also indicated that pure dephasing (time T_2^*) was negligible and the plasmon dephasing (time T_2) was dominated by nonradiative decay (time T_1) into single-particle excitations [14].

In the other single-particle study [49], 40-nm gold nanoparticles embedded in polymer films of various thicknesses showed a strong dependence of the plasmon resonance frequency on the particle environment. In particular, the plasmon absorption red-shifted with increasing film thickness. In the thinner films, the particles were assumed to be partially surrounded by air, which accounted for the change in the local environment. Using a scanning near-field optical microscope (SNOM), the near-field transmission spectra of individual gold nanoparticles with an average size of 40 nm were measured [50]. The dephasing times extracted from homogeneous linewidths were around 8 fs. Deviations in linewidths observed for individual particles were attributed to variations in the local environment. These examples clearly show that effects due to the surrounding medium as well as interparticle coupling are the issues that can be best addressed using single-particle spectroscopy.

Another approach to circumvent the inhomogeneous broadening is to use nonlinear optical techniques [16,41,51]. For example, Heilweil and Hochstrasser [41] found that both T_1 and T_2 are shorter than 48 fs for 20-nm colloidal gold nanoparticle solutions. Lamprecht et al. [51] measured T_2 for lithographically fabricated 200-nm gold and silver nanoparticles using a second-order nonlinear optical autocorrelation in the femtosecond regime. They obtained dephasing times of 6 fs for gold nanoparticles and 7 fs for silver nanoparticles. Liau et al. [16] combined nonlinear optical measurements (two-pulse, second-order interferometric autocorrelation) with high-spatial-resolution imaging to study optical responses from single silver nanoparticles. For a 75-nm particle, they obtained a dephasing time T_2 of 10 fs.

This brief overview shows that the available experimental results on plasmon dephasing (obtained by different spectroscopic techniques) indicate that coherent plasmon oscillations decay on ultrafast timescales of a few femtoseconds.

IV. NONRADIATIVE DECAY OF THE SURFACE PLASMON RESONANCE AFTER LASER EXCITATION

The use of femtosecond laser pulses to study electron dynamics in thin metal films [52] and metal nanoparticles [17–23] allows one to selectively excite the electrons and then to follow energy relaxation due, for example, to interactions with the lattice phonons. The size dependence of the electron–phonon coupling is an important issue in the problem of electron energy relaxation. An enhanced electron-surface scattering was already shown to have an effect on the plasmon bandwidth and, hence, on the dephasing time T_2. The same effect can, in principle, influence energy relaxation, leading to more efficient energy losses in smaller nanoparticles.

Experimentally, the excitation of the electron gas in metal nanoparticles leads to a bleach of the surface plasmon absorption band, as shown in Fig. 5a for 15-nm spherical gold nanoparticles (excitation wavelength is 400 nm) [20–22]. This bleach is caused by a decrease in the amplitude (damping) as well as by broadening of the surface plasmon resonance due to effective heating of electrons by laser pulses. The transient absorption spectra in Fig. 5a can be described in terms of a difference between the plasmon absorption at the increased electron temperature (following the laser excitation) and the plasmon absorption at room temperature (before the laser excitation). The latter spectrum (dotted line in Fig. 5a) shows that the plasmon bleach maximum

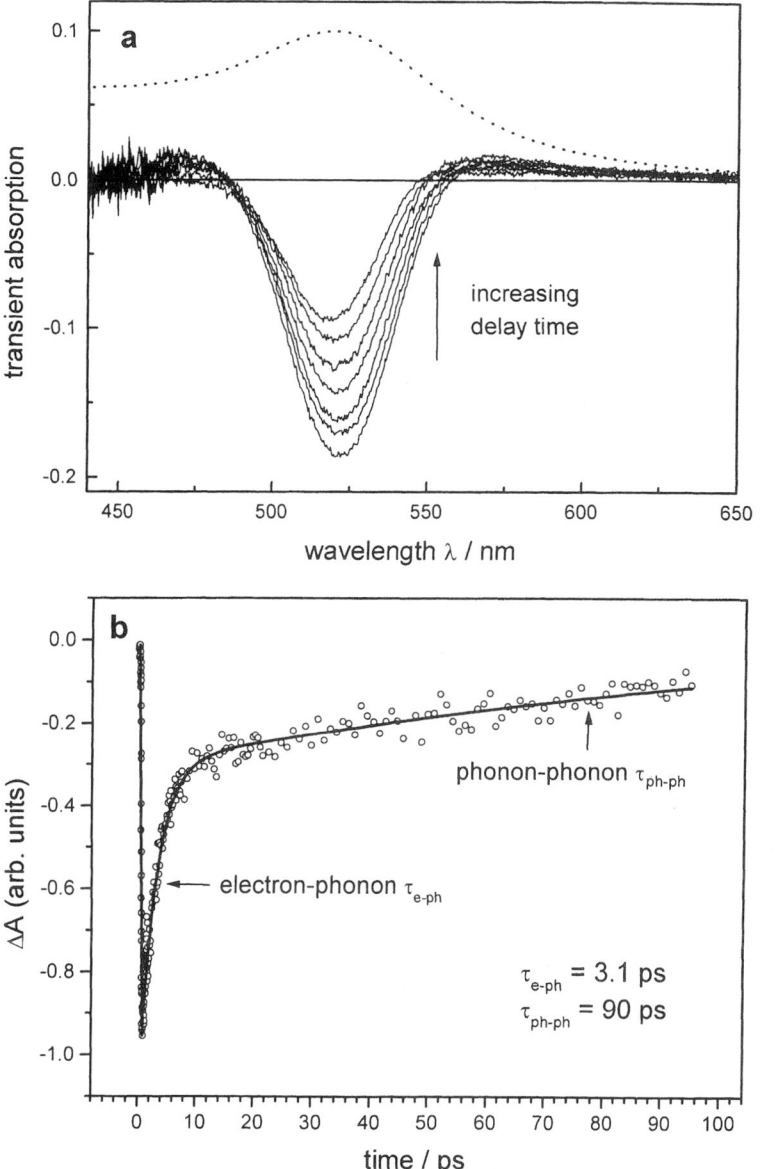

Figure 5 (a) Femtosecond transient absorption spectra of 15-nm spherical gold nanoparticles in aqueous solution taken at different delay times after excitation with 400-nm femtosecond laser pulses (solid lines). Also shown is the ground-state absorption spectrum (dotted line). (b) The recovery dynamics of a plasmon bleach at 520 nm is fit to a two-exponential decay. Two time constants derived from this fit are attributed to electron–phonon (3.1 ps) and phonon–phonon (90 ps) interactions.

is located at the position of the ground-state plasmon absorption maximum. The recovery of the bleach is due to the energy exchange between the photoexcited electrons, the nanoparticle lattice vibrations, and the molecule vibrations in the surrounding solvent [20–22]. The dynamics of the plasmon bleach recovery at 520 nm (maximum of the plasmon band bleach) shows a biexponential decay with time constants of 3.1 and 90 ps. The initial fast decay was attributed to the electron–phonon coupling (energy exchange between photoexcited electrons and lattice vibrations) and was analyzed according to the two-temperature model [53]. This first step in the relaxation leads to thermal equilibrium between the electrons and phonons. The second step involves the cooling of the nanoparticle by energy exchange with the surrounding medium on the ~100-ps timescale [20–22]. Especially the electron–phonon relaxation time is strongly dependent on the laser pump power increasing with increasing pump power. This is due to the temperature-dependent electronic heat [20–22] capacity, and relaxation times as short as 700 fs to 1 ps have been measured for very weak excitation [21].

The size and shape dependence of the plasmon bleach dynamics is analyzed in Fig. 6, which compares the data obtained for 15- and 48-nm spherical gold nanoparticles and gold nanorods with an aspect ratio of 3.8 [22]. The fact that the dynamics for all these samples are very similar indicates that the electron–phonon coupling is size and shape independent. These results are consistent with experiments carried out by Hartland and co-workers [21], who studied spherical gold nanoparticles in the size range between 2 and 120 nm. The time constant obtained for nanoparticles are comparable to those measured for bulk metals using similar time-resolved, transient-absorption techniques [52].

The fact that energy relaxation dynamics in gold nanoparticles are size independent is rather surprising considering that the mean free path of electrons is only ~50 nm. In particles smaller than 50 nm, the electron-surface scattering should become more frequent than the electron-lattice scattering. Therefore, it was thought that the enhanced electron-surface interactions should give rise to a pronounced size dependence of the electron cooling dynamics in particles smaller than the electron mean free path. However, this effect is not observed experimentally (see, e.g., Fig. 6). Using a theory developed by Belotskii and Tomchuk [54], it can be shown that the coupling constant between the electrons and acoustic and capillary modes of the surface phonons is determined by the ratio of the electron concentration to the metal density [55]. Because of a small number of valence electrons in gold (one valence electron) and a large gold atomic mass (Au = 197), the electron-surface scattering contribution to the overall rate of electron energy losses is small (<10%) [55]. Physically, this result means that electron-surface collisions are primarily elastic and, therefore, although they lead to changes

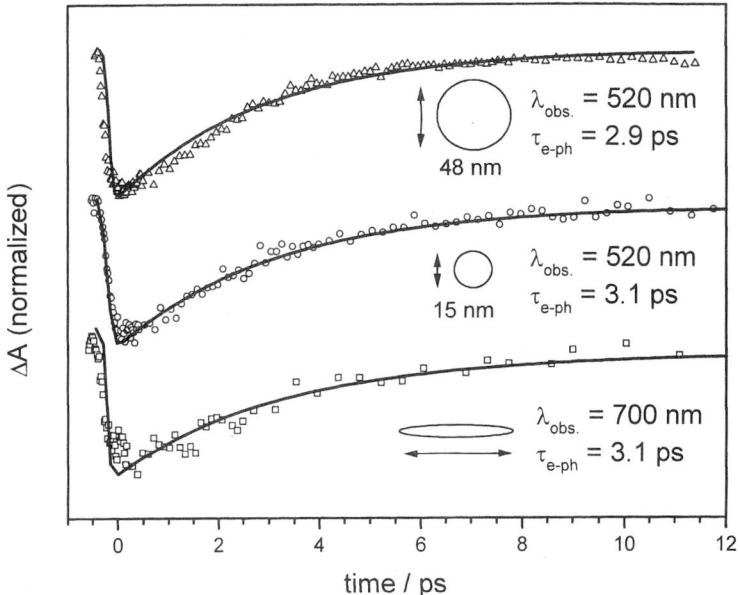

Figure 6 Electron–phonon dynamics in 48- and 15-nm spherical gold nanoparticles and gold nanorods with a mean aspect ratio of 3.8. These data suggest that the electron–phonon coupling is independent of the size and shape of gold nanoparticles.

in the momentum (and hence phase), they do not change the energy of the excited electrons.

V. PHOTOTHERMAL SHAPE CHANGES OF GOLD NANORODS

Large extinction coefficients of gold nanoparticles and their small masses result in a significant increase of the lattice temperature even at low laser excitation powers [several nanojoules (nJ) per pulse]. If the excitation intensity is increased above a few microjoules level, changes of the nanoparticle shape can be observed [26–30]. Figure 7 shows TEM images of gold nanorods before (Fig. 7a) and after (Fig. 7b) exposure to 50-μJ femtosecond laser pulses at 800 nm. The particles do not precipitate after laser exposure but, rather, a selective heating of gold nanorods in the aqueous solution leads to transformation of rodlike particles into particles of almost spherical shape. A comparison of the particle volume before and after laser exposure is shown

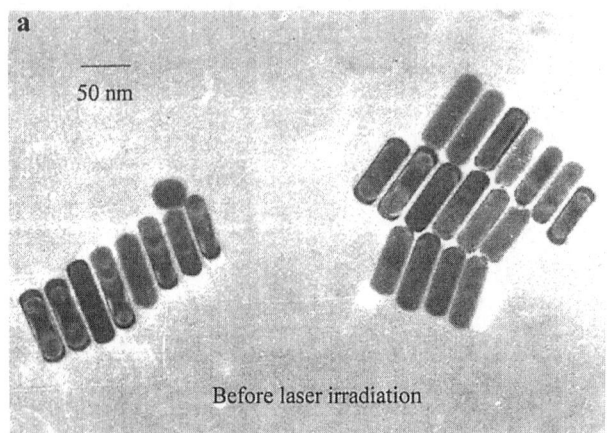

Before laser irradiation

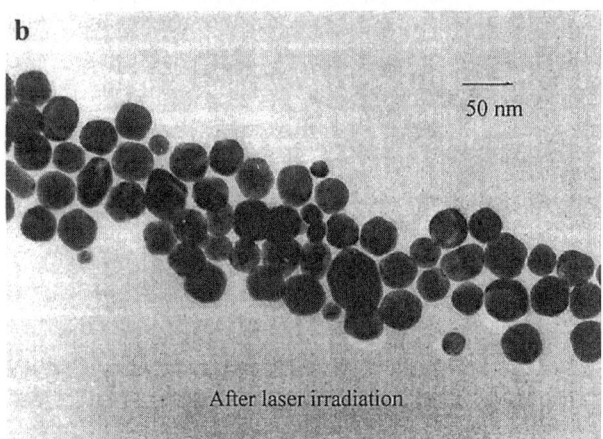

After laser irradiation

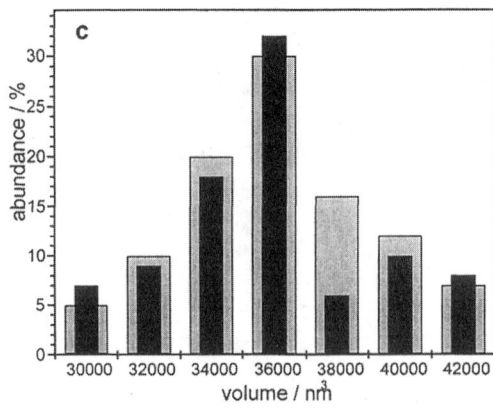

in Fig. 7c. Because the average volume distribution of the spheres is comparable to that of the original nanorods, it can be concluded that each nanorod transforms or "isomerizes" into a spherical particle without a significant fragmentation into smaller particles [26,27].

The observed shape changes can be attributed to thermal process in which the electrons heat the lattice after intense laser irradiation [26,27]. The lattice temperature reaches the melting point and a shape transformation into a more thermodynamically stable spherical shape occurs before the lattice loses its energy to the surrounding medium. Due to a large amount of solvent, the temperature increase for the solution is minimal compared to the local temperature of the nanorods.

The shape changes can be monitored by monitoring the absorption spectra of colloidal solutions, as illustrated in Fig. 8 [28]. This experiment was carried out by placing 400 µL of a gold nanorod solution into a cylindrical cuvette, which was rotated so that a different portion of the solution was exposed to each laser pulse. The wavelength of the femtosecond pulses was 820 nm and the pulse energies were 20 and 0.5 µJ (Figs. 8a and 8b, respectively). The spectra were taken in time intervals of several seconds, which corresponded to several thousands of laser pulses (the laser repetition rate was 1 kHz). One can see that the absorption intensity of the longitudinal surface plasmon band at 800 nm decreases with increasing exposure time (i.e., the number of laser pulses), whereas the absorption band at 515-nm becomes more intense. The longitudinal surface plasmon mode is characteristic of particles of nonspherical shapes (e.g., cylinders) [8]. Therefore, the decrease in the intensity of this mode indicates a transformation of rodlike particles into spheres, which is consistent with results of the TEM studies. The increase in the amplitude of the 515-nm mode observed during this transformation indicates that the extinction coefficient for the surface plasmon absorption in spherical nanoparticles is larger than that for the transverse surface plasmon mode in nanorods.

In order to answer the question of whether this shape transformation requires a single or multiple successive laser pulses, we plot the absorption intensity at the maximum of the longitudinal surface plasmon absorption as a function of the number of laser pulses (see insets to Figs. 8a and 8b). The solid lines are exponential fits of the depletion of the longitudinal surface plasmon absorption. The characteristic constant $\tau_{1/e}$ corresponds to the time that is

---

Figure 7 Gold nanorods on a TEM grid deposited before (a) and after (b) irradiation with 820-nm femtosecond laser pulses. (c) Comparison of the volume distribution of the nanorods prior to laser irradiation with that of the spherical nanoparticles obtained after irradiation.

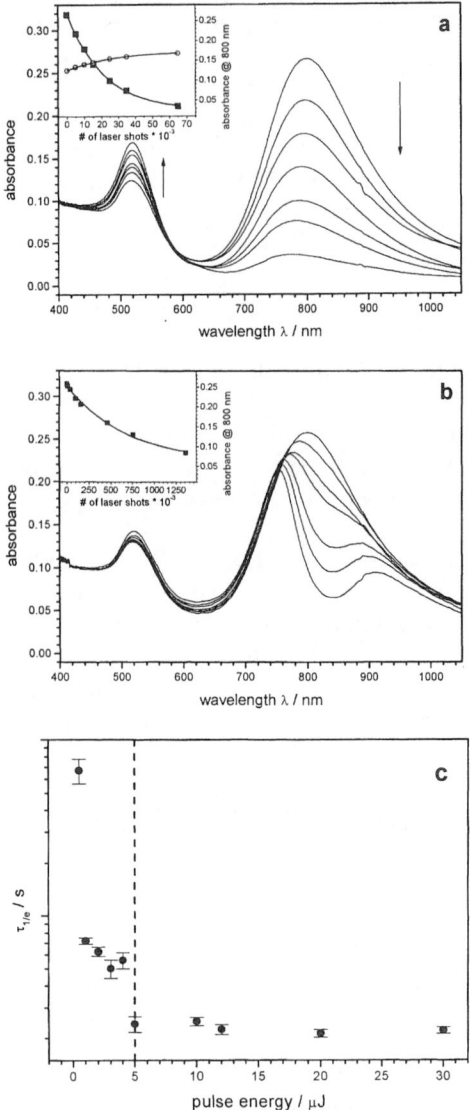

Figure 8 Ultraviolet–visible absorption spectra of a solution of gold nanorods with a mean aspect ratio of 4.1 taken after exposure to an increasing number of femtosecond laser pulses. The intensity of the longitudinal surface plasmon absorption at 800 nm decreases with increasing exposure time; the intensity of the transverse mode at 520 nm increases as shown in the inset. The laser wavelength is 820 nm and the pulse energy is (a) 20 μJ and (b) 0.5 μJ. (c) Plot of the time (number of laser pulses) required to decrease the absorption intensity to 1/e of its initial value as a function of the pulse energy.

required to reduce the number of nanorods to $1/e$ of its initial value. For high laser intensities (Fig. 8a), $\tau_{1/e}$ correlates well with the time that characterizes the increase in the absorption intensity at 515 nm. This result indicates that the absorption of a single laser pulse transforms all nanorods present in the excitation volume into nanodots [28]. Because the spheres do not absorb at 820 nm after the shape transformation is complete, the same particle cannot absorb another laser pulse.

For low-intensity excitation at 820 nm (Fig. 8b), a spectral "hole" at the excitation wavelength is burned in a broad absorption band associated with the longitudinal surface plasmon resonance [27,28]. The generation of this spectral "hole" indicates a considerable inhomogeneous broadening of the longitudinal surface plasmon band in rods, especially compared to a relatively small contribution from sample inhomogeneity into the plasmon linewidth for samples of spherical nanoparticles (see above discussion). These measurements also allow us to establish the minimal energy required to induce shape changes for an ensemble of nanorods in solution [28]. A plot of $\tau_{1/e}$ as a function of the laser pulse energy is shown in Fig. 8c. Time $\tau_{1/e}$ is independent of the laser pulse energy above 5 μJ. For smaller pulse energies, $\tau_{1/e}$ increases significantly, suggesting that there exists a threshold energy for the complete melting of all nanorods within the broad distribution with a single laser pulse. The reason for the constant value of $\tau_{1/e}$ above the threshold pulse energy of 5 μJ is that the energy in excess of the amount required for the shape transformation simply heats the same particles to a higher temperature without inducing further shape changes. Furthermore, the effect of extra energy (extra pump pulses) is also reduced because of the reduced absorption at the position of the longitudinal surface plasmon due to the shape change.

Using this spectroscopically determined value of 5 μJ together with the optical density of the sample and the concentration of nanorods in the excitation volume, one can calculate the energy required to melt a single gold nanorod [28]. For nanorods with a mean aspect ratio of 4.1, a melting energy of 65 fJ is found. On the other hand, assuming bulk thermodynamic properties of the gold nanorods, one can calculate the minimum melting energy, which yields a value of 16 fJ. As the nanorods are much larger than the size range for which a decrease of the bulk melting temperature has been observed (<5 nm), this comparison is in reasonable qualitative agreement.

Below the threshold pulse energy of 5 μJ, it is only possible to convert a part of nanorods from the ensemble into particles of other shapes. Only the nanorods that strongly absorb at the excitation wavelength experience shape transformation, as can be seen from the appearance of a spectral "hole" in the absorption spectrum [27,28]. The increase in the number of laser pulses that are required to reduce the nanorod concentration to $1/e$ of its initial value indicates that the particles are likely reshaped by absorbing several successive laser pulses while being mixed in the rotating cylindrical cell. A partial melting of the nanorods only slightly reduces their longitudinal plasmon absorption and, hence, the absorption of another laser pulse is possible.

This partial melting of gold nanorods is illustrated in Fig. 9, which shows the TEM images before (Fig. 9a) and after (Fig. 9b) irradiation with low-energy femtosecond pulses [27]. These images indicate that mostly shorter and wider nanorods are present in the final solution. This is confirmed quantitatively by the statistical analysis of the nanorod distributions before (gray columns) and after (black columns) laser exposure, as illustrated in Fig. 9c. The mean aspect ratio of the nanorods decreases from 4.1 to 2.6. Although some nanorods are transformed into nanodots (complete particle melting), the main effect is a decrease in the nanorod length (from 44 to 35 nm) and a simultaneous increase in the nanorod width (from 11 to 13 nm). This result indicates a gentle reshaping of the gold nanorods into shorter and wider particles by partial surface melting and surface reconstruction [27]. Femto-second laser pulses with appropriate energies and wavelengths can, therefore, be used to narrow a relatively wide nanorod size distribution typical for existing chemical procedures.

In order to understand the mechanism of the shape transformation, we have performed high-resolution TEM (HR-TEM) studies [29]. Colloidal gold nanorods were exposed to laser pulses with pulse energies below the threshold needed for a complete melting of the nanorods. The laser-induced structural changes were then followed by HR-TEM. Freshly prepared gold nanorods before laser irradiation were found to have the {100}, {111}, and {110} facets and contain no volume dislocations, stacking faults, or twins [56]. The relatively unstable {110} facet is, however, usually absent in the spherical nanodots prepared either by chemical methods or by photothermal melting of the nanorods. Spherical nanoparticles are dominated by {111} and {100} facets with shapes of truncated octahedral, icosahedral, and decahedral [56]. The {111} and {100} facets are the lower-energy faces of gold. In order to reduce the strain associated with the sphericallike particle shape and to accommodate the presence of only {111} and {100} facets, the nanodots must contain planar defects. An example of a multiple twinned particle is an icosahedral, which consists of 20 tetrahedra with {111} facets.

Figure 9 Transmission electron microscopic images taken before (a) and after (b) exposure to low-energy femtosecond laser pulses (0.5 μJ). (c) Comparison of the aspect ratio distribution of gold nanorods before (gray columns) and after (black columns) laser irradiation showing that the mean aspect ratio of the nanorods decreases from 4.1 to 2.6.

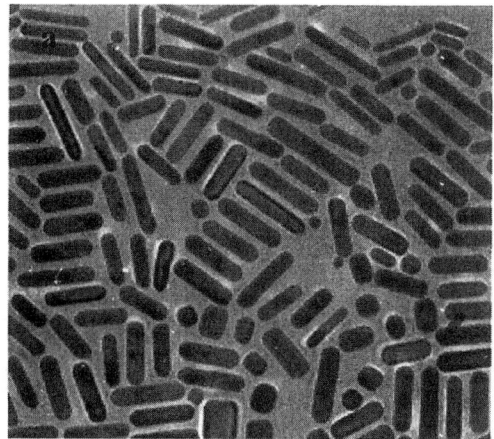

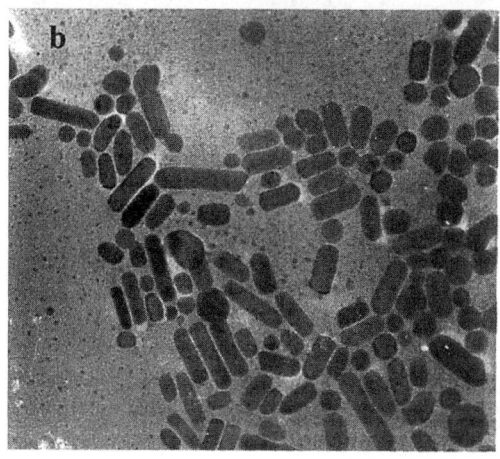

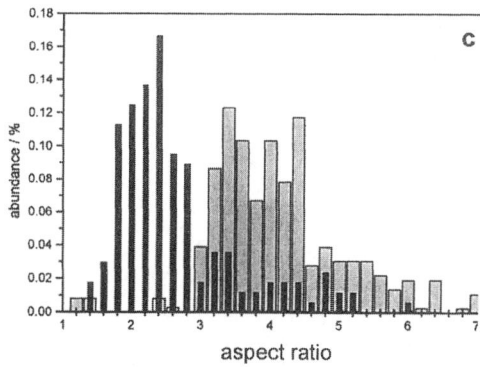

Figure 10 shows HR-TEM images of gold nanorods before (Fig. 10a) and after (Fig. 10b–10d) exposure to femtosecond laser pulses having an increasing amount of energy [29]. The as-prepared nanorod in Fig. 10a shows no defects sites, as explained earlier. The nanorods in Figs. 10b and 10c show point defects and twins, respectively. Finally, the nanodot obtained after the complete shape transformation shows multiple twinning (Fig. 10d). A possible mechanism of the rod-to-sphere shape transformation process is the following [29]. The as-prepared nanorods are defect-free single crystals. The sides of the rod are enclosed by {110} and {100} facets and its growth direction is [001]. The small {111} facets are present, but only at the corners. While being illuminated by laser pulses, point defects are first created in the body of the nanorods and they serve as the nuclei for the formation of twins and stacking faults. The twin is formed by two crystals with a specific orientation, which is only possible if local melting takes place. This suggests that the melting first takes place at the defect sites in order to form a twinned crystal. Then, surface diffusion must occur simultaneously in order to enhance the growth of the twinned crystal. This process is thermodynamically driven by the fact that the lower surface energy {111} face gains the area while the area of the {110} face slowly decreases. A continuing growth of the twinned crystal finally eliminates the unstable {110} surface. Finally, the entire particle is enclosed by the more stable {111} and/or {100} faces. Melting, therefore, occurs not only on the surface of the nanorods but also inside the particle by creating defect sites. This is a mechanism which is quite different from what one would expect for melting of bulk materials.

So far, we have reviewed the results on the rod-to-sphere shape transformation induced by femtosecond laser pulses. However, interesting differences are observed when nanosecond pulses are used instead [26,27]. In particular, it was found that the energy threshold for the complete melting of gold nanorods was lower for femtosecond pulses than for pulses of a nanosecond duration [27]. This result is illustrated in Figs. 11a and 11b, which show nanoparticles produced by irradiation with femtosecond (Fig. 11a) and nanosecond (Fig. 11b) laser pulses. In this experiment, the same pulse energies, beam sizes, and excitation wavelengths (~800 nm) were used to allow a direct comparison. The dominant particle shape after nanosecond irradiation

---

Figure 10 Series of TEM images of gold nanorods taken at different stages of the rod-to-sphere shape transformation. The amount of laser energy absorbed by the particles increases from top to bottom. The image in (a) shows a defect-free gold nanorod before laser irradiation. The particle in (b) has point defects inside the particle and (c) shows an example of a nanorod with a twin defect. The spherical nanoparticles obtained after the shape transformation show multiple twinning (d).

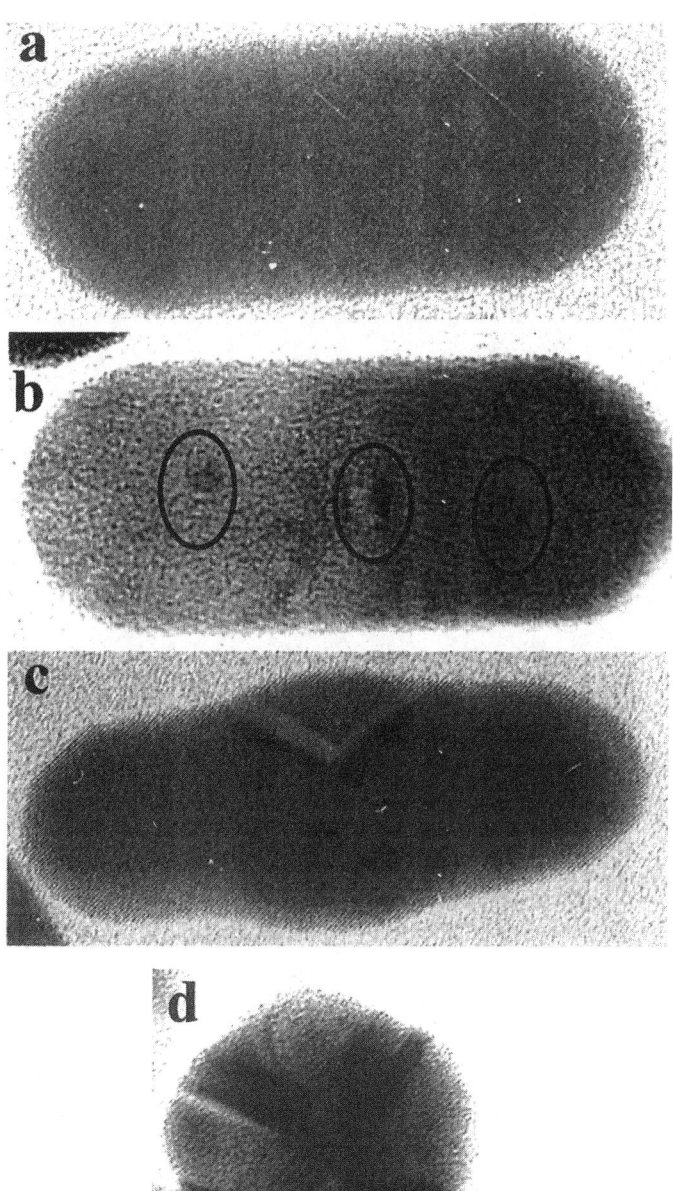

100 Femtosecond Pulses   7 Nanosecond Pulses

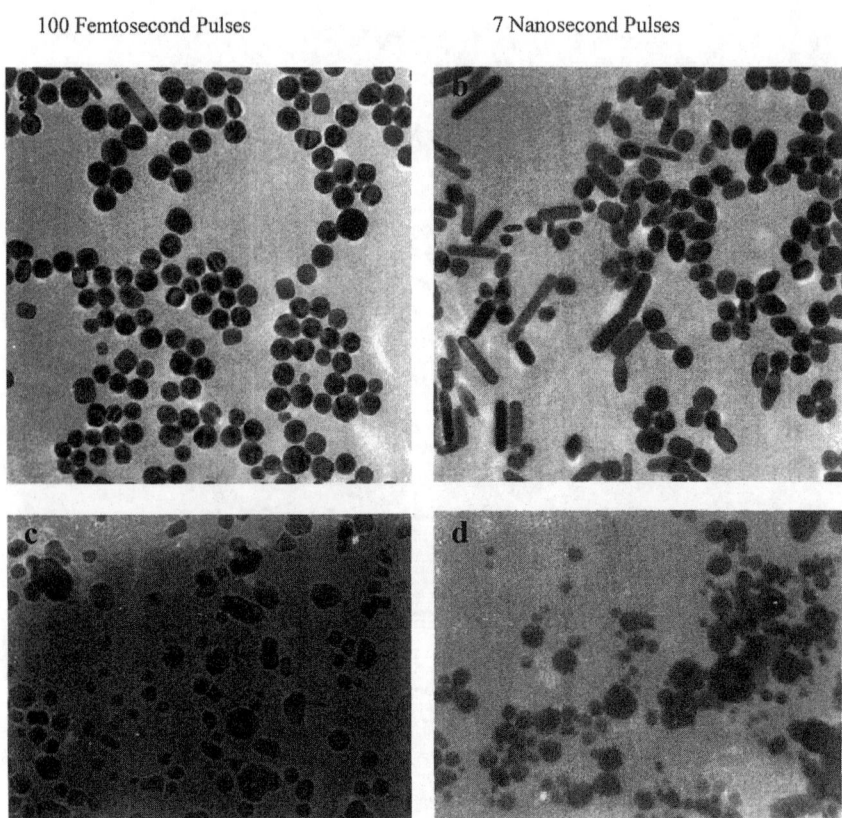

Figure 11 Transmission electron microscopic images of a gold nanorods solution with a mean aspect ratio of 4.1 after exposure to the same energy femtosecond (a) and nanosecond (b) laser pulses. (c) and (d) are TEM images of nanoparticles produced by irradiation with high-energy femtosecond and nanosecond laser pulses, respectively.

(Fig. 11b) is a "φ-shape," indicating only partial melting compared to the nanodots (Fig. 11a) produced by femtosecond pumping [27,57]. This comparison leads to the conclusion that femtosecond laser pulses are more effective in inducing a complete shape transformation than nanosecond pulses. This observation can be explained by the fact that with femtosecond laser pulses, the energy is rapidly absorbed by the nanorods and the melting then occurs after heating the lattice via electron–phonon interactions. In the case of nanosecond pumping, cooling of nanorods (time constant ~100 ps) can compete with heating induced by relatively long 7-ns laser pulses [26,27].

Melting into spheres having a comparable volume to the original nanorods or partial melting (shorter and wider rods for femtosecond pulses and "φ-shaped" particles for nanosecond pulses) are, however, not the only products observed. When higher energy pulses are used, a fragmentation of the nanorods into small spheres can be observed [26,27]. Figures 11c and 11d compare the different particle shapes obtained after fragmentation of the gold nanorods using femtosecond (Fig. 11c) and nanosecond (Fig. 11d) laser pulses. In the case of irradiation with femtosecond pulses, mostly irregular shaped nanoparticles are formed (Fig. 11c), whereas the particles are nearly spherical after fragmentation caused by nanosecond pulses (Fig. 11d). Fragmentation with femtosecond pulses can be explained by a rapid explosion of the nanorods caused by their multiphoton ionization. The repulsion of the accumulated positive charges on the particles after multiphoton ionization leads to fragmentation [27,58]. Because the initial excitation energy is released directly into the solvent by means of photoejected electrons, the lattice of the nanorods never becomes hot, which explains the irregular particle shapes seen in Fig. 11c. In the case of fragmentation with nanosecond pulses (Fig. 11d), more regular spherical shapes of the fragments are observed. This was attributed to the absorption of additional photons by the hot lattice within the nanosecond pulse duration [26,27]. Fragmentation would therefore not necessarily lead to the total cooling of the fragmented parts, but atomic rearrangement into a more spherical shape is possible. Basically, hot particles fragment in the case of nanosecond laser pulses.

Fragmentation is not only observed for gold nanorods but also for spherical metal nanoparticles [58–63]. Irradiation of gold particles with 532 nm nanosecond laser pulses was found to cause fragmentation, as reported by Koda et al. [59,60]. This was explained in terms of the slow heat release of the deposited laser energy into the surrounding solvent, which, in turn, leads to melting and vaporization of the nanoparticles, as estimated from the deposited laser energy and the absorption cross section. As the nanoparticles cannot cool off as fast as they are heated, they fragment into smaller nanodots. On the other hand, in a study of silver nanoparticles that were irradiated with 355-nm picosecond laser pulses, Kamat et al. [58] proposed that strong laser excitation causes the ejection of photoelectrons. As the particles become positively charged, the repulsion between the charges then leads to fragmentation. Mafune et al. [61–63] combined the creation of gold nanoparticles by laser ablation in aqueous solution with subsequent fragmentation in order to produce size-selected gold nanoparticles. The ablation was carried out with 1064-nm laser light, whereas the fragmentation was induced by 532-nm light, which was in resonance with the gold nanoparticle absorption. These examples together with the above-presented results demonstrate that the laser pulse width is an important factor that strongly influences the size and shape of the final irradiation product. At the same time, these studies show the possibilities for "machining" nanostructured materials using pulsed laser light.

An interesting question is how long it takes to melt a gold nanorod. To address this question, time-resolved pump-probe spectroscopy was used to monitor the disappearance of the longitudinal surface plasmon absorption [30]. As the longitudinal surface plasmon absorption is characteristic of the rodlike shape, the rise time for the permanent bleaching of this band corresponds to the melting time of nanorods. This experiment was carried out in a flow cell with a large sample volume in order to assure that the overall percentage of nanorods destroyed by laser irradiation remained small enough that a new probe volume contained mainly unexposed nanorods [30] (these precautions allowed a continuous averaging of the transient absorption). The results of these measurements are shown in Fig. 12 for a rod sample with an

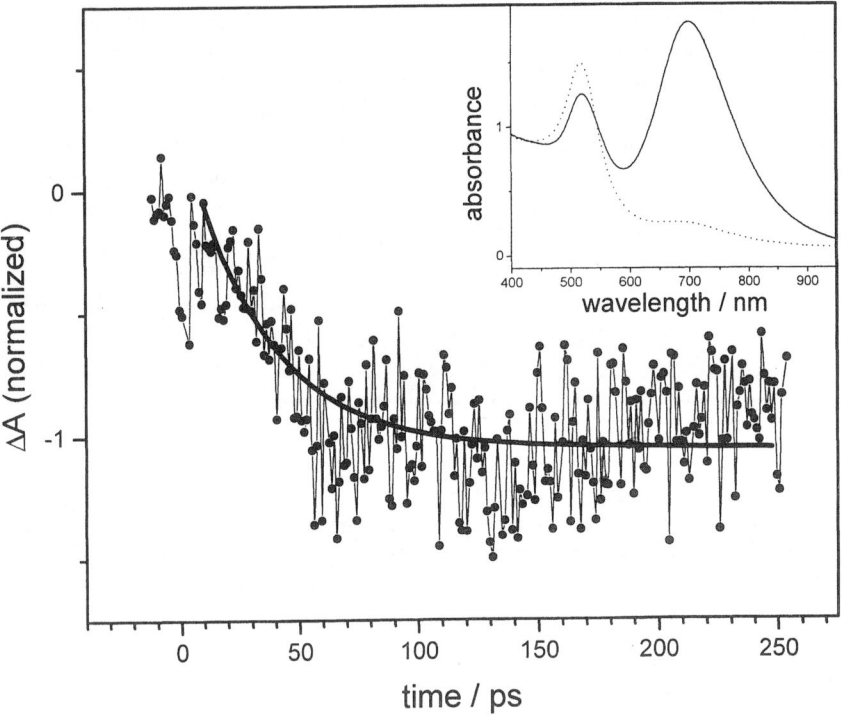

Figure 12 Transient absorption dynamics of the rod-to-sphere shape transformation recorded at 790 nm for the excitation at 400 nm. The rise time for the permanent bleaching of the longitudinal surface plasmon band (see inset) corresponds to the "melting" (transformation) time. The gold nanorods in this example had an aspect ratio of 2.9.

average aspect ratio of 2.9. The excitation wavelength was 400 nm and the monitoring wavelength was 790 nm. The laser pulse energy (9 μJ) was above the threshold for a complete rod-to-sphere transformation [30]. The fact that nanodots were indeed produced in this experiment was verified independently by TEM measurements on a nanorod sample that was irradiated until a complete bleaching of the longitudinal surface plasmon band (see inset). Figure 12 shows the buildup of a permanent bleach signal at the position of the longitudinal resonance, which is associated with the shape transformation from a rod to a sphere. An exponential fit to the experimental data yields a time constant of 30 ± 5 ps, which represents the nanorod-to-nanodot transformation time [30]. No significant dependence of the photoinduced isomerization dynamics on the rod aspect ratio was detected for four samples, with aspect ratios ranging from 1.9 to 3.7 [30]. It is interesting that the measured transformation time is close to the physical limit imposed on the rate of melting by the longitudinal speed of sound [64,65]. For example, it would take ~20 ps for a melting front to travel from one to another end of a 60 nm rod if we assume that the front speed is equal to the speed of sound in gold.

VI. SUMMARY AND CONCLUSIONS

Colloidal metal nanoparticles show a characteristic surface plasmon absorption with extinction coefficients exceeding those for organic laser dyes by several orders of magnitude. Large absorption cross sections are combined with the ease for spectral tunability over the entire visible and near-infrared spectral ranges achieved through the size, shape, and composition control. Furthermore, collective plasmon oscillations lead to a strong field enhancement in the proximity of metal nanoparticles. All of these properties make metal nanoparticles very attractive for a wide range of optical and optoelectronic applications.

The surface plasmon absorption is a small-particle effect observed in the size range from a few nanometers to tens or a few hundreds of nanometers. For particles discussed in this chapter, the energy separation between electronic levels is much smaller than the thermal energy and, therefore, quantum size effects are not significant. Quantum confinement becomes important for sizes below ~2 nm. For such "ultrasmall" particles, the plasmon absorption is strongly damped or even completely suppressed.

A critical length scale for metal nanoparticles is the electron mean free path. A reduction of the particle size below this critical length leads to the enhanced electron-surface scattering, which has a direct effect on the dephasing of the coherent electron motion and, hence, the width of the surface plasmon resonance. Surprisingly, the electron energy relaxation dynamics following the laser excitation appear to be independent of the nanoparticle size and shape and resemble those in bulk metals. This observation indicates that electron-surface collisions are primarily elastic and do not change the electron's energy.

An interesting effect of shape transformation of gold nanorods into spherical nanoparticles is observed in colloidal solutions irradiated with intense laser pulses. The size/shape of nanoparticles produced by irradiation depends on both the energy and the duration of laser pulses. Under appropriate excitation conditions, a narrowing of the nanorod distribution can be achieved with femtosecond laser pulses. This result implies that laser irradiation can be used to selectively reshape metallic nanoparticles (nanoscale laser "machining"). However, it also imposes some limitations on the use of metal nanoparticles in applications that require high laser powers.

One possible direction in the research on shape transformation is the study of laser-induced shape changes for colloidal metal nanoparticles in different media/solvents. These studies will provide new insights into the mechanisms for energy transfer from "hot" particles to the surrounding medium. The timescales of heating and cooling should determine the possible shape changes as was discussed in this chapter by comparing results for femtosecond and nanosecond excitations. Furthermore, the study of single metal nanoparticles could help to overcome problems associated with the polydispersity of colloidal samples. Large extinction coefficients make single-particle studies possible, which was already demonstrated by absorption and scattering measurements on single silver and gold nanoparticles of different sizes and shapes.

ACKNOWLEDGMENTS

We wish to acknowledge our collaborators who contributed to this work: Z. L. Wang, C. Burda, B. Nikoobakht, and M. B. Mohamed. The support of the National Science Foundation, Division of Materials Research (grant No. 0138391) is greatly appreciated.

REFERENCES

1. Henglein, A. J. Phys. Chem. 1993, 97, 8457.
2. Brus, L.E. Appl. Phys. A 1991, 53, 465.
3. Alivisatos, A.P. Science 1996, 271, 933.
4. Mulvaney, P. Langmuir 1996, 12, 788.

5. Storhoff, J.J.; Mirkin, C.A. Chem. Rev. 1999, *99*, 1849.
6. Heath, J.R.; Shiang, J.J. Chem. Soc. Rev. 1998, *27*, 65.
7. Faraday, M. Phil. Trans. 1857, *147*, 145.
8. Kreibig, U.; Vollmer, M. *Optical Properties of Metal Clusters*; Berlin: Springen-Verlag, 1995.
9. Kerker, M. *The Scattering of Light and Other Electromagnetic Radiation*; New York: Academic Press, 1969.
10. Bohren, C.F.; Huffman, D.R. *Absorption and Scattering of Light by Small Particles*; New York: Wiley, 1983.
11. Mie, G. Ann. Phys. 1908, *25*, 329.
12. Kreibig, U.; Genzel, U. Surface Sci. 1985, *156*, 678.
13. Kreibig, U.; von Fragstein, C. Z. Phys. 1969, *224*, 307.
14. Soennichsen, C.; Franzl, T.; Wilk, G.; von Plessen, G.; Feldmann, J.; Wilson, O.; Mulvaney, P. Phys. Rev. Lett. 2002, *88*, 77,402.
15. Link, S.; El-Sayed, M.A. J. Phys. Chem. B 1999, *103*, 4212.
16. Liau, Y.H.; Unterreiner, A.N.; Chang, Q.; Scherer, N.F. J. Phys. Chem. B 2001, *105*, 2135.
17. Voisin, C.; Del Fatti, N.; Christofilos, D.; Vallee, F. J. Phys. Chem. B 2001, *105*, 2264.
18. Bigot, J.-Y.; Halte, V.; Merle, J.C.; Danois, A. Chem. Phys. 2000, *251*, 181.
19. Stagira, S.; Nisoli, M.; De Silvesti, S.; Stella, A.; Tognini, P.; Cheyssac, P.; Kofman, R. Chem. Phys. 2000, *251*, 259.
20. Perner, M.; Bost, P.; von Plessen, G.; Feldmann, J.; Becker, U.; Mennig, M.; Schmidt, H. Phys. Rev. Lett. 1997, *78*, 2192.
21. Hodak, J.H.; Henglein, A.; Hartland, G.V. J. Phys. Chem. B 2000, *104*, 9954.
22. Link, S.; El-Sayed, M.A. J. Phys. Chem. B 1999, *103*, 8410.
23. El-Sayed, M.A. Acc. Chem. Res. 2001, *34*, 257.
24. Hodak, J.H.; Henglein, A.; Hartland, G.V. J. Chem. Phys. 1999, *111*, 8613.
25. Del Fatti, N.; Voisin, C.; Chevy, F.; Vallee, F.; Flytzanis, C. J. Chem. Phys. 1999, *110*, 11,484.
26. Link, S.; Burda, C.; Mohamed, M.B.; Nikoobakht, B.; El-Sayed, M.A. J. Phys. Chem. A 1999, *103*, 1165.
27. Link, S.; Burda, C.; Nikoobakht, B.; El-Sayed, M.A. J. Phys. Chem. B 2000, *104*, 6152.
28. Link, S.; El-Sayed, M.A. J. Chem. Phys. 2001, *114*, 2362.
29. Link, S.; Wang, Z.L.; El-Sayed, M.A. J. Phys. Chem. B 2000, *104*, 7867.
30. Link, S.; Burda, C.; Nikoobakht, B.; El-Sayed, M.A. Chem. Phys. Lett. 1999, *315*, 12.
31. Creighton, J.A.; Eadon, D.G. J. Chem. Soc. Faraday Trans. 1991, *87*, 3881.
32. Link, S.; Wang, Z.L.; El-Sayed, M.A. J. Phys. Chem. B 1999, *103*, 3529.
33. van der Zande, B.M.I.; Bohmer, M.R.; Fokkink, L.G.J.; Schonenberger, C. J. Phys. Chem. B 1997, *101*, 852.
34. Yu, Y.; Chang, S.; Lee, C.; Wang, C.R.C. J. Phys. Chem. B 1997, *101*, 6661.
35. Mohamed, M.B.; Ismael, K.Z.; Link, S.; El-Sayed, M.A. J. Phys. Chem. B 1998, *102*, 9370.

36. Papavassiliou, G.C. Prog. Solid State Chem. 1979, *12*, 185.
37. Alvarez, M.M.; Khoury, J.T.; Schaaff, T.G.; Shafigullin, M.N.; Vezmer, I.; Whetten, R.L. J. Phys. Chem. B 1997, *101*, 3706.
38. Schaaff, T.G.; Shafigullin, M.N.; Khoury, J.T.; Vezmer, I.; Whetten, R.L.; Cullen, W.G.; First, P.N.; Wing, C.; Ascensio, J.; Yacaman, M.J. J. Phys. Chem. B 1997, *101*, 7885.
39. Hostetler, M.J.; Wingate, J.E.; Zhong, C.J.; Harris, J.E.; Vachet, R.W.; Clark, M.R.; Londono, J.D.; Green, S.J.; Stokes, J.J.; Wignall, G.D.; Glish, G.L.; Porter, M.D.; Evans, N.D.; Murray, R.W. Langmuir 1998, *14*, 17.
40. Ashcroft, N.W.; Mermin, N.D. *Solid State Physics*; Philadelphia: Saunders College, 1976.
41. Heilweil, E.J.; Hochstrasser, R.M. J. Chem. Phys. 1985, *82*, 4762.
42. Wood, D.M.; Ashcroft, N.W. Phys. Rev. B 1982, *25*, 6255.
43. Kraus, W.A.; Schatz, G.C. J. Chem. Phys. 1983, *79*, 6130.
44. Yannouleas, C.; Broglia, R.A. Ann. Phys. 1992, *217*, 105.
45. Brack, M. Rev. Mod. Phys. 1993, *65*, 677.
46. Bonacic-Koutecky, V.; Fantucci, P.; Koutecky, J. Chem. Rev. 1991, *91*, 1035.
47. Palpant, B.; Prevel, B.; Lerme, J.; Cottancin, E.; Pellarin, M.; Treilleux, M.; Perez, A.; Vialle, J.L.; Broyer, M. Phys. Rev. B 1998, *57*, 1963.
48. Persson, N.J. Surface Sci. 1993, *281*, 153.
49. Itoh, T.; Asahi, T.; Masuhara, H. Jpn. J. Appl. Phys. 2002, *41*, L76.
50. Klar, T.; Perner, M.; Grosse, S.; von Plessen, G.; Spirkl, W.; Feldmann, J. Phys. Rev. Lett. 1998, *80*, 4249.
51. Lamprecht, B.; Leitner, A.; Aussenegg, F.R. Appl. Phys. B 1999, *68*, 419.
52. Sun, C.K.; Vallee, F.; Acioli, L.H.; Ippen, E.; Fujimoto, J.G. Phys. Rev. B 1994, *50*, 15,337.
53. Anisimov, L.; Kapeliovich, B.L.; Perel'man, T.L. Sov. Phys. JETP 1975, *39*, 375.
54. Belotskii, E.D.; Tomchuk, P.M. Int. J. Electron 1992, *73*, 915.
55. Hodak, J.H.; Henglein, A.; Hartland, G.V. J. Chem. Phys. 2000, *112*, 5942.
56. Wang, Z.L.; Mohamed, M.B.; Link, S.; El-Sayed, M.A. Surface Sci. 1999, *440*, L809.
57. Chang, S.; Shih, C.; Chen, C.; Lai, W.; Wang, C.R.C. Langmuir 1999, *15*, 701.
58. Kamat, P.V.; Flumiani, M.; Hartland, G.V. J. Phys. Chem. B 1998, *102*, 3123.
59. Kurita, H.; Takami, A.; Koda, S. Appl. Phys. Lett. 1998, *72*, 789.
60. Takami, A.; Kurita, H.; Koda, S. J. Phys. Chem. B 1999, *103*, 1226.
61. Mafune, F.; Kohno, J.; Takeda, Y.; Kondow, T. J. Phys. Chem. B 2002, *106*, 7575.
62. Mafune, F.; Kohno, J.; Takeda, Y.; Kondow, T. J. Phys. Chem. B 2001, *105*, 9050.
63. Mafune, F.; Kohno, J.; Takeda, Y.; Kondow, T.; Schwabe, H. J. Phys. Chem. B 2001, *105*, 5114.
64. Stuart, B.C.; Feit, M.D.; Herman, S.; Rubenchik, A.M.; Shore, B.W.; Perry, M.D. Phys. Rev. B 1996, *53*, 1749.
65. Sokolowski-Tinten, K.; Bialkowski, J.; Cavalleri, A.; von der Linde, D.; Oparin, A.; Meyer-ter-Vehn, J.; Anisimov, S.I. Phys. Rev. Lett. 1998, *81*, 224.

12

Time-Resolved Spectroscopy of Metal Nanoparticles

Gregory V. Hartland
University of Notre Dame, Notre Dame, Indiana, U.S.A.

I. INTRODUCTION

In the past decade, time-resolved laser spectroscopy has proven to be a powerful technique for probing the fundamental properties of metal particles [1–6]. In these experiments, a pump laser pulse is used to excite the electron distribution, and a second probe laser pulse monitors how the system returns to equilibrium. The major aim of these studies has been to obtain information about how confinement affects the couplings between the electrons and between the electrons and phonons. This work has recently been reviewed by several authors [1–6] and our current understanding is that confinement effects are relatively small for electron–phonon coupling (at least for noble metals like Ag and Au) and can be qualitatively described by considering how the electrons couple to surface phonon modes [7–9]. An unexpected result from these experiments is the observation that ultrafast laser excitation can actually coherently excite the acoustic phonon modes of the particles [11–13]. The aim of this chapter is to review these experiments—specifically, to describe the assignment and excitation mechanism for the vibrational modes and (more importantly) discuss what new physics can be learned from these measurements.

A simple picture of these experiments is that the pump laser selectively excites the electrons. The absorbed energy flows into the phonon modes on a picosecond timescale: The exact time depends on the electron–phonon coupling constant, the electronic heat capacity, and the pump laser power [1,2, 9,14]. This results in an increase in the lattice temperature, which causes the particles to expand. Because the heating time is extremely rapid, the phonon mode that correlates to the expansion coordinate can be impulsively excited [12,13]. For spherical metal particles, the symmetric breathing mode is excited. This mode contributes to the signal in transient absorption experiments because it changes the volume of the particles, which changes the dielectric function of the metal and, therefore, the optical response of the system [11,13]. This description of the excitation mechanism represents an indirect coupling between the electrons and the coherently excited phonon modes. A significant question that will be addressed in this review is whether the electrons can directly couple to the symmetric breathing mode [15,16]. Other issues that will be examined are how the period of the vibrational modes depends on the size and shape of the particles and what happens for bimetallic particles—in particular, particles composed of two metals with significantly different elastic properties [17].

In addition to the vibrational studies, electron–phonon coupling in bimetallic particles will be discussed. Compared to noble metal particles, much less is known about the photophysics of multicomponent metal particles or particles of transition metals, even though these materials are important components of catalysts [18]. At present, one of the limiting factors for studying these materials is that it is difficult to make multicomponent particles with arbitrary combinations of metals that have a well-defined structure (e.g., core–shell compared to alloyed) and are in a controlled environment. In this chapter, our recent ultrafast studies of electron–phonon coupling in Pt–Au core–shell nanoparticles will be reviewed [19]. These particles were prepared using radiation chemistry techniques. The experimental results show that the effective electron–phonon coupling constant in a bimetallic particle is an average of the electron–phonon coupling constants of the pure metals, weighted by the density of electronic states. This is an intuitive result that was not at all obvious to us until we did these experiments!

II. EXPERIMENTAL: SYNTHESIS AND LASER TECHNIQUES

The key factors in these studies (and really all studies) are to have high-quality, well-characterized metal particle samples and to perform experiments with high sensitivity. For most of the experiments described in this section, the particles used were in the mesoscopic regime (10–100 nm in diameter). An extremely versatile and controlled way of producing these materials is to use radiation chemistry [20]. All noble metal ions and many electronegative metal ions can be reduced in aqueous solution by γ-radiation. The radiation is primarily absorbed by the solvent, generating reducing species (free radicals, aqueous electrons, and hydrogen atoms) and oxidizing $\cdot$OH radicals. These species will attack any dissolved substances. To achieve complete reduction of the metal ions in solution, an alcohol is added to scavenge the •OH radicals. This produces organic radicals, which act as reductants. In most cases, it is useful to eliminate the primary radicals from water radiolysis (i.e., only produce organic radicals). This can be achieved by irradiating under an atmosphere of nitrous oxide. The hydrated electrons are scavenged by N_2O to produce •OH radicals, which then react with the alcohol.

Radiation chemistry can be used to produce small ($\sim$2 nm) noble metal particles in solution [21], but for our purposes, its most important use is to grow existing nanoparticles into larger ones [22]. In a typical growth experiment, a solution containing Au seed particles, $KAu(CN)_2$, methanol, and N_2O is γ-irradiated, producing 1-hydroxymethyl radicals. These radicals cannot reduce Au–I complexes in solution, due to the large free energy of formation of a free Au atom; however, they can react with the Au particles by electron transfer. Once the colloidal particles have accumulated enough electrons, they are able to reduce $Au(CN)_2^-$ directly onto their surface [23]. Irradiation is carried out until all of the Au–I complex is reduced, and the final size of the particles is simply determined by the amount of gold complex used [21,22]. The particles produced by this technique have an extremely narrow size distribution (< 7%, *vida infra*). This technique can also be used to prepare core–shell metal particles. For example, when $Ag(CN)_2^-$ is used instead of the gold complex, a Ag shell is formed around the seed particle [24]. Bimetallic particles with Au, Ag, or Pt cores and shells of Au, Ag, Hg, Cd, Pb, or Pt have been produced by radiation chemistry [25–29]. Note that for particles containing non-noble metals, air must be rigorously excluded during the synthesis and photophysical measurements. The Au seeds are typically made using the conventional citrate reduction method [30]. Some of the data discussed here involves platinum particles, which were made by reduction of $PtCl_2(H_2O)_2$ by hydrogen, with citrate as a stabilizer. The Pt particles produced had a narrow size distribution and mainly consist of cubic and cuboctahedral crystals [31]. The rod-shaped particles discussed in Section III were supplied by Professor Mulvaney (The University of Melbourne, Australia) and were synthesized using the electrochemical technique first described by Wang and co-workers [32,33].

The laser system used in the transient absorption experiments is a regeneratively amplified Ti: sapphire laser (Clark-MXR, CPA-1000, with a SpectraPhysics Millennia Vs pump laser). This system produces pulses in the 780–820-nm range with a full width at half-maximum of 120–150 fs (sech^2 deconvolution) and an energy of 400–500 µJ per pulse. The output is split by a 90 : 10 beam splitter to create the pump and probe beams. The timing between the two laser pulses is controlled by a stepper motor-driven translation stage (Newport, UTM150PP.1). For the experiments described below, the pump was frequency doubled in a 1-mm B- Barium Borate (BBO) crystal before the sample. Visible probe laser pulses were obtained by gently focusing the 10% portion of the output from the regen into a 3-mm sapphire window to generate a single-filament, white-light continuum. The pump and probe beams were spatially overlapped at the sample, and the transmitted probe intensity was monitored by a Si–PIN photodiode (Thorlabs, PA150). The probe laser wavelength was selected by a Jobin–Yvon Spex H-10 monochromator placed after the sample. A normalization scheme using gated integration, an analog division circuit, and lock-in detection of the signal was used to reduce noise from fluctuations in the probe laser [14]. Normalization and a stable white light beam are crucial for the success of the experiments described in Section III, as they allow good signal-to-noise data to be collected at relatively low pump laser power. For some of our experiments, the metal particle samples were flowed through a 3-mm-path-length sample cell; however, the air-sensitive samples were kept in sealed cuvettes (2-mm path length).

III. COHERENT EXCITATION OF ACOUSTIC VIBRATIONAL MODES

Transient absorption data for ~50-nm-diameter Au particles collected with 550-nm and 510-nm probe laser pulses are presented in Fig. 1. These two probe wavelengths lie on the red and blue sides, respectively, of the gold plasmon band, which is shown in the insert of Fig. 1. The plasmon band is a collective oscillation of the conduction band electrons and occurs at ~525 nm for the particles in Fig. 1. The initial bleach signal (negative ΔA) is due to the hot electrons created by laser excitation, which cause a strong broadening of the plasmon band [1,2,14]. The decay time of the bleach provides information about the electron–phonon coupling constant for the particles. The modulations are due to the coherently excited vibrational modes. Note that the period and damping times of the modulations are identical for the two scans, but the phases are very different: The modulations at 510 nm are completely out of phase with those at 550 nm. This implies that the modulations arise from a periodic shift in the *position* of the plasmon band [9, 11–13].

The period of the modulations depends on the average size of the particles. Figure 2 shows the frequency (in cm^{-1} units) versus the inverse of the radius, for a series of transient absorption experiments with different size Au particles (all in aqueous solution). The frequency changes from 1 to 12 cm^{-1} as the radius changes from 60 to 4 nm [13]. Also shown in Fig. 2 are the results from continuum mechanics calculations for the symmetric breathing mode of a sphere. These calculations, which were first performed over 100 years ago,

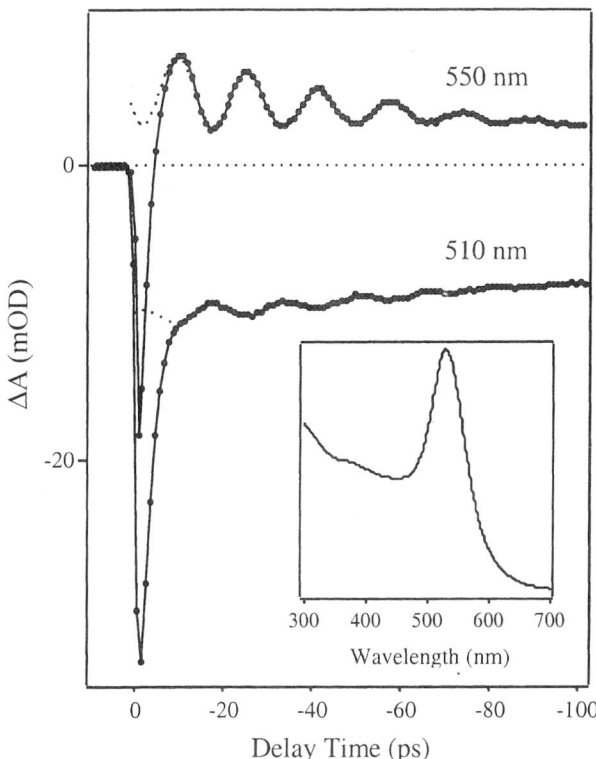

Figure 1 Transient absorption data for ~50-nm-diameter Au particles collected with 510-nm and 550-nm probe laser wavelengths. The insert shows the ultraviolet–visible absorption spectra of the particles.

predict that the frequency should be inversely proportional to the radius R [34,35]:

$$\bar{v}(\text{cm}^{-1}) = \frac{\eta c_1}{2\pi c R} \tag{1}$$

where c_l is the longitudinal speed of sound in Au, c is the speed of light, and η is an eigenvalue that depends on the transverse and longitudinal speeds of sound. For the symmetric breathing mode, η is the smallest root of $\eta \cot \eta = 1 - \eta^2/4\delta^2$, where $\delta = c_t/c_l$, where c_t is the tranverse speed of sound. The excellent agreement between the calculated and experimental results confirms that the modulations are due to the symmetric breathing mode [13].

The modulations are damped because the samples are polydisperse: Different-sized particles have different vibrational periods [13]. Assuming

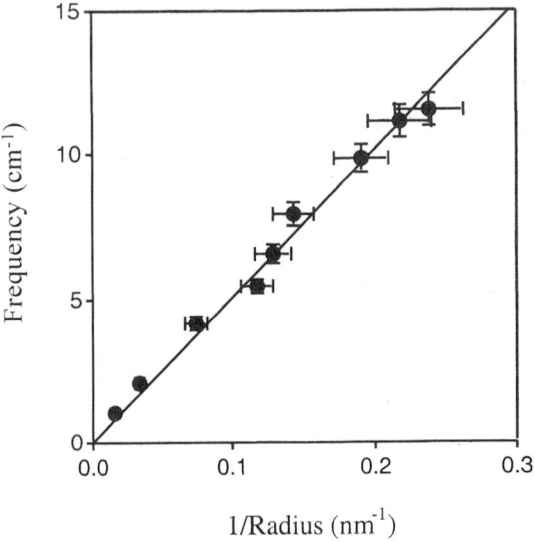

Figure 2 Frequency of the acoustic vibrational mode versus the inverse radius of the particles: (•) experimental values determined from transient absorption data; (—) calculated values using Eq. (1) and the bulk speeds of sound for Au.

that the particles have a normal size distribution and that the average radius $\overline{R}$ is much larger than the standard deviation σ_R, the signal in our experiment can be expressed as [36]

$$S(t) = \cos\left(\frac{2\pi}{\overline{T}}t + \phi\right)e^{-t^2/\tau^2} \tag{2}$$

where ϕ is the phase of the modulation, $\overline{T} = 2\pi\overline{R}/\eta c_l$ is the period corresponding to the average radius of the sample, and the damping time is given by $\tau = \overline{R}\,\overline{T}/\sqrt{2}\pi\sigma_R$. Fits to the transient absorption data using Eq. (2) are included in Fig. 1. This analysis yields an average period of $\overline{T} = 15.9 \pm 0.1$ ps, which corresponds to an average radius of $\overline{R} = 24.2 \pm 0.1$ nm for this sample. The damping time is $\tau = 51 \pm 1$ ps, which implies a standard deviation of $\sigma_R = 1.70 \pm 0.05$ nm. In comparing size distributions, it is common to normalize the standard deviation by the average radius. For our sample, we find $\sigma_R/\overline{R} = 7.0 \pm 0.2\%$, which corresponds to an extremely high-quality, monodisperse sample. In principle, the decay of the modulations could contain contributions from energy relaxation to the environment. Thus, the value of σ_R derived from this analysis is strictly an upper limit to the true width of the distribution (i.e., $\sigma_R/\overline{R} \leq 7.0\%$). Note that the number of oscillations in the transient absorption data depends on τ/T, which is inversely proportional to $\sigma_R/\overline{R}$. Thus, the sample quality can be judged simply by the number of beats in the experimental traces.

Time-resolved experiments were performed over a range of wavelengths for the sample in Fig. 1. The transient absorption versus wavelength data at different times was then analyzed to determine the position and width of the plasmon band versus time [36]. In this analysis, the Au plasmon band was modeled as a Lorentzian function. The peak position versus time data was then be converted to radius versus time through the Mie theory expression for the absorption cross section of small particles. Specifically, assuming that the dielectric function of the metal is dominated by free-electron (Drude model) contributions, it can be shown that the maximum of the plasmon band occurs at [37].

$$\omega_{max} = \frac{\omega_p}{\sqrt{1 + 2\varepsilon_m}}, \tag{3}$$

where ε_m is the dielectric constant of the medium and ω_p is the plasma frequency. The plasma frequency is given by $\omega_p = (ne^2/\varepsilon_0 m_e)^{1/2}$, where n is the electron density, m_e is the effective mass of the electrons, and ε_0 and e have their usual meanings. Equation (3) shows that the frequency of the plasmon band maximum is proportional to $\sqrt{n}$. This proportionally allows the peak wavelength versus time data to be converted into size versus time. The results of these calculations are shown in Fig. 3. The key points to note are that (1) laser excitation produces an overall increase in the radius of the particles of $\sim0.4\%$, just after the electrons and phonons have reached equilibrium (~10 ps); (2) the oscillations in the radius have an amplitude that is $\sim50\%$ of the overall size increase.

A simple way to understand how the beat signal is generated is to consider the lattice heating process. The pump laser deposits energy in the electronic degrees of freedom, which subsequently flows into the phonon modes on a picosecond timescale. This increases the temperature of the nuclei, which causes the particles to expand. Because the lattice heating is faster than the phonon mode that correlates with the expansion coordinate, the nuclei cannot respond instantaneously. Thus, following laser excitation, the nuclei will start to move along the expansion coordinate (the symmetric breathing mode) and pick up momentum. When they reach the equilibrium radius of the hot particles, their inertia will cause them to overshoot. The elastic properties of the particle then provide a restoring force that makes the nuclei stop and reverse their motions. The competition between the impulsive kick from the rapid laser-induced heating and the restoring force from the elastic response of the particles causes the nuclei to "ring" around the new equilibrium radius—the value of which is determined by the temperature rise in the particles.

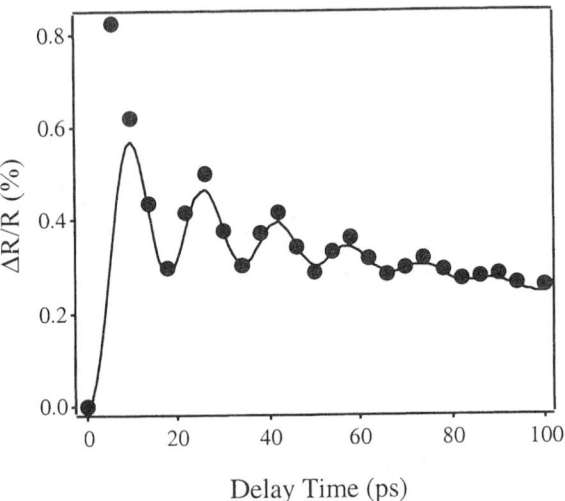

Figure 3 Change in radius ($\Delta R/R$) versus time for ~50-nm-diameter Au particles after ultrafast laser excitation: (•) experimental values determined from the transient absorption data; (—) calculated using the harmonic oscillator model (see text for details).

This behavior is analogous to a harmonic oscillator [12,13,15]. Consider a ball at the minimum of a parabolic well. If the potential minimum is suddenly changed, the ball will initially be displaced from the new minimum. The ball will subsequently move toward the new minimum, pick up momentum, and overshoot, exactly as described earlier. The amount that the ball overshoots depends on the initial displacement and the friction in the system. This is shown schematically in Fig. 4. Note that this simple model predicts that the amount of overexpansion should be approximately the same as the overall increase in the radius due to lattice heating.

A coherent response in the nuclei can also be induced by direct coupling between the expansion coordinate and the hot electrons [15,16]. This coupling arises from the pressure exerted by the hot electrons, [38,39] and dies off rapidly as the electrons equilibrate with the lattice. In terms of the harmonic oscillator model described earlier, the hot electron pressure effect is equivalent to a transient force that gives the nuclei an instantaneous kick at $\tau = 0$ [15]. As shown in Fig. 4, this generates a very different response compared to lattice heating. First, the nuclei oscillate around the original equilibrium radius (not the radius of the hot particles). Second, the oscillations induced by hot electron pressure are $90°$ out of phase with those from heating the nuclei

Nuclear Heating   Hot-Electron Pressure

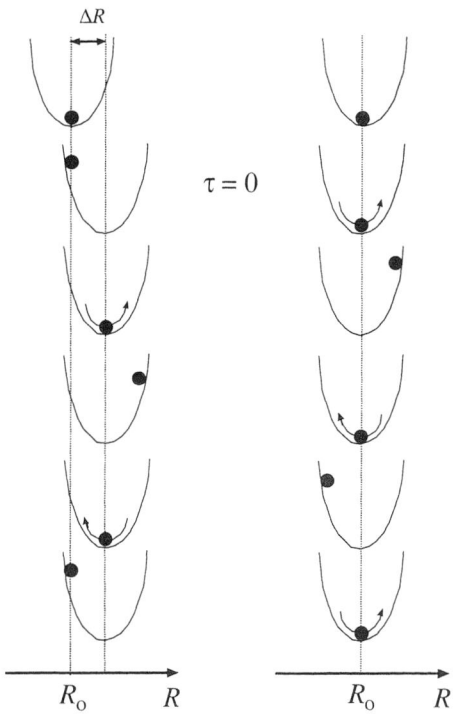

Figure 4 Harmonic oscillator model for the generation of coherent vibrational motion in metal nanoparticles by nuclear heating (left) or by hot electron pressure (right).

(strictly, a 90° phase difference only applies for delta-function excitation of the electrons and the nuclei).

    The relative contributions of these two effects can be calculated through the thermal stresses from heating the electrons (σ_e) or the lattice (σ_l) [38,39]:

$$\sigma_e = -\frac{2}{3}\Delta E_e(T_e) \tag{4a}$$

$$\sigma_l = -\gamma_l \Delta E_l(T_l) \tag{4b}$$

where γ_l is the Grüneisen parameter for the lattice ($\gamma_l = 2.7$ for Au) and ΔE_e and ΔE_l are the increase in energy of the electrons and lattice, respectively, which depend on T_e and T_l, the electronic and lattice temperatures. The time dependence of T_e and T_l can be calculated by the "two-temperature" model.

In this model, the electrons couple to the phonon modes in a way that depends on their temperature difference [40]:

$$C_e(T_e)\frac{\partial T_e}{\partial t} = -g(T_e - T_l) + \frac{E_0}{\sqrt{\pi}\sigma_L}\exp\left[\frac{-t^2}{\sigma_L^2}\right] \tag{5a}$$

and

$$C_1\frac{\partial T_l}{\partial t} = g(T_e - T_l) - \frac{T_l - 298}{\tau_s} \tag{5b}$$

In these equations, g is the electron–phonon coupling constant, $C_e(T_e) = \gamma T_e$ and C_l are the heat capacities of the electrons and the lattice, respectively, and τ_s is the timescale for energy transfer to the solvent. The last term in Eq. (5a) represents the laser pulse: E_0 corresponds to the amount of energy absorbed by the sample and σ_L gives the pulse width [36]. Equations (5a) and (5b) can be solved numerically if E_0 and τ_s are specified (all the other parameters are known).

Once T_e and T_l have been determined, the response of the system to heating can be calculated by the following harmonic oscillator equation:

$$\frac{d^2R}{dt^2} + \frac{2}{\tau_d}\frac{dR}{dt} + \left(\frac{2\pi}{T}\right)^2[R - R_0] = F_e(T_e) + F_l(T_l) \tag{6}$$

where T is the period the phonon mode that correlates with expansion, R_0 is the initial radius, τ_d is a phenomenological constant that accounts for the decay of the modulations, and $F_e(T_e)$ and $F_l(T_l)$ are the forces due to heating the electrons and the lattice, which are proportional to σ_e and σ_l, respectively. The results of these calculations are shown as the solid line in Fig. 3. The values of T, R_0, τ_d, and τ_s were chosen to match the data in Figs. 1 and 3. The value of E_0 (the absorbed energy) was determined from the experimental conditions: the laser spot size, the energy per pulse, and the absorbance of the sample at 400 nm. For our experiments $E_0 = (7 \pm 1) \times 10^3$ J/mol, which produces an overall temperature rise in the lattice of $\Delta T_l = 290°C$ just after the electrons and phonons have reached equilibrium (i.e., before any heat dissipaton to the solvent). Note that the only adjustable parameters in these calculations are τ_d and τ_s.

There are several points to note about the data in Fig. 3. First, the amplitude (the amount of overexpansion and compression) and the phase of the oscillations are in excellent agreement for the calculations and the experiments. The amplitude is mainly determined by the lattice heating contribution to the driving force for the expansion. Thus, the agreement between the calculated and experimental amplitudes shows that all of the absorbed energy goes into expansion. In contrast, the phase is sensitive to both the electronic and lattic contributions. Neglecting hot electron pressure effects in Eq. (6) produces a phase difference of 45° between the experimental and calculated $\Delta R/R$ versus time traces. Thus, hot electron pressure effects play a significant role in launching the coherant vibrational motion of the particles. This conclusion is consistent with that from recent time-resolved studies of Ag particles in a solid matrix by Perner et al. [15] and Voisin and co-workers [16].

IV. GOLD NANORODS

Studies of the properties of nonspherical metal particles are extremely important, as these materials are likely to be vital components in nanoelectronics devices [41]. Extensive work has been done on the optical properties of cylindrical and ellipsoidal particles. For example, it is well known that these particles show two plasmon bands, corresponding to oscillation of the conduction-band electrons along the longitudinal or transverse directions [37]. Time-resolved studies of these systems have been performed by the El-Sayed group [42–44], who studied electron–phonon coupling and laser-induced melting in Au nanorods, and by Perner et al. who examined the coherently excited vibrational modes of Ag ellipses [15]. In their work, Perner et al. observed that the period of the beat signal depended on the relative orientation of the ellipse with the laser polarization vector. The measured periods are approximately given by $2d/c_l$, where c_l is the longitudinal speed of sound and d is either the length or width of the ellipse— depending on whether the probe laser interrogates the longitudinal or transverse plasmon band, respectively. They attributed this signal to expansin and contraction along the major or minor axis of the ellipse.

Figure 5 shows initial results from our ultrafast laser studies of Au nanorods [45] Modulations with a period on the order of 50–70 ps can be clearly seen in the transient absorption data. The inset in Fig. 5 shows the measured period versus the length of the rod. The straight line corresponds to $T = 2L/c_R$, where c_R is the speed of Rayleigh surface waves in Au and L is the length of the rod. This is approximately the period expected for a Rayleigh surface wave that propagates along a cylinder. The exact value of the period depends on the probe laser wavelength, which implies that different probe wavelengths interrogate different length rods in the sample. The dependence of the measured period on the laser wavelength has been analyzed to obtain information about the homogeneous and inhomogeneous contributions to the spectra of the rods [45]. The quantitative differences between our results ($T = 2L/c_R$) and those of Perner et al. ($T = 2L/c_l$) show that the vibrational dynamics of nonspherical particles is very complex. It is not at all clear what

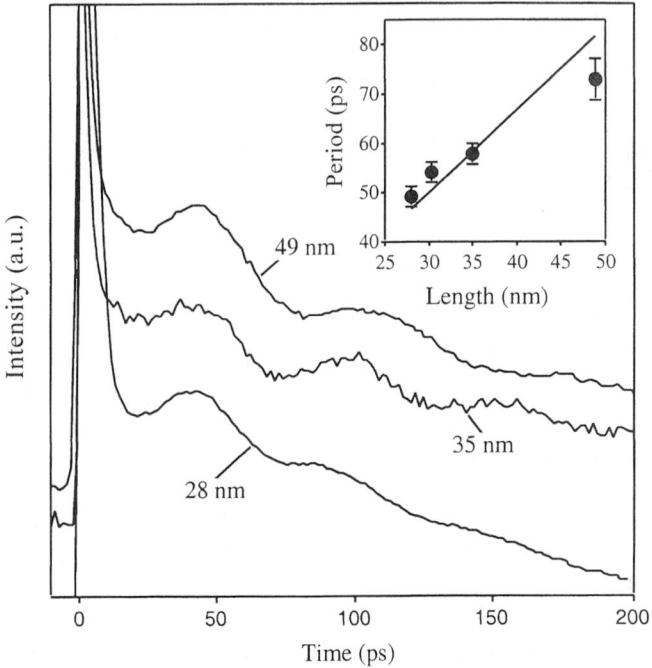

Figure 5 Transient bleach data for different aspect ratio nanorods. The experiments were performed with 400-nm pump pulses and with the probe laser tuned to the maximum of the longitudinal band for each sample. The width of the rods was ~10 nm and their length is given in the figure. The inset shows the measured period versus the rod length, for the samples investigated to date.

modes will be excited and how their frequencies depend on the size and shape of the particles.

V. BIMETALLIC NANOPARTICLES

Bimetallic particles are extremely important in catalysis [18]; however, their photophysics has not received anywhere near the attention that has been given to single-component particles. Thus, at this stage, very little is known about the properties of these materials—in particular, whether one can simply the average of the properties of the individual components or whether a more complex treatment is needed. Figure 6 shows the ultraviolet–visible (UV-vis) absorption spectra of Pt–Au core–shell particles with the same of Pt core and

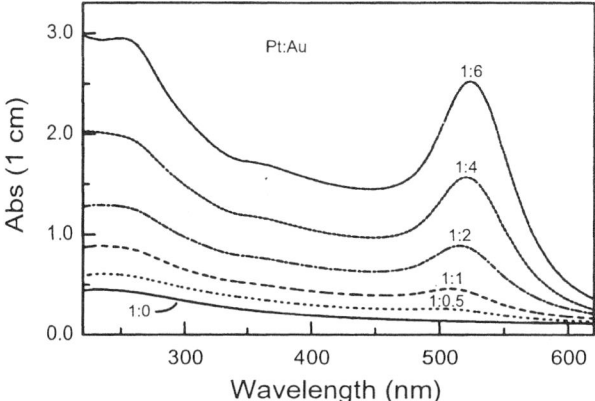

Figure 6 Ultraviolet–visible absorption spectra of Pt–Au core–shell nanoparticles. The molar ratios of Pt : Au are given in the figure. The plasmon band for these composite particles occurs at 520 nm, close to the plasmon band of pure Au.

different amounts of Au [29]. There are two important things to note about these spectra. First, Pt has a very weak absorption in the visible/near-UV region. Second, particles with a >1:1 ratio of Au to Pt have a strong plasmon band at 520 nm, which is very close to the position for pure Au (see Fig. 1). This is typical of core–shell particles: After the addition of enough shell material, the UV-vis absorption spectra tends to resemble that of particles of the shell.

Transient bleach data for Pt–Au core–shell and pure Au particles are presented in Fig. 7a. These experiments were performed with 400-nm pump laser pulses, with the probe laser tuned to the peak of the plasmon band [19]. The bleach recovery corresponds to relaxation of the hot electrons by energy exchange with the phonon modes. The most noticeable feature of these experiments is that the relaxation time is much faster for the core–shell particles compared to the pure Au particles. Similar results are observed for Au_{core}–Pt_{shell} particles [19].

Figure 7b shows the characteristic timescale for electron–phonon coupling plotted again the mole fraction of Au. The presence of even a small amount of Pt drastically shortens the relaxation time. The dashed line shows the effective time constant for electron–phonon coupling calculated by averaging the time constants for Pt and Au:

$$\frac{1}{\tau_{eff}} = \frac{(1 - \alpha_{Au})}{\tau_{Pt}} + \frac{\alpha_{Au}}{\tau_{Au}} \tag{7}$$

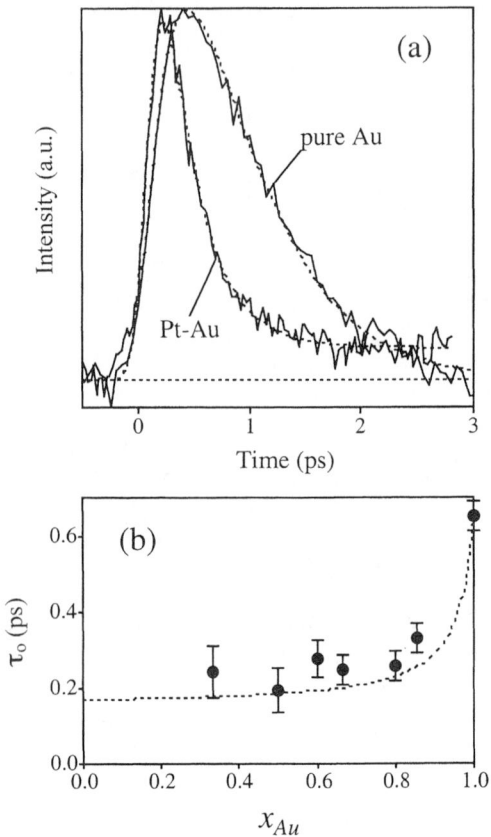

Figure 7 (a) Transient bleach data for pure Au and Pt–Au core–shell nano-particles (1 : 0.5 molar ratio of Pt : Au) recorded with 400-nm pump and visible probe laser pulses. Note that the dynamics are much faster for the bimetallic particles. (b) Characteristic timescale for electron–phonon coupling versus mole fraction of Au (x_{Au}) for Pt–Au core–shell nanoparticles. The dashed line is the time constant calculated using Eqs. (7) and (8).

where α_{Au} is the fraction of Au electronic states, which is given by

$$\alpha_{Au} = \frac{x_{Au}\rho(\varepsilon_F)_{Au}}{x_{Au}\rho(\varepsilon_F)_{Au} + (1 - x_{Au})\rho(\varepsilon_F)_{Pt}} \qquad (8)$$

In Eq. (8), x_{Au} is the mole fraction of Au, and $\rho(\varepsilon_F)_{Au}$ and $\rho(\varepsilon_F)_{Pt}$ are the density of electronic states at the Fermi level for Au and Pt. In this calcu-lation, the values of $\rho(\varepsilon_F)_{Au}$ and $\rho(\varepsilon_F)_{Pt}$ were taken from the literature [46],

τ_{Au} = 0.65 ps was determined from transient bleach experiments [9], and τ_{Pt} = 0.17 ps was calculated from electrical conductivity data [47]. The calculations and experiments are in excellent agreement. Essentially perfect agreement can be obtained by allowing the time constant for Pt to vary. This yields a characteristic electron–photon coupling time for Pt of 0.22 ps (in close agreement with the value calculated from the conductivity of Pt). Note

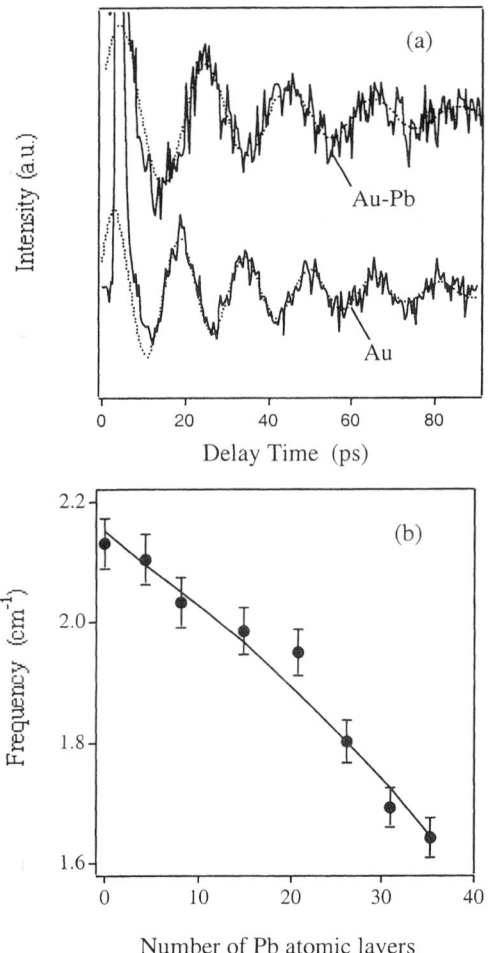

Figure 8 (a) Transient absorption data for pure Au particles (47 nm in diameter) and the Au particles coated with ~40 monolayers of Pb. (b) Period versus Pb coverage for the Au–Pb core–shell nanoparticles. The line shows the results from continuum mechanics calculations for these particles.

that the particles with >80% Au have spectra that closely resemble pure Au (see Fig. 6) but have a dynamic response that is controlled by Pt. This occurs because (1) the electron–phonon coupling constant for Pt is much larger than that for Au and (2) Pt has a much larger density of states at the Fermi level than Au [19].

Coherently excited acoustic vibrational modes can also be observed for bimetallic particles. Pt and Au have very similar elastic modulii; therefore, the vibrational periods observed for Pt–Au core–shell particles are very similar to those for pure Au particles of the same size. A more interesting situation arises where the two metals have different elastic properties. Figure 8a shows transient bleach data for pure Au particles and for the Au particles coated with ∼40 monolayers of Pb, where modulations due to the coherently excited breathing mode of the particle can be seen [17]. Clearly, the addition of the Pb has a strong affect on the period.

Figure 8b shows the measured frequency (in cm⁻¹ units) versus the number of Pb atomic layers for all the samples investigated. It is important to note that the decrease in the frequency with increasing Pb content cannot be reproduced by using Eq. (3) and average values of c_l and c_t, that is, the interface between the core and shell must be explicitly taken into account [17]. The solid line in Fig. 8b shows the results of continuum mechanics calculations for the breathing mode of a core–shell particle. In these calculations, the displacement and normal stress were set to be continuous at the interface, and the bulk elastic properties of Au and Pb were used, details of the theory are given in Ref. 48. The experimental results and calculations are in excellent agreement. This shows that continuum mechanics gives an excellent description of the elastic properties of core–shell nanoparticles. Furthermore, even at a few atomic layers (<10), the elastic properties of Pb are the same as that for bulk Pb.

VI. SUMMARY AND CONCLUSIONS

Time-resolved laser spectroscopy is a powerful tool for studying the fundamental properties of metal nanoparticles. Our work has concentrated on (1) examining how the coupling between the electrons and phonons depends on the size and composition of the particles [9,10] and (2) studies of coherently excited acoustic vibrational modes in nanoparticles. We have previously shown that the timescale for electron–phonon coupling in Au does not depend on size for particles in the 2–200-nm size region [2,9]. For bimetallic particles, our transient absorption experiments show that the electron–phonon coupling constant is an average of the coupling constants for the individual metals, weighted by the density of electronic states for each [19]. For the specific system studied—Pt–Au core–shell particles—the steady-state optical properties are determined by the Au shell, but the dynamics are dominated by the Pt core. This occurs because Pt has a much larger density of electronic states than Au [19].

The rapid electronic and lattice heating that accompanies ultrafast laser excitation can also coherently excite the phonon mode that correlates with the expansion coordinate of the particles. For spherical particles, the symmetric breathing mode is observed in transient absorption experiments. The measured periods exactly match the predictions of continuum mechanics calculations [12,13]. In addition, the amplitude and phase of the modulations are in excellent agreement with model calculations, where the expansion coordinate is treated as an harmonic oscillator [12,13,15,16] and the driving force for expansion arises from the thermal stresses created by heating the electrons and the nuclei [38,39]. These results show that both direct and indirect coupling between the laser excited electrons and the symmetric breathing mode are important. For rod-shaped particles, ultrafast laser excitation generates a Rayleigh surface wave, which has a frequency that depends on the length of the rod [45]. This observation is very different compared to recent results from Perner and co-workers, who studied Ag ellipses in a solid matrix [15]. This shows that the vibrational dynamics of nonspherical particles is very complex. Coherently excited vibrational modes can also be observed for bimetallic particles. For core–shell particles composed of materials with very different elastic properties, the interface between the core and the shell must be explicitly taken into account in continuum mechanics calculations of the breathing mode [17,48].

ACKNOWLEDGMENTS

The work described in this chapter was supported by the National Science Foundation by grants No. CHE98-16164 and CHE02-36279. The author is extremely grateful to his students for performing the work (primarily Jose Hodak), to Professor Arnim Henglein and Professor Paul Mulvaney for providing the metal particle samples, and to Professor John Sader for performing the contiuum mechanics calculations for the core–shell nanoparticles.

REFERENCES

1. Link, S.; El-Sayed, M.A. J. Phys. Chem. B 1999, *103*, 8410.
2. Hodak, J.H.; Henglein, A.; Hartland, G.V. J. Phys. Chem. B 2000, *104*, 9954.

3. Del Fatti, N.; Vallée, F.; Flytzanis, C.; Hamanaka, Y.; Nakamura, A. Chem. Phys. 2000, *251*, 215.

4. Stagira, S.; Nisoli, M.; De Silvestri, S.; Stella, A.; Tognini, P.; Cheyssac, P.; Kofman, R. Chem. Phys. 2000, *251*, 259.

5. Voisin, C.; Del Fatti, N.; Christofilos, D.; Vallée, F. J. Phys. Chem. B 2001, *105*, 2264.

6. El-Sayed, M.A. Acc. Chem. Res. 2001, *34*, 257.

7. Belotskii, E.D.; Tomchuck, P.M. Int. J. Electron. 1992, *73*, 915.

8. Stella, A.; Nisoli, M.; De Silvestri, S.; Svelto, O.; Lanzani, G.; Cheyssac, P.; Kofman, R. Phys. Rev. B 1996, *53*, 15497.

9. Hodak, J.H.; Henglein, A.; Hartland, G.V. J. Chem. Phys. 2000, *112*, 5942.

10. Nisoli, M.; De-Silvestri, S.; Cavalleri, A.; Malvezzi, A.M.; Stella, A.; Lanzani, G.; Cheyssac, P.; Kofman, R. Phys. Rev. B 1997, *55*, R13,424.

11. Hodak, J.H.; Martini, I.; Hartland, G.V. J. Chem. Phys. 1998, *108*, 9210.

12. Del Fatti, N.; Voisin, C.; Chevy, F.; Vallée, F.; Flytzanis, C. J. Chem. Phys. 1999, *110*, 11,484.

13. Hodak, J.H.; Henglein, A.; Hartland, G.V. J. Chem. Phys. 1999, *111*, 8613.

14. Hodak, J.H.; Martini, I.; Hartland, G.V. J. Phys. Chem. B 1998, *102*, 6958.

15. Perner, M.; Gresillon, S.; Marz, J.; von Plessen, G.; Feldmann, J.; Porsten-dorfer, J.; Berg, K.J.; Berg, G. Phys. Rev. Lett. 2000, *85*, 792.

16. Voisin, C.; Del Fatti, N.; Christofilos, D.; Vallee, F. Appl. Surface Sci. 2000, *164*, 131.

17. Hodak, J.H.; Henglein, A.; Hartland, G.V. J. Phys. Chem. B 2000, *104*, 5053.

18. Schmid, G., Ed. *Clusters and Colloids: From Theory to Application.* Weinheim: VCH, 1994.

19. Hodak, J.H.; Henglein, A.; Hartland, G.V. J. Chem. Phys. 2001, *114*, 2760.

20. Henglein, A. J. Phys. Chem. 1993, *97*, 5457.

21. Henglein, A. Langmuir 1999, *15*, 6738.

22. Henglein, A.; Meisel, D. Langmuir 1998, *14*, 7392.

23. Henglein, A.; Lilie, J. J. Am. Chem. Soc. 1981, *103*, 1059.

24. Mulvaney, P.; Giersig, M.; Henglein, A. J. Phys. Chem. 1993, *97*, 7061.

25. Katsikas, L.; Gutierrez, M.; Henglein, A. J. Phys. Chem. 1996, *100*, 11,203.

26. Henglein, A.; Brancewicz, C. Chem. Mater. 1997, *9*, 2164.

27. Henglein, A. J. Phys. Chem. B 2000, *104*, 2201.

28. Henglein, A.; Giersig, M. J. Phys. Chem. B 2000, *104*, 5056.

29. Henglein, A. J. Phys. Chem. B 2000, *104*, 6683.

30. Enustun, B.V.; Turkevich, J. J. Am. Chem. Soc. 1963, *85*, 3317.

31. Ahmadi, T.S.; Wang, Z.L.; Green, T.C.; Henglein, A.; El-Sayed, M.A. Science 1996, *272*, 1924.

32. Yu, Y.Y.; Chang, S.S.; Lee, C.L.; Wang, C.R.C. J. Phys. Chem. B 1997, *101*, 6661.

33. Chang, S.S.; Shih, C.W.; Chen, C.D.; Lai, W.C.; Wang, C.R.C. Langmuir 1999, *15*, 701.

34. Lamb, H. Proc. Lond. Math. Soc. 1882, *13*, 189.

35. Dubrovskiy, V.A.; Morochnik, V.S. Izv. Earth Phys. 1981, *17*, 494.

36. Hartland, G.V. J. Chem. Phys. 2002, *116*, 8042.
37. Kriebig, U.; Vollmer, M. *Optical Properties of Metal Clusters.* Berlin: Springer-Verlag, 1995.
38. Gusev, V.E. Opt. Commun. 1992, *94*, 76.
39. Tas, G.; Maris, H.J. Phys. Rev. B 1994, *49*, 15,046.
40. Belotskii, E.D.; Tomchuk, P.M. Surface Sci. 1990, *239*, 143.
41. Heath, J.R. Pure Appl. Chem. 2000, *72*, 11.
42. Link, S.; Burda, C.; Mohamed, M.B.; Nikoobakht, B.; El-Sayed, M.A. Phys. Rev. B 2000, *61*, 6086.
43. Link, S.; Burda, C.; Nikoobakht, B.; El-Sayed, M.A. J. Phys. Chem. B 2000, *104*, 6152.
44. Link, S.; El-Sayed, M.A. J. Chem. Phys. 2001, *114*, 2362.
45. Hartland, G.V.; Hu, M.; Wilson, O.; Mulvaney, P.; Sader, J.E.; Giersig, M. J. Phys. Chem. B 2002, *106*, 743.
46. Gosavi, S.; Marcus, R.A. J. Phys. Chem. B 2000, *104*, 2067.
47. Achcroft, N.W.; Mermin, N.D. *Solid State Physics.* Orlando, FL: Harcourt Brace, 1976.
48. Sader, J.E.; Hartland, G.V.; Mulvaney, P. J. Phys. Chem. B 2002, *106*, 1399.

Index